PROBLEMS IN ORGANIC STRUCTURE DETERMINATION
A Practical Approach to NMR Spectroscopy

PROBLEMS IN ORGANIC STRUCTURE DETERMINATION

A Practical Approach to NMR Spectroscopy

Roger G. Linington
UC Santa Cruz, USA

Philip G. Williams
University of Hawaii, USA

John B. MacMillan
U.T. Southwestern, Texas, USA

CRC Press
Taylor & Francis Group
Boca Raton London New York

CRC Press is an imprint of the
Taylor & Francis Group, an **informa** business

CRC Press
Taylor & Francis Group
6000 Broken Sound Parkway NW, Suite 300
Boca Raton, FL 33487-2742

© 2016 by Taylor & Francis Group, LLC
CRC Press is an imprint of Taylor & Francis Group, an Informa business

No claim to original U.S. Government works

Printed in Canada on acid-free paper
Version Date: 20150909

International Standard Book Number-13: 978-1-4987-1962-9 (Paperback)

This book contains information obtained from authentic and highly regarded sources. Reasonable efforts have been made to publish reliable data and information, but the author and publisher cannot assume responsibility for the validity of all materials or the consequences of their use. The authors and publishers have attempted to trace the copyright holders of all material reproduced in this publication and apologize to copyright holders if permission to publish in this form has not been obtained. If any copyright material has not been acknowledged please write and let us know so we may rectify in any future reprint.

Except as permitted under U.S. Copyright Law, no part of this book may be reprinted, reproduced, transmitted, or utilized in any form by any electronic, mechanical, or other means, now known or hereafter invented, including photocopying, microfilming, and recording, or in any information storage or retrieval system, without written permission from the publishers.

For permission to photocopy or use material electronically from this work, please access www.copyright.com (http://www.copyright.com/) or contact the Copyright Clearance Center, Inc. (CCC), 222 Rosewood Drive, Danvers, MA 01923, 978-750-8400. CCC is a not-for-profit organization that provides licenses and registration for a variety of users. For organizations that have been granted a photocopy license by the CCC, a separate system of payment has been arranged.

Trademark Notice: Product or corporate names may be trademarks or registered trademarks, and are used only for identification and explanation without intent to infringe.

Visit the Taylor & Francis Web site at
http://www.taylorandfrancis.com

and the CRC Press Web site at
http://www.crcpress.com

DEDICATION

"This book is dedicated to Professors R. J. Andersen, W. Fenical, W. H. Gerwick, T. F. Molinski, and the late R. E. Moore who taught us this craft."

TABLE OF CONTENTS

Acknowledgements	ix
Introduction	xi
Section 1: Structure Confirmation for Simple Molecules	1
Section 2: Structure Confirmation for Complex Molecules	91
Section 3: Determination of Unknown Structures for Simple Molecules	309
Section 4: Determination of Unknown Structures for Complex Molecules	457
Section 5: Determination of Relative Configurations Using NMR Methods	637
Section 6: Complex Unknown Natural Product Structure Determination	709
Appendix A: NMR Solvent Reference Chemical Shifts	731
Appendix B: Problem Notes	732
Appendix C: Problem Hints	734
Appendix D: Distribution of Experiment Types	741
Appendix E: Distribution of Functional Groups	745
Appendix F: Selected ^{15}N NMR Chemical Shifts	749
About the Authors	751
About the Spectrometer	753
About the Cover	755

ACKNOWLEDGEMENTS

This work would not have been possible without the efforts of many people. This book contains spectra from 118 small molecules and well over 500 total experiments, all of which were acquired under the expert technical skill of Wes Y. Yoshida of the University of Hawaii. We are indebted to Wes for his significant contribution to this project, including his dedicated efforts to obtain data of the highest quality, and his seemingly limitless patience with our shifting requests.

We thank our colleagues William Chain, Jef DeBrabander, Thomas Hemscheidt, Scott Lokey, Joseph Ready and Uttam Tambar for the generous gifts of chemicals that allowed us to include such a wide diversity of small molecules.

We have benefited greatly from conversations about the design of this project with members of the natural products community, many of whom reviewed portions of this book and provided valuable suggestions. In particular we would like to thank Professors Raymond Andersen (U. British Columbia), Marcy Balunas (U. Connecticut), William Gerwick (Scripps Institution of Oceanography, U. California San Diego), Harald Gross (U. Tubingen), Brian Murphy (U. Illinois-Chicago), Philip Proteau (Oregon State U.) and Ryuichi Sakai (Hokkaido U.) for their valuable feedback. We also thank the many students and postdocs who provided comments and worked through many of these problems: Jehad Almaliti, Judith Bauer, Jingqiu Dai, Amanda Fenner, Joshua Gurr, Isaiah Gomez, Henrik Harms, Mian Huang, Norbert Kirchner, Shangwen Luom, Henrike Miess, Bailey Miller, Nathan Moss, Michael Mullowney, Ram Neupane, Stephen Parrish, Anam Shaikh, and Chen Zhang, Nathaniel Oswald, David Brumley, Dominic Colosimo and Peng Fu.

Dr. Karen S. MacMillan deserves a special thanks, as we took advantage of her attention to detail to carry out proofreading of the text and quality control of the final spectra.

The staff at CRC press has been extremely helpful in bringing the final product to completion. In particular we owe a great deal to Hilary LaFoe who has been very supportive of this project since its inception.

Finally, we hope that you find the problems as challenging, intellectually stimulating, and enjoyable as we do, and that this text will sharpen your skills in the area of organic structure determination.

Roger Linington
Philip Williams
John MacMillan

November 2015

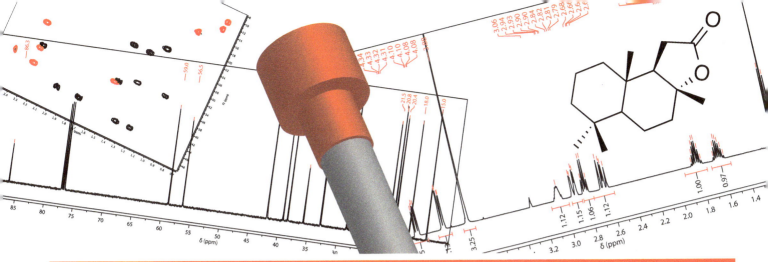

Problems in Organic Structure Determination

OBJECTIVE

The goal of this book is to train undergraduates, graduates and professional scientists in structure elucidation through a problem-based learning approach. The book consists entirely of NMR-based problems, and is split into six sections that train users in the art of structure elucidation using incrementally more challenging compounds and complex NMR experiments. The book stems from our collective research and teaching experiences in the field of natural products chemistry and a desire to share these experiences with others. It is intended to complement existing NMR structure determination textbooks (see below for recommended texts and reference material) by providing a large set of experimentally acquired NMR data, complete with the usual combination of impurities, noise and other issues that illustrate the common challenges encountered when working in the area of small molecule structure elucidation.

Fundamentally, we believe that small molecule structure determination can only be mastered by solving problems and that, despite the prevalence and utility of modern 2D NMR experiments, this requires a solid understanding of all of the information imparted from NMR analysis, including chemical shifts, coupling constants, and integrations. This consideration has driven the design of this text with its high-resolution expansions of 1D and 2D spectra, readily available raw NMR data files and the inclusion of a section on stereochemical analysis where careful consideration of coupling constants in conjunction with ROESY or NOESY data is of critical importance.

We envision that users of this book will already possess a basic knowledge of NMR structure determination, equivalent to having completed the standard sophomore organic chemistry sequence, and so begin Section 1 from this starting point. The gradation of problems throughout the text is such that it provides a large number of entry level problems that build gradually in difficulty, culminating with a number of highly complex natural product structure determination challenges in Section 6. The distribution of difficulty throughout this range is suitable for upper division undergraduates learning the fundamentals of NMR structure elucidation, while also providing sufficient challenging problems to be valuable for graduate students, post-doctoral scholars or professional scientists looking to improve their skills in this area.

Problems are designed to be solved without IR, UV or other supplemental analytical data as in general it has been our experience that the corresponding information can be extracted with higher resolution from the NMR data provided based on chemical shift considerations or analysis of the 2D NMR data. This 'NMR-centric' approach to structure elucidation mirrors the strategy most commonly employed in our own laboratories, and the natural products and organic chemistry communities at large. Any users wishing to supplement this text with IR, UV or MS spectra should have little difficulty doing so, since the majority of compounds used are commercially available at reasonable prices. A full list of CAS numbers is provided at the end of the answer key. Finally, with the exception of ^{15}N NMR chemical shift values listed in Appendix F, tables of chemical shift and coupling data are not included in this text as this information is easily available and well summarized in the references listed below or on the internet.

BIBLIOGRAPHY

Tables of Data

Breitmaier, E. (1987). Carbon-13 NMR spectroscopy : high-resolution methods and applications in organic chemistry and biochemistry. New York, New York : VCH Publishers.

Pretsch, E., P. Bühlmann, et al. (2009). Structure determination of organic compounds: tables of spectral data. Berlin, Springer.

Witanowski, M. (1973). Nitrogen NMR. London, New York, Plenum Press.

Structure Determination

Crews, P., J. Rodríguez, et al. (2010). Organic structure analysis. New York, Oxford University Press.

Hoye, T. R. and H. Zhao (2002). "A method for easily determining coupling constant values: an addendum to "A practical guide to first-order multiplet analysis in ^{1}H NMR spectroscopy." The Journal of Organic Chemistry 67 (12): 4014-4016.

Macomber, R. S. (1998). A complete introduction to modern NMR spectroscopy. New York, New York : Wiley.

Silverstein, R. M., F. X. Webster, et al. (2005). Spectrometric identification of organic compounds. Hoboken, N.J., Wiley.

Simpson, J. H. (2012). Organic structure determination using 2-D NMR spectroscopy: a problem-based approach. Boston, Elsevier/AP.

NMR Experiments & Theory

Berger, S. and S. Braun (2004). 200 and more NMR experiments. A practical course. 3. Edition, Wiley-VCH.

Claridge, T. D. W. (2008). High-resolution NMR techniques in organic chemistry. Amsterdam, Pergamon.

Keeler, J. (2010). Understanding NMR spectroscopy. Chichester, U.K, John Wiley and Sons.

Processing NMR spectra

Burrow, T. E., R. G. Enriquez, et al. (2009). "The signal/noise of an HMBC spectrum can depend dramatically upon the choice of acquisition and processing parameters." Magnetic Resonance in Chemistry 47(12): 1086-1094.

Reynolds, W. F. and R. G. Enríquez (2002). "Choosing the best pulse sequences, acquisition parameters, postacquisition processing strategies, and probes for natural product structure elucidation by NMR spectroscopy." Journal of Natural Products 65(2): 221-244.

INTRODUCTION

Common NMR Impurities

Fulmer, G. R., A. J. M. Miller, et al. (2010). "NMR chemical shifts of trace impurities: common laboratory solvents, organics, and gases in deuterated solvents relevant to the organometallic chemist." Organometallics 29(9): 2176-2179.

Gottlieb, H. E., V. Kotlyar, et al. (1997). "NMR chemical shifts of common laboratory solvents as trace impurities." The Journal of Organic Chemistry 62(21): 7512-7515.

LAYOUT

This book is designed to lead users from an initial starting point using simple examples typically found in most undergraduate organic chemistry textbooks, to the final complex natural products that represent the pinnacle of organic structure analysis. Within this continuum of difficulty, users will encounter many of the functional groups commonly found in organic chemistry, starting with simple alkyl amines, and building up to terminal alkynes, epoxides, polycyclic structures, chiral compounds and fluorine-containing molecules. In addition, users will be gradually introduced to all of the standard 1D- and 2D-NMR experiments, as well as some valuable but less common advanced techniques including selectively decoupled proton spectra, 1D-TOCSY, 1D-NOESY, HSQC-TOCSY, and ^{15}N-HMBC experiments. By working progressively through these problems it is hoped that users will develop mastery over the intricacies and complexities of structure elucidation. By the end of the book it is expected that users will be able to tackle even the most challenging and complex structure elucidation problems, and be able to incorporate data from a wide array of experiment types.

This book is divided into six sections. Sections 1 and 2 focus on connecting structures to spectra, and are designed to develop an understanding of how to analyze 1D and 2D datasets in an integrated fashion, while emphasizing the information available from each of the most commonly used NMR experiments. Sections 3 and 4 involve the *de novo* structure determination of unknowns of progressively increasing difficulty and will introduce users to an extended range of 1D- and 2D-NMR experiments. Section 5 focuses on one of the most challenging areas of small molecule structure determination, namely configurational analysis. Finally, section 6 contains problems derived from complex natural products for those wanting more experience with these most difficult analytical objectives.

HOW TO USE THIS BOOK

Several features have been added to this book in order to make it easier to use. First, because most problems are several pages long, copies of the ^{1}H and ^{13}C NMR spectra have been reproduced on a single page for each problem. These single page 1D spectra are available for download at http://www.crcpress.com/product/isbn/9781498719629. and are aimed to eliminate the need to constantly flip back and forth between the 1D and 2D spectra while solving complex problems. Second, each spectrum is labeled with the experiment type (e.g. gCOSY) to avoid confusion. Third, for instructors looking for problems that use specific combinations of NMR experiments appendix D includes a table describing the experiments used for each problem. In a similar vein, appendix E contains a summary of the functional groups present in each problem, for those wishing to gain experience with particular structural types (e.g. fluorine-containing molecules). Finally, select problems marked with an "H" contain hints about the problem in appendix C to provide small clues to aid in solving the problem for instances in which users are having difficulty reaching the final solution.

COMPOUND SELECTION

In reality, the compounds included in this book represent only a subset of those originally examined which met certain criteria. With the exception of a handful of problems in section 6, all compounds are commercially available for the reason mentioned earlier. Compounds were used as received, without further purification, since small impurities are the norm rather than the exception in most NMR spectra. Throughout the book, compounds are referred to by their supplier-given names, even when IUPAC conventions would dictate an alternative, but CAS numbers are available at the end of the answer key for the entire data set. Representatives of all of the major functional groups encountered in organic chemistry, including compounds incorporating fluorine, have been incorporated in the data set.

DATA ACQUISITION

Data collection was performed on a Varian Unity Inova 500 MHz NMR spectrometer equipped with a switchable 5 mm room temperature probe. This instrument was selected because it represents a hardware configuration and magnetic field-strength accessible to most labs, and the data are therefore representative of the types of data users will encounter in their daily research activities. Most samples were prepared with ~20 mg of compound in Norrell 507-HP 5 mm NMR tubes with 670 µL of deuterated NMR solvent.

ELECTRONIC DATA

In addition to the printed spectra provided in the main text, we have provided all of the original NMR data files from the spectrometer as downloadable files from the book website (http://www.crcpress.com/product/isbn/9781498719629). Provision of these files permits users to process the data using any of a number of commercially available NMR processing packages, and to explore the spectra in more detail than is permitted from the static printed layouts. We feel that this will be particularly valuable for problems later in the book where identifying key 2D correlations can be difficult without careful examination of the data. However, in all cases (with the exception of section 6) the printed spectra are of sufficient detail to permit the problems to be solved from the textbook. Consequently these electronic files should be seen as an additional resource, rather than a requirement for solving these structures. For section 6 problems, pdf copies of the full processed datasets are available for download at the website listed above.

DATA LAYOUT

All spectra in the book have been standardized as follows:

- All problems contain a ^{1}H NMR spectrum from 0.4 to 8.6 ppm with peak picking at the center of each multiplet with labels provided in parts per million (ppm). Expansions of the ^{1}H NMR spectrum are included for each problem with detailed peak picking in hertz (Hz). Solvent and water signals are excluded from peak picking.

- Integrations on ^{1}H NMR spectra are set with a non-exchangeable proton to the lowest integer value (i.e. if the molecule contains a methine then the lowest integer set to 1; if the molecule does not contain a methine, but does have a methylene, then the lowest integer is set to 2). If the molecule is symmetrical then the integration is set to the lowest integer value, rather than the absolute number of protons contributing to each signal.

- All problems contain a ^{13}C NMR spectrum from 5 to 215 ppm with peak picking provided in ppm.

- DEPT-135 spectra are phased using the standard convention so that carbons bearing odd numbers of protons are positive, while those bearing even numbers of protons are negative.

- We have utilized a multiplicity-edited gHSQC throughout this book. For problems that contain this experiment cross peaks deriving from CH_2 groups are depicted in blue, and cross peaks derived from CH and CH_3 groups are depicted in red.

- Minor impurities are neither labeled nor peak picked; however, significant impurities are labeled with an "i" in both the 1D and 2D spectra.

- There are a number of problems that include a ^1H–^{15}N HMBC experiment. Consistent with IUPAC's 2001 recommendations, we have used nitromethane as the reference molecule for all ^{15}N experiments. For reference, relevant ^{15}N chemical shifts are provided in appendix F.

ANSWERS & WORKED SOLUTIONS

Balancing some practical limitations with a desire to provide feedback to users, we have opted to provide all answers as a PDF that can be downloaded from the publisher's website at http://www.crcpress.com/product/isbn/9781498719629. In general, the answer key includes the chemical structure annotated with corresponding ^1H and ^{13}C chemical shift assignments in blue and red text respectively, and ^{15}N chemical shifts in orange where available. While we have made every attempt to fully assign all proton and carbon chemical shifts, on occasion NMR signals cannot be unequivocally assigned using the data provided. In those cases interchangeable assignments are labeled with a "*".

We have also provided signal multiplicities and J_{HH} and J_{CF} values for the majority of problems. The J_{HH} values included in the answers were obtained by using the peak picking in Hz on the ^1H NMR spectral expansions, always starting from the furthest upfield portion of the multiplet, provided the resolution was adequate. As this is real data, the proton multiplets are not always perfectly symmetrical, and there is a certain amount of experimental error in determining the magnitude of the proton-proton couplings; so, expect small deviations from our J_{HH} values if you do not follow our convention or if you process and peak pick the raw NMR data yourself. Note that in cases where signals display non-first order coupling, such as situations where magnetic inequivalence is observed, we have not assigned coupling constant values. Finally, worked solutions to certain problems - denoted by the W icon on their first page of each problem - are also available in the answer key.

ANCILLARY MATERIAL

As described above, this book is accompanied by a number of electronic resources, accessible through the book website at http://www.crcpress.com/product/isbn/9781498719629. These resources include:

1) A complete set of original fids for all compounds, including both 1D and 2D spectra. These files are arranged in folders by problem number, making them suitable for use as take-home exam questions and homework problems for institutions that have access to NMR processing software.

2) A full color answer key, including ^1H and ^{13}C chemical shifts and coupling constants for all atoms. NMR data tables are included for the more complex structures. Three-dimensional depictions and key nOe correlations are presented where appropriate for the relative configurational problems in section 5.

3) Complete pdf layouts for all complex natural product problems from section 6. It is anticipated that most users will solve these complex structures using the electronic files, however for those users who do not have access to appropriate NMR processing software, full print layouts are available as pdfs for problems 111 - 130.

4) Condensed ^1H and ^{13}C spectra. To aid users in solving complex problems containing

many pages of expansions, we have generated the ^{1}H and ^{13}C spectra on a single page for all problems, in order to reduce the need to flip back and forth between spectra in the book.

Section 1

Section 1 of this book is geared towards introducing relative newcomers to NMR-based structure elucidation, as well as serving as a review of the fundamentals of NMR analysis for those with previous experience.

LEARNING OBJECTIVES

- Familiarize you with assigning ^{1}H and ^{13}C chemical shifts in gradually more complex molecules.
- Analyze and predict multiplicity patterns in ^{1}H NMR spectra in order to obtain and interpret coupling constants.
- Integrate data from different NMR experiments and apply this knowledge to the assignment of chemical structure.

EXPERIMENTS INCLUDED

^{1}H, ^{13}C, DEPT-135, gCOSY, gHSQC and gHMBC.

Note: All gHSQC experiments are multiplicity edited with cross peaks deriving from CH_2 groups depicted in blue, and cross peaks deriving from CH and CH_3 groups depicted in red.

TYPES OF MOLECULES

This section contains most of the basic functional groups found in small molecules including: alcohols, amines, halides, olefins, carbonyls, aromatics and heteroaromatics.

SPECTRUM LAYOUT CONVENTIONS

The solvent used for each problem is indicated in the question, but solvent signals are not annotated on the spectra. A list of solvents and their ^{1}H and ^{13}C chemical shifts can be found in appendix A. In general, only signals deriving from the molecule are peak picked on the ^{1}H and ^{13}C spectra. Minor impurities are left unannotated, while major impurities are marked with a lower case "i".

STRATEGIES FOR SUCCESS

An important objective of this chapter is to become adept at the basic aspects of NMR structural elucidation, particularly chemical shift and coupling constant analyses. Problems 1 and 25, which focus on predicting these values, emphasize these concepts and will help you develop an appreciation of the trends. Equally important, especially as the complexity of the problems increases, is to make sure you evaluate ALL the data before reaching a conclusion, and to be sure that all of the available data are in agreement with your proposed answer.

LEGEND

Spectrum annotations:

i = impurity.

W = worked problem in the answer key.

N = technical note about the data. For example: *"This spectrum is missing one exchangeable proton."*

H = hint to assist in solving the problem. For example: *"This molecule would have an IR stretch at 2240 cm^{-1}."*

SECTION 1 Problem 1

a) Functional groups with large electronegativities result in a downfield shift in the chemical shifts of adjacent atoms. For example, listed below are the proton chemical shifts for the methylene groups in ppm.

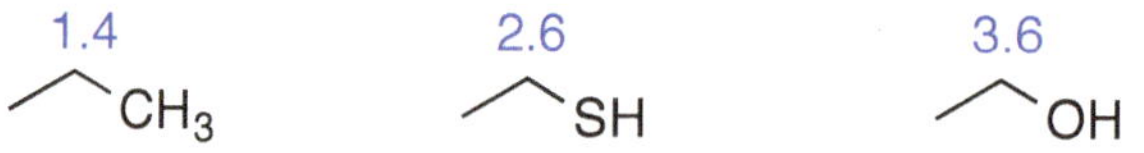

$$\text{CH}_2\text{CH}_3 \quad 1.4 \qquad \text{CH}_2\text{SH} \quad 2.6 \qquad \text{CH}_2\text{OH} \quad 3.6$$

Based on this trend, rank order the following molecules in terms of chemical shift (most downfield to most upfield), and estimate the proton chemical shift of the methylene groups (CH_2) in each molecule.

Hint: It would be helpful to consult a periodic table to compare the relative positions of the groups above in relation to the magnitude of the difference.

$$\text{CH}_2\text{Cl} \qquad \text{CH}_2\text{NH}_2 \qquad \text{CH}_2\text{F} \qquad \text{CH}_2\text{Si}$$

b) Ethyl acetate has the proton chemical shift values shown below:

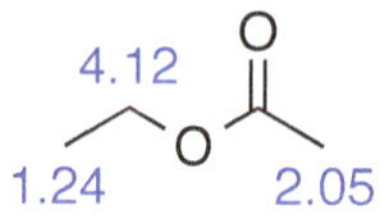

Ethyl acetate: OCH$_2$ = 4.12, CH$_3$ (ethyl) = 1.24, CH$_3$ (acetyl) = 2.05

Estimate the chemical shifts of the proton signals for the following structures:

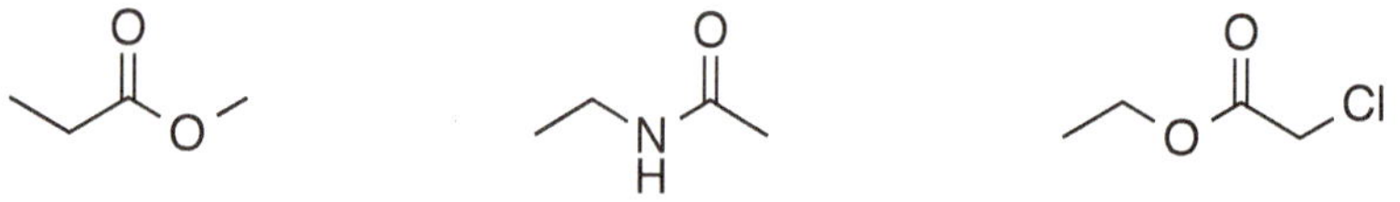

(propionate anion, N-ethyl acetamide, ethyl chloroacetate)

c) For monosubstituted aromatic rings, electron withdrawing groups shift aromatic protons downfield while electron donating groups have the opposite effect. For example, listed below are the proton chemical shifts for the ortho, meta and para protons of a series of monosubstituted benzene rings.

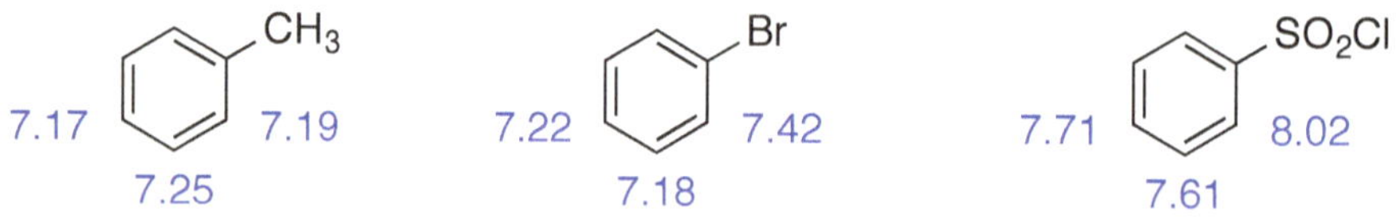

- Toluene (CH$_3$): ortho 7.17, meta 7.25, para 7.19
- Bromobenzene (Br): ortho 7.42, meta 7.18, para 7.22
- Benzenesulfonyl chloride (SO$_2$Cl): ortho 8.02, meta 7.61, para 7.71

Rank order the following compounds based on the chemical shift of the ortho proton (most downfield to most upfield) and then estimate the proton chemical shifts of the ortho, meta and para proton.

(ethylbenzene CH$_2$CH$_3$, benzyl chloride CH$_2$Cl, anisole OCH$_3$, nitrobenzene NO$_2$)

d) Carbon chemical shifts of carbonyls are helpful in structure elucidation. In general, ketones are the furthest downfield, followed by aldehydes, acids, esters and amides. Furthermore, conjugation to a double bond or aromatic ring causes a 6 - 10 ppm shift upfield.

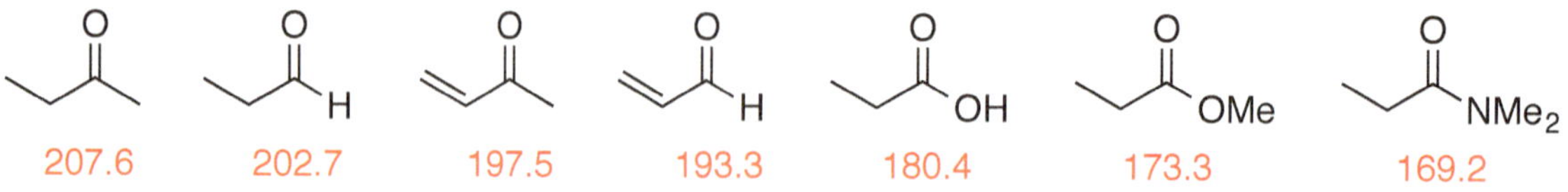

Ketone	Aldehyde	Enone	Enal	Acid	Ester	Amide
207.6	202.7	197.5	193.3	180.4	173.3	169.2

Estimate the chemical shifts of the following carbonyl carbons:

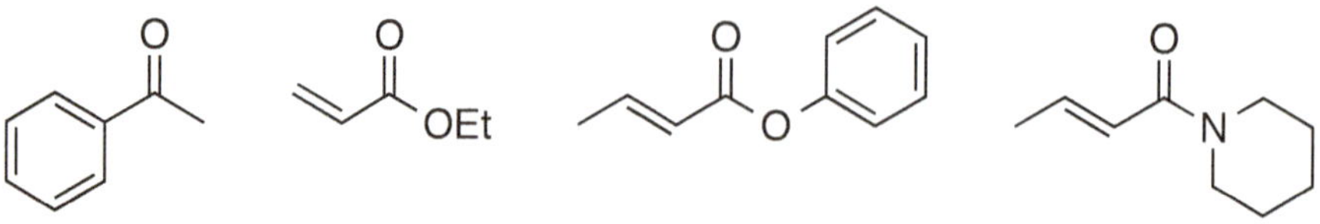

(acetophenone, ethyl acrylate, phenyl acrylate, acryloyl piperidine)

SECTION 1 Problem 2

Problems in Organic Structure Determination

Using the ¹H spectrum below, assign all proton resonances to the structure of *N*-methylbutylamine.

Spectrum acquired in CDCl₃ at 500 MHz.

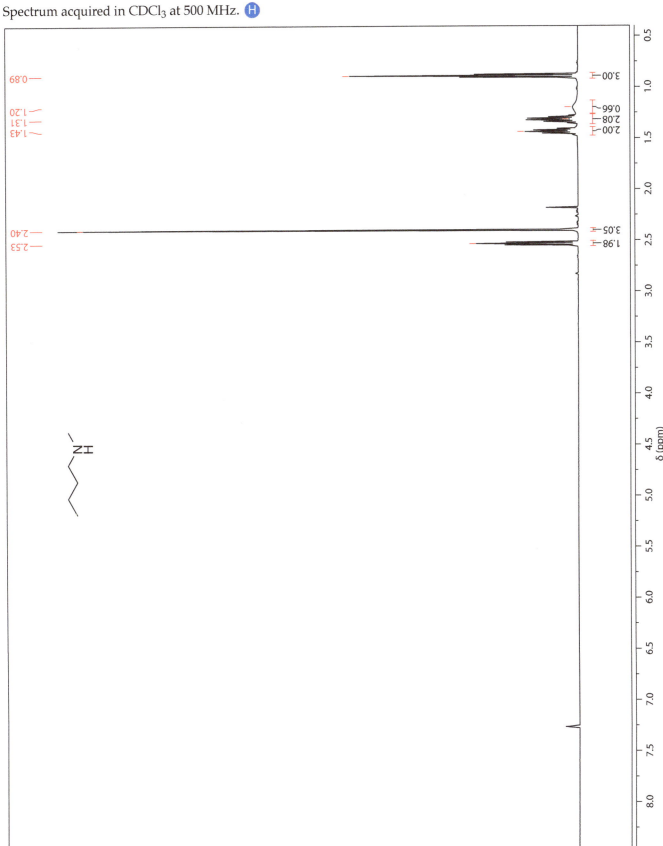

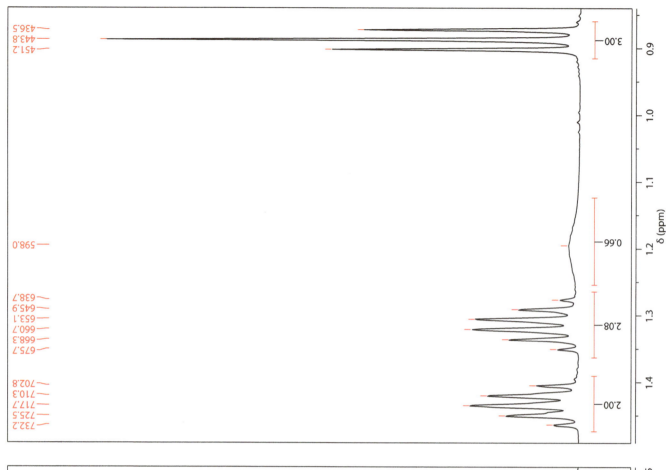
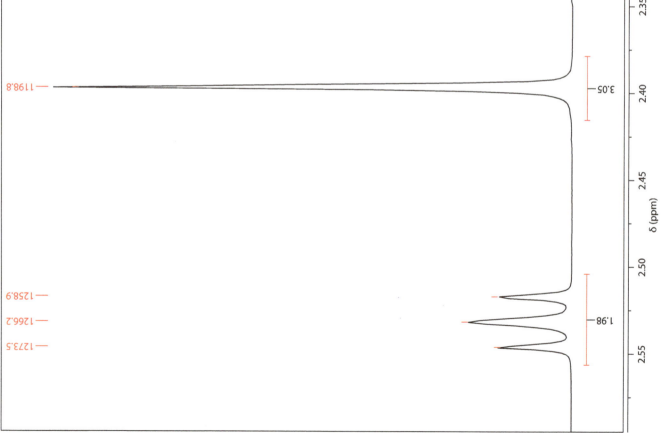

SECTION 1 Problem 3

Using the ¹H spectrum, assign all proton resonances to the structure of ethyl-3-chloropropionate below.

Spectrum acquired in CDCl₃ at 500 MHz.

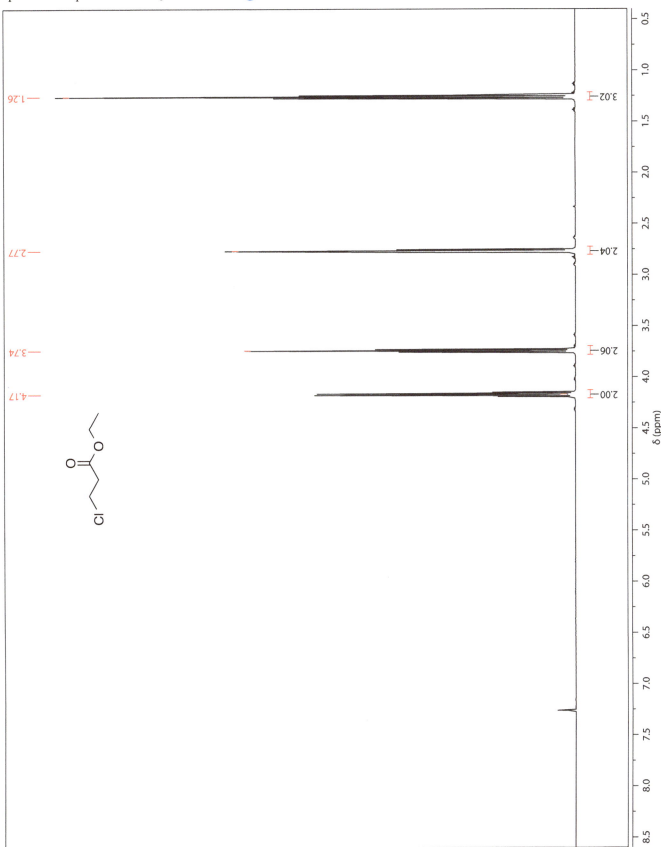

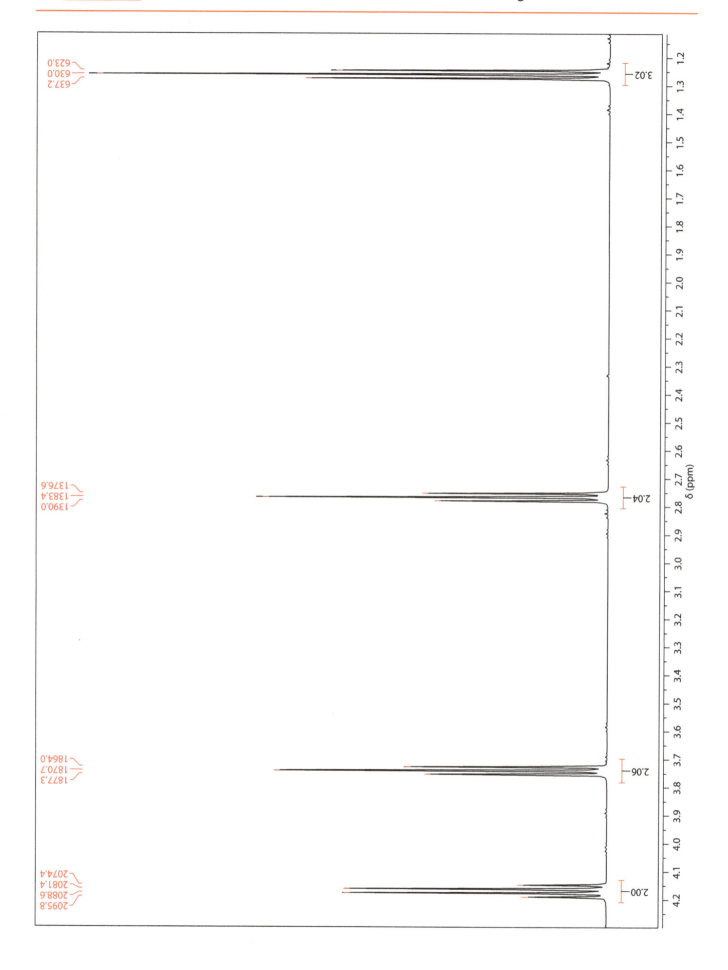

SECTION 1 Problem 4

Using the ¹H spectrum, assign all proton resonances to the structure of *p*-anisaldehyde dimethyl acetal below. Spectrum acquired in CDCl₃ at 500 MHz.

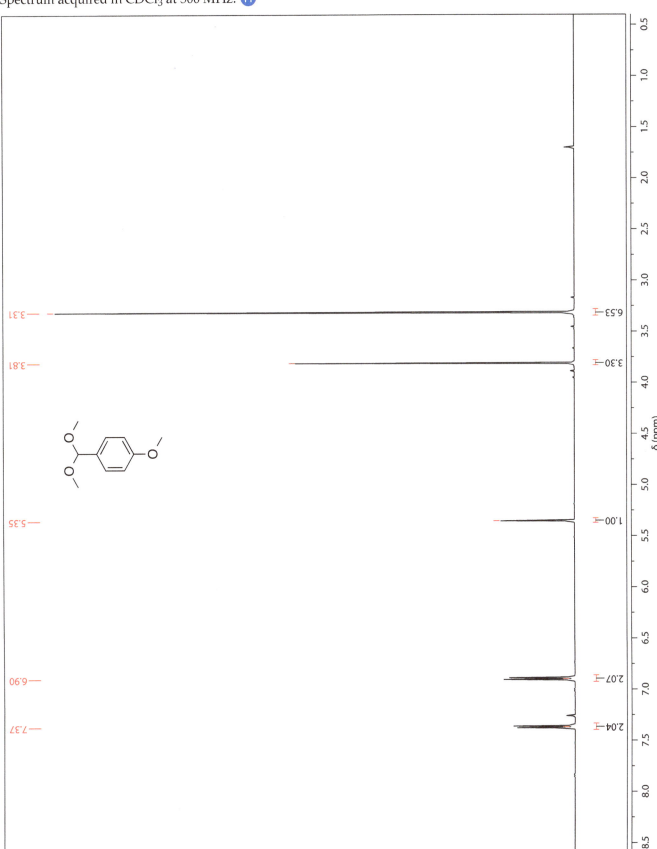

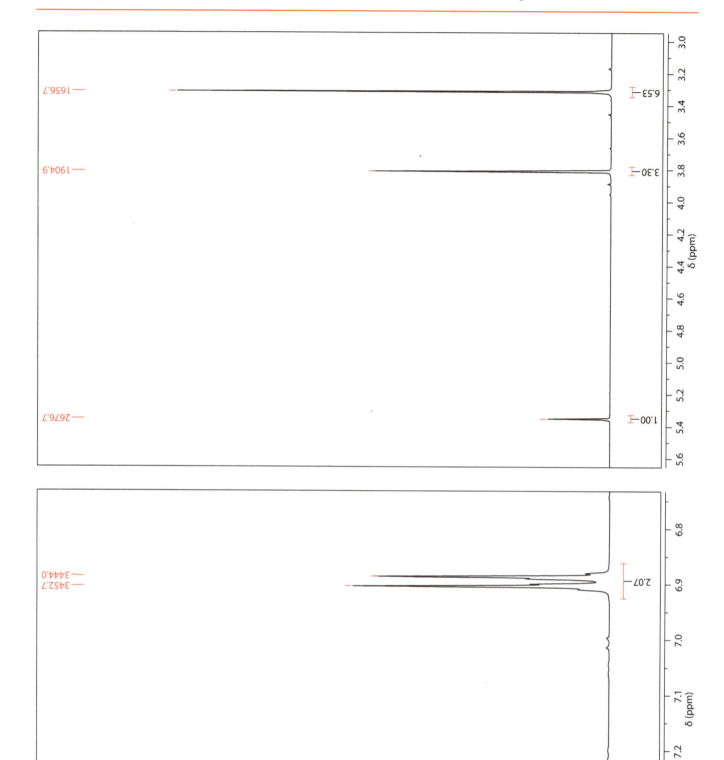

SECTION 1 Problem 5

Problems in Organic Structure Determination

Which of the following compounds is the correct structure for the spectra below?

Spectra acquired in CDCl₃ at 500 MHz (¹H) or 125 MHz (¹³C).

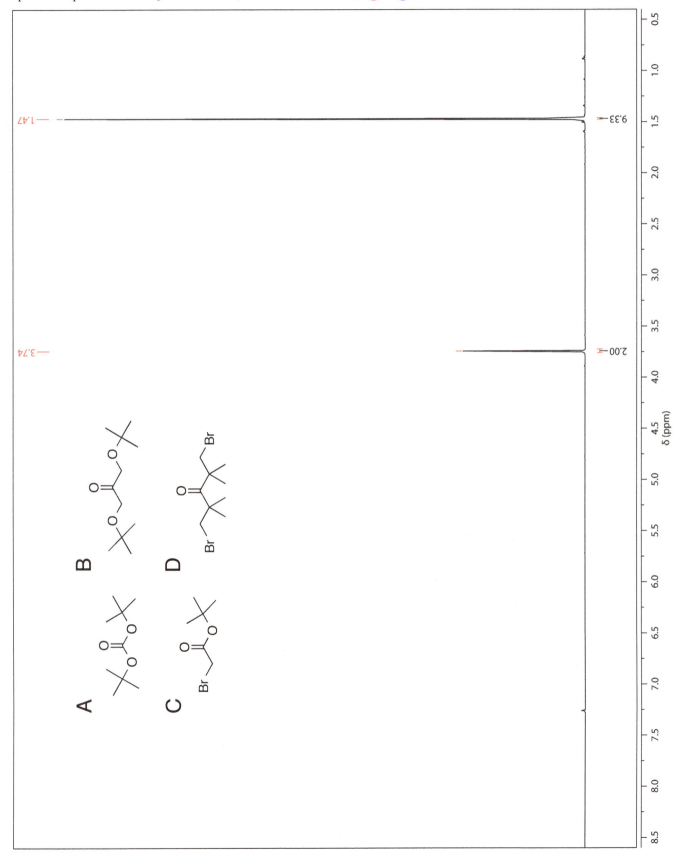

SECTION 1 Problem 5

¹³C NMR peaks: 166.4, 83.0, 27.9, 27.8 ppm

Expansion (26.3–29.6 ppm): 27.9, 27.8 ppm

SECTION 1 Problem 6 — Problems in Organic Structure Determination

Which of the following compounds is the correct structure for the spectra below?

Spectra acquired in (CD₃)₂SO at 500 MHz (¹H) or 125 MHz (¹³C).

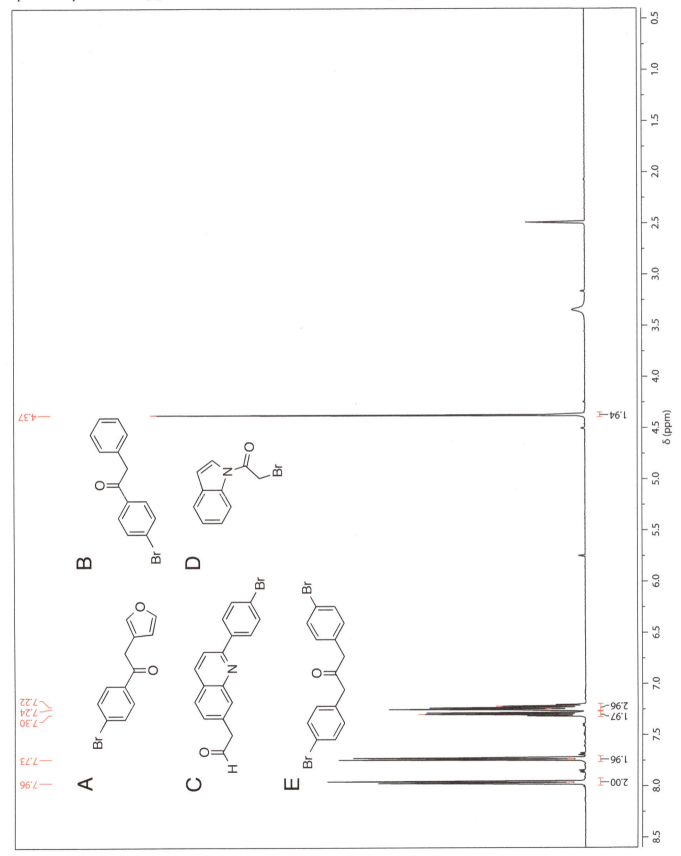

SECTION 1 Problem 6

SECTION 1 Problem 6

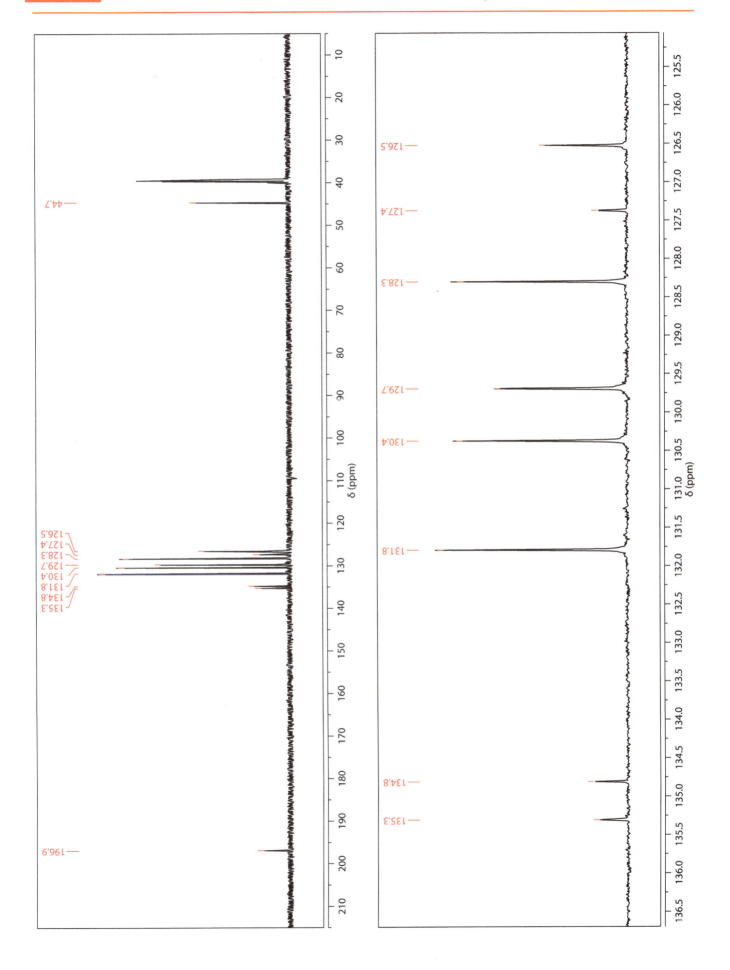

SECTION 1 Problem 7

Problems in Organic Structure Determination

Which of the following compounds is the correct structure for the spectra below?

Spectra acquired in CD$_3$OD at 500 MHz (^{1}H) or 125 MHz (^{13}C).

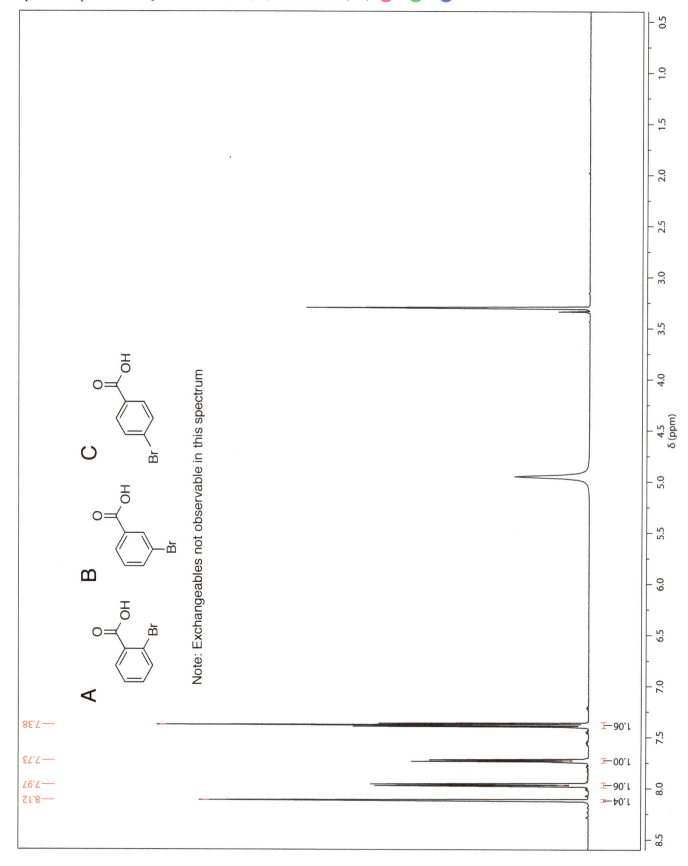

SECTION 1 Problem 7

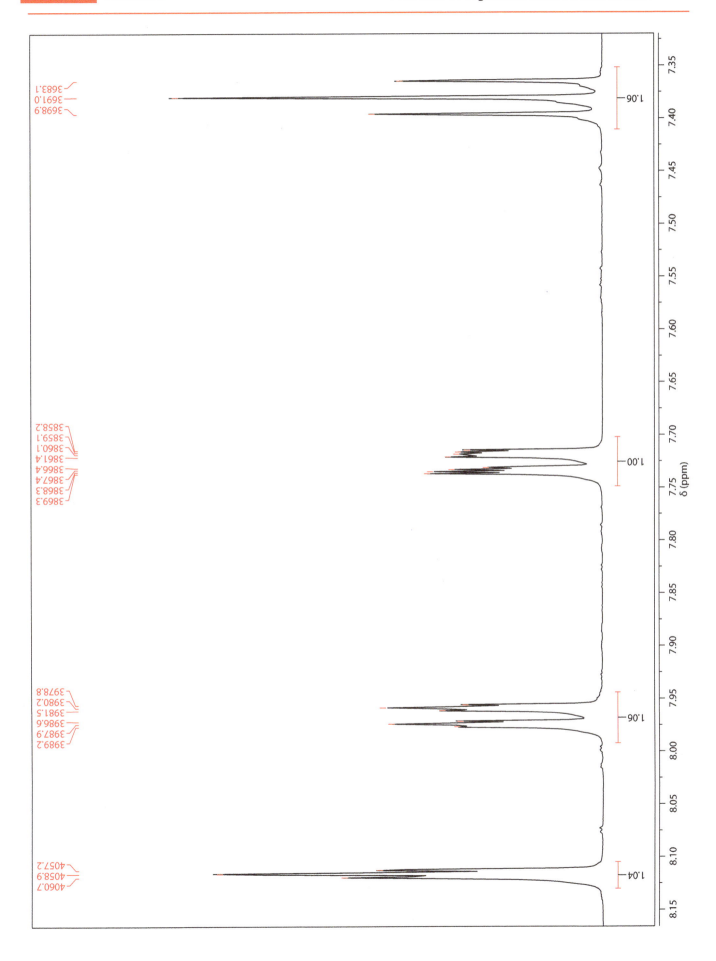

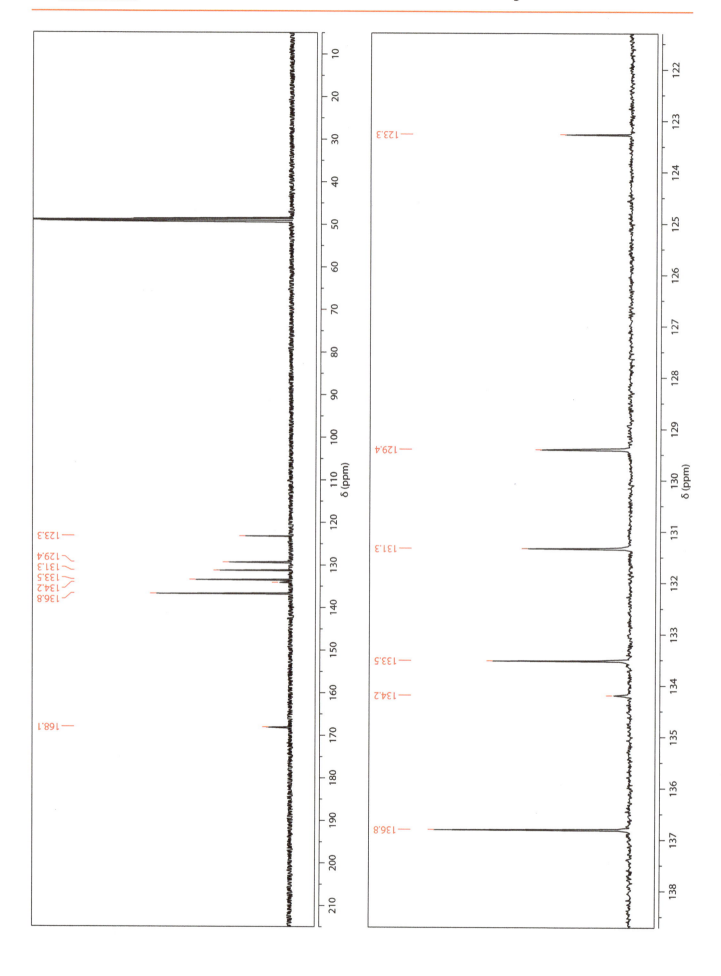

SECTION 1 Problem 8 — Problems in Organic Structure Determination

Using the ¹H and ¹³C spectra, assign all proton and carbon resonances to the structure of diethyl methylmalonate below.
Spectra acquired in CDCl₃ at 500 MHz (¹H) or 125 MHz (¹³C).

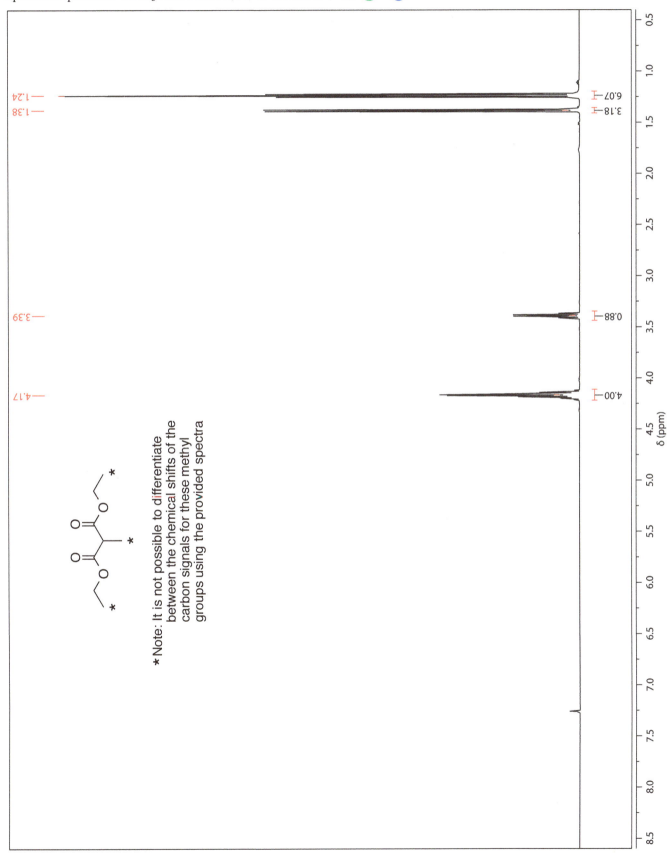

Problem 8

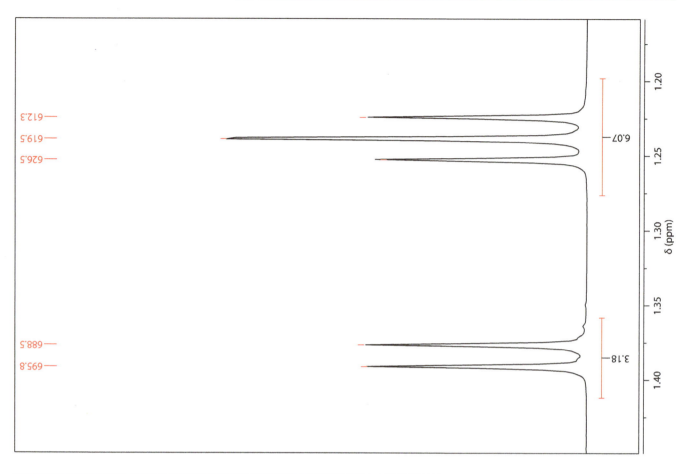

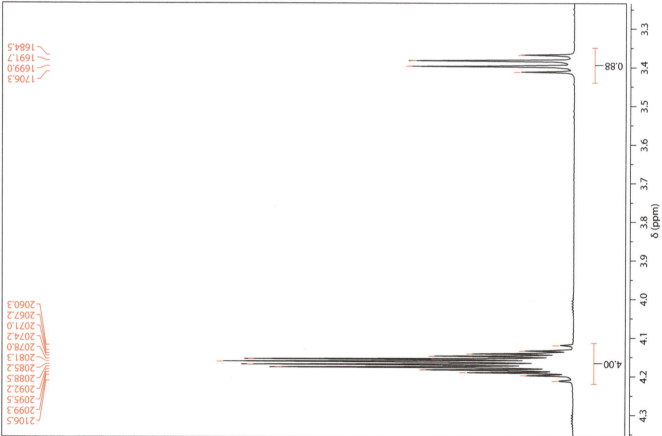

SECTION 1 Problem 8

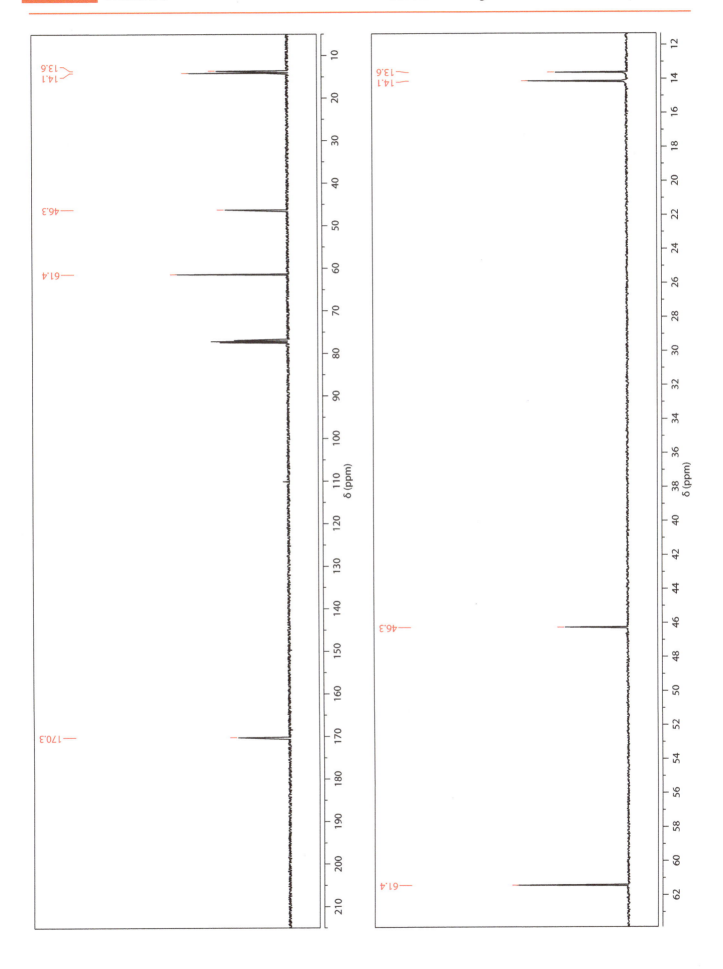

SECTION 1 Problem 9

Which of the following compounds is the correct structure for the spectra below?

Spectra acquired in CD$_3$OD at 500 MHz (^{1}H) or 125 MHz (^{13}C).

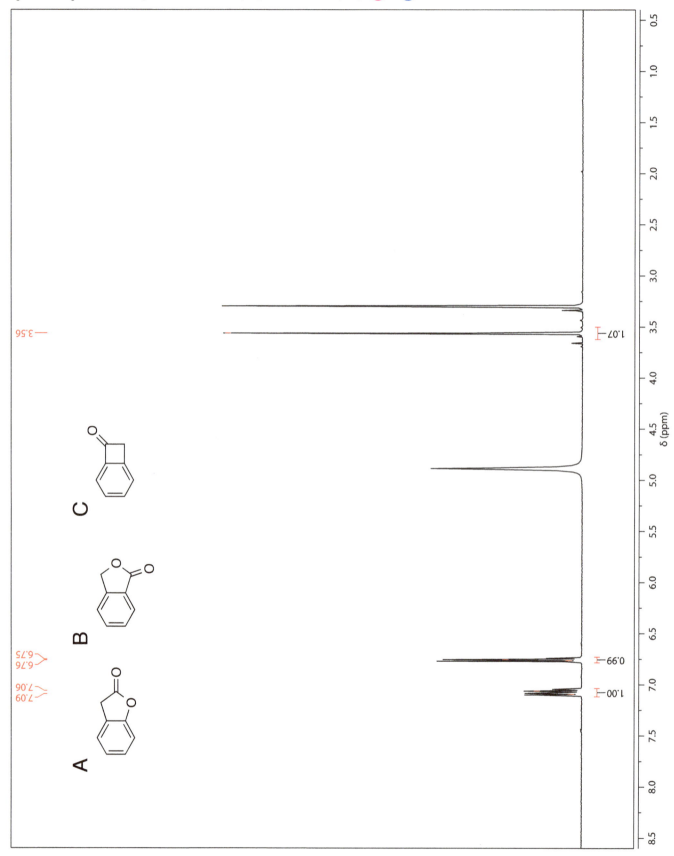

Problem 9

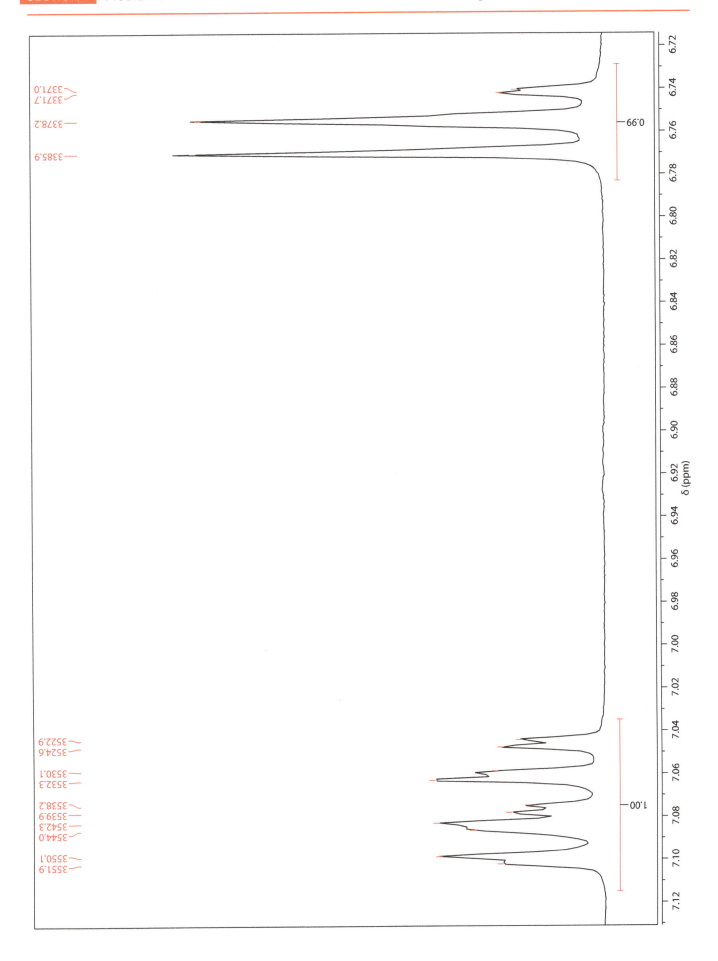

SECTION 1 Problem 9

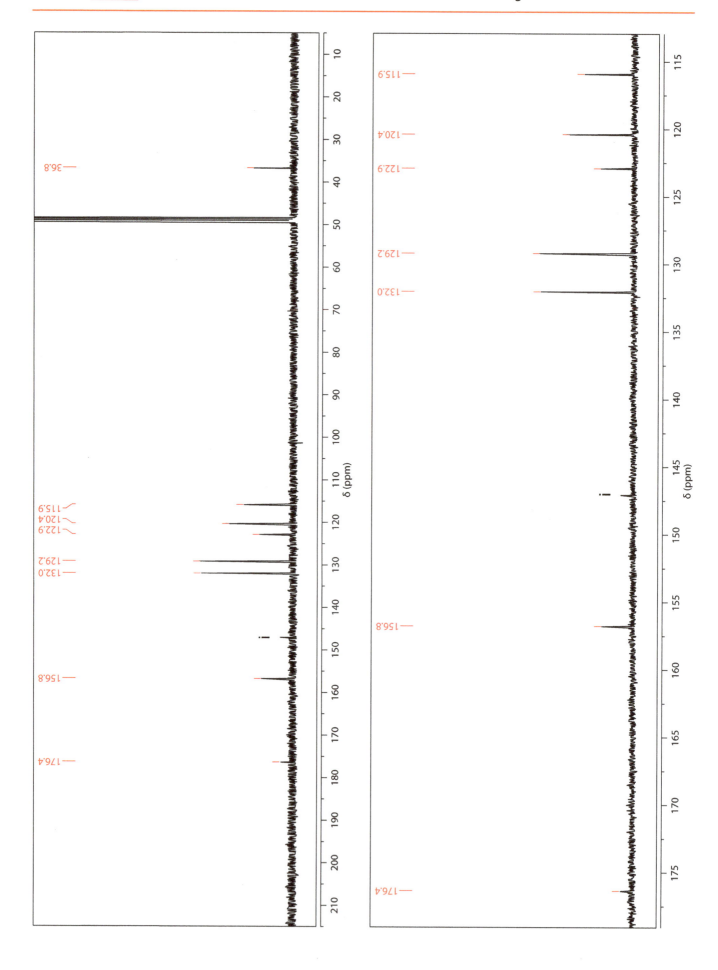

SECTION 1 Problem 10 — Problems in Organic Structure Determination

Which of the following compounds is the correct structure for the spectra below?

Spectra acquired in CDCl₃ at 500 MHz (¹H) or 125 MHz (¹³C). Ⓦ Ⓗ

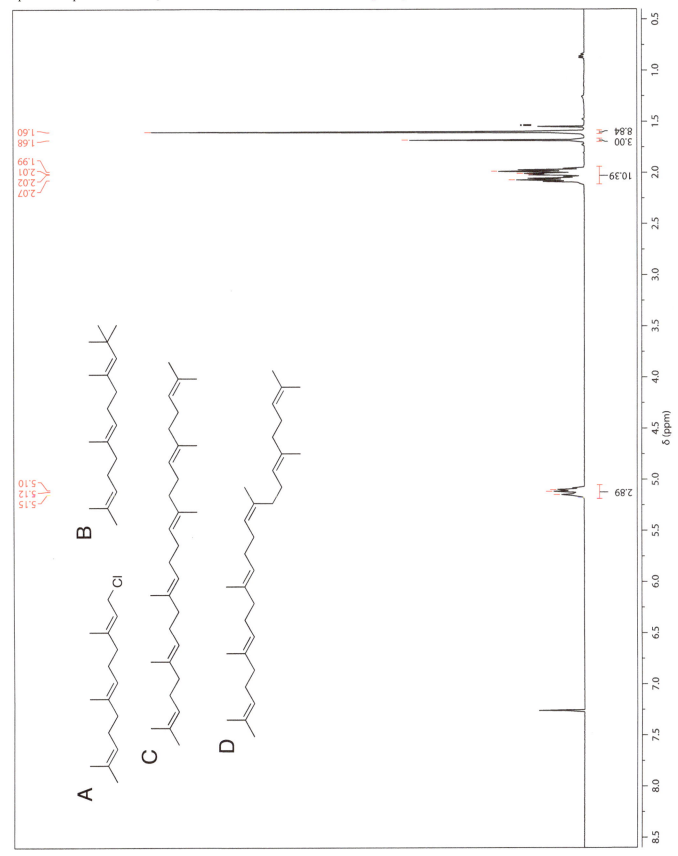

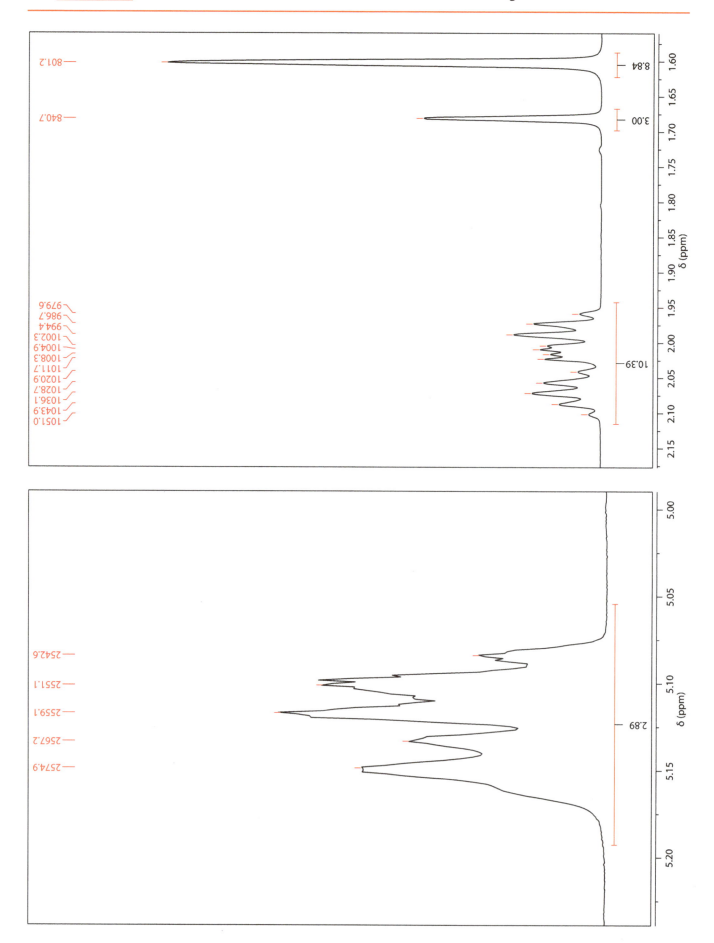

SECTION 1 Problem 10

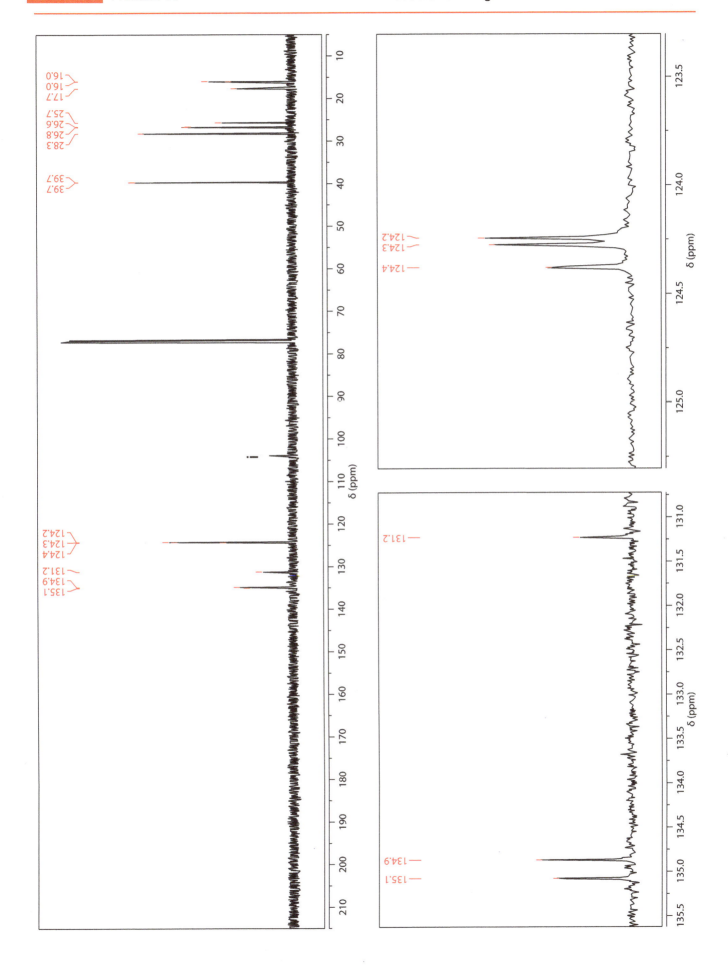

SECTION 1 Problem 10

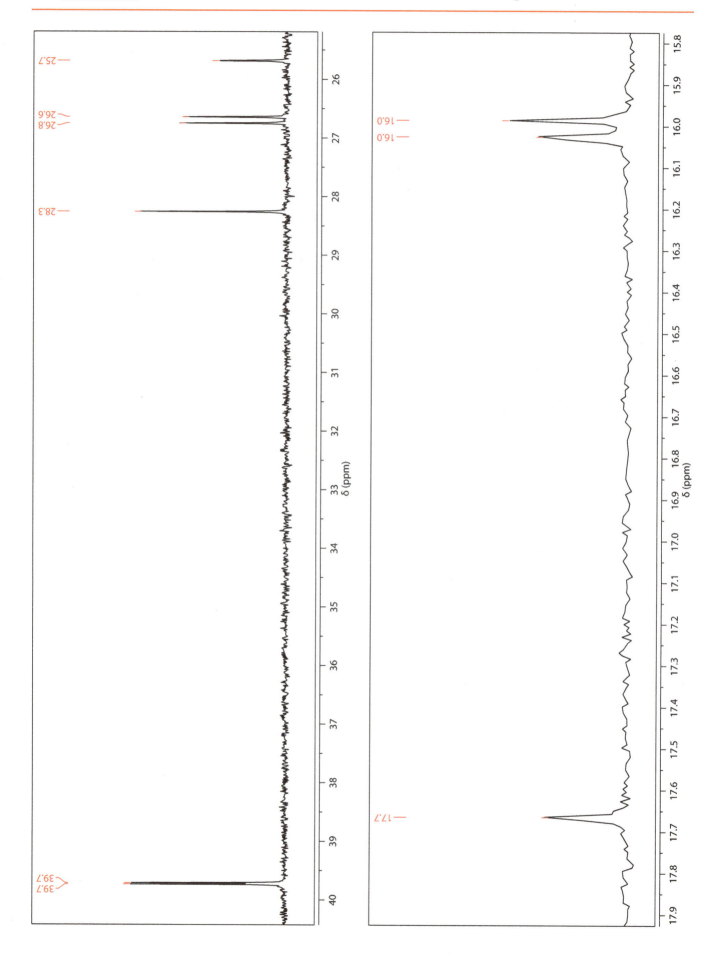

Problem 10

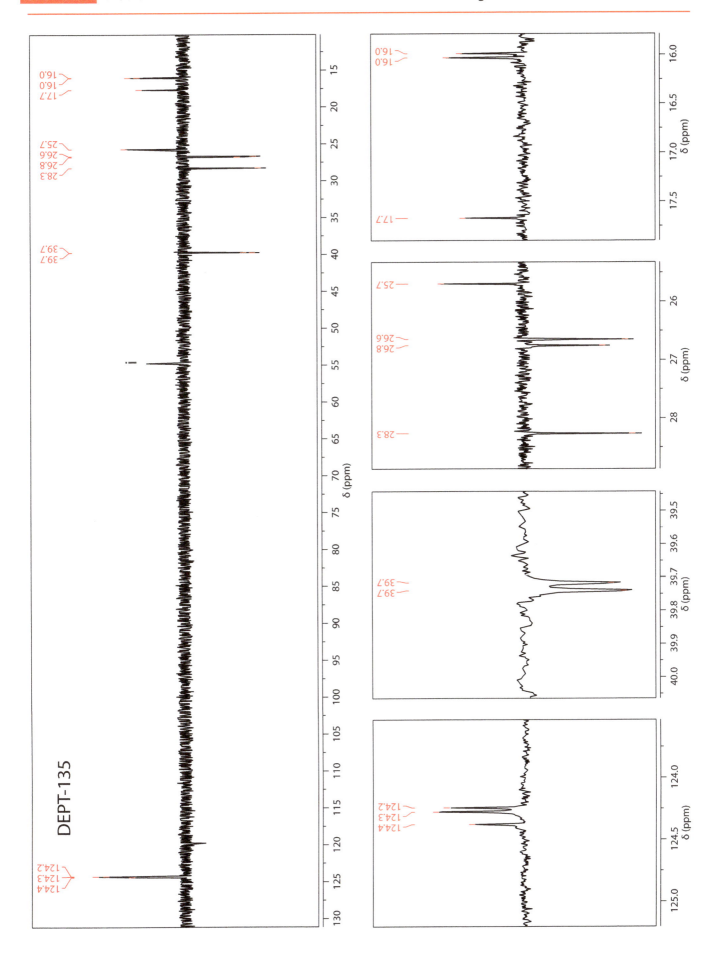

SECTION 1 Problem 11

Problems in Organic Structure Determination

Which of the following compounds is the correct structure for the spectra below?

Spectra acquired in CDCl₃ at 500 MHz (¹H) or 125 MHz (¹³C).

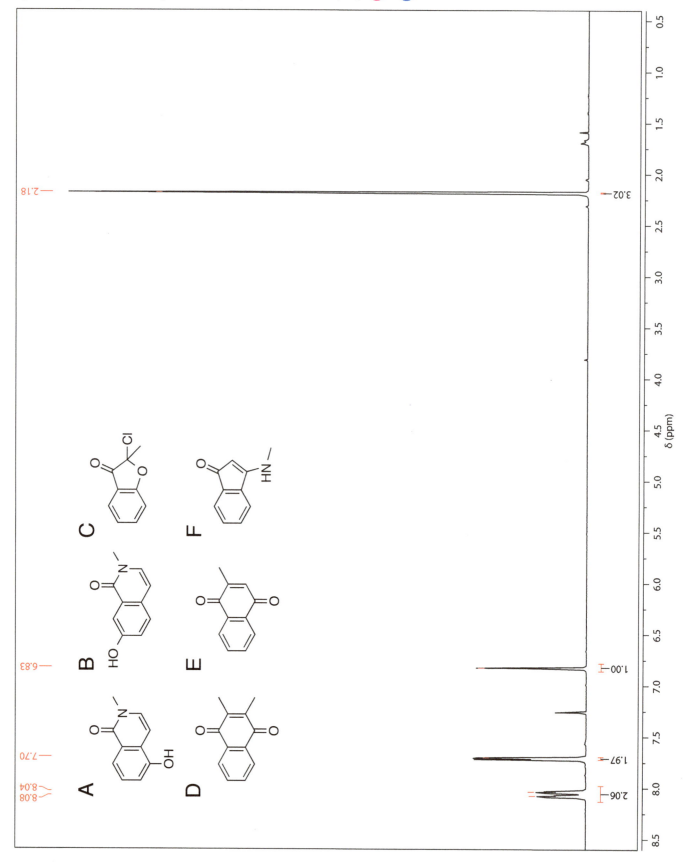

Problem 11

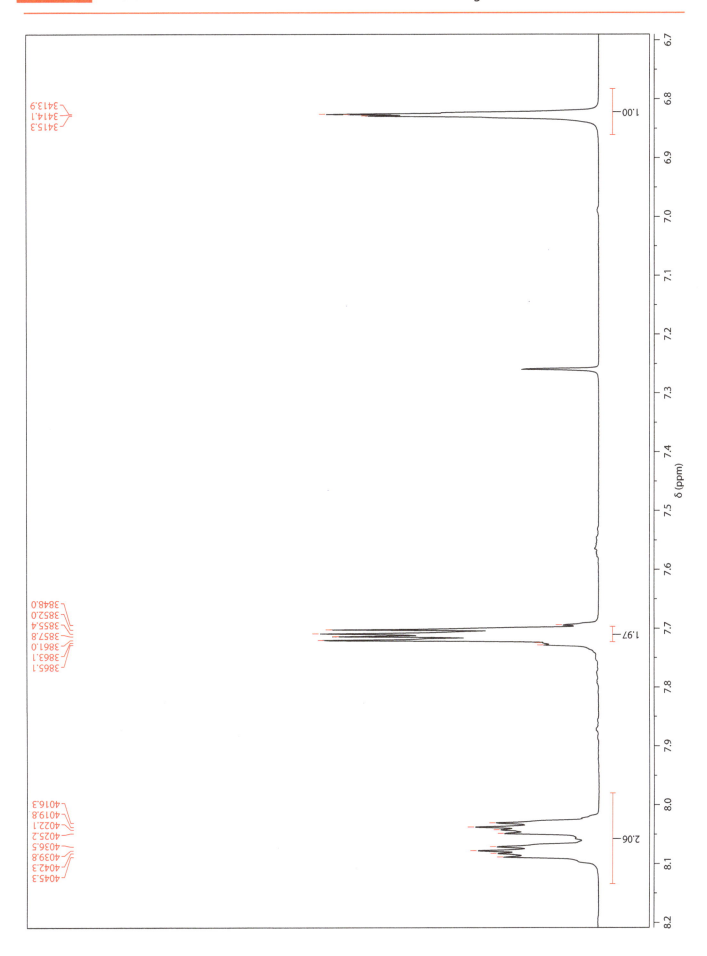

Problem 11

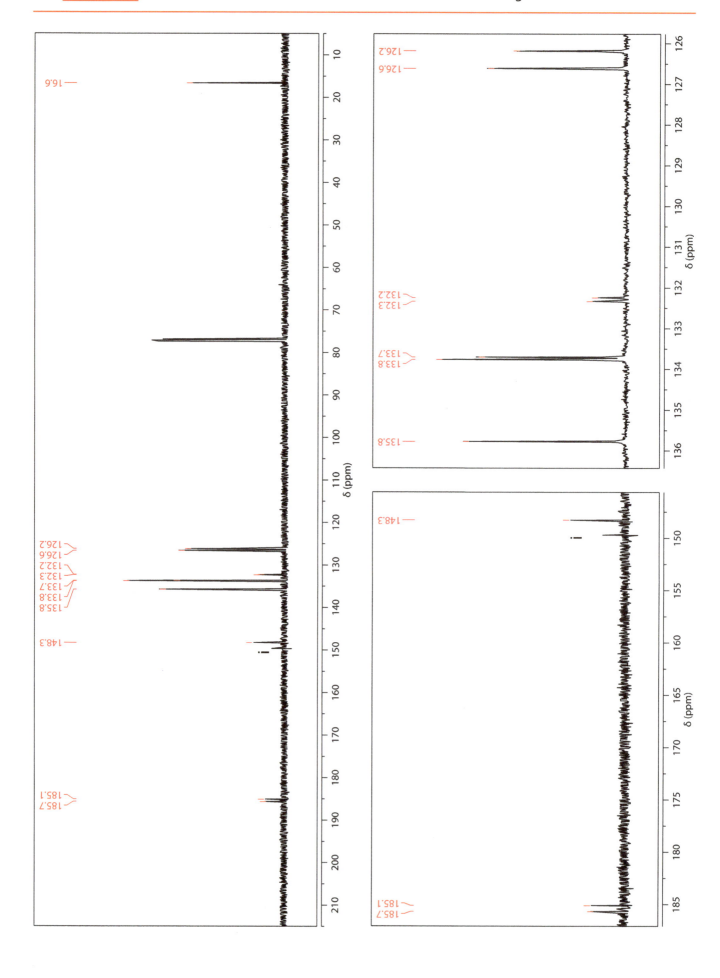

SECTION 1 Problem 12 — Problems in Organic Structure Determination

Which of the following compounds is the correct structure for the spectra below?

Spectra acquired in CDCl₃ at 500 MHz (¹H) or 125 MHz (¹³C).

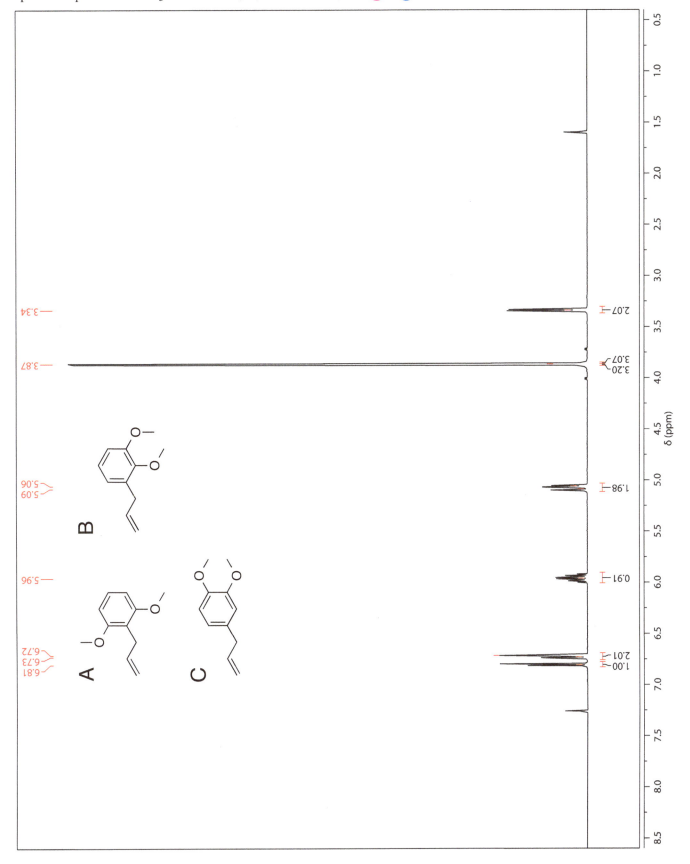

Problem 12

SECTION 1 Problem 12

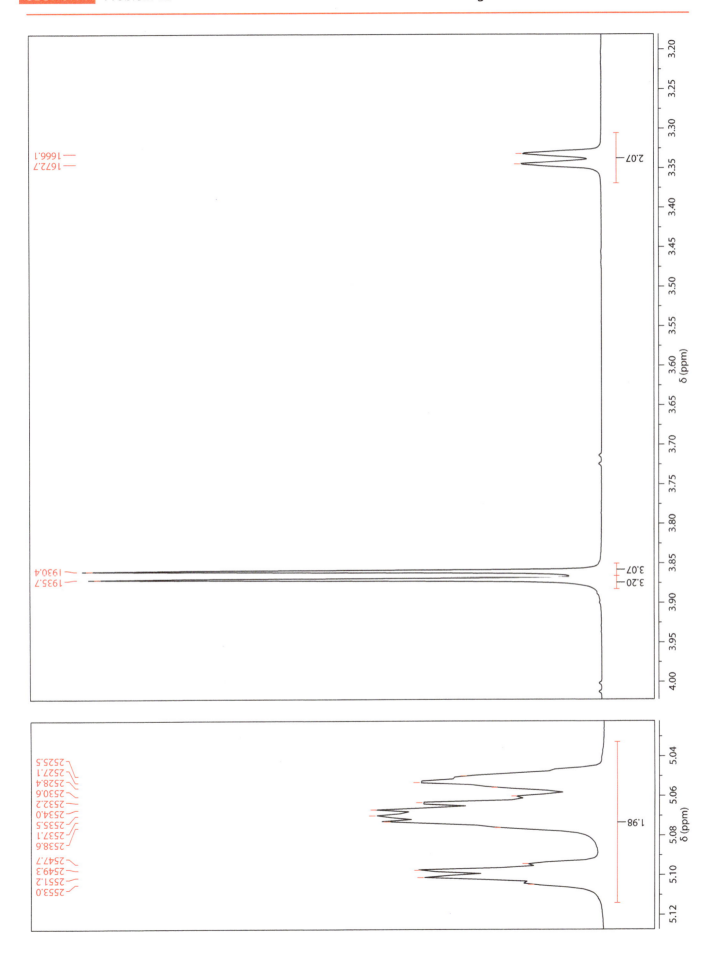

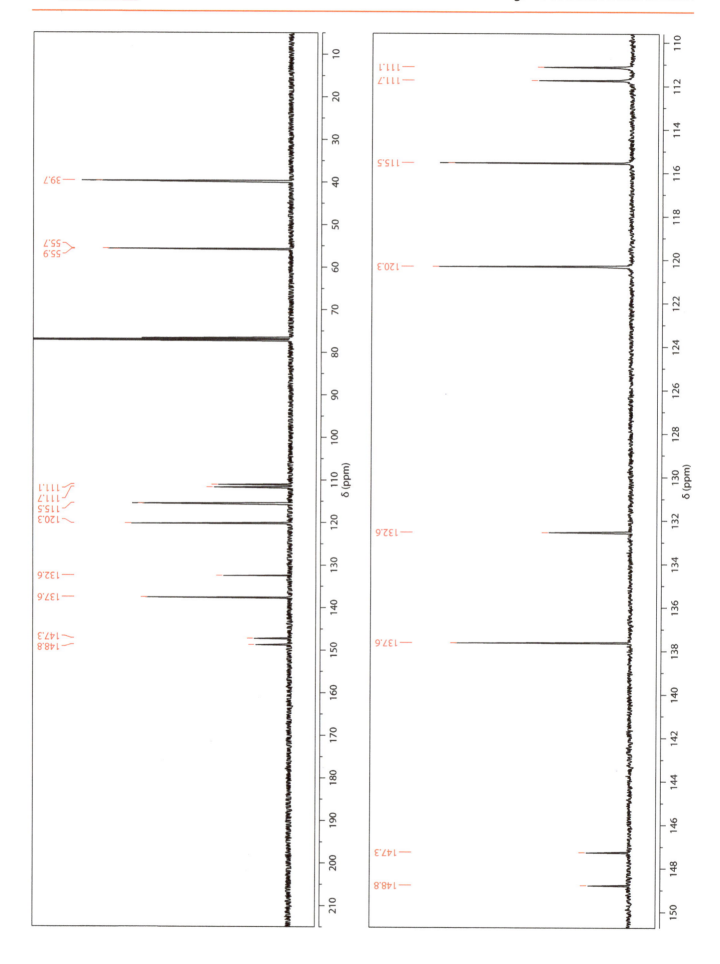

SECTION 1 Problem 13

Using the ^{1}H, ^{13}C and gHSQC spectra, assign all proton and carbon resonances to the structure of 4-penten-1-ol below and assign all proton coupling constants.
Spectra acquired in CDCl$_3$ at 500 MHz (^{1}H) or 125 MHz (^{13}C).

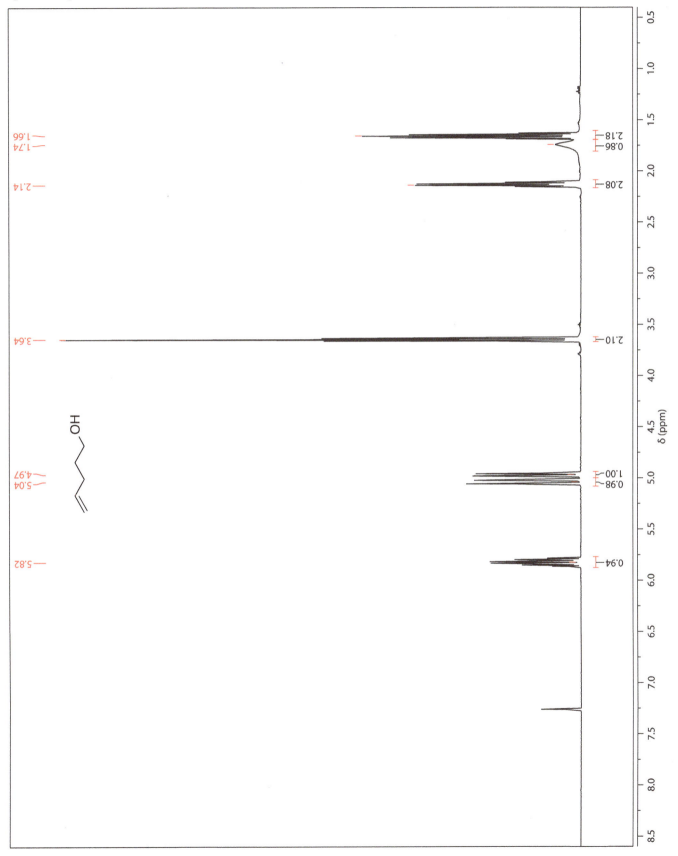

SECTION 1 Problem 13

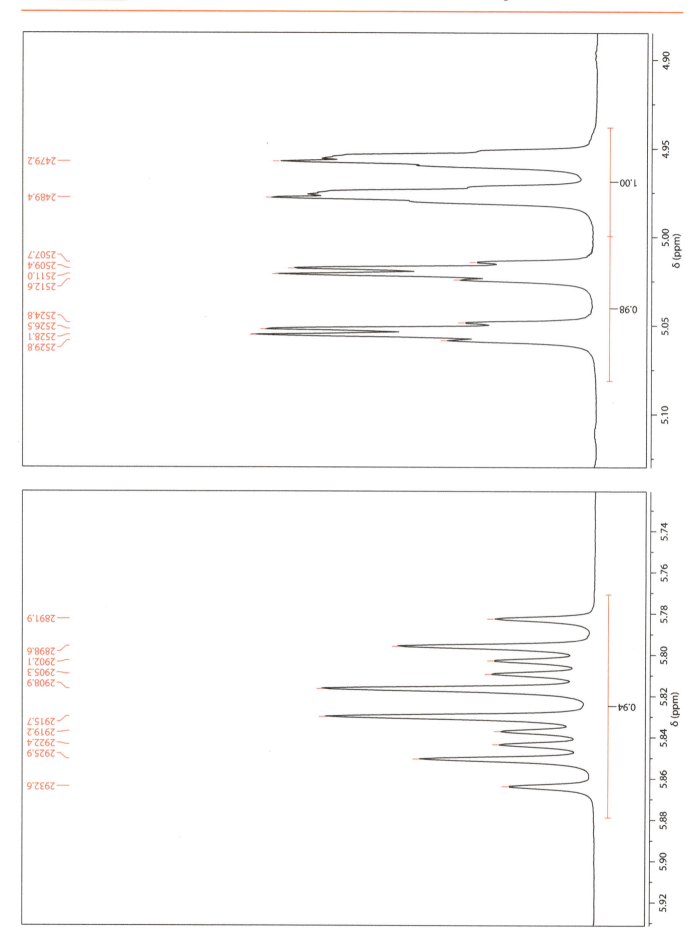

SECTION 1 Problem 13

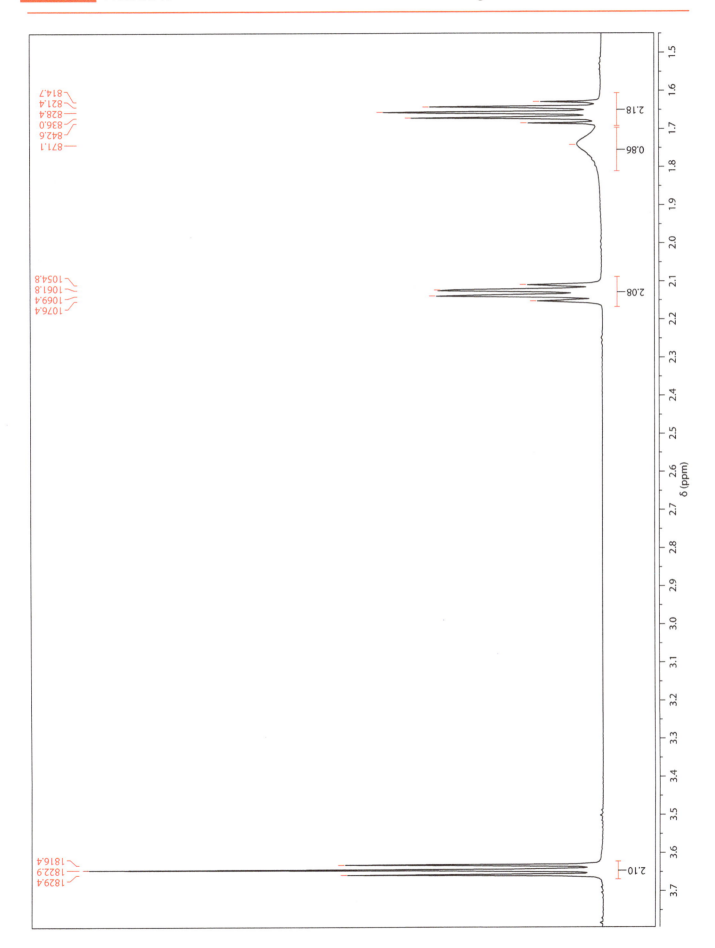

SECTION 1 Problem 13

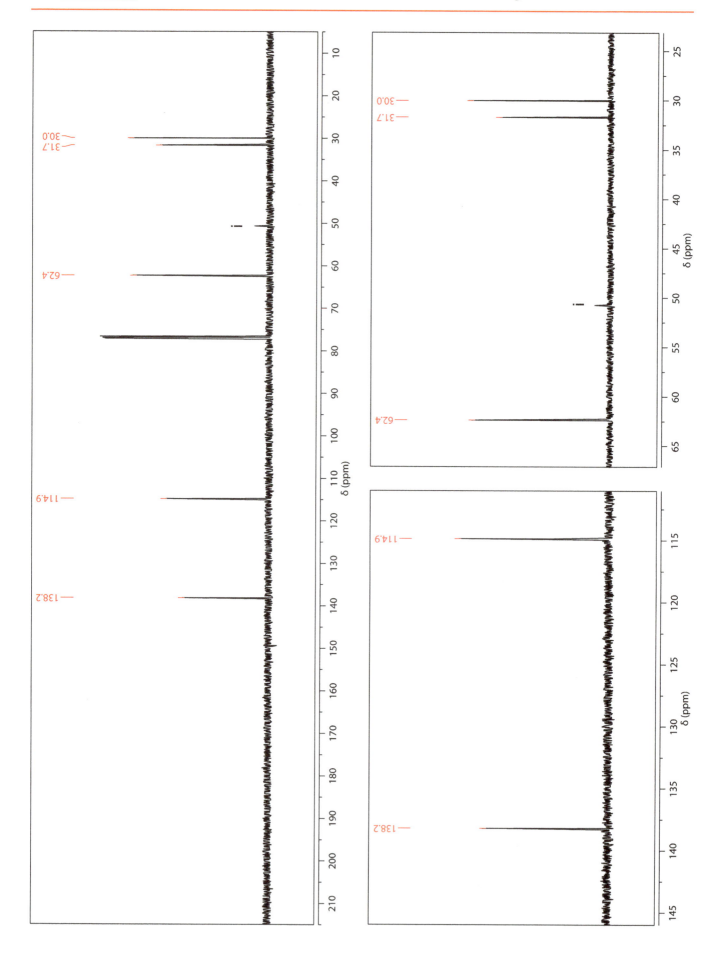

Problem 13

Problems in Organic Structure Determination

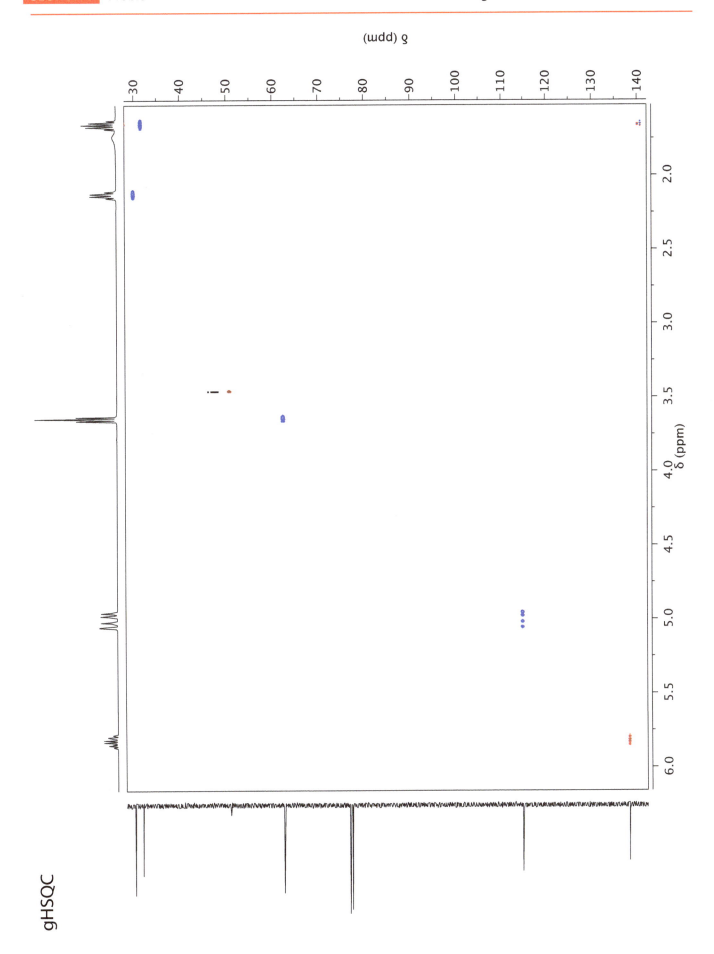

gHSQC

Problem 14

Using the ¹H, ¹³C and gHSQC spectra, assign all proton and carbon resonances to the structure of 5-bromosalicylaldehyde below and assign all proton coupling constants.
Spectra acquired in CDCl₃ at 500 MHz (¹H) or 125 MHz (¹³C).

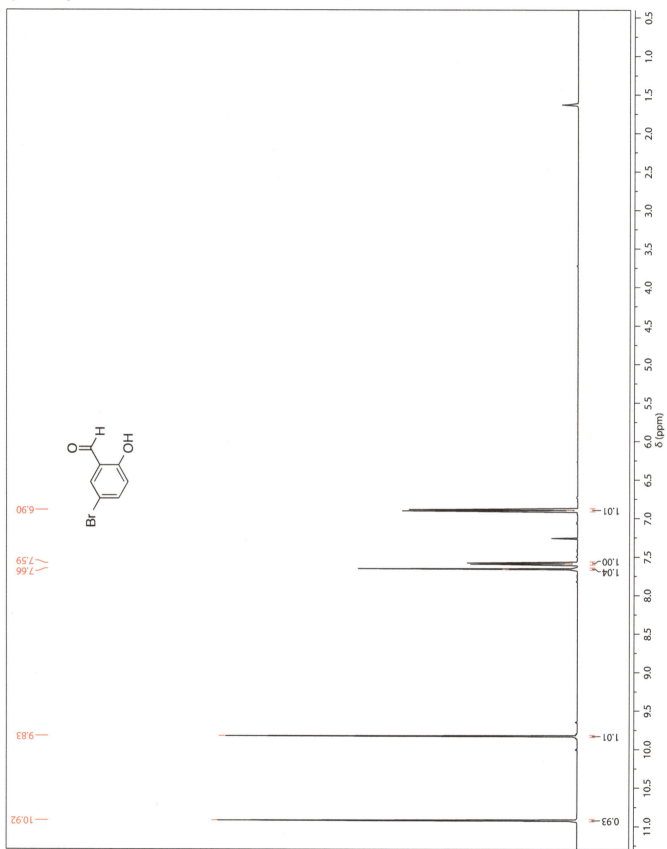

SECTION 1 Problem 14

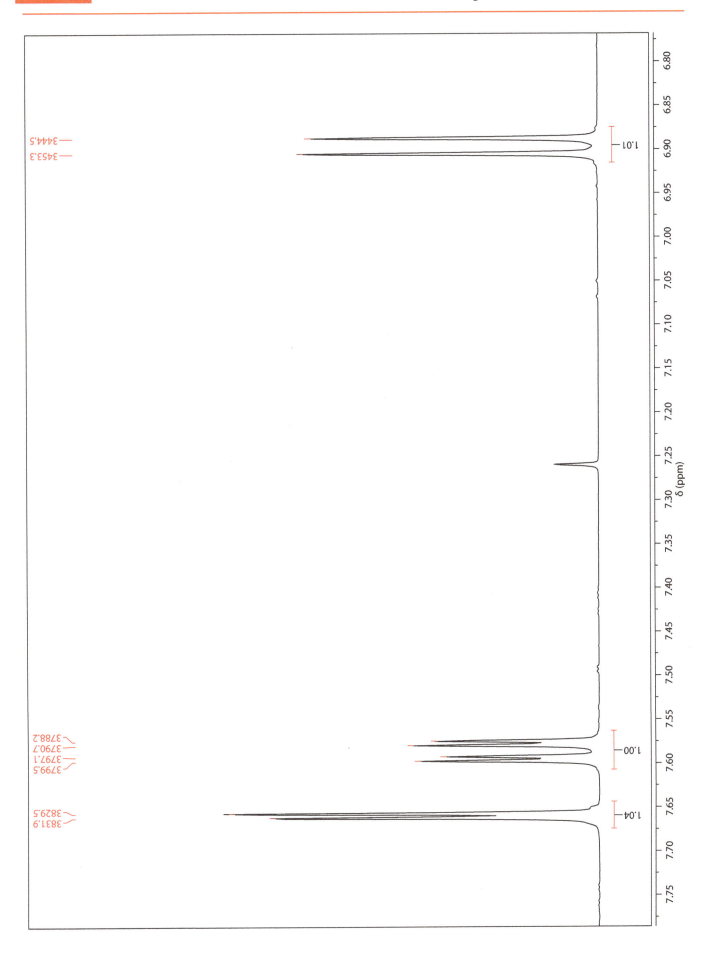

Problem 14

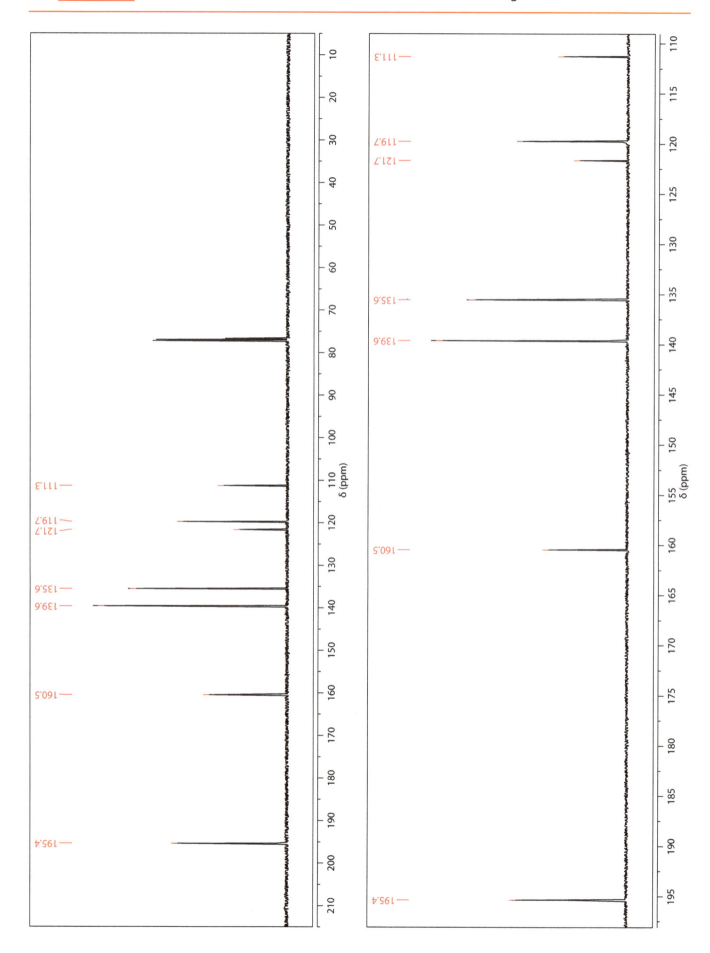

SECTION 1 Problem 14

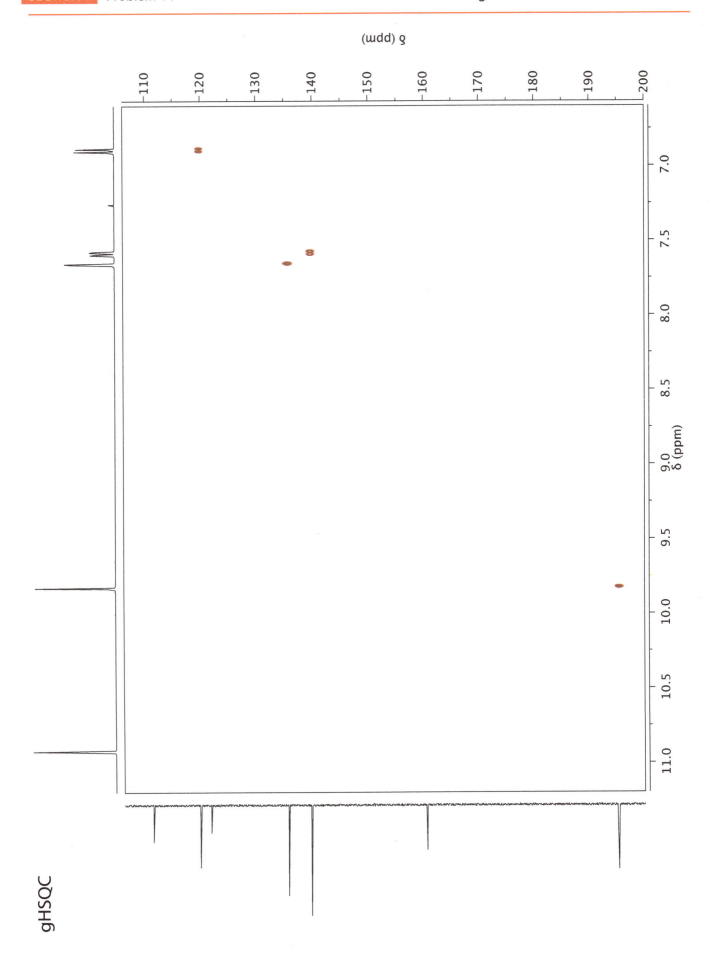

gHSQC

SECTION 1 Problem 15

Which of the following compounds is the correct structure for the spectra below?

Spectra acquired in $(CD_3)_2SO$ at 500 MHz (^1H) or 125 MHz (^{13}C).

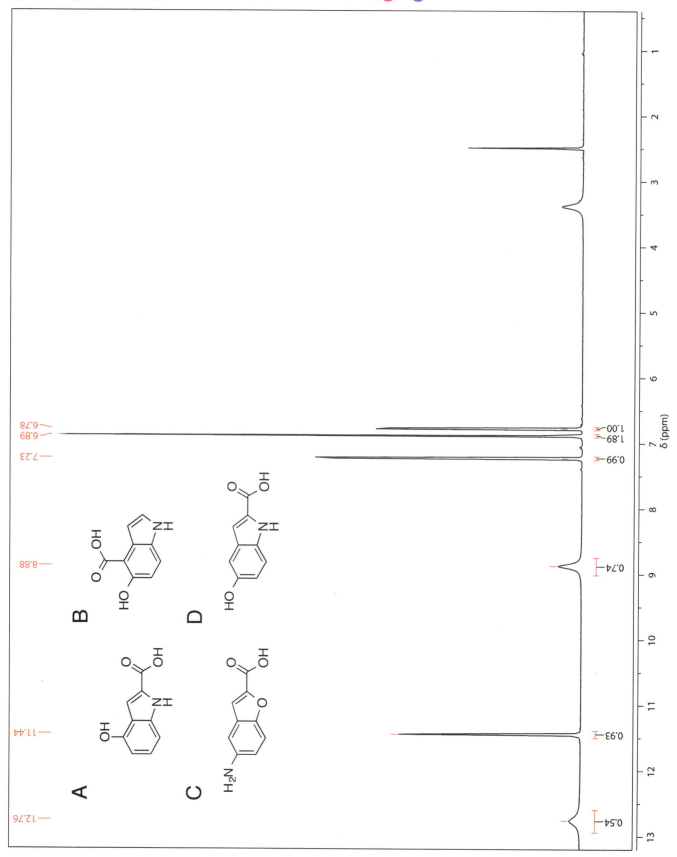

SECTION 1 Problem 15 — Problems in Organic Structure Determination

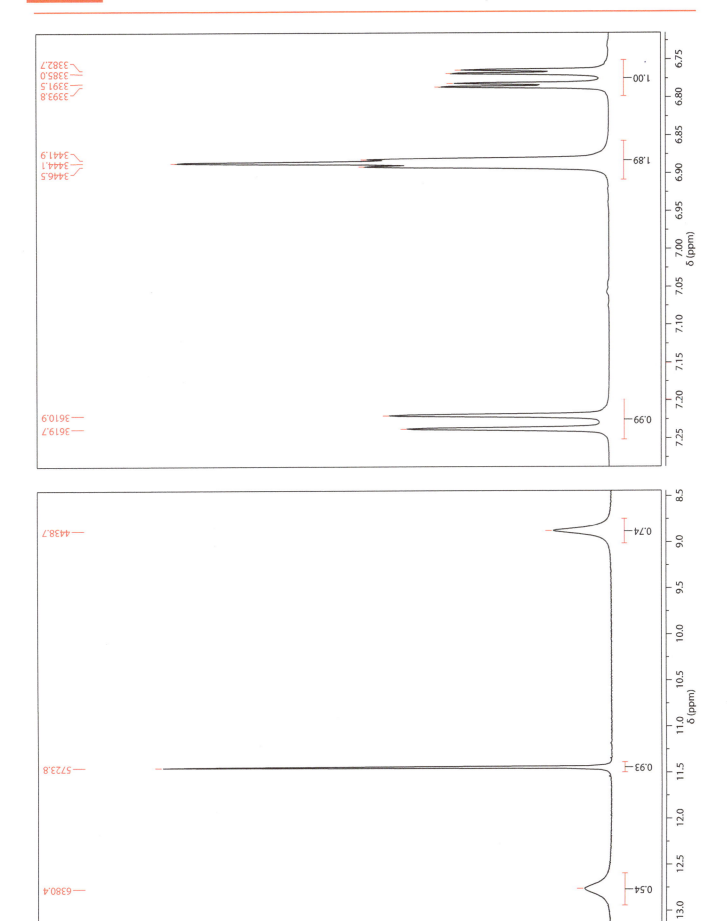

SECTION 1 Problem 15

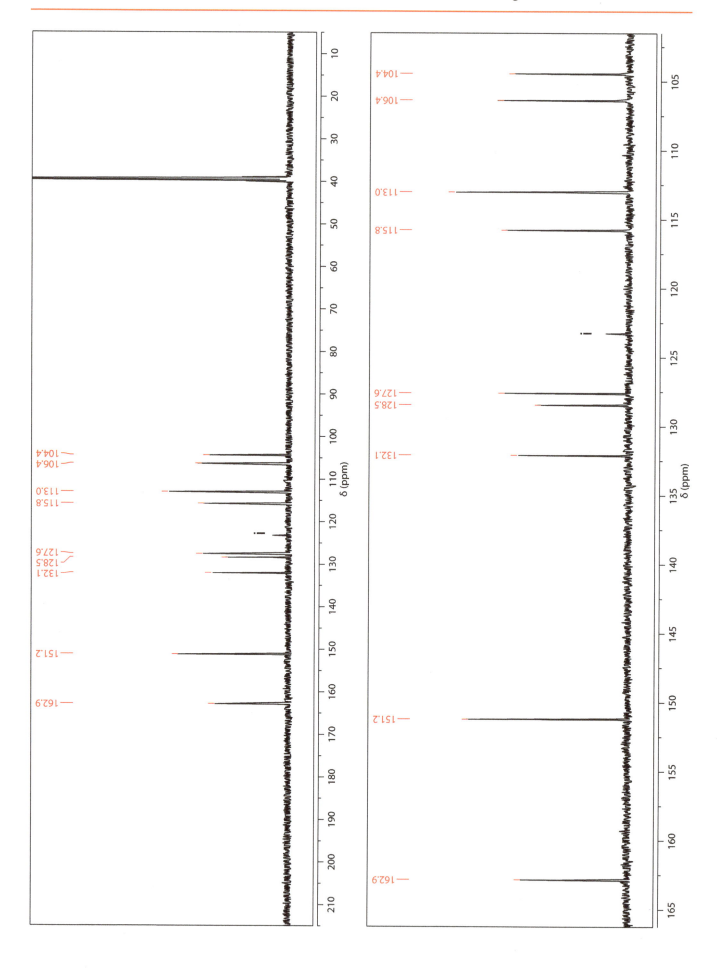

SECTION 1 Problem 16 — Problems in Organic Structure Determination

Using the ¹H, ¹³C, COSY and gHSQC spectra, assign all proton and carbon resonances to the structure of 3-bromopyridine below and assign all proton coupling constants.
Spectra acquired in CDCl₃ at 500 MHz (¹H) or 125 MHz (¹³C).

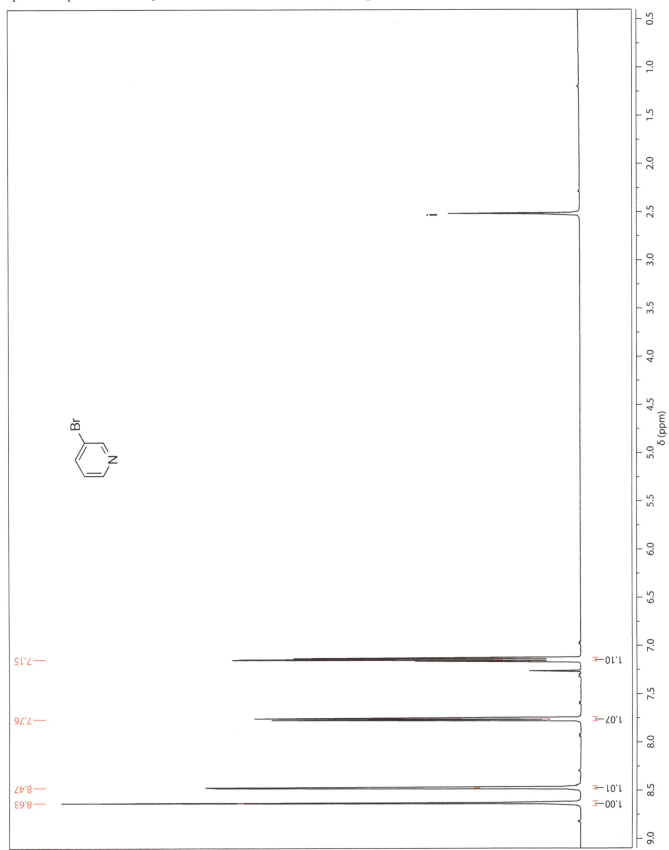

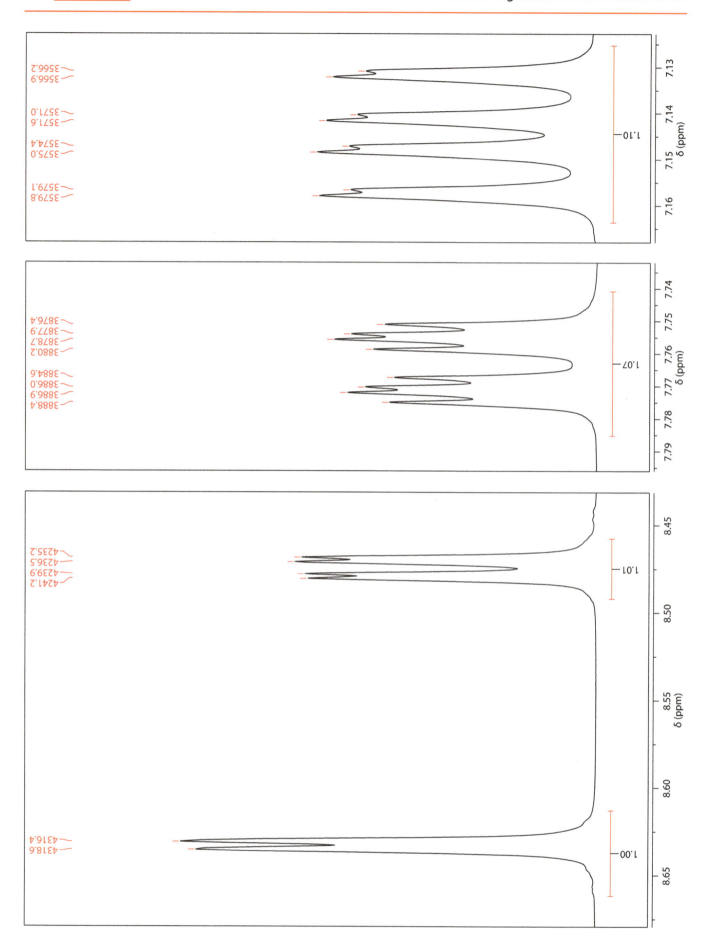

SECTION 1 Problem 16

Problems in Organic Structure Determination

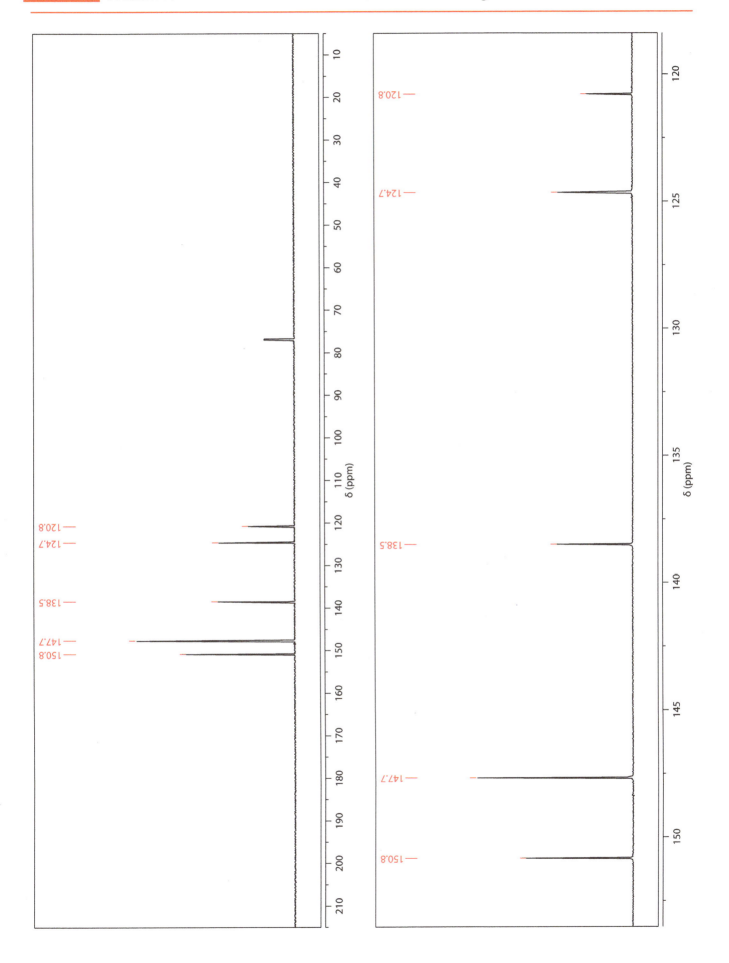

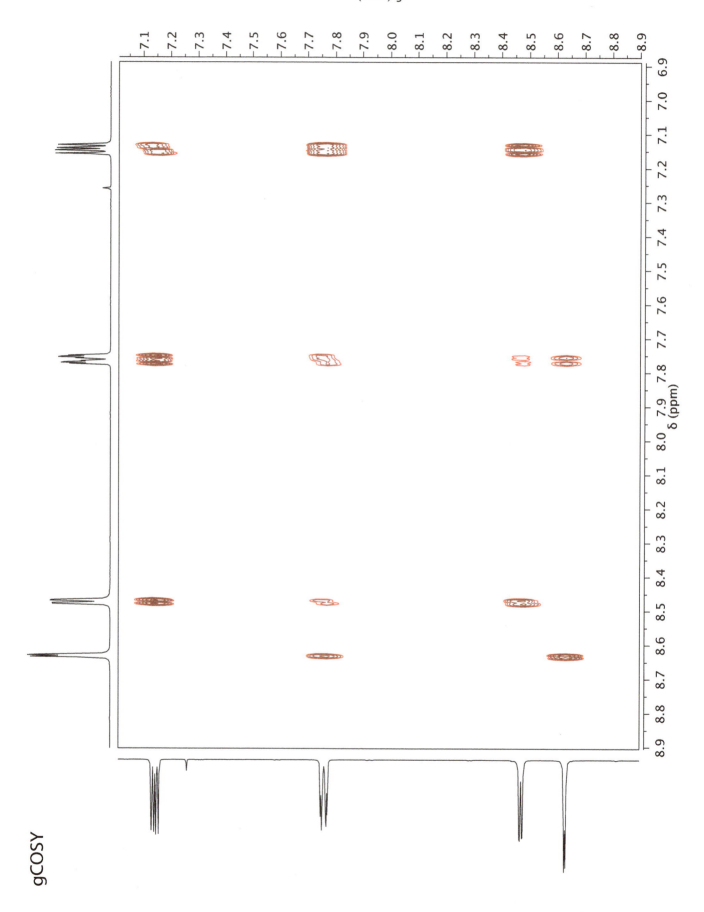

Problem 16

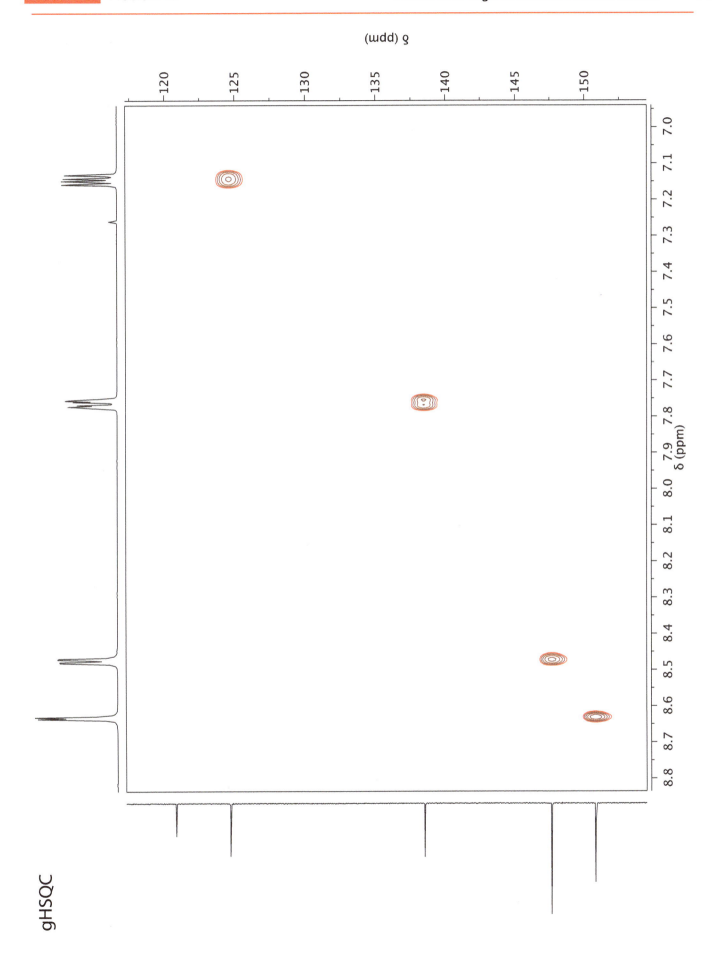

SECTION 1 Problem 17

Using the ^1H, ^{13}C, gCOSY and gHSQC spectra, assign all proton and carbon resonances to the structure of 2-hydroxyisocaproic acid below.
Spectra acquired in $CDCl_3$ at 500 MHz (^1H) or 125 MHz (^{13}C).

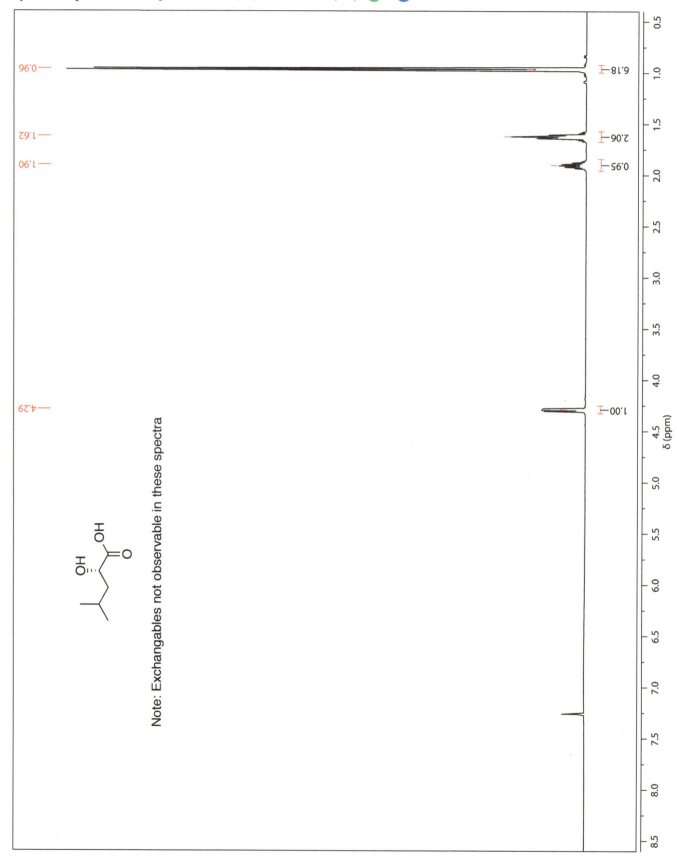

SECTION 1 Problem 17

Problems in Organic Structure Determination

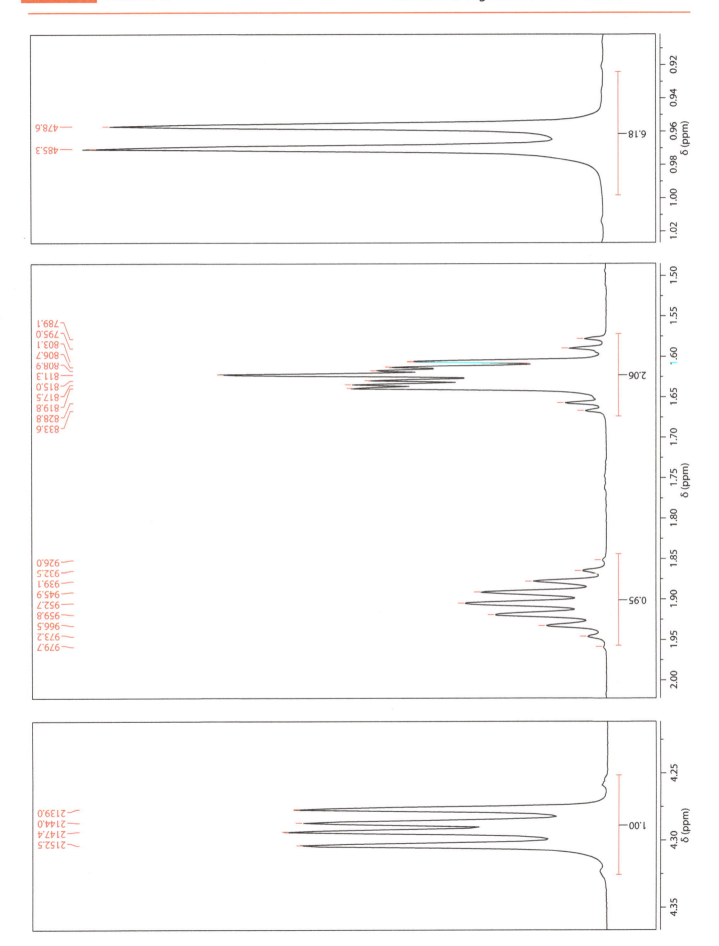

SECTION 1 Problem 17

Problems in Organic Structure Determination

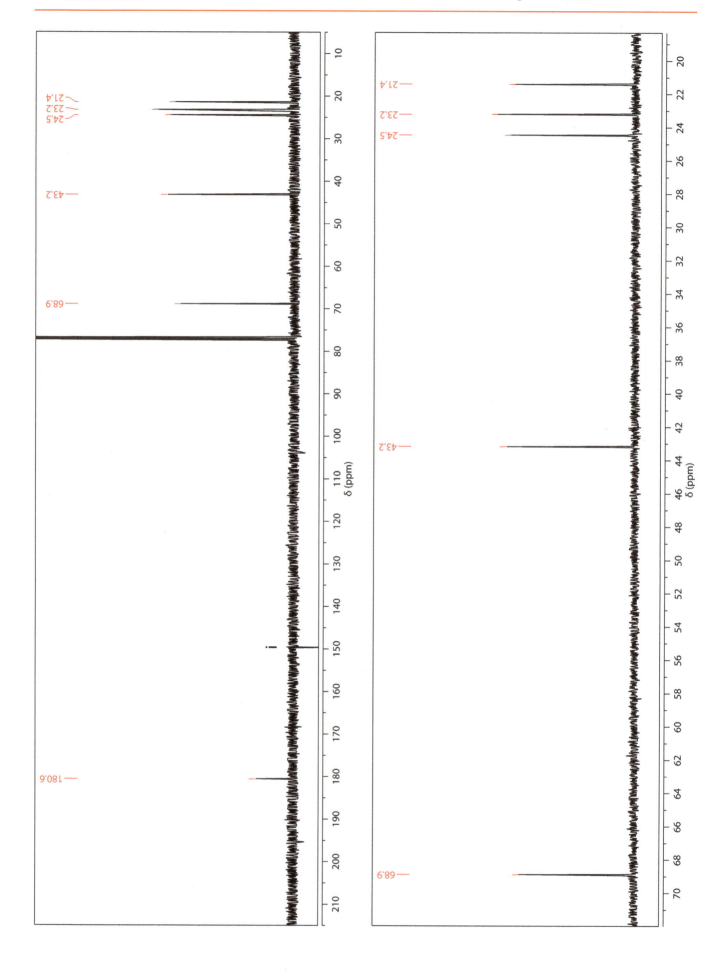

SECTION 1 Problem 17

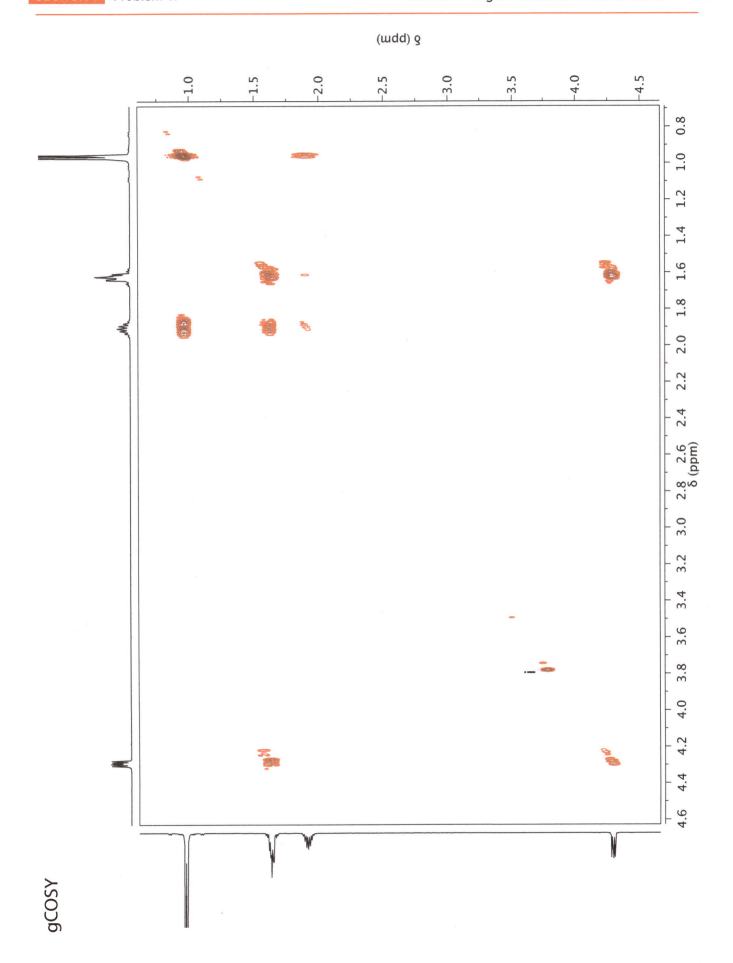

gCOSY

Problem 17

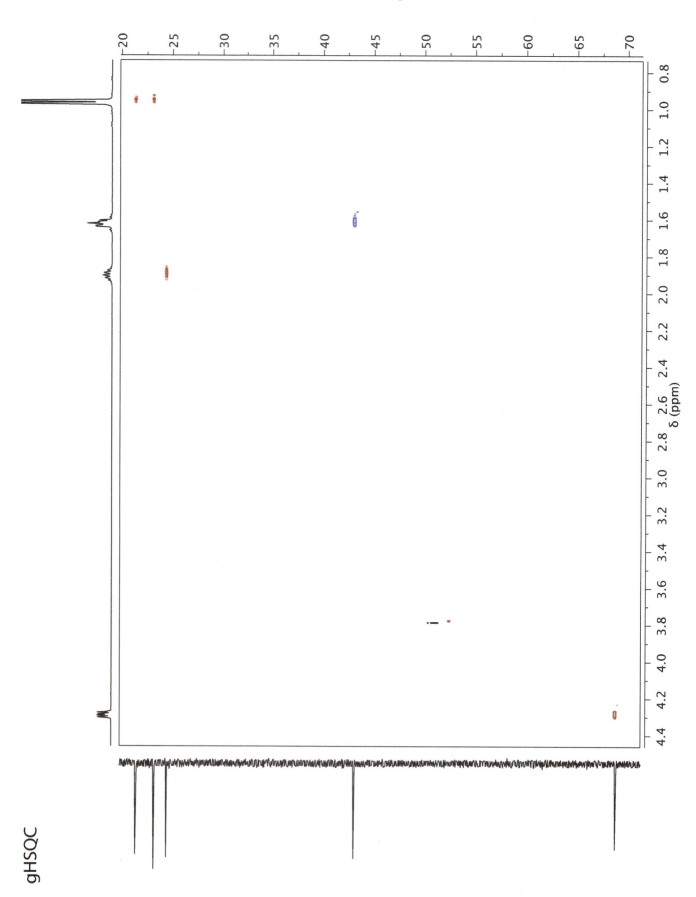

SECTION 1 Problem 18 — Problems in Organic Structure Determination

Using the ^{1}H, ^{13}C, gCOSY and gHSQC spectra, assign all proton and carbon resonances to the structure of 7-octenoic acid below and assign all proton coupling constants.
Spectra acquired in CDCl$_3$ at 500 MHz (^{1}H) or 125 MHz (^{13}C).

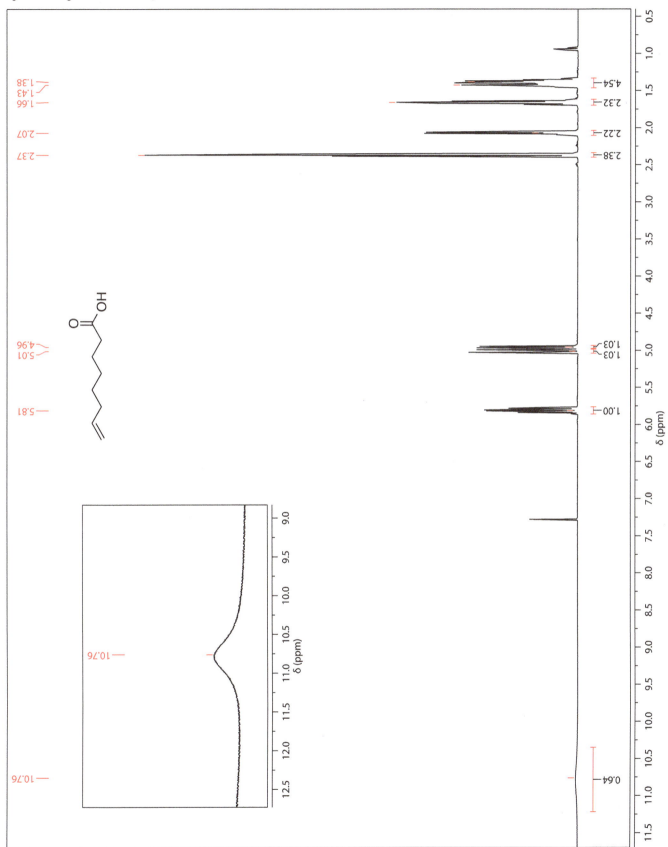

SECTION 1 Problem 18

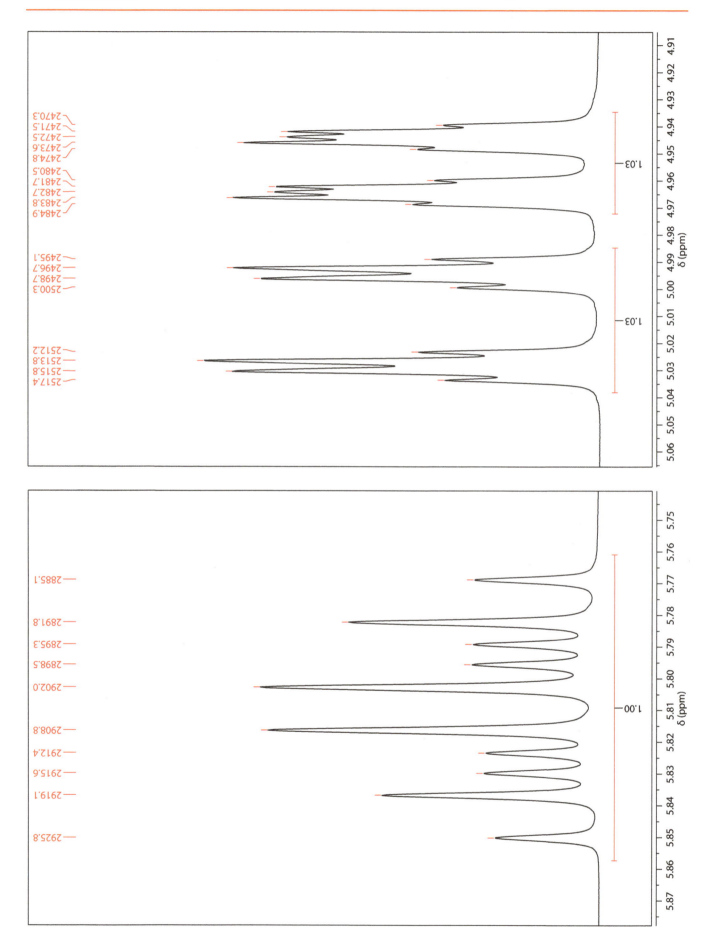

SECTION 1 Problem 18

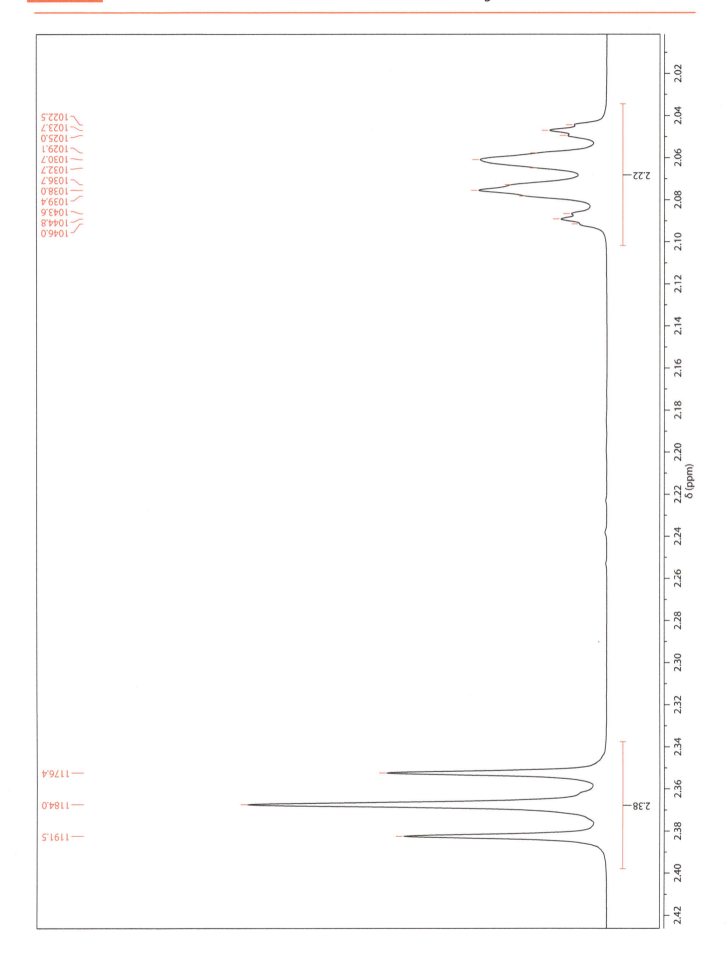

SECTION 1 Problem 18

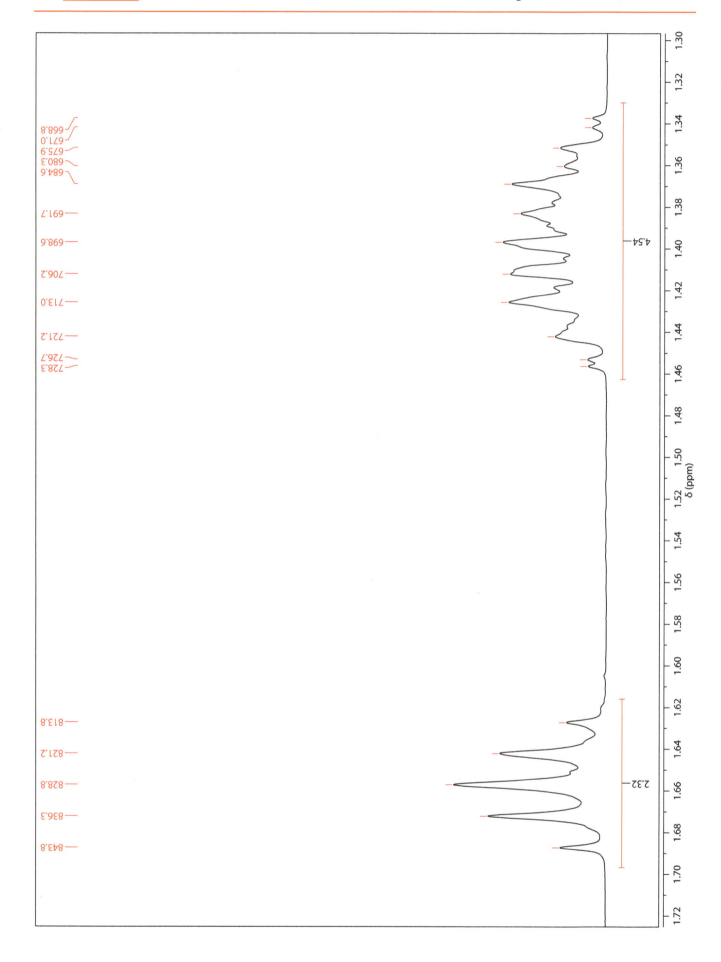

SECTION 1 Problem 18

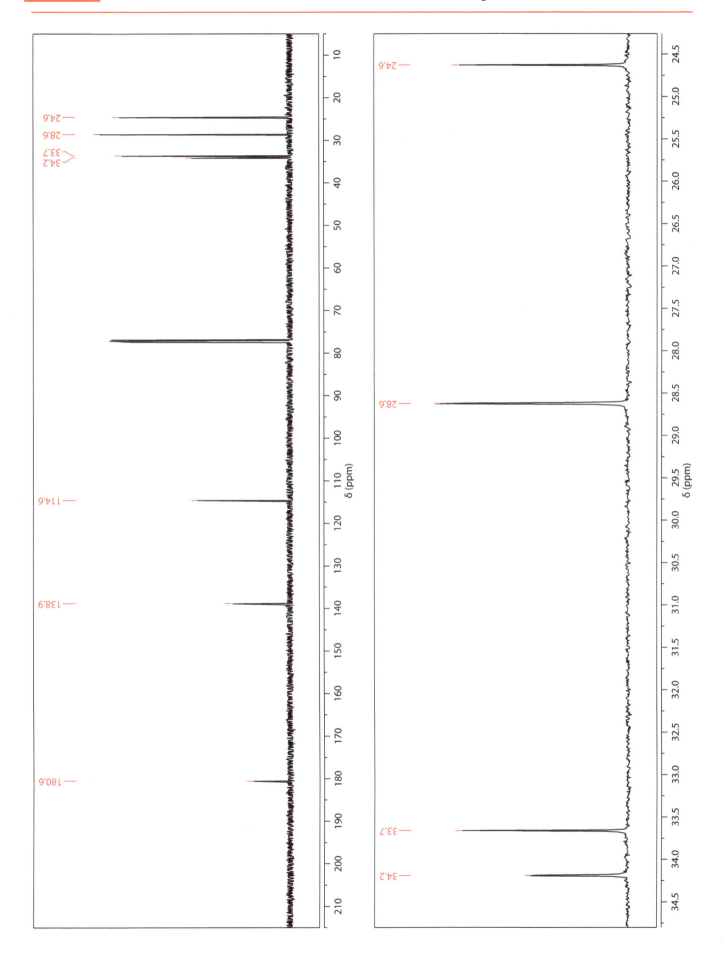

Problem 18

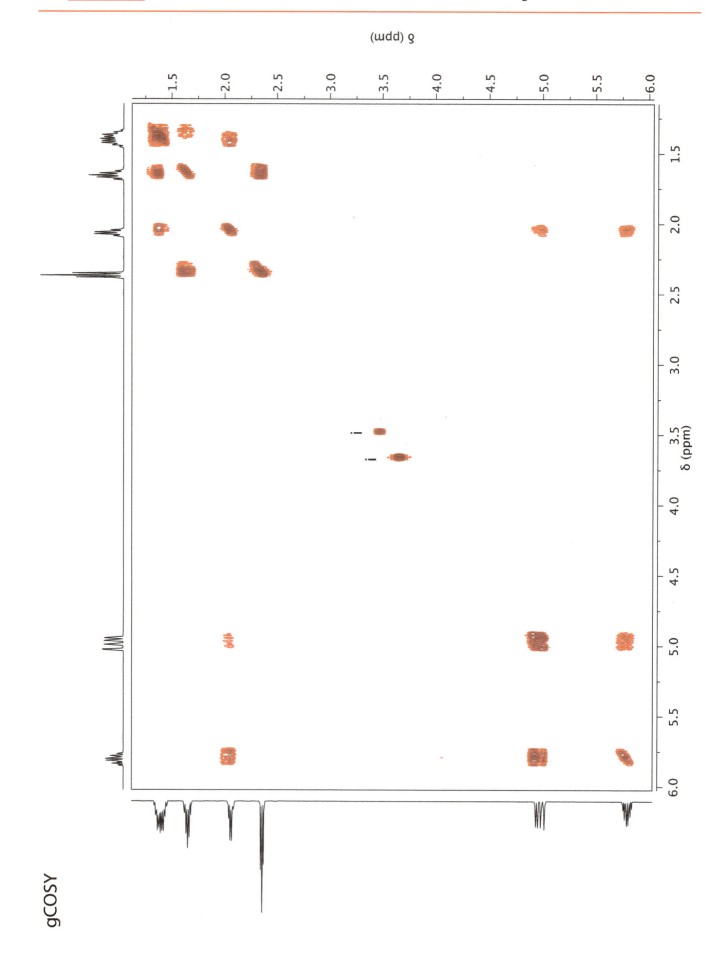

gCOSY

SECTION 1 Problem 18

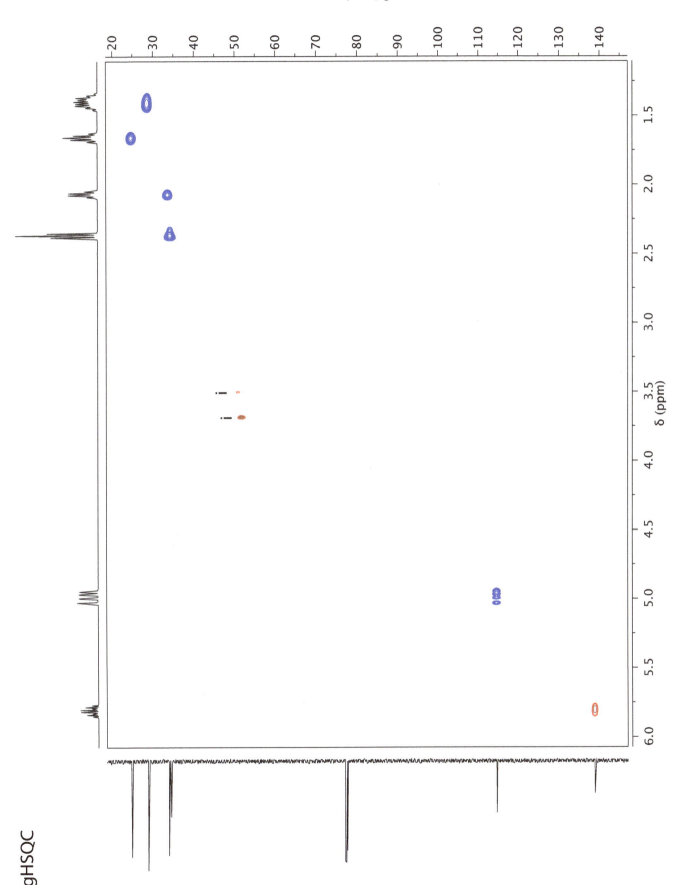

gHSQC

Problem 18

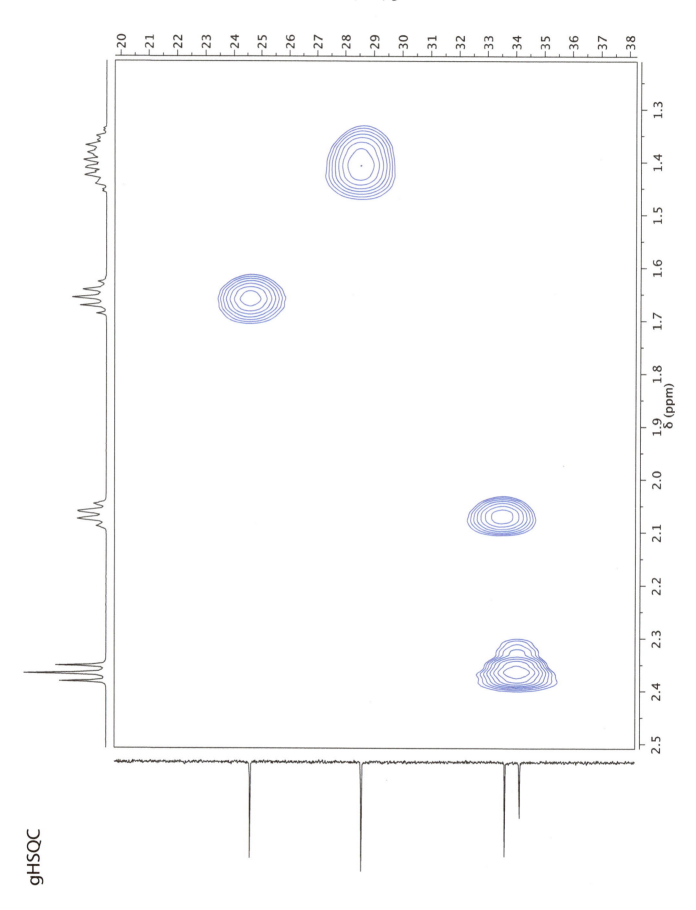

gHSQC

SECTION 1 Problem 19

Problems in Organic Structure Determination

Using the ¹H, ¹³C, gCOSY and gHSQC spectra, assign all proton and carbon resonances to the structure of benzyl-3-bromopropyl ether below.
Spectra acquired in CDCl₃ at 500 MHz (¹H) or 125 MHz (¹³C).

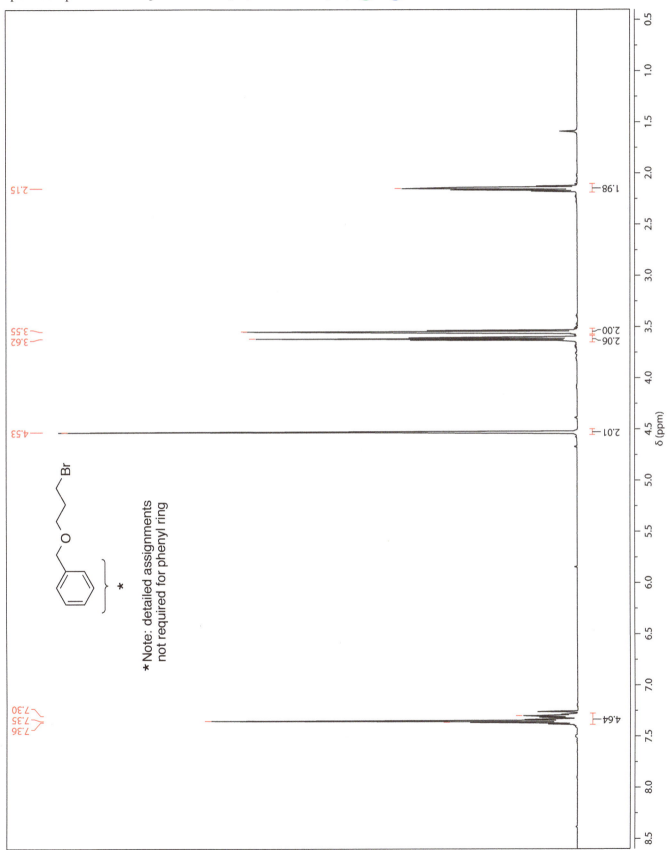

SECTION 1 Problem 19

SECTION 1 Problem 19 — Problems in Organic Structure Determination

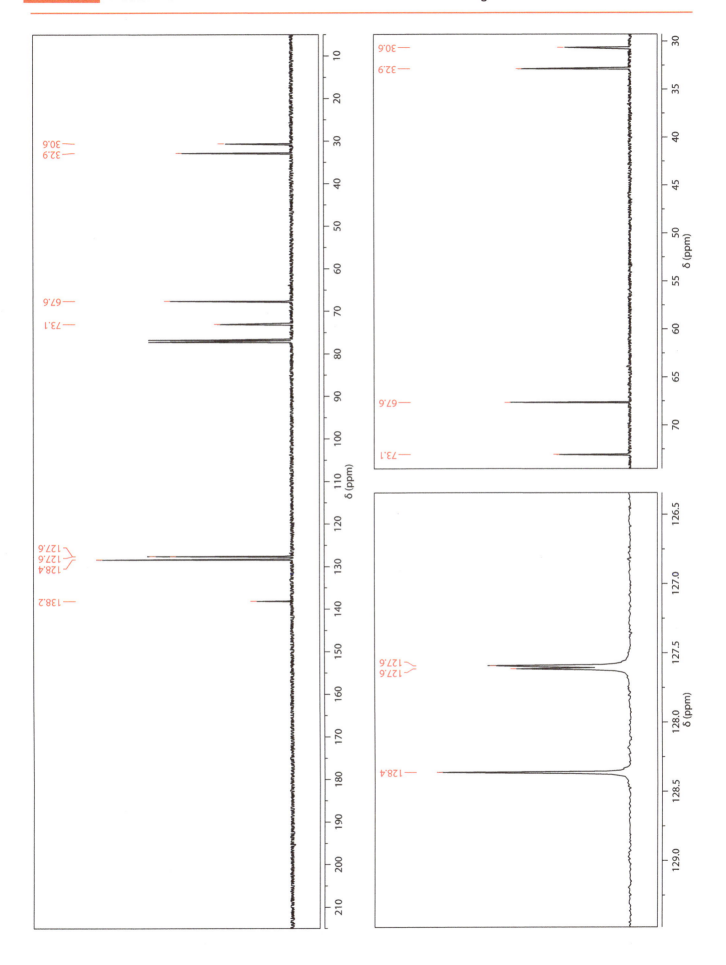

Problem 19

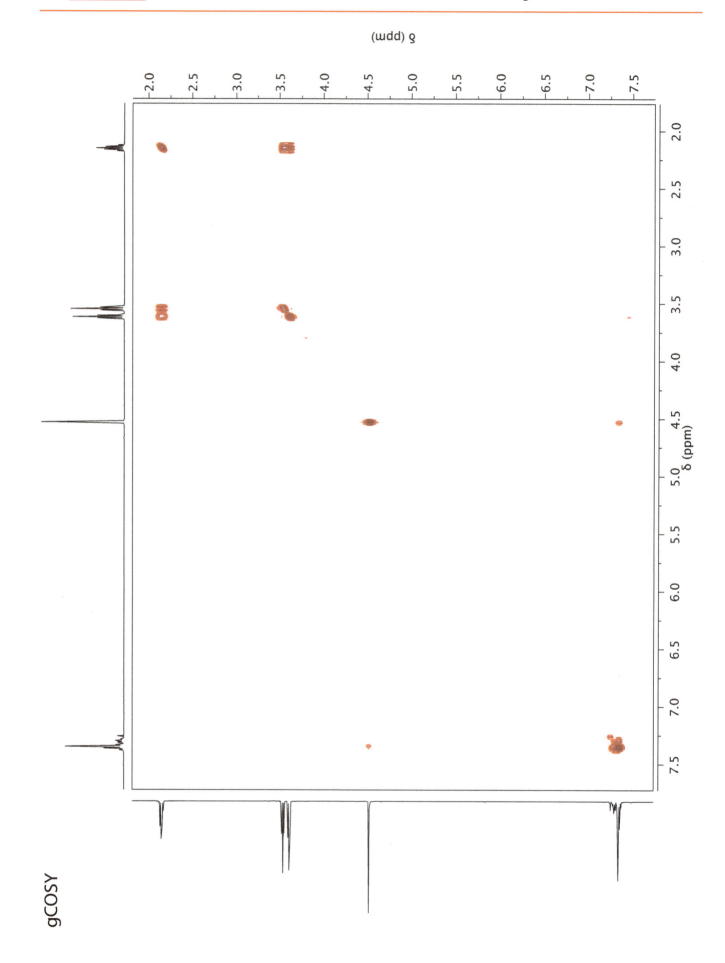

gCOSY

Problem 19

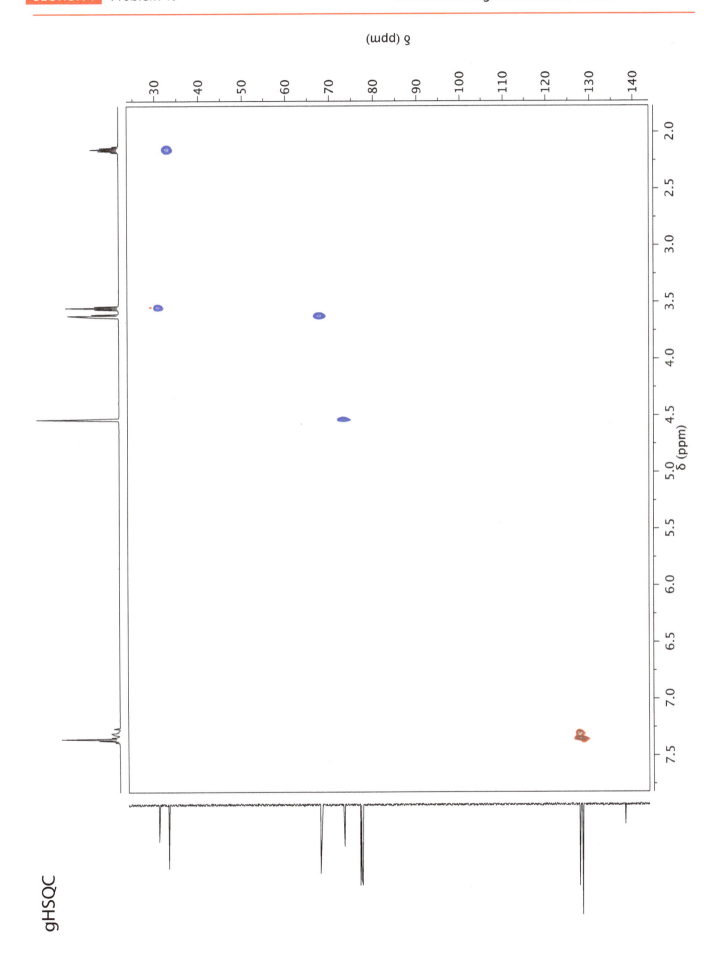

Problem 20

Which of the following compounds is the correct structure for the spectra below?

Spectra acquired in CDCl₃ at 500 MHz (¹H) or 125 MHz (¹³C).

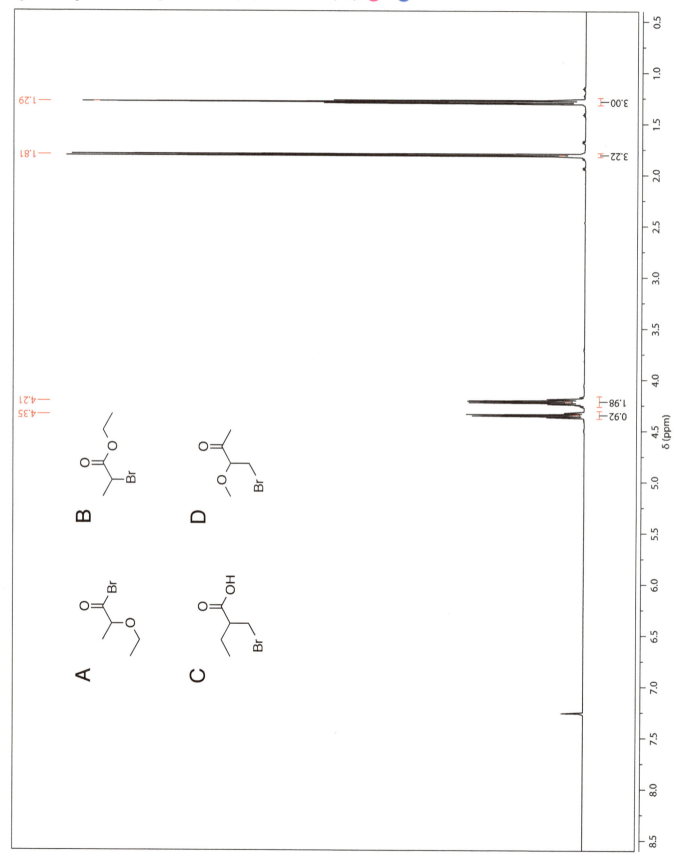

SECTION 1 Problem 20

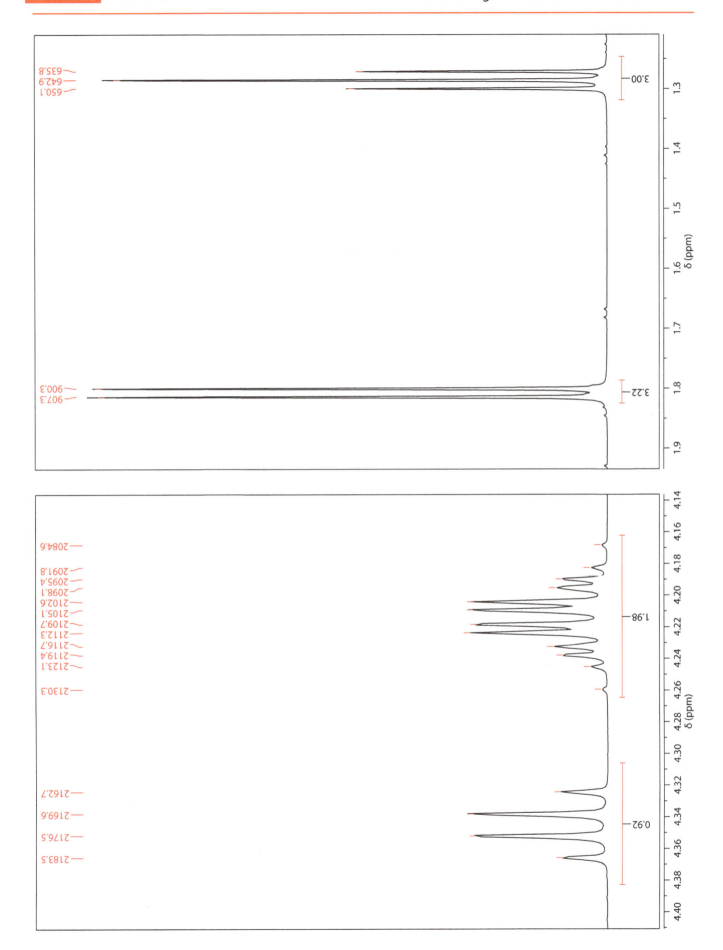

Problem 20

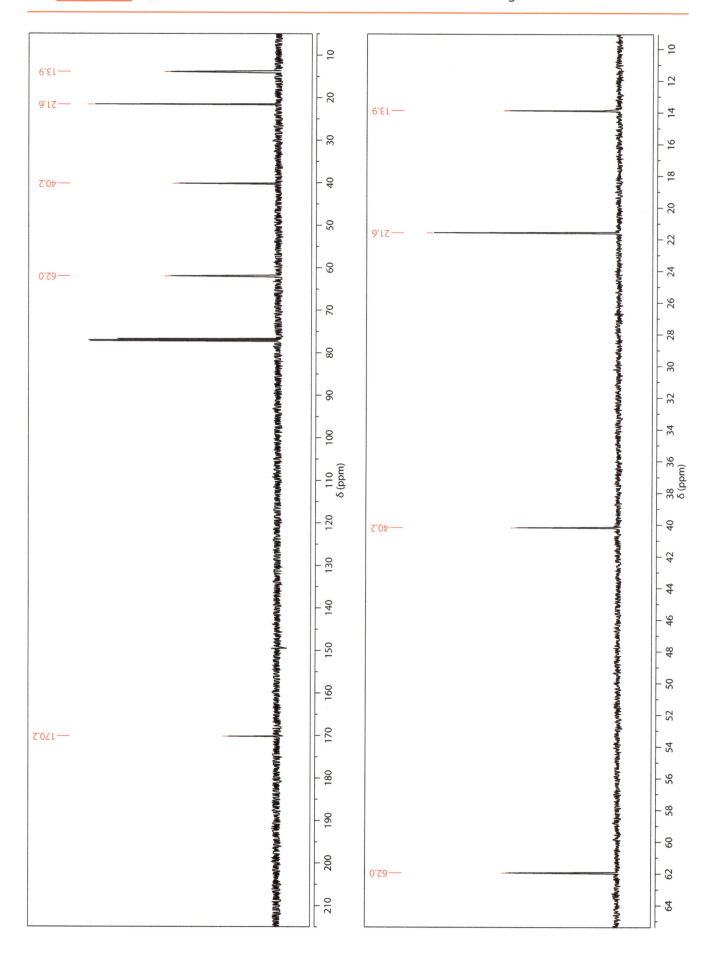

SECTION 1 Problem 21 — Problems in Organic Structure Determination

Which of the following compounds is the correct structure for the spectra below?

Spectra acquired in CD$_3$OD at 500 MHz (^{1}H) or 125 MHz (^{13}C).

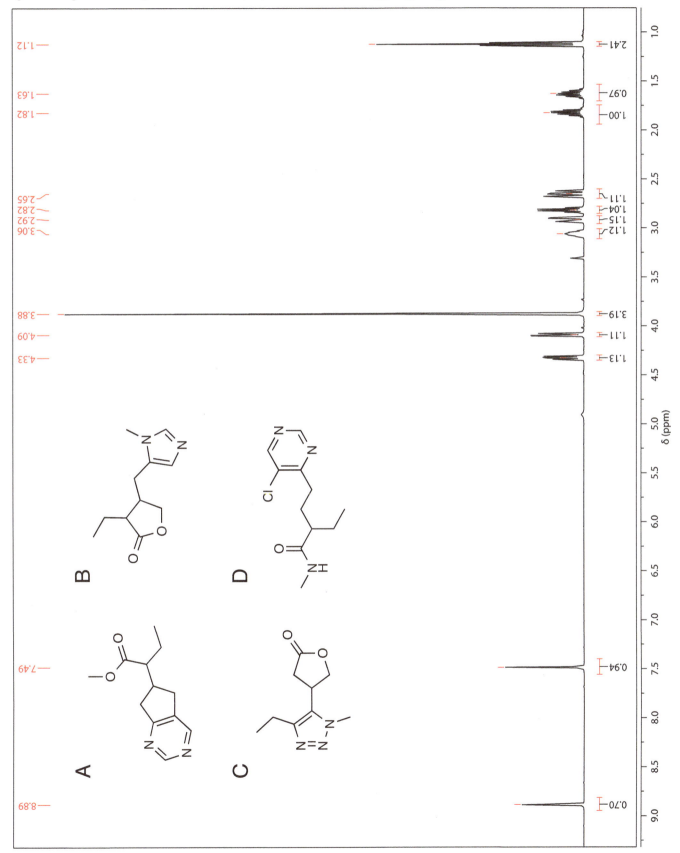

SECTION 1 Problem 21

Problems in Organic Structure Determination

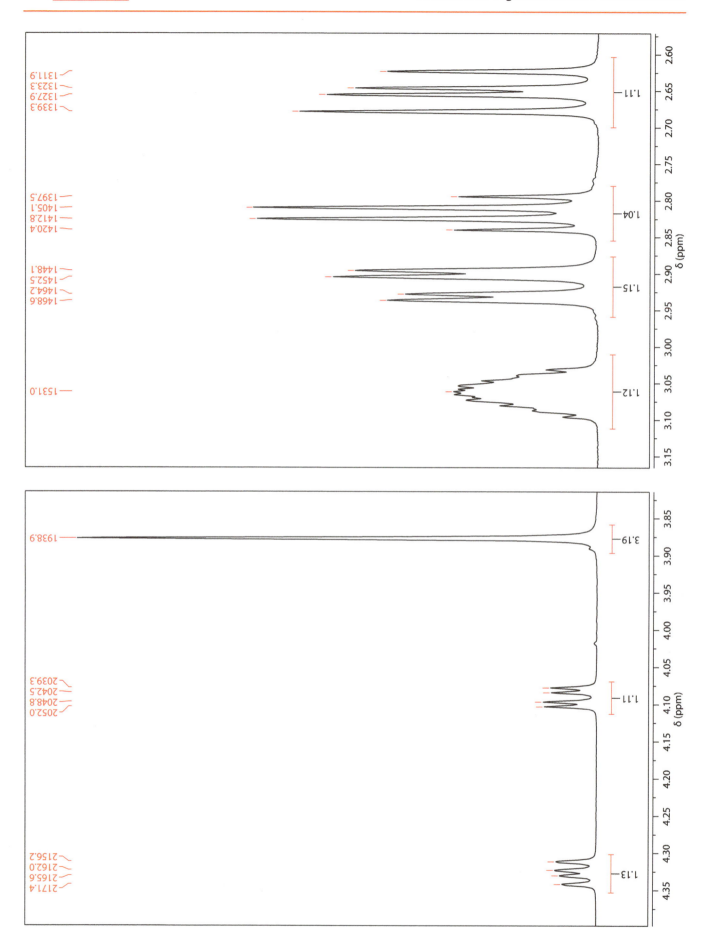

SECTION 1 Problem 21

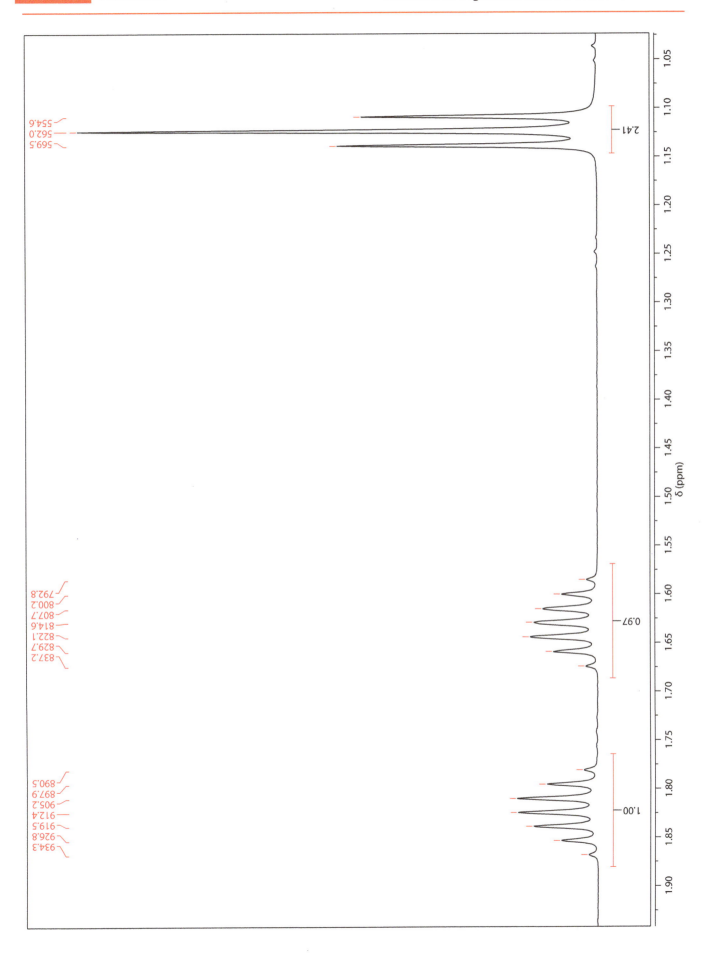

SECTION 1 Problem 21

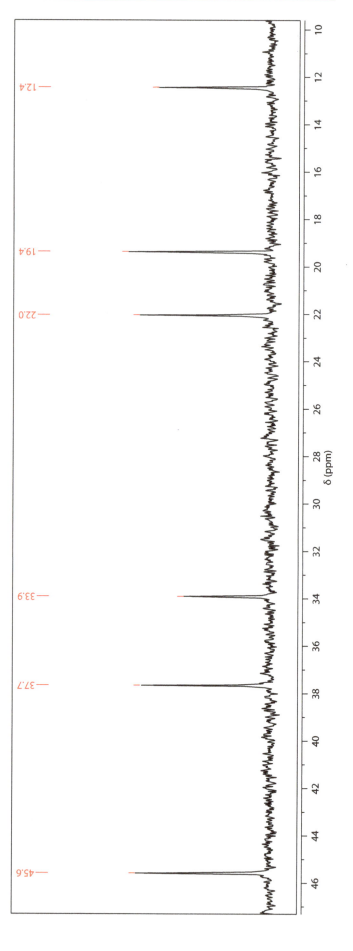

SECTION 1 Problem 22 — Problems in Organic Structure Determination

Identify the isomer of aminobenzhydrol that provides the following spectra. Assign all proton and carbon resonances. **Note:** detailed assignments not required for the phenyl ring.
Spectra acquired in CDCl₃ at 500 MHz (^{1}H) or 125 MHz (^{13}C).

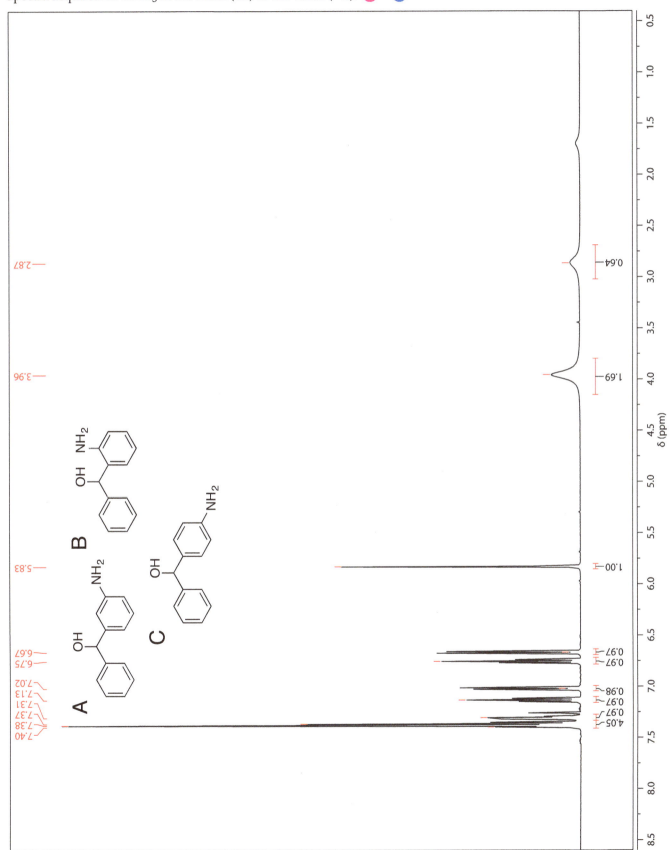

SECTION 1 Problem 22

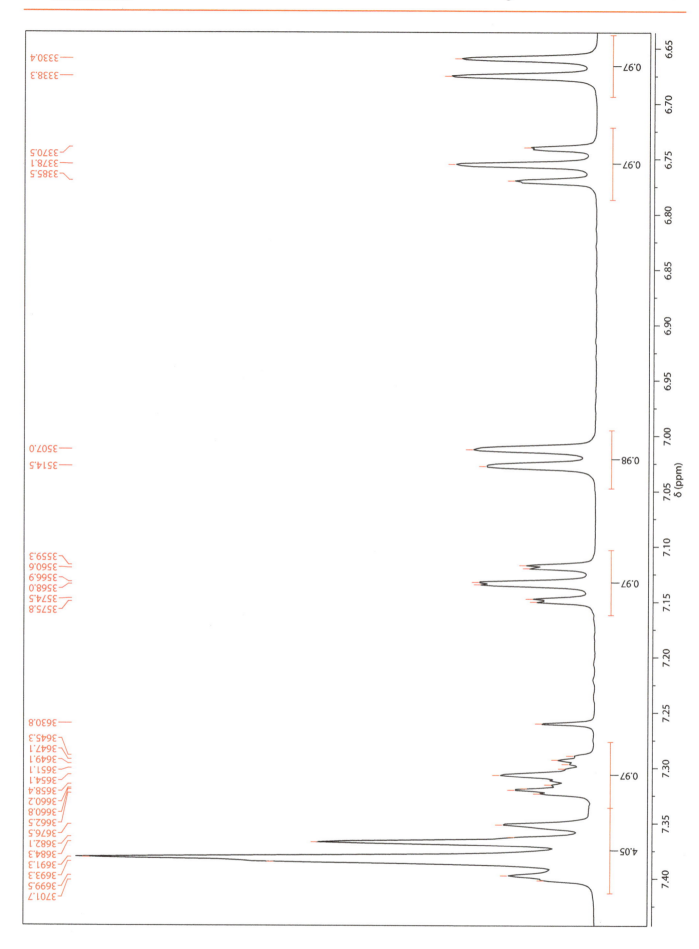

SECTION 1 Problem 22

Problems in Organic Structure Determination

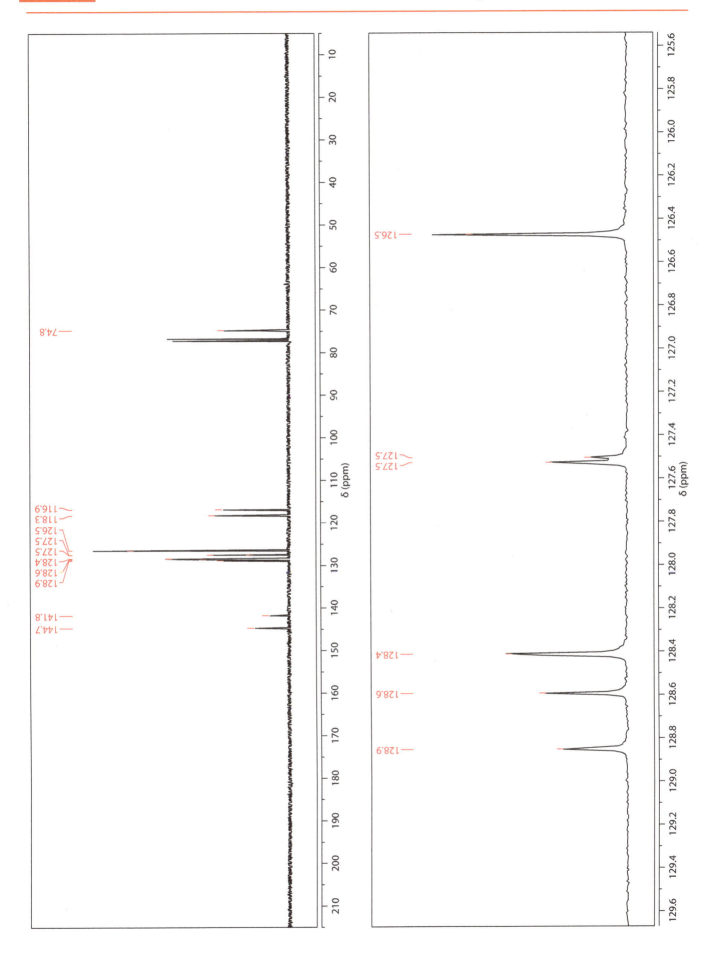

SECTION 1 Problem 23

Which of the following compounds is the correct structure for the spectra below?

Spectra acquired in CDCl$_3$ at 500 MHz (^{1}H) or 125 MHz (^{13}C).

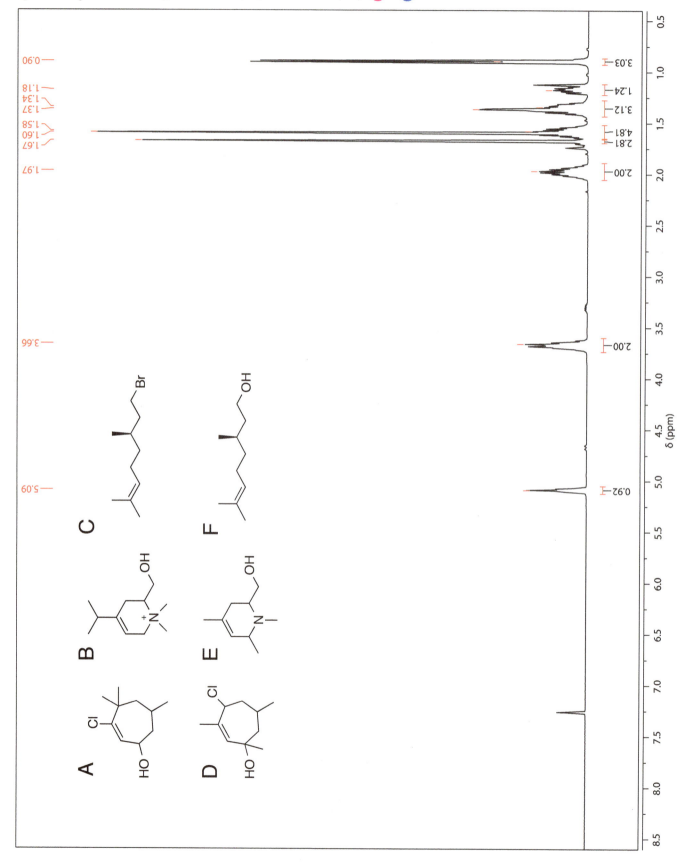

SECTION 1 Problem 23

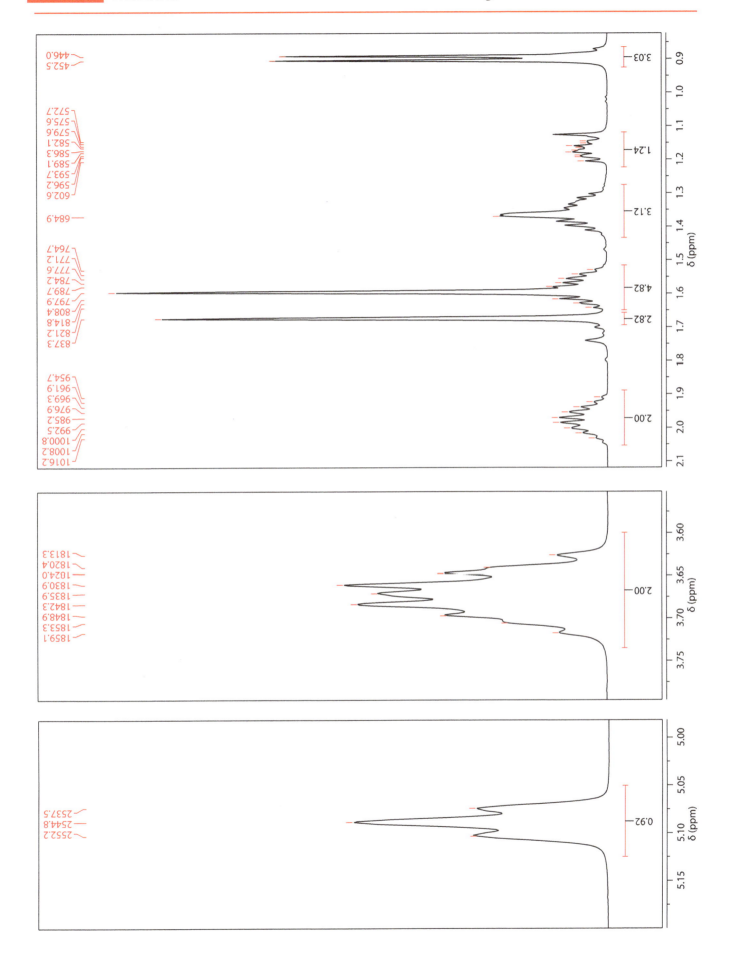

Problem 23

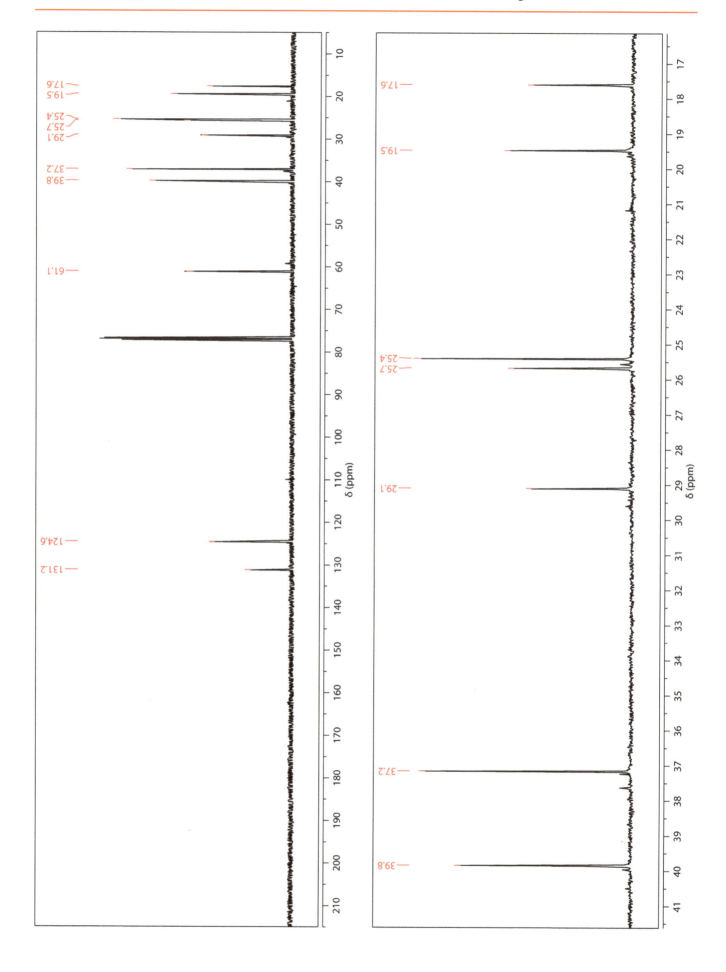

SECTION 1 Problem 24

Using the ¹H, ¹³C, gHSQC and gHMBC spectra, assign all proton and carbon resonances to the structure of neocuproine below and assign all proton coupling constants.
Spectra acquired in CDCl₃ at 500 MHz (¹H) or 125 MHz (¹³C).

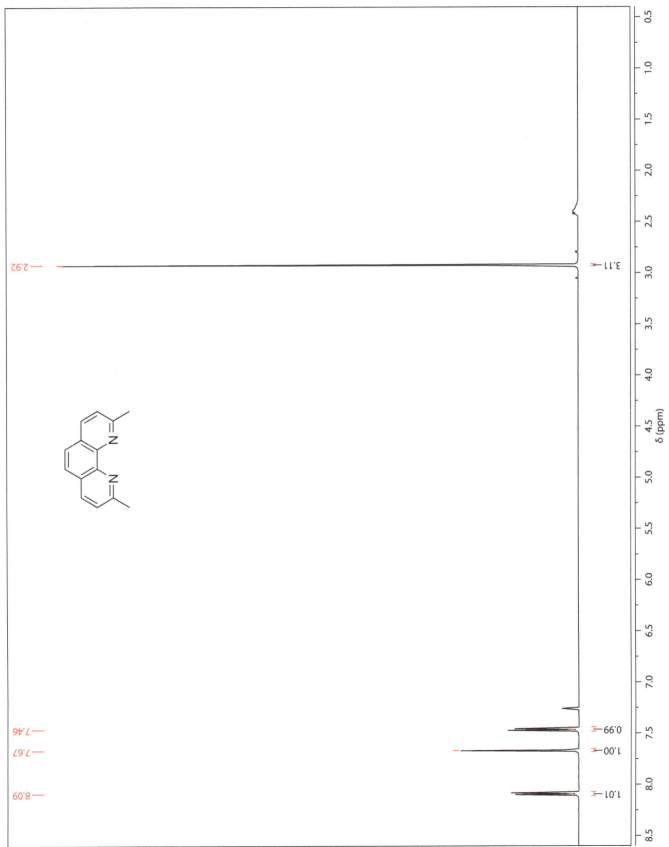

Problem 24

Problem 24

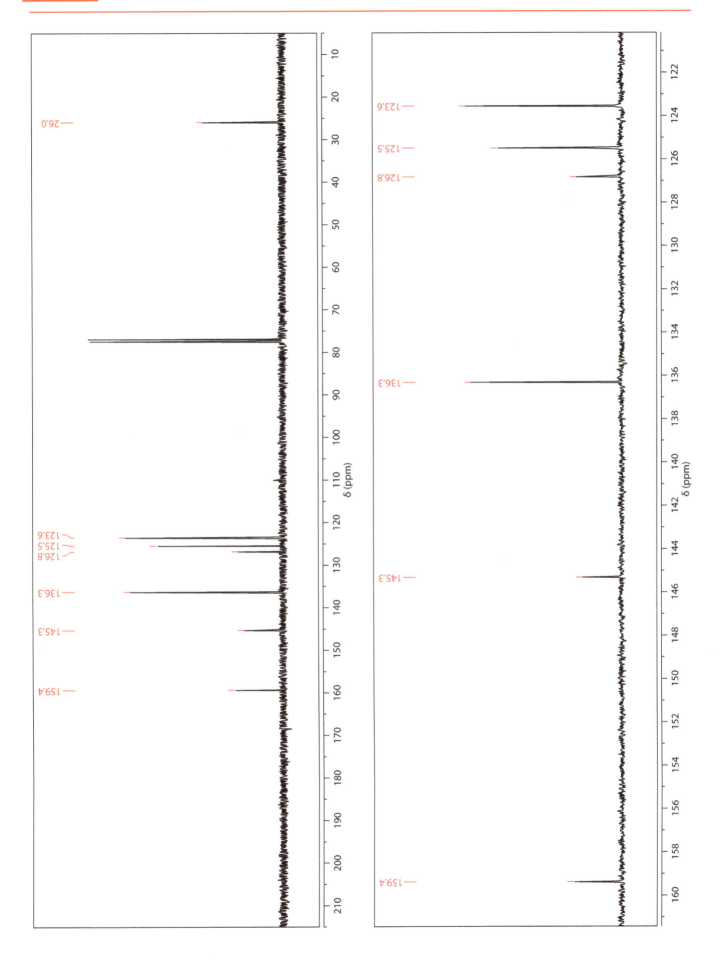

SECTION 1 Problem 24

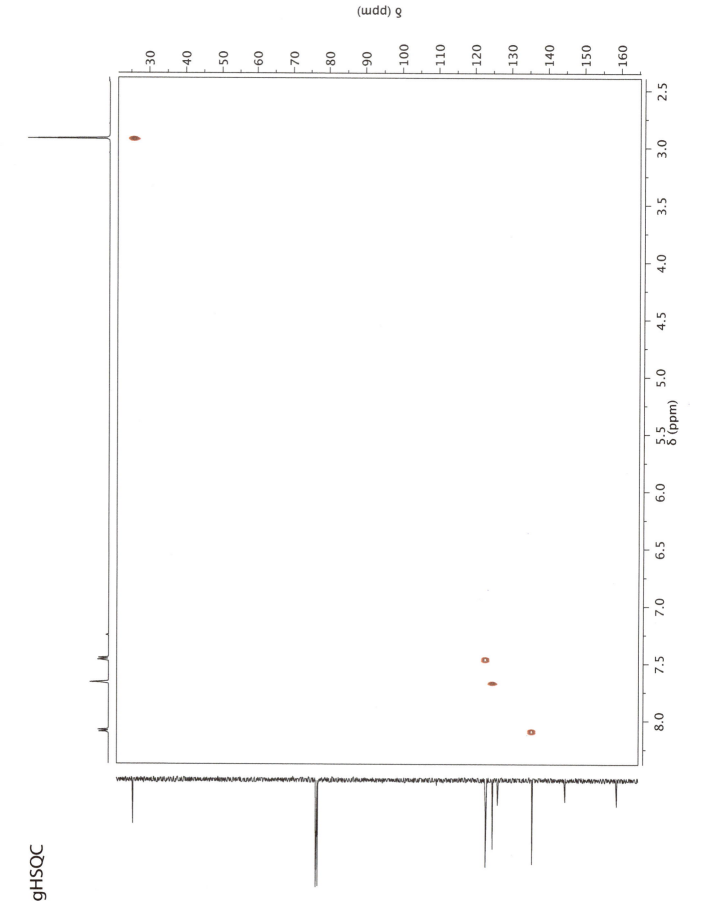

gHSQC

SECTION 1 Problem 24

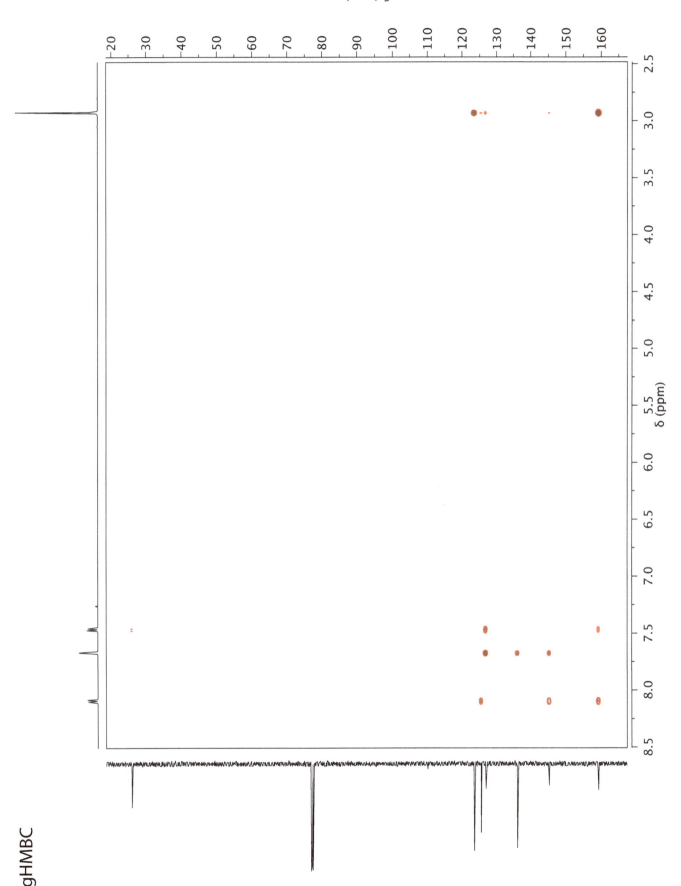

gHMBC

Problem 24

gHMBC

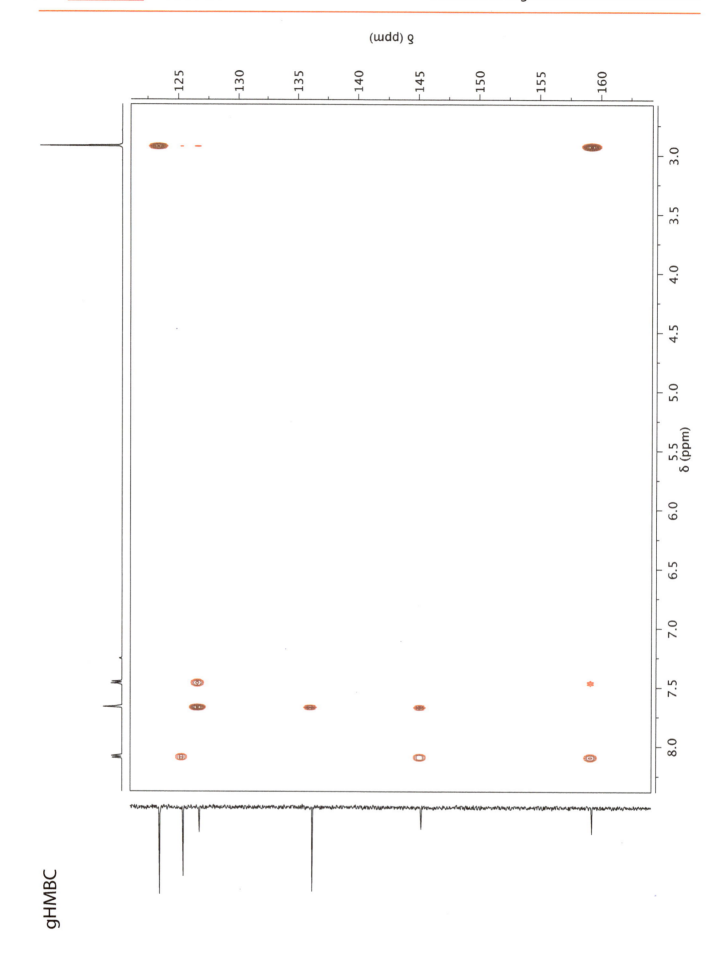

SECTION 1 Problem 25

Predict the ¹H and ¹³C spectra for the compounds listed below, including chemical shifts, multiplicities and integrals for all signals.
Spectra predicted at 500 MHz.

Section 2

Section 2 is geared towards those with previous experience with NMR structure elucidation. The problems build upon those in section 1 and focus on integrating data from 1D- and 2D-NMR experiments to assign spectra to given structures.

LEARNING OBJECTIVES

- Develop skills required for assigning ^{1}H and ^{13}C chemical shifts to molecules of increasing complexity.
- Analyze and predict multiplicity patterns for signals containing complicated fine structure in order to derive coupling constant values.
- Integrate data from a broad array of different NMR experiments, including an increasing use of 2D-NMR spectra, and apply this knowledge to the assignment of chemical structures.

EXPERIMENTS INCLUDED

^{1}H, ^{13}C, DEPT-135, gCOSY, gHSQC, gHMBC, 1D TOCSY, ^{1}H selective decoupling and gHSQC-TOCSY.

TYPES OF MOLECULES

This section contains most of the basic functional groups commonly found in small molecules (alcohols, amines, halides, olefins, carbonyls, aromatics and heteroaromatics), as well as a number of more unusual functional groups (complex heteroaromatics, fluorine, cyclopropanes). Most molecules in this section also contain one or more stereogenic centers.

LEGEND

Spectrum annotations:
  i = impurity.

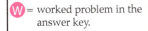 = worked problem in the answer key.

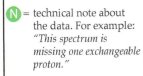 = technical note about the data. For example: *"This spectrum is missing one exchangeable proton."*

H = hint to assist in solving the problem. For example: *"This molecule would have an IR stretch at 2240 cm^{-1}."*

SPECTRUM LAYOUT CONVENTIONS

The solvent used for each problem is indicated in the question, but solvent signals are not annotated on the spectra. A list of solvents and their ^{1}H and ^{13}C chemical shifts can be found in appendix A. Only signals deriving from the molecule are peak picked on the ^{1}H and ^{13}C spectra. Minor impurities are left unannotated, while major impurities are marked with a lower case "i".

STRATEGIES FOR SUCCESS

A key goal of this chapter is to develop expertise in the use of sets of NMR spectra in concert for the assignment of complex structures. In many cases these problems can only be solved by extracting and integrating information from multiple spectra. As the complexity of the problems increase it is therefore critical to make sure you evaluate ALL the data before reaching your conclusion.

Problem 26

Using the ^{1}H and ^{13}C spectra assign all proton and carbon resonances to the structure of (1-ethoxycyclopropoxy)trimethylsilane.
Spectra acquired in CDCl$_3$ at 500 MHz (^{1}H) or 125 MHz (^{13}C).

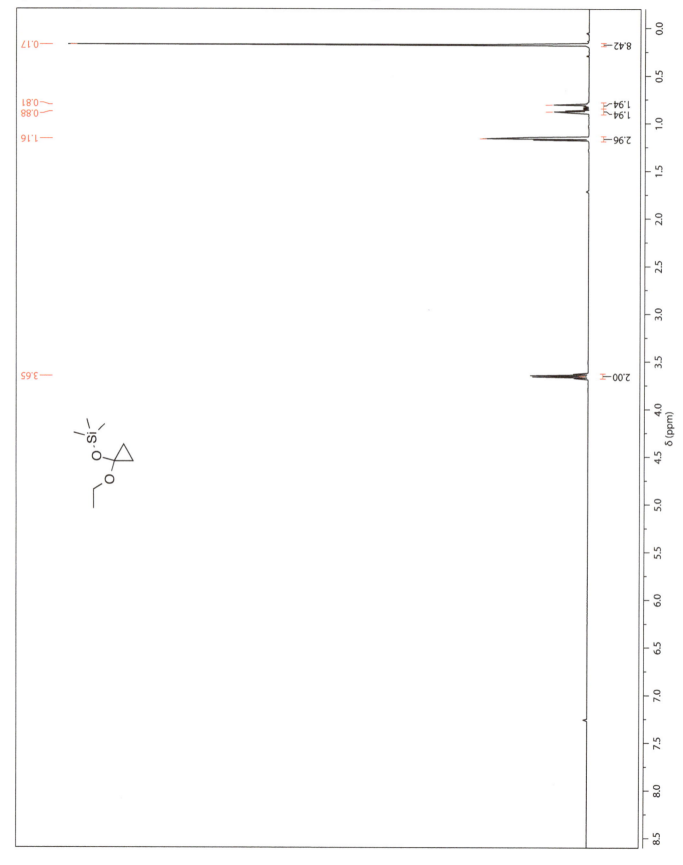

SECTION 2 Problem 26 — Problems in Organic Structure Determination

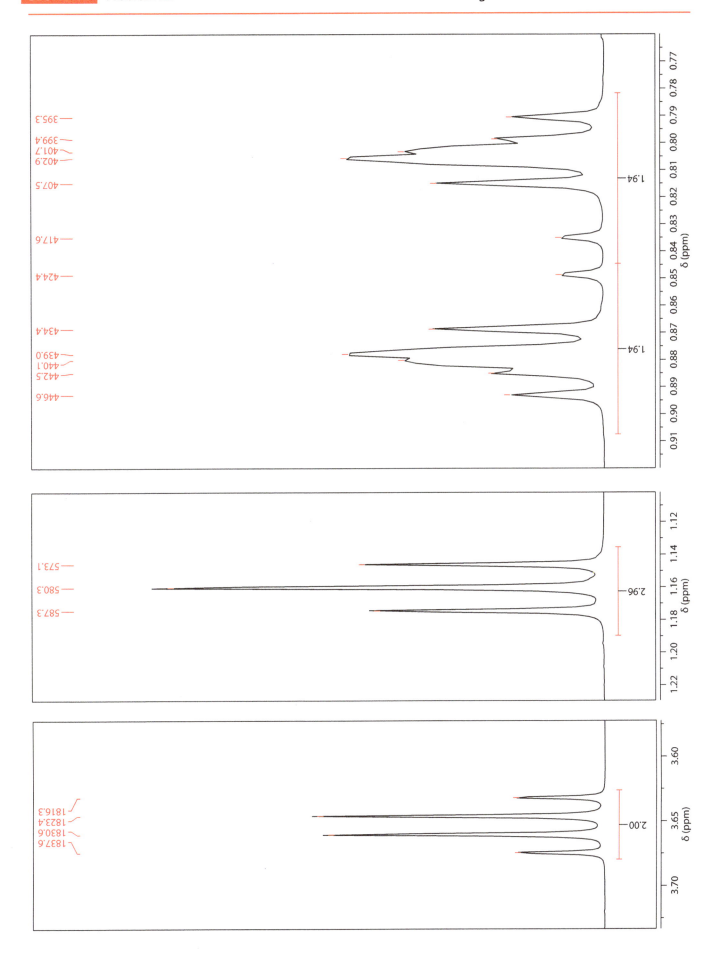

SECTION 2 Problem 26

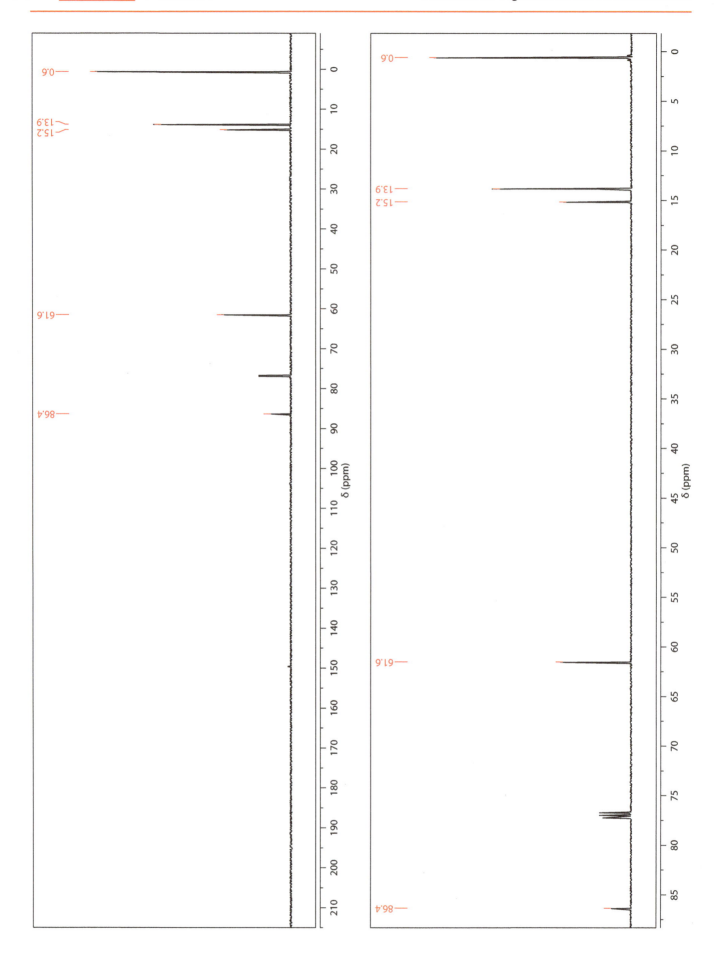

Problem 27

Using the ^{1}H and ^{13}C spectra assign all proton and carbon resonances to the structure of L-malic acid.

Spectra acquired in CD_3OD at 500 MHz (^{1}H) or 125 MHz (^{13}C).

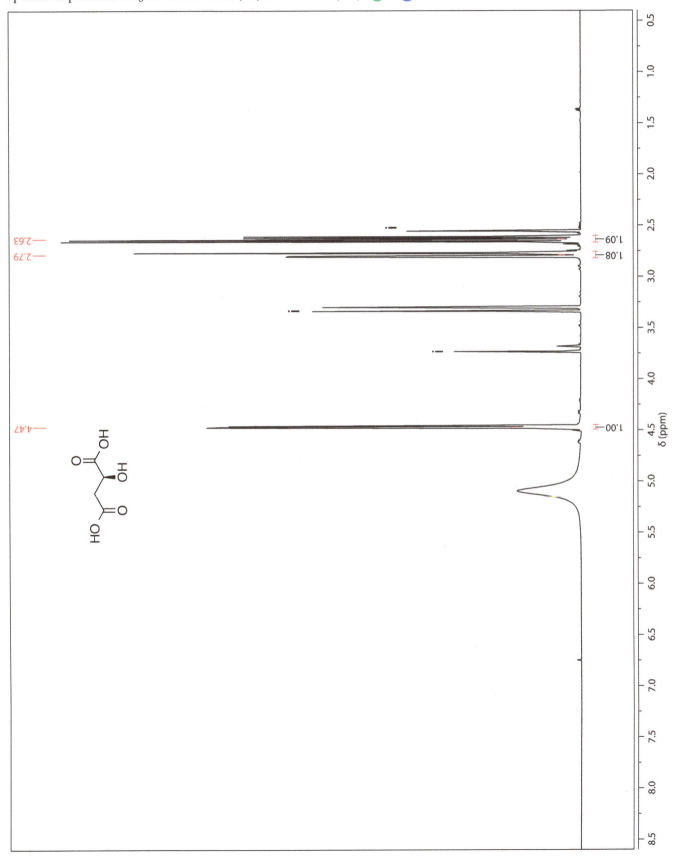

SECTION 2 Problem 27

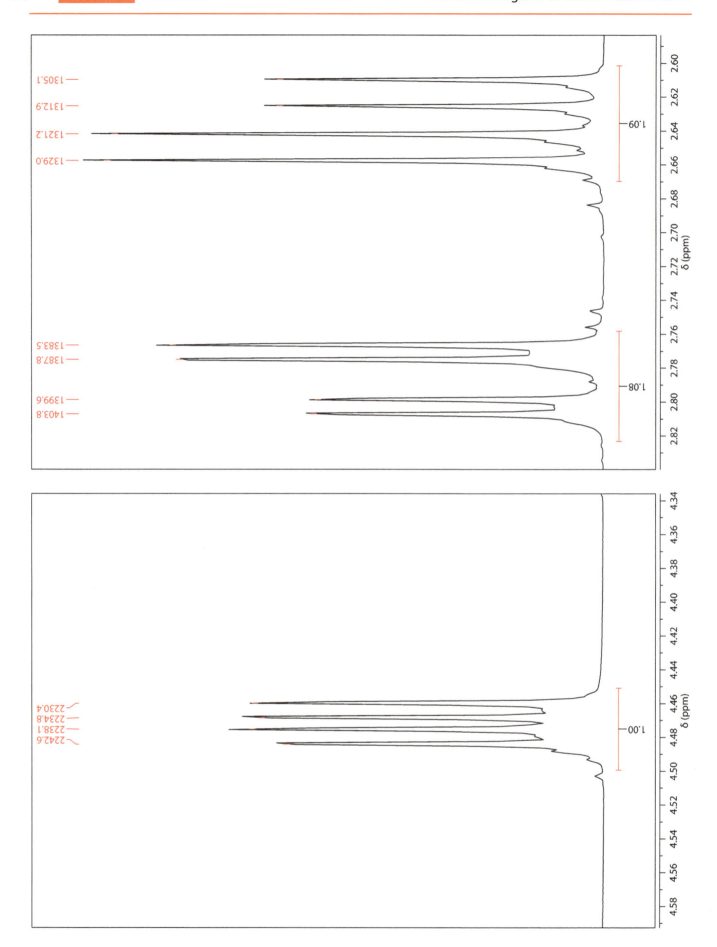

SECTION 2 Problem 27

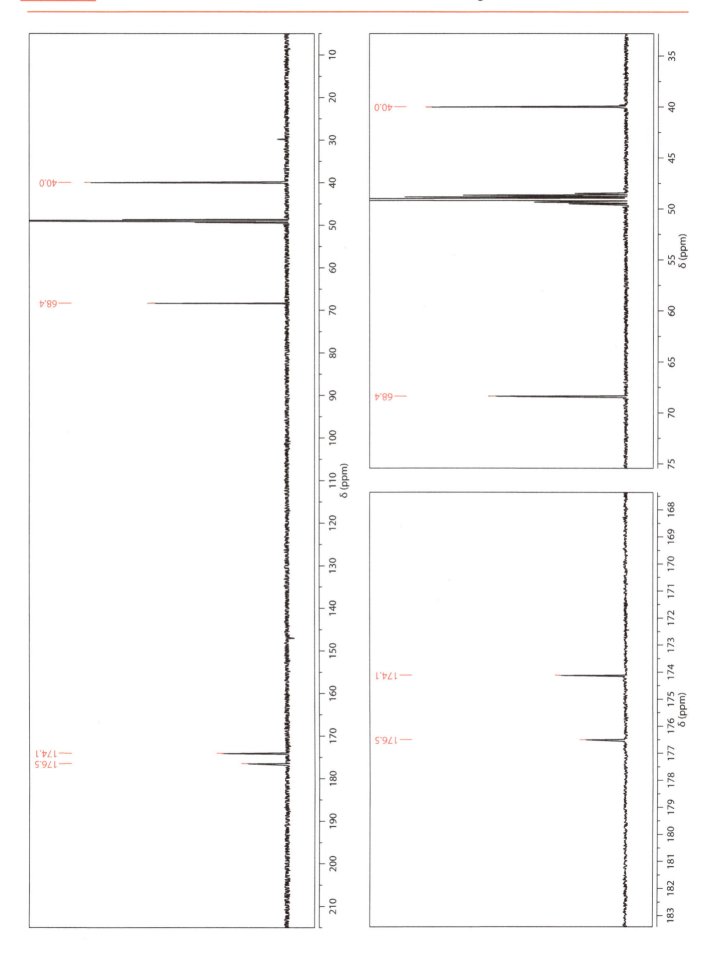

Problem 28

Which of the following compounds is the correct structure for the spectra below?

Spectra acquired in CDCl₃ at 500 MHz (¹H) or 125 MHz (¹³C).

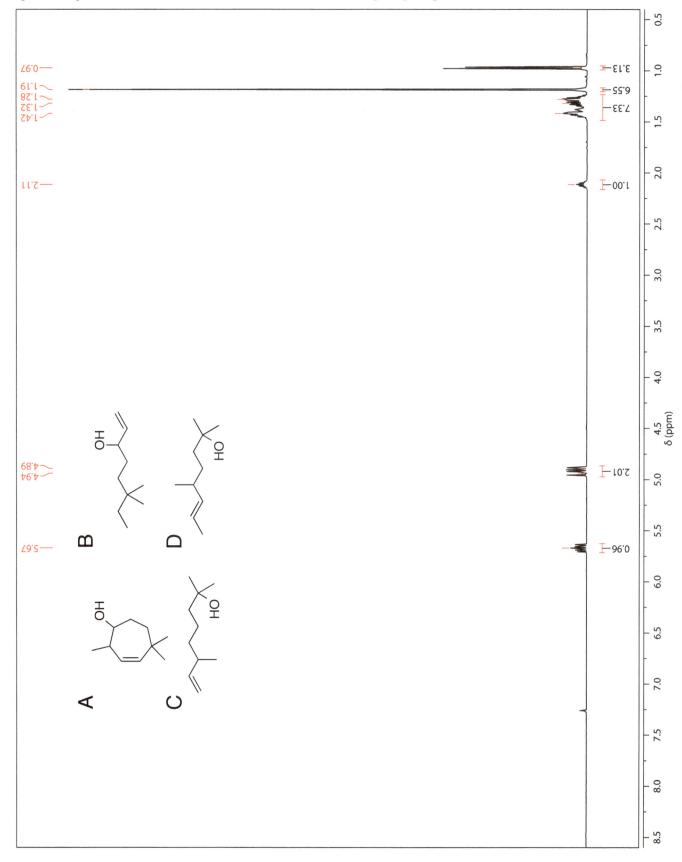

SECTION 2 Problem 28

Problems in Organic Structure Determination

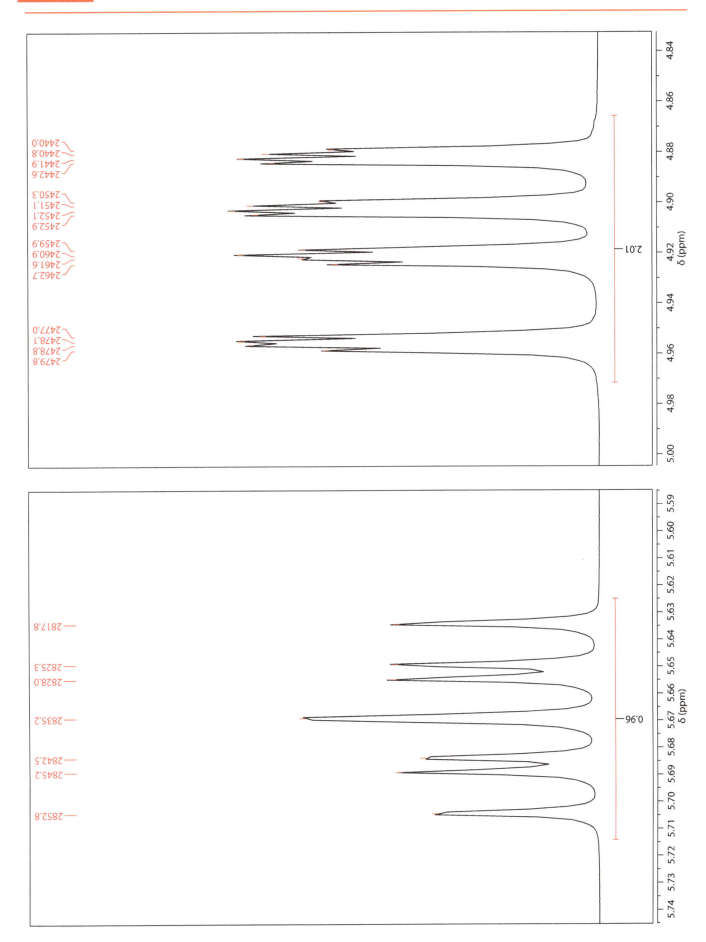

Problem 28

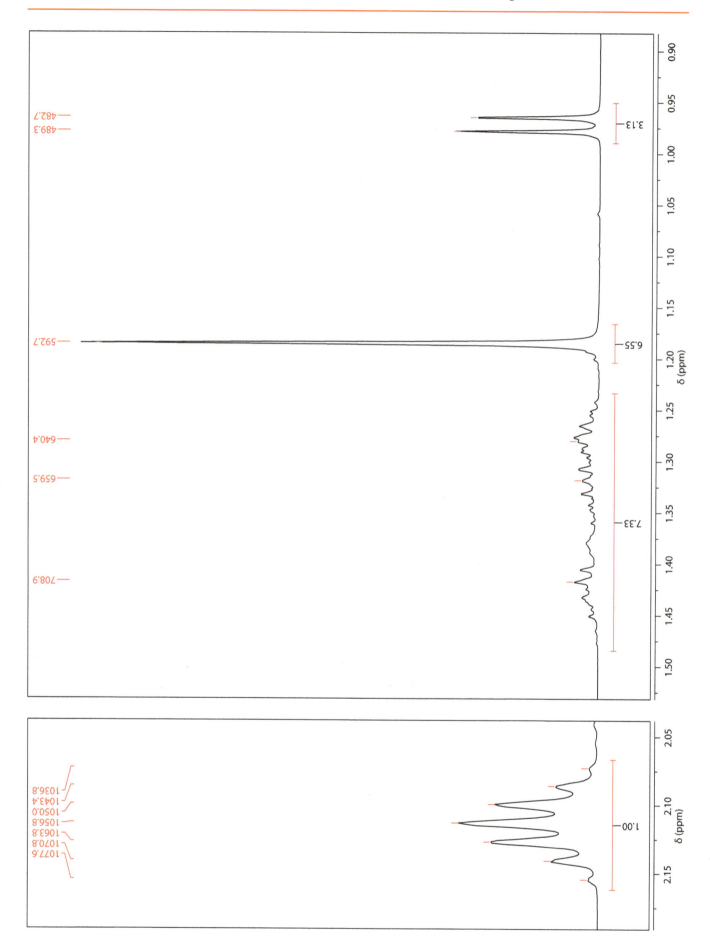

SECTION 2 Problem 28

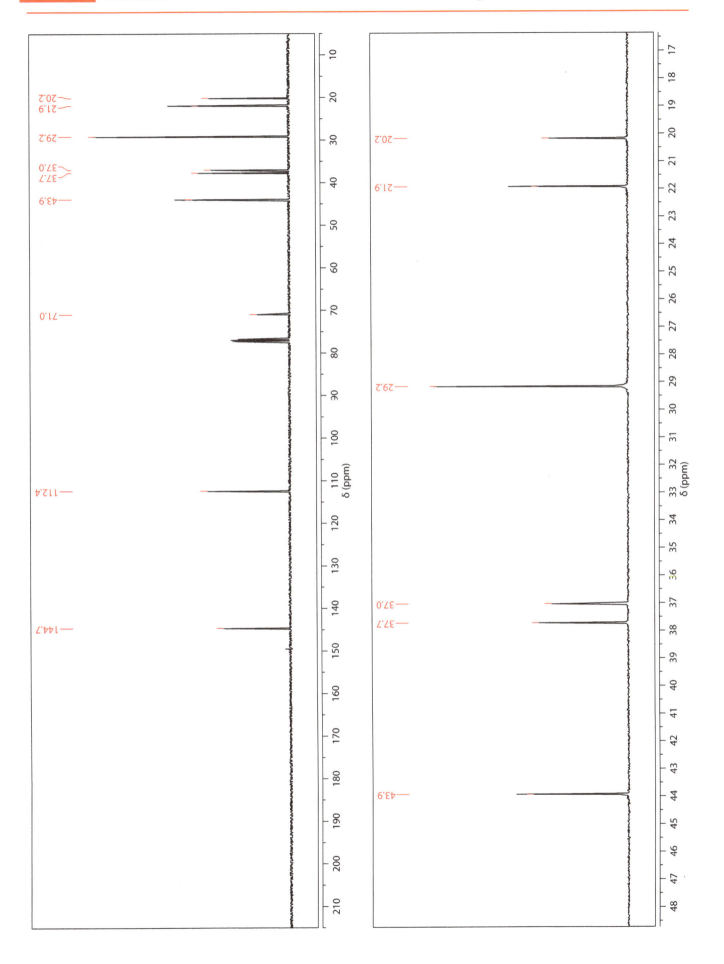

Problem 28

gCOSY

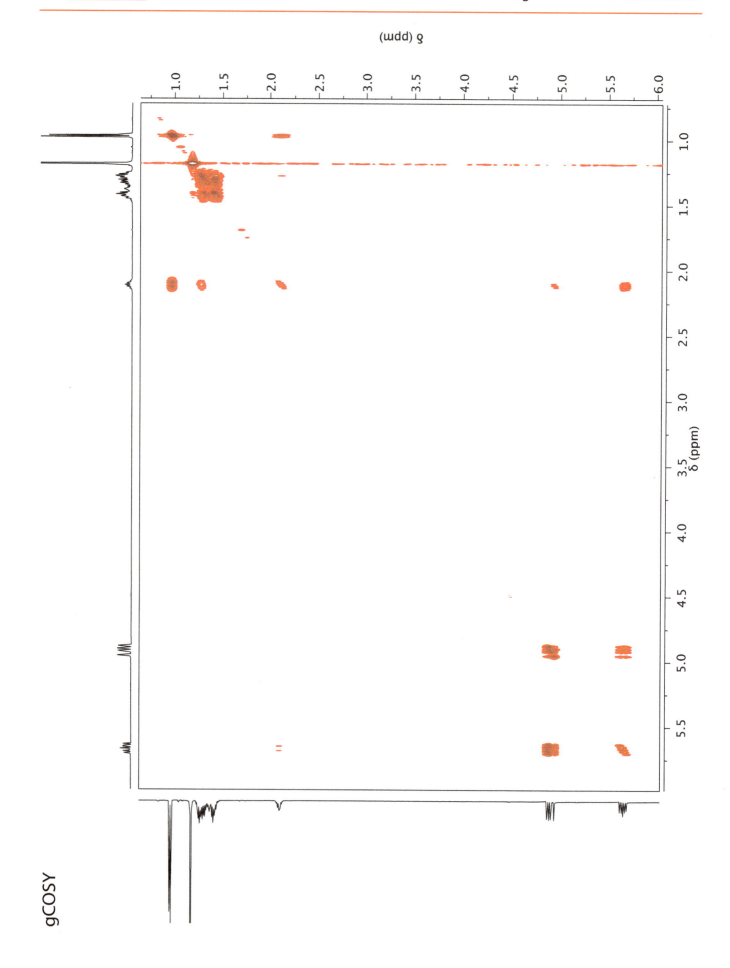

SECTION 2 Problem 28

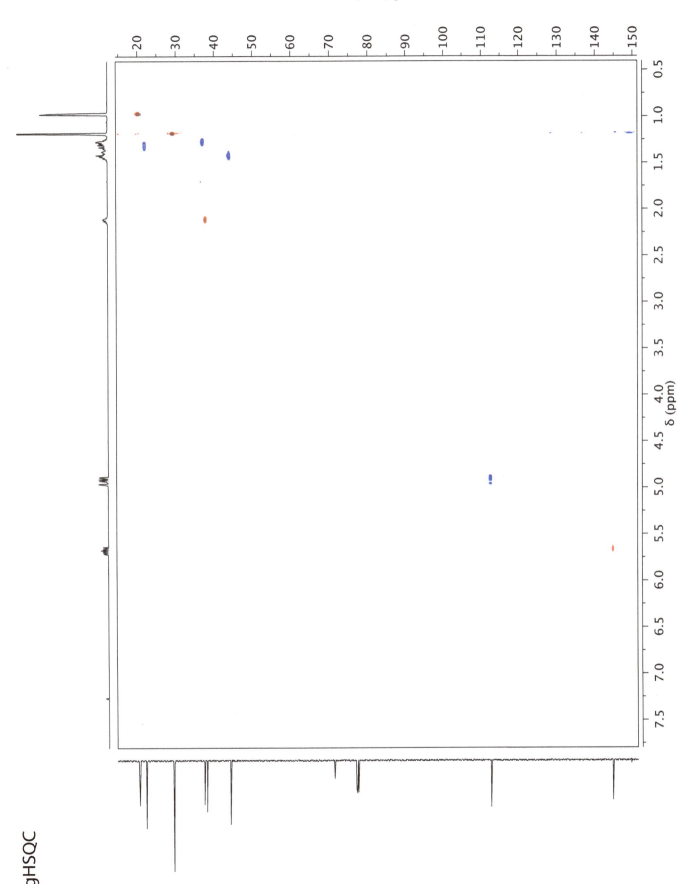

gHSQC

Problem 28

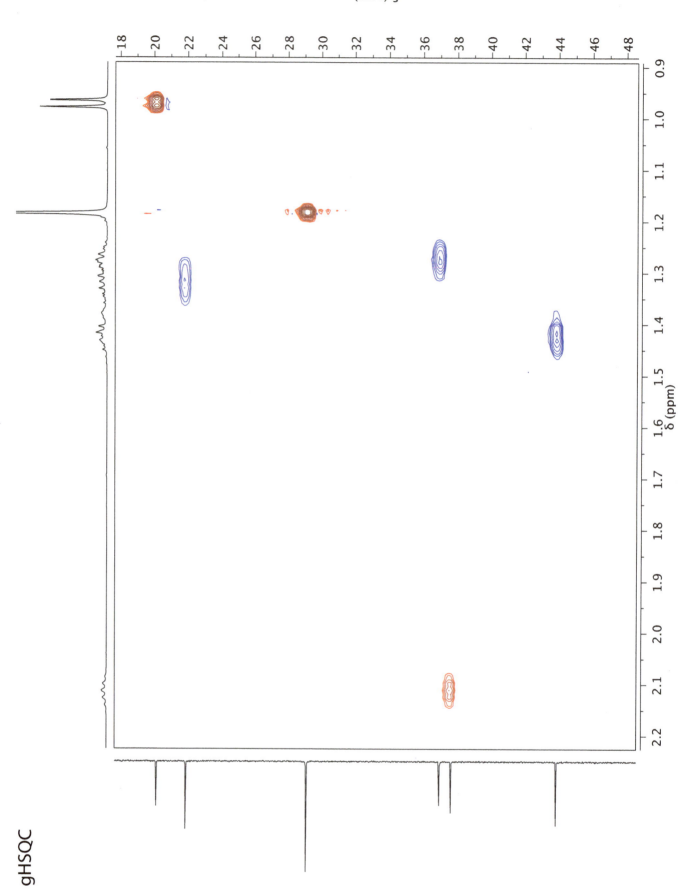

gHSQC

SECTION 2 Problem 29

Problems in Organic Structure Determination

Which of the following compounds is the correct structure for the spectra below?

Spectra acquired in CDCl$_3$ at 500 MHz (^{1}H) or 125 MHz (^{13}C).

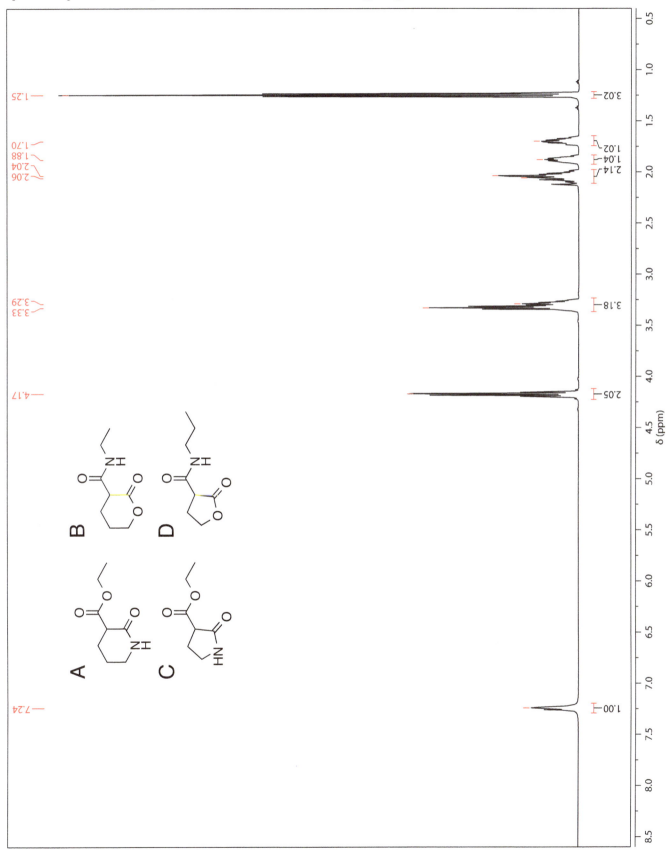

Problem 29

SECTION 2 Problem 29 — Problems in Organic Structure Determination

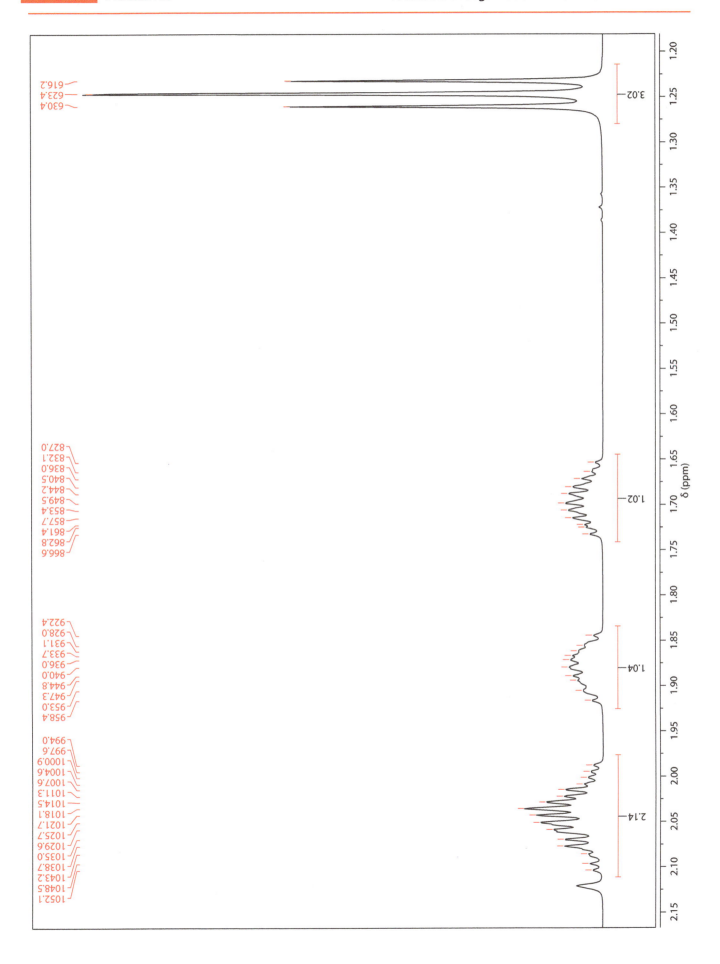

Problem 29

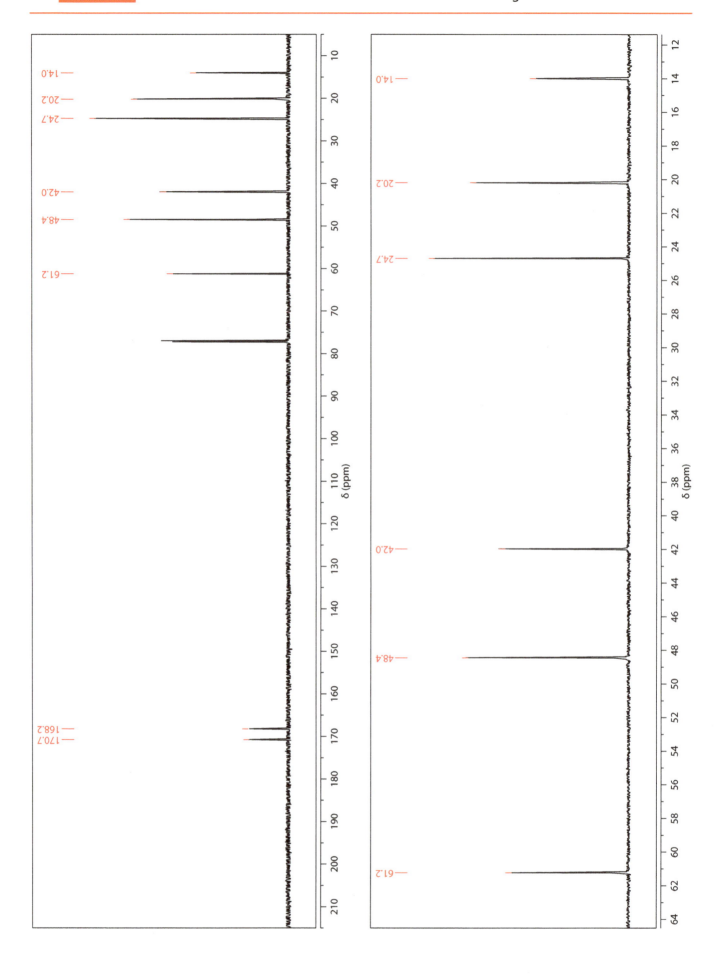

SECTION 2 Problem 29

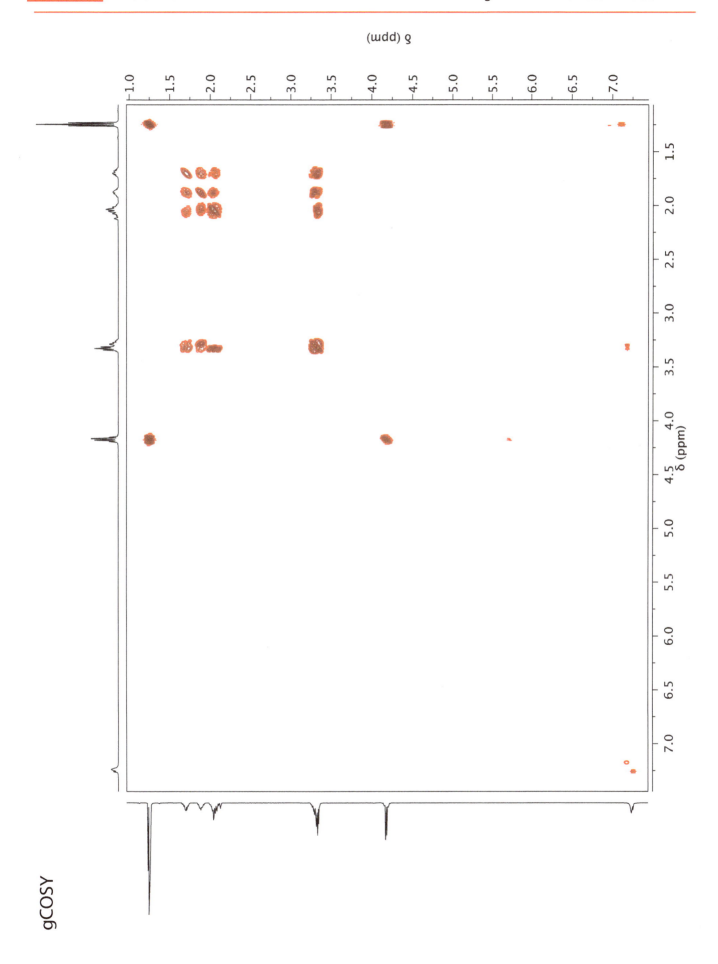

gCOSY

Problem 29

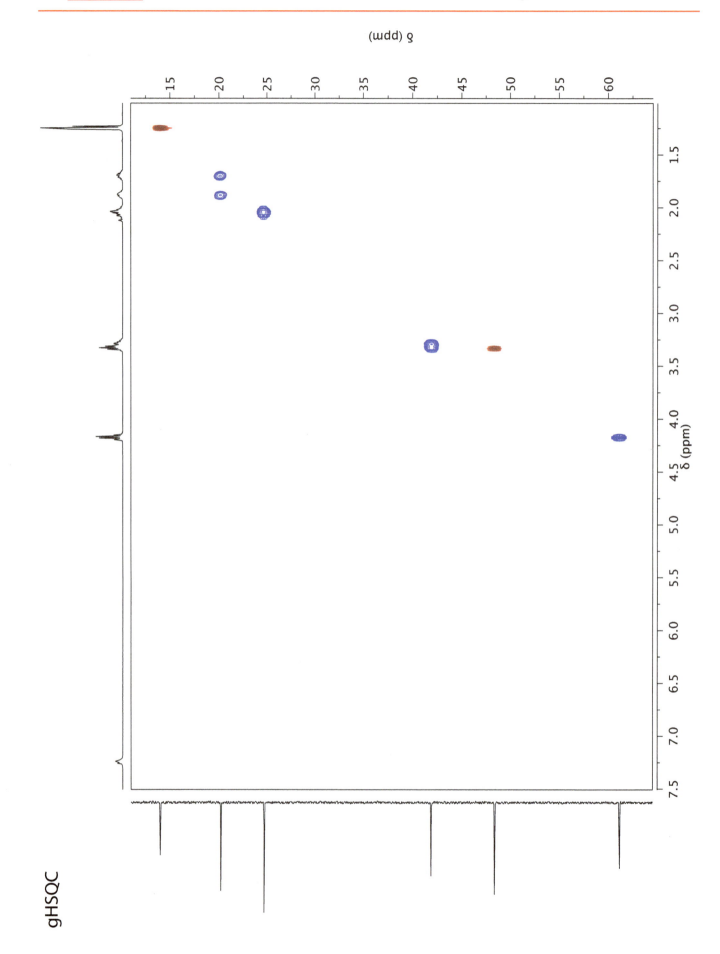

gHSQC

SECTION 2 Problem 29

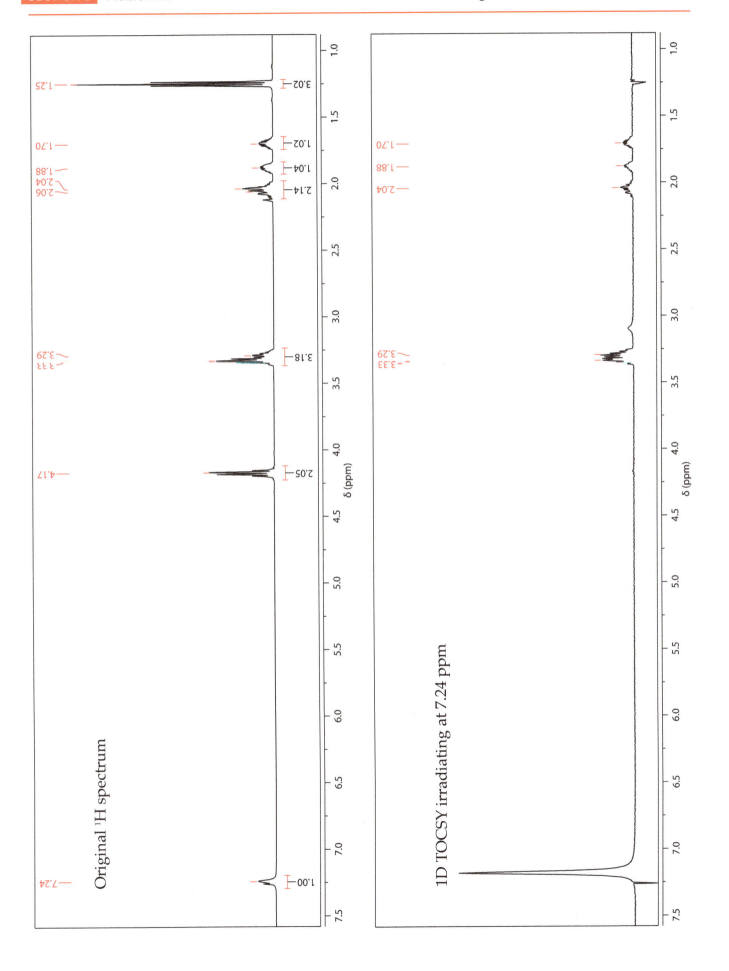

Using the ¹H, ¹³C, gCOSY and gHSQC spectra, assign all proton and carbon resonances to the structure of D-mannitol below and assign all proton coupling constants.
Spectra acquired in (CD₃)₂SO at 500 MHz (¹H) or 125 MHz (¹³C).

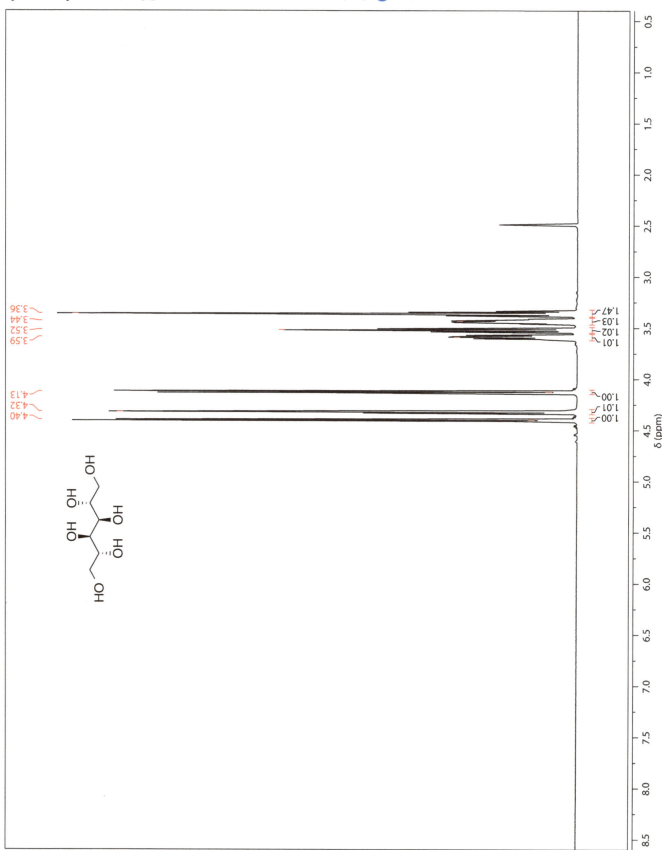

SECTION 2 Problem 30

Problems in Organic Structure Determination

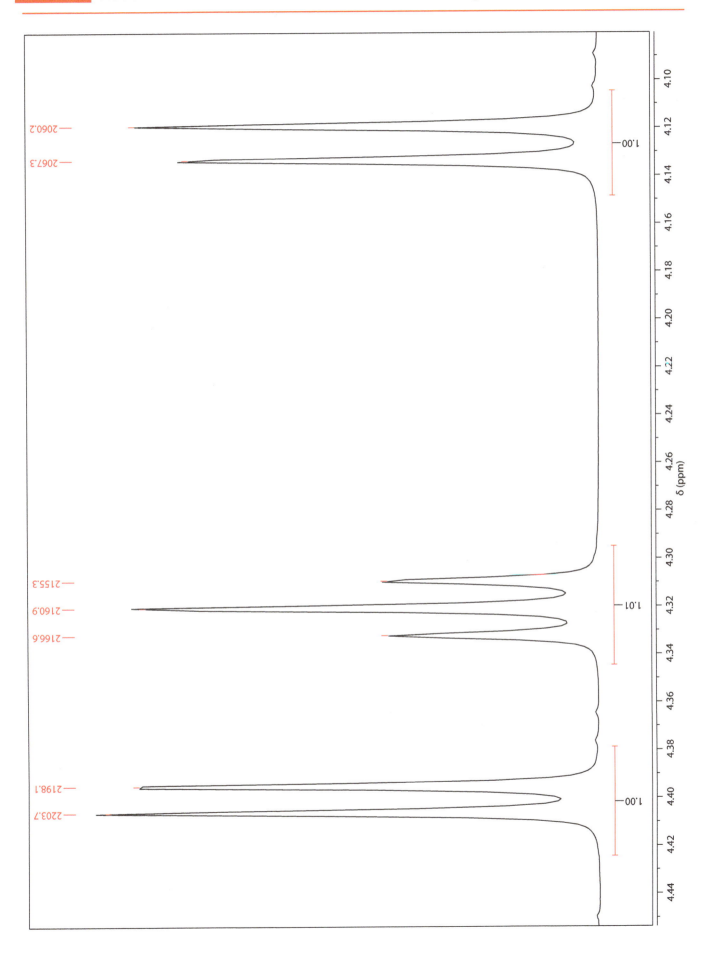

Problem 30

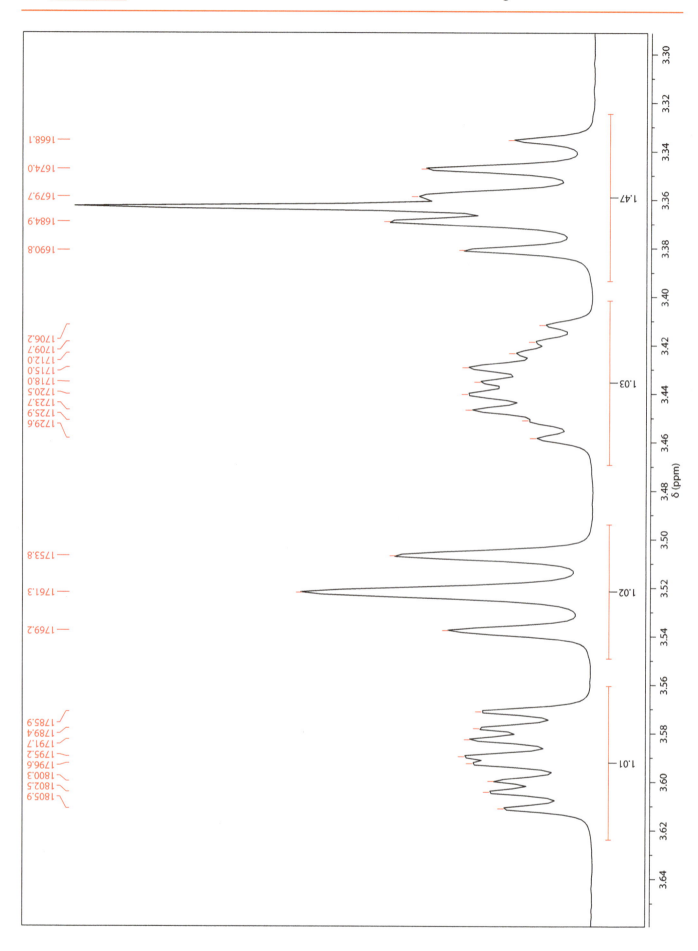

SECTION 2 Problem 30

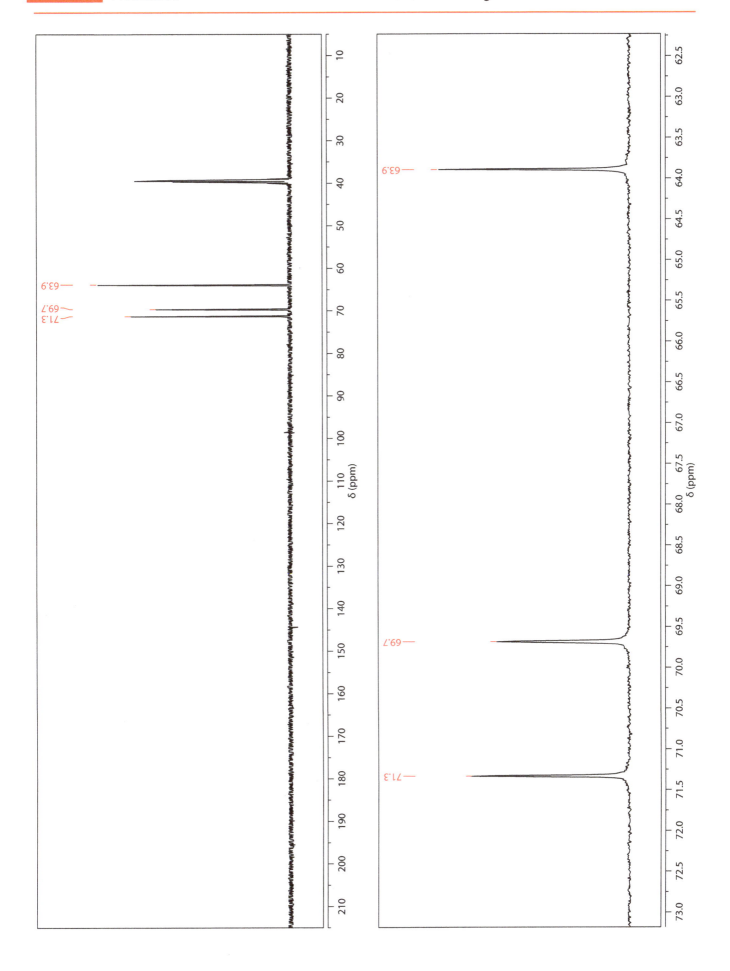

Problem 30

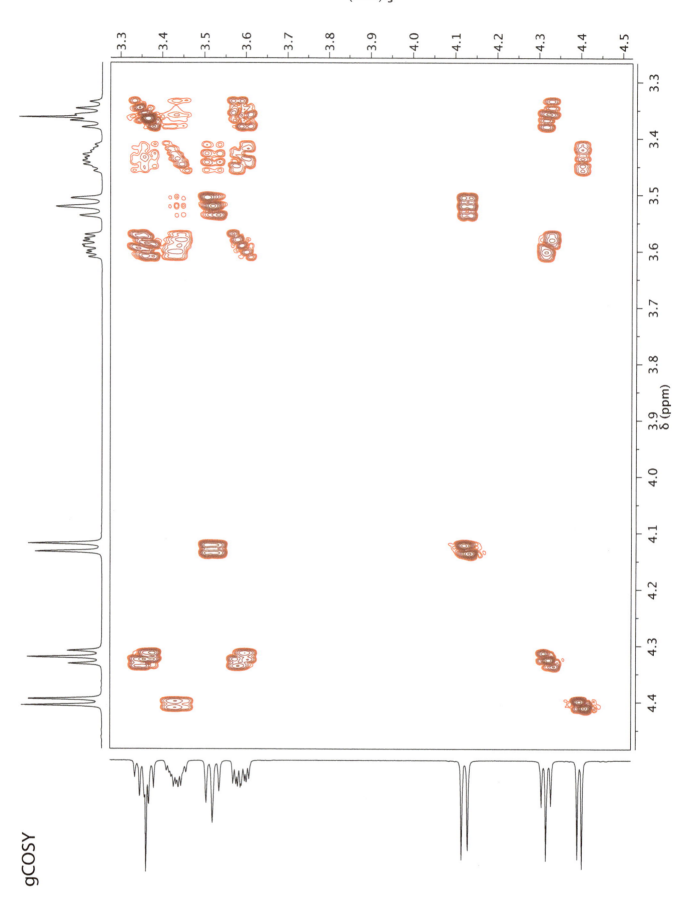

gCOSY

Problem 30

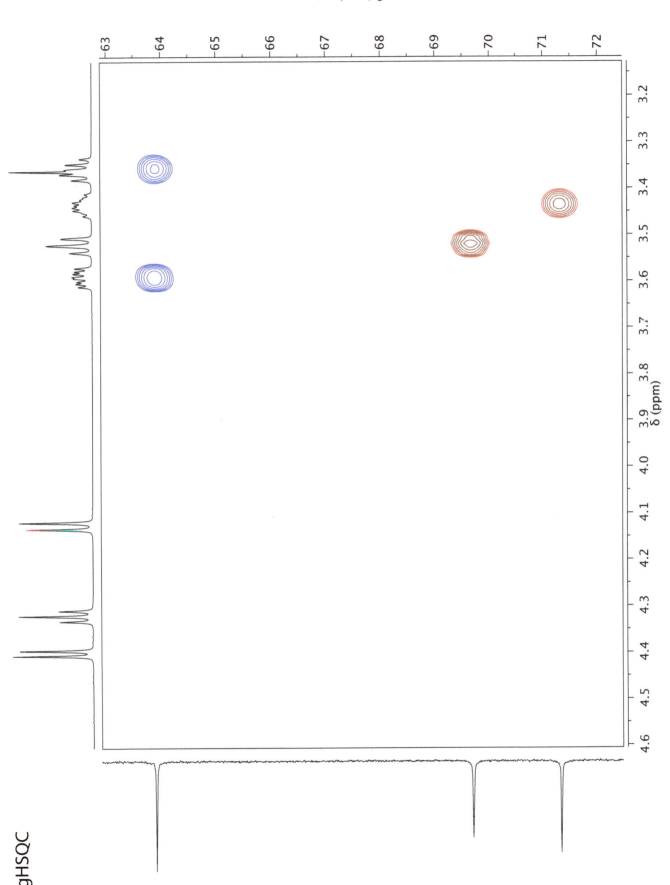

gHSQC

Using the ¹H, ¹³C, gCOSY and gHSQC spectra, assign all proton and carbon resonances to the structure below and assign all proton coupling constants.
Spectra acquired in CD₃OD at 500 MHz (¹H) or 125 MHz (¹³C).

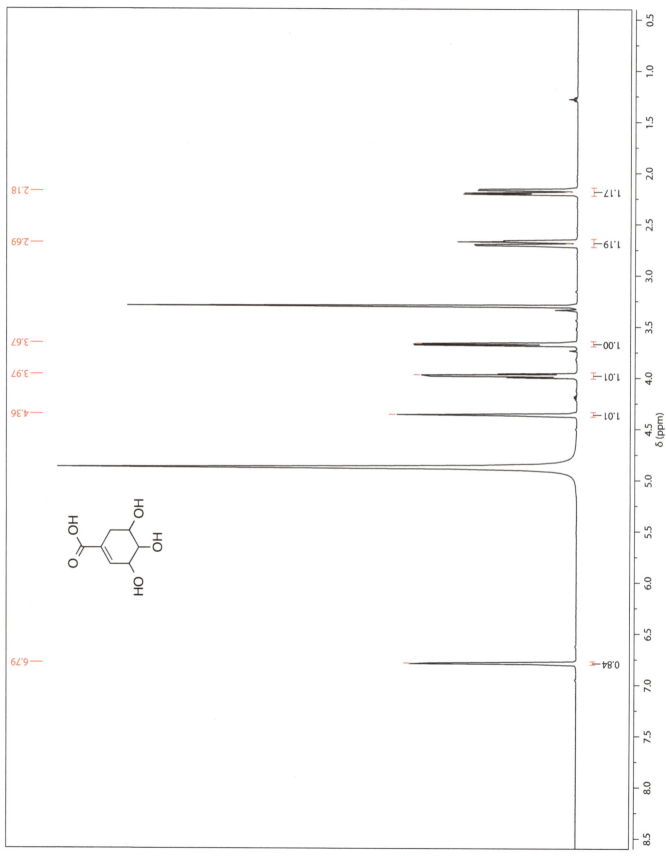

SECTION 2 Problem 31

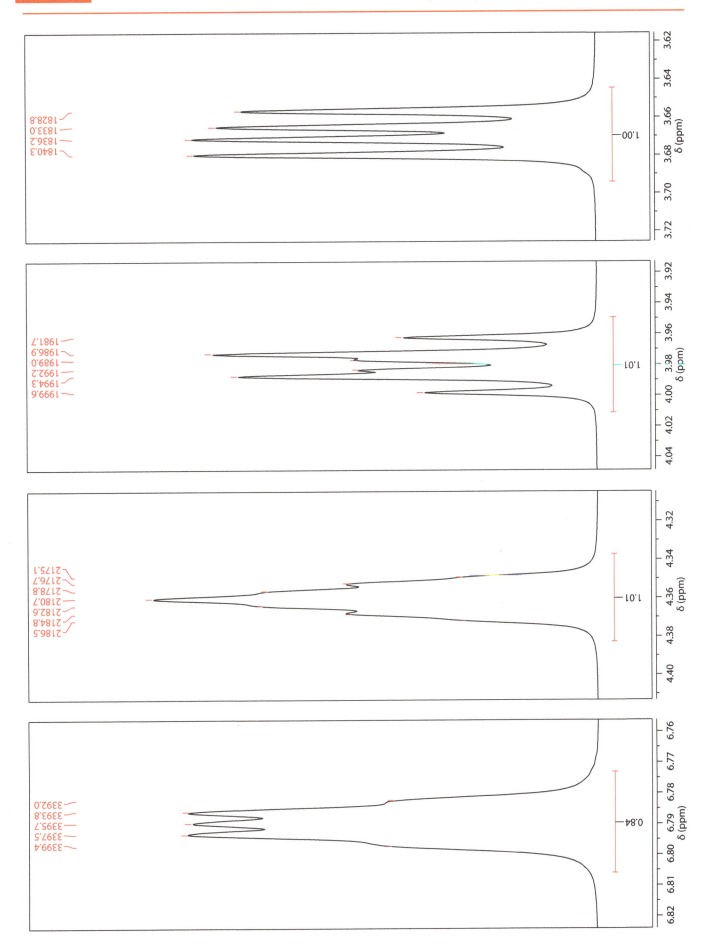

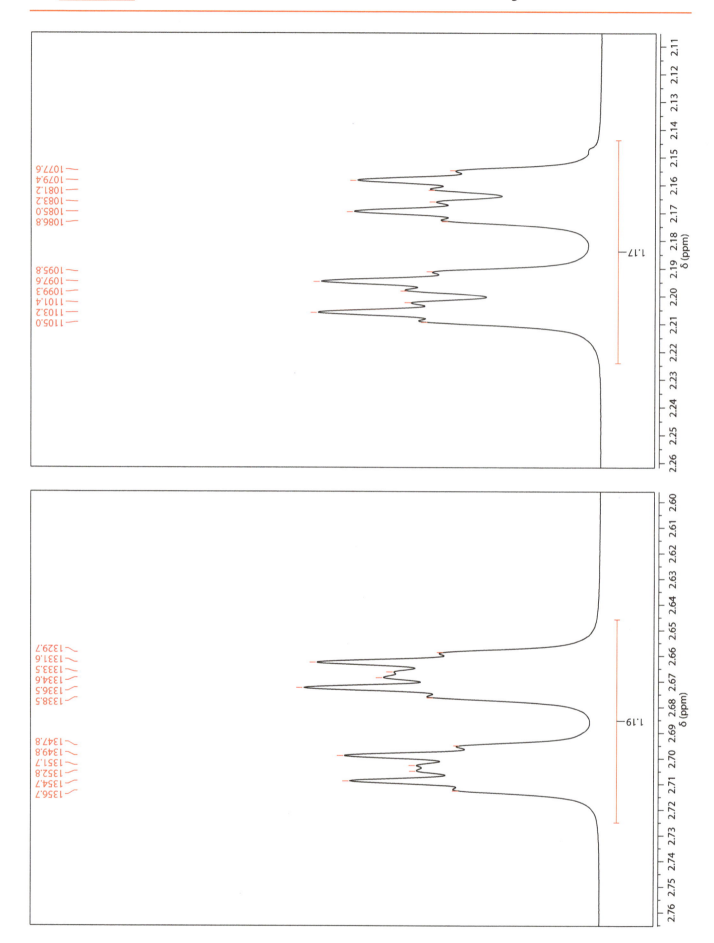

SECTION 2 Problem 31

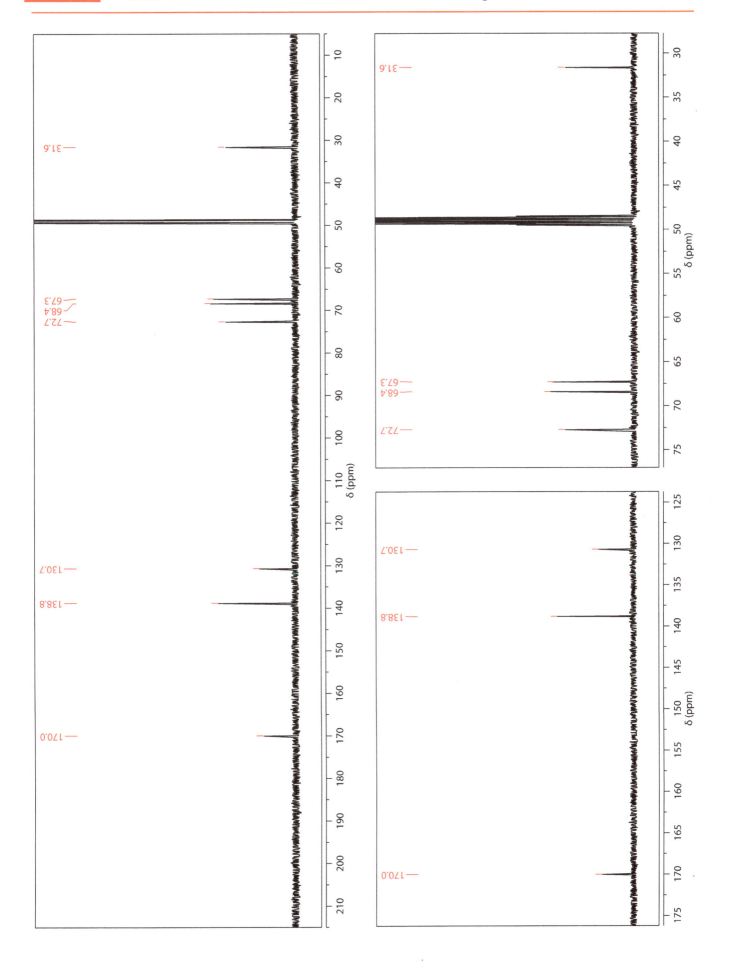

Problem 31

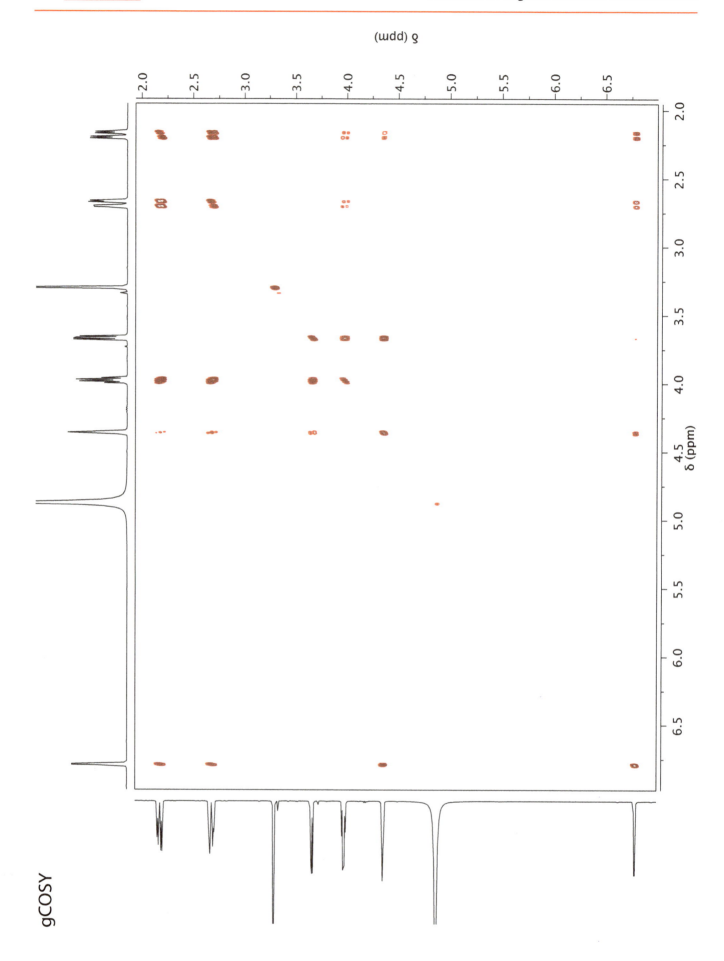

gCOSY

Problem 31

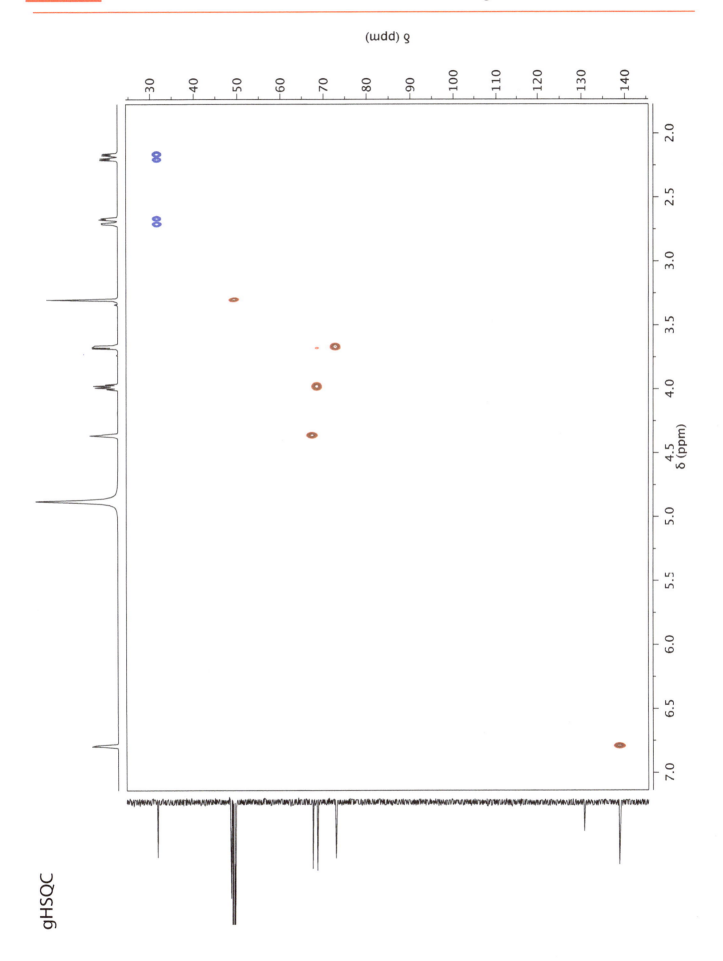

gHSQC

Using the ¹H, ¹³C, gCOSY and gHSQC spectra, assign all proton and carbon resonances to the structure of L-menthol below and assign all coupling constants.
Spectra acquired in CDCl₃ at 500 MHz (¹H) or 125 MHz (¹³C).

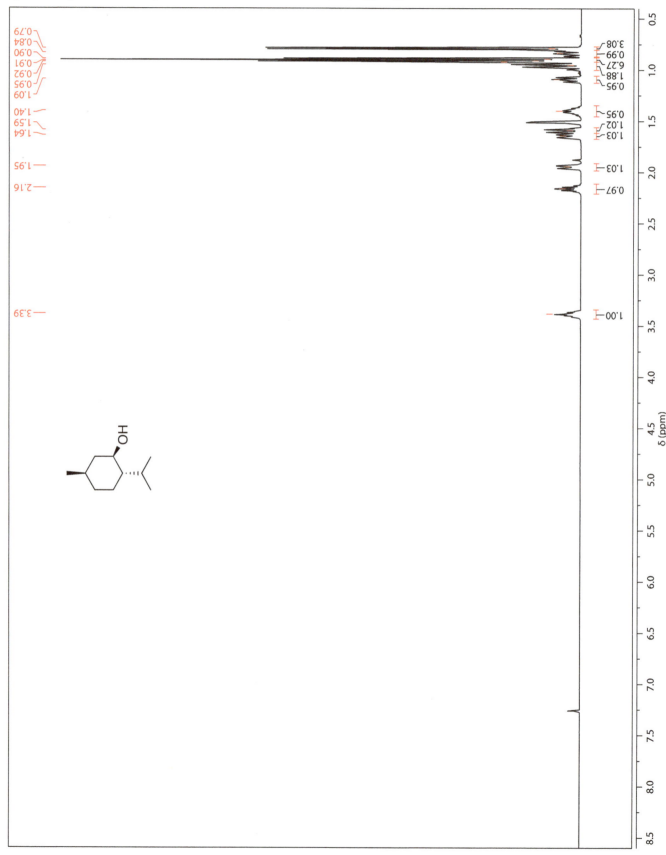

SECTION 2 Problem 32

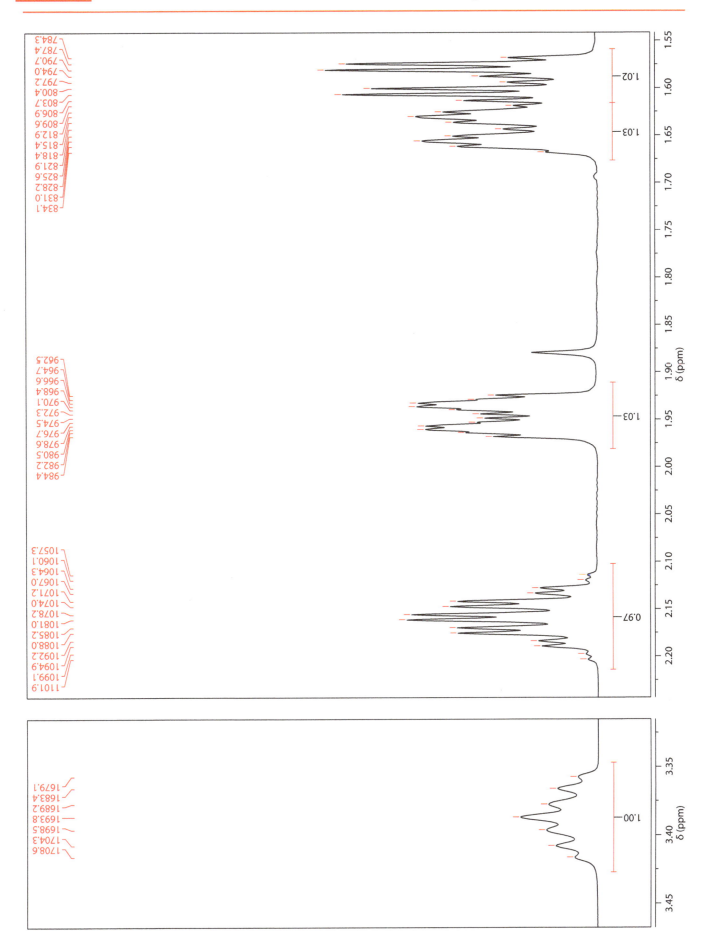

Problem 32

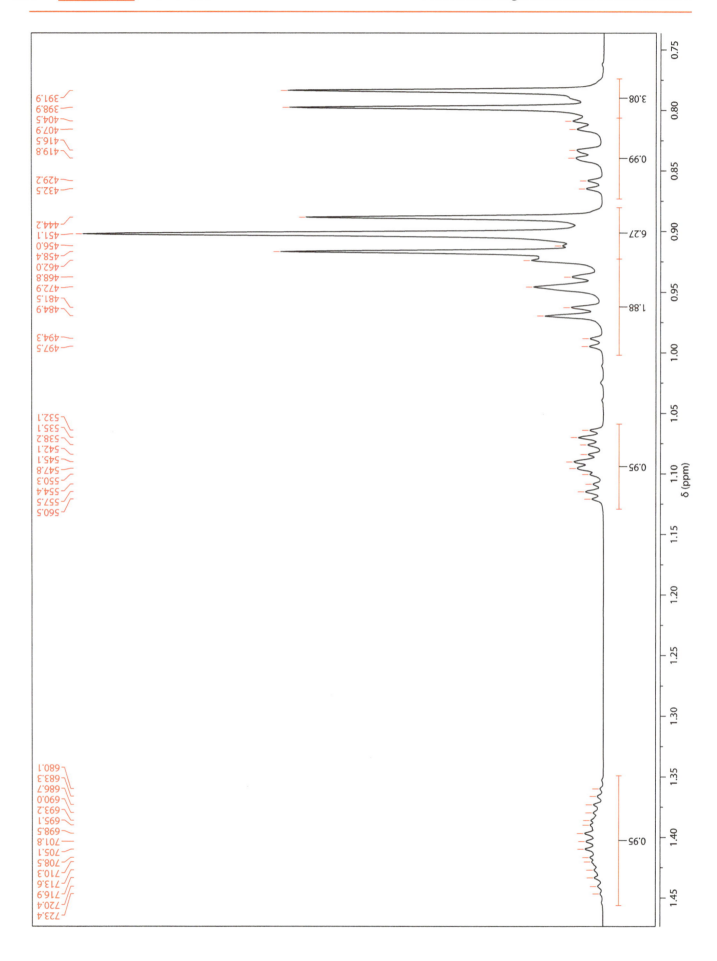

SECTION 2 Problem 32

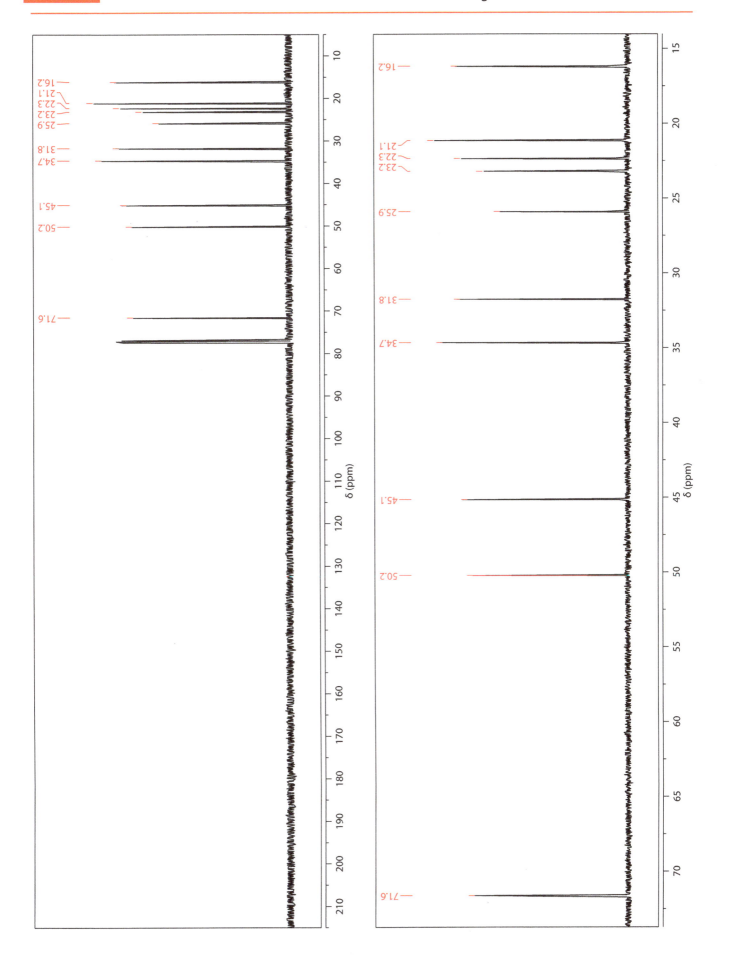

Problem 32

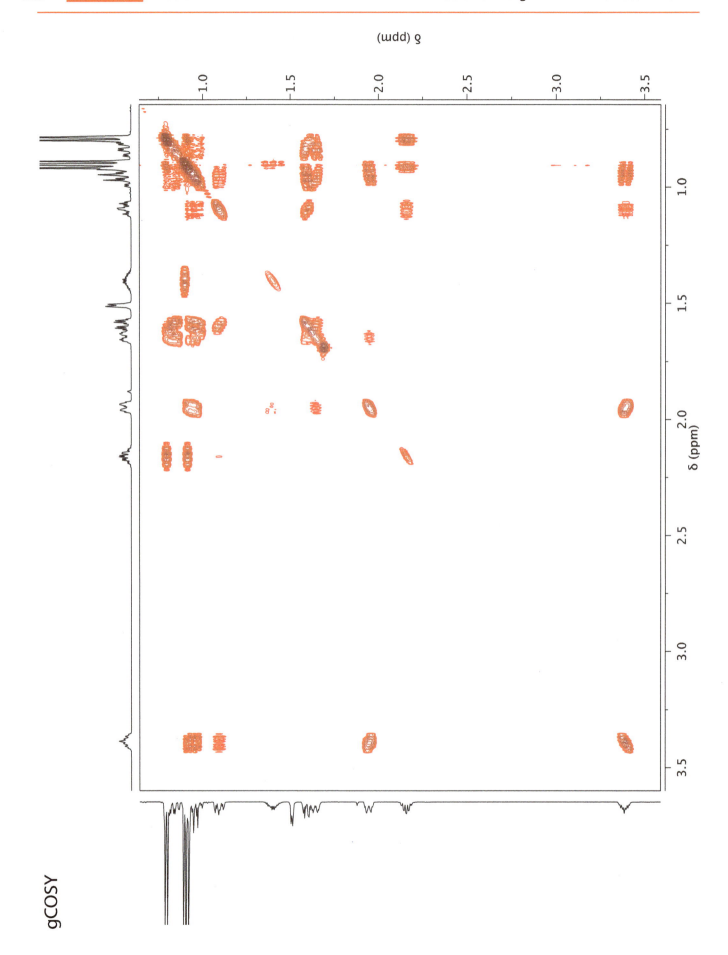

gCOSY

Problem 32

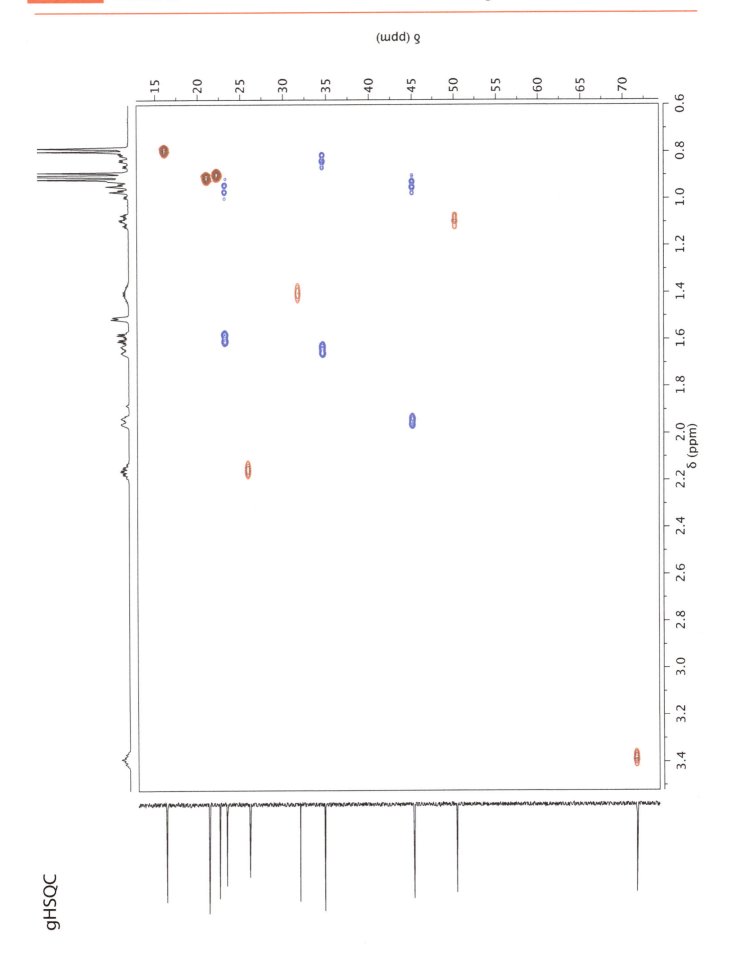

gHSQC

SECTION 2 Problem 33

Problems in Organic Structure Determination

Which of the following compounds is the correct structure for the spectra below?

Spectra acquired in CDCl₃ at 500 MHz (^{1}H) or 125 MHz (^{13}C).

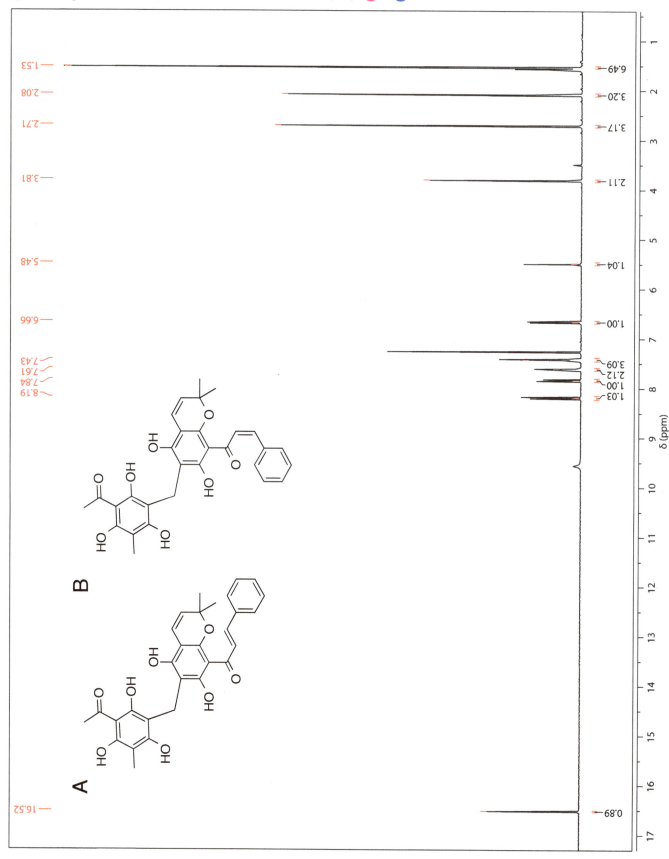

Problem 33

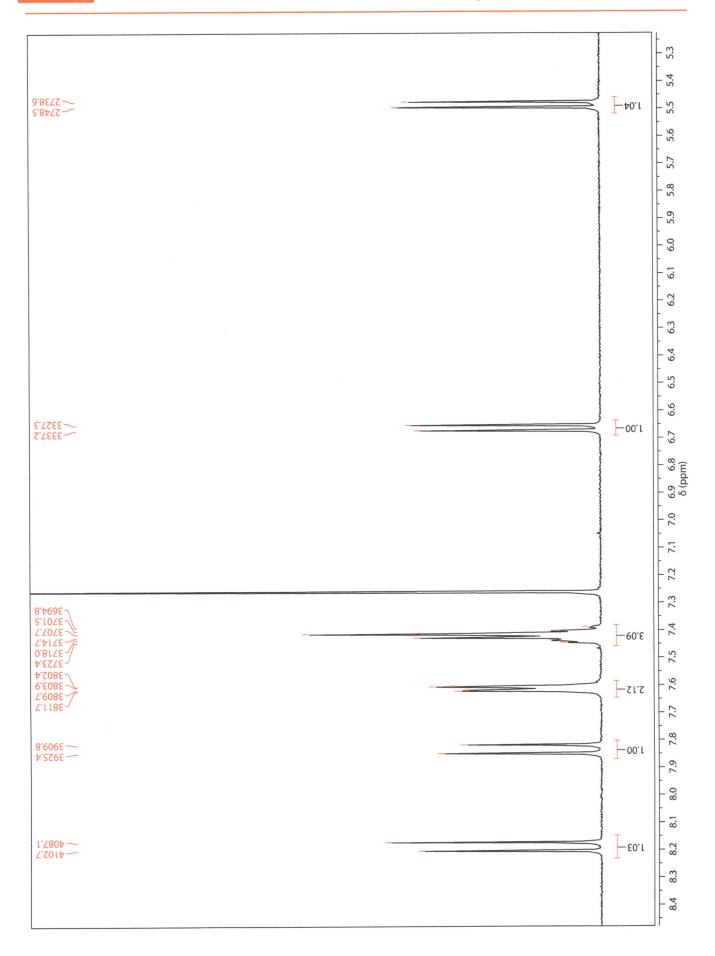

Problem 33

gCOSY

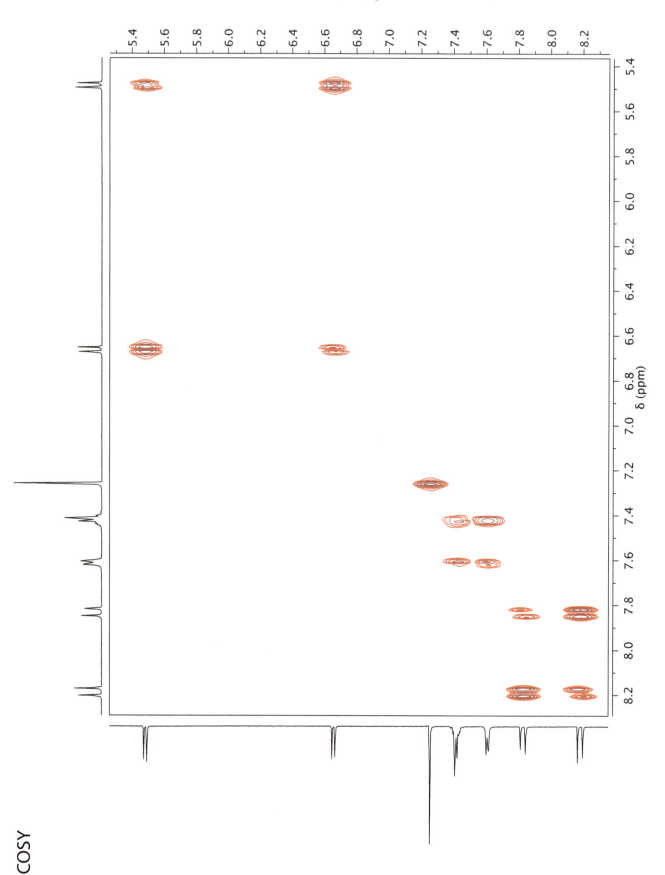

SECTION 2 Problem 34

Using the ¹H, ¹³C, gCOSY, gHSQC and gHMBC spectra, assign all proton and carbon resonances to the structure of 7-benzyloxyindole below.
Spectra acquired in CDCl₃ at 500 MHz (¹H) or 125 MHz (¹³C).

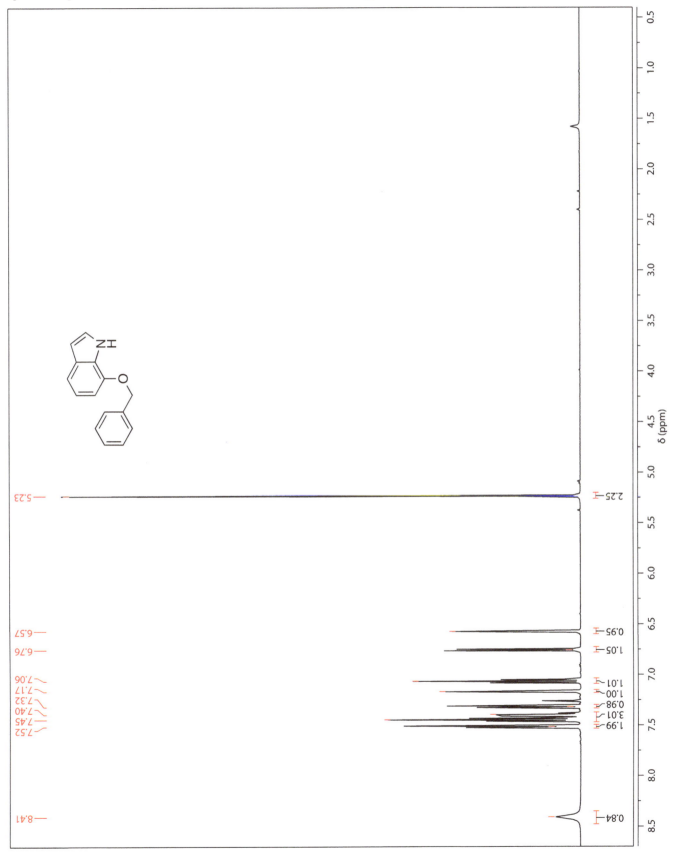

SECTION 2 Problem 34

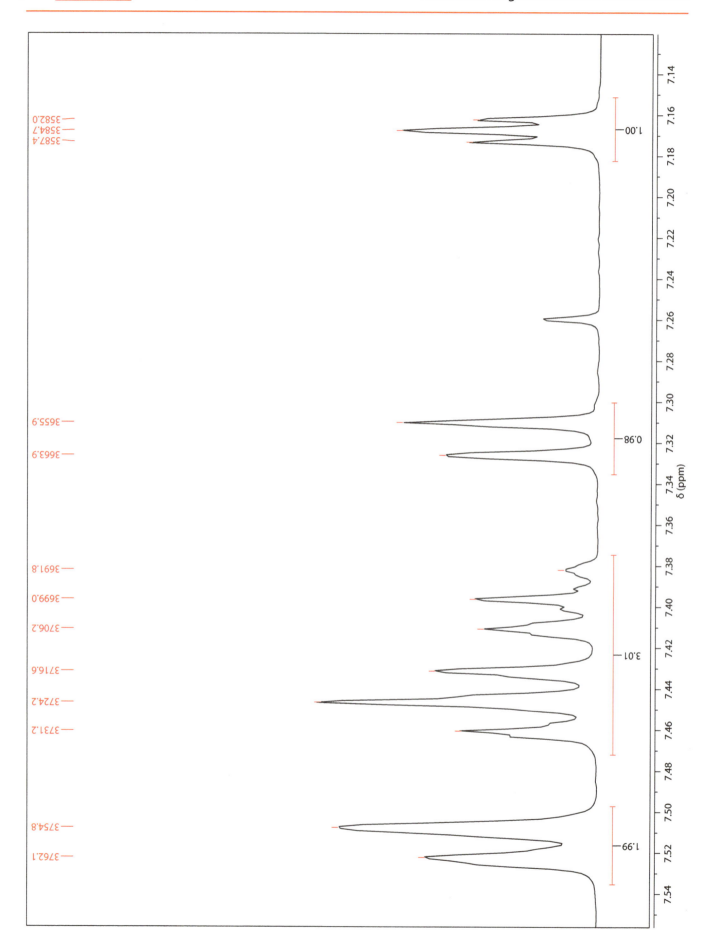

Problem 34

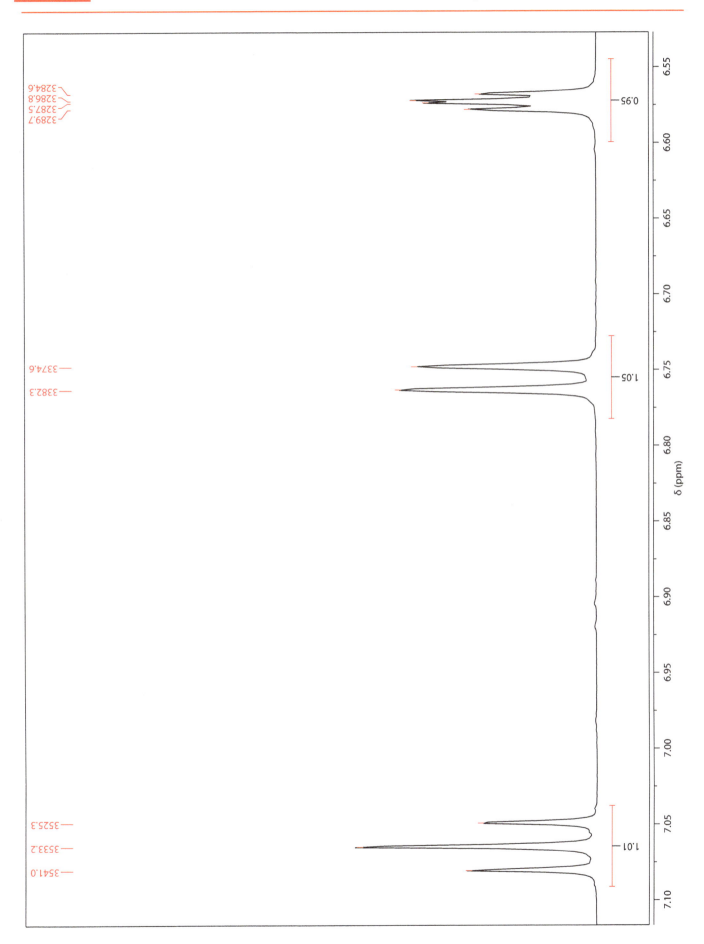

Problem 34

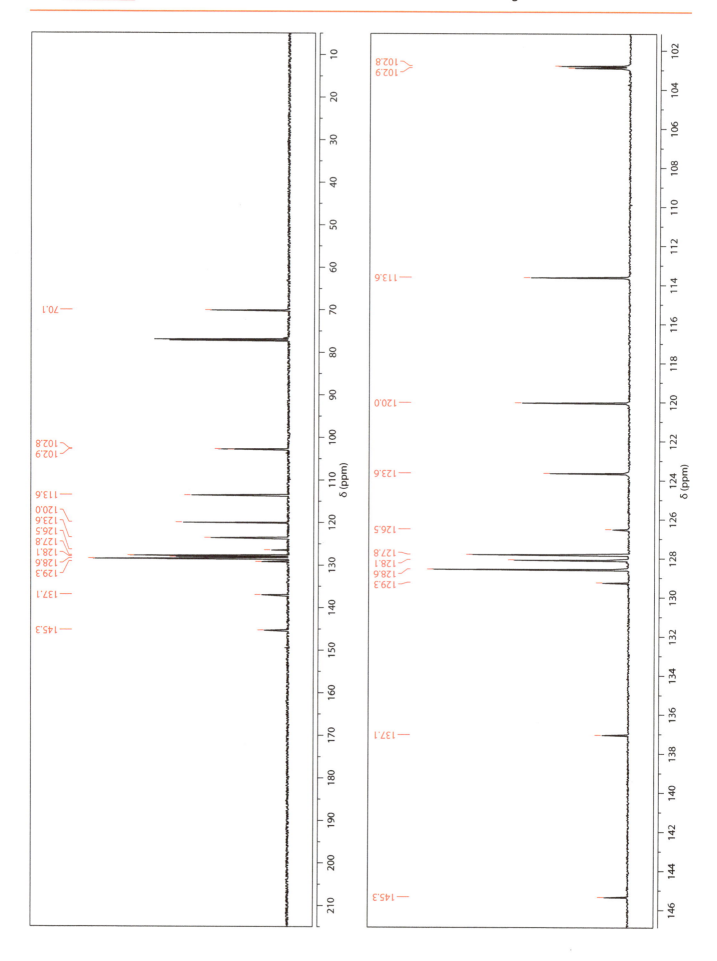

SECTION 2 Problem 34

Problems in Organic Structure Determination

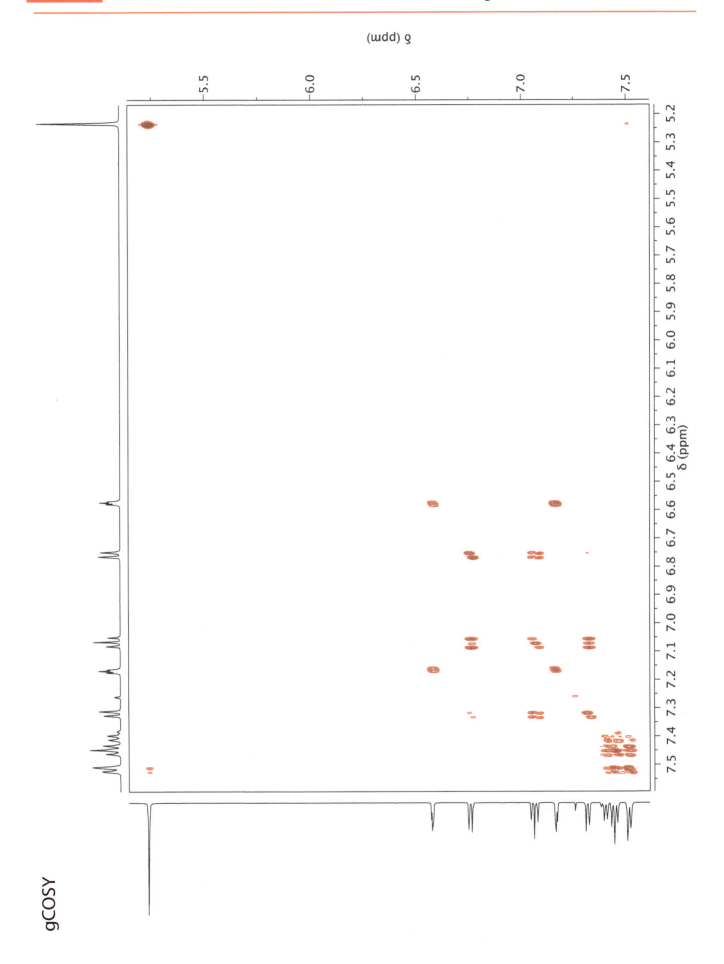

gCOSY

Problem 34

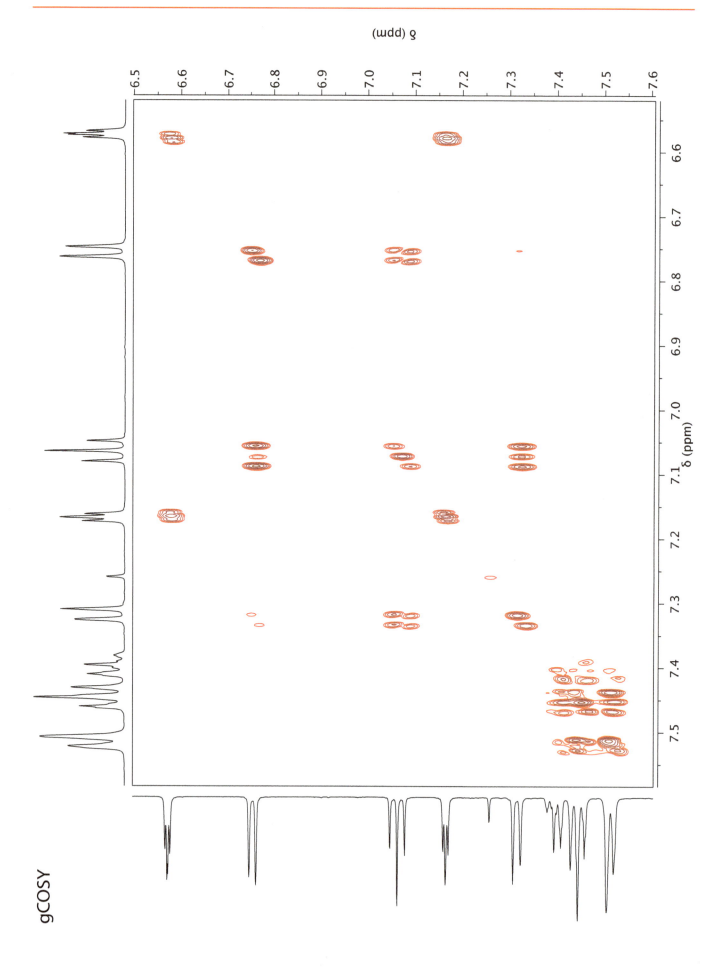

gCOSY

SECTION 2 Problem 34

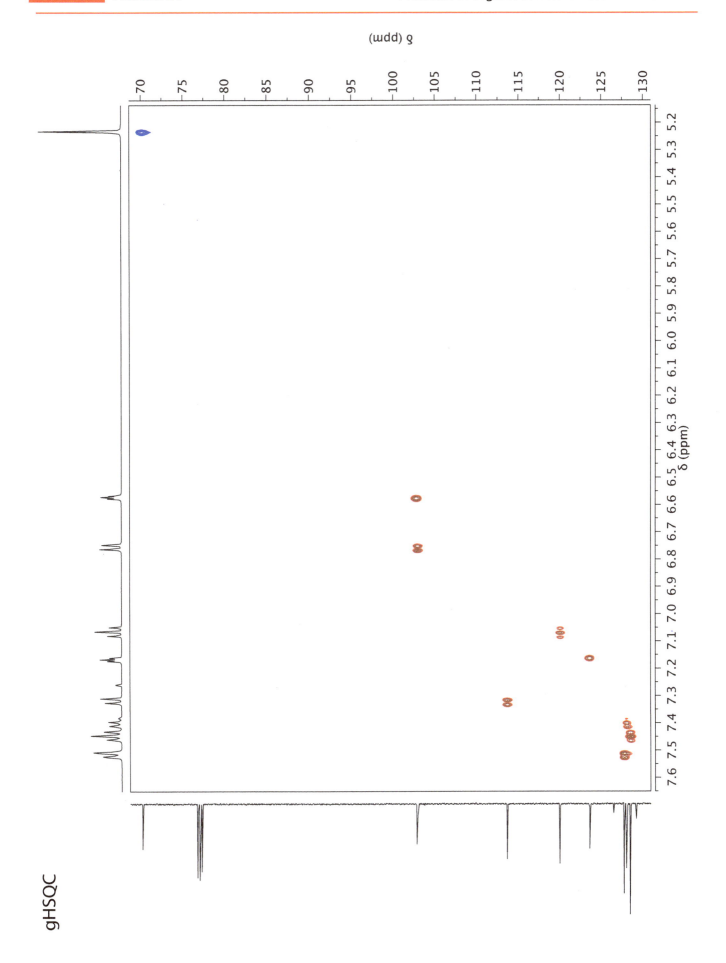

gHSQC

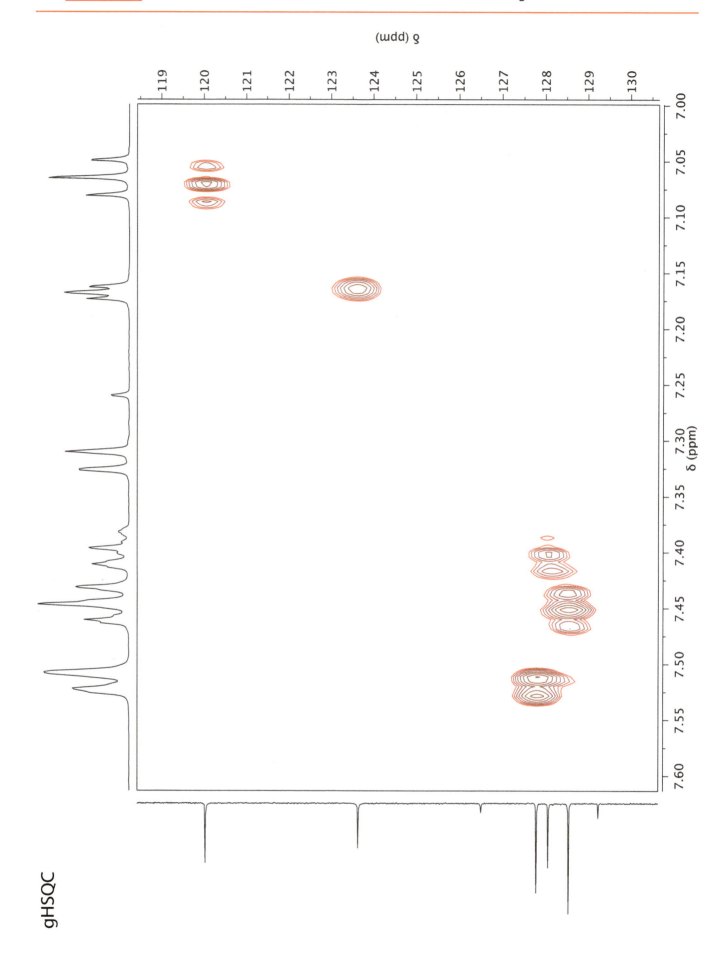

SECTION 2 Problem 34

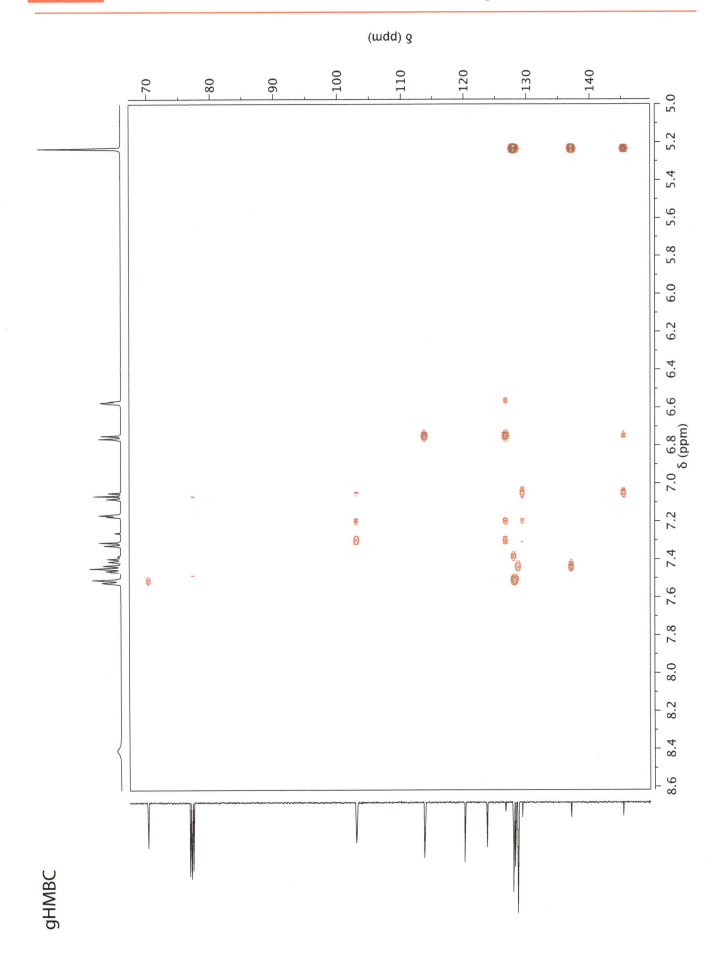

gHMBC

Problem 34

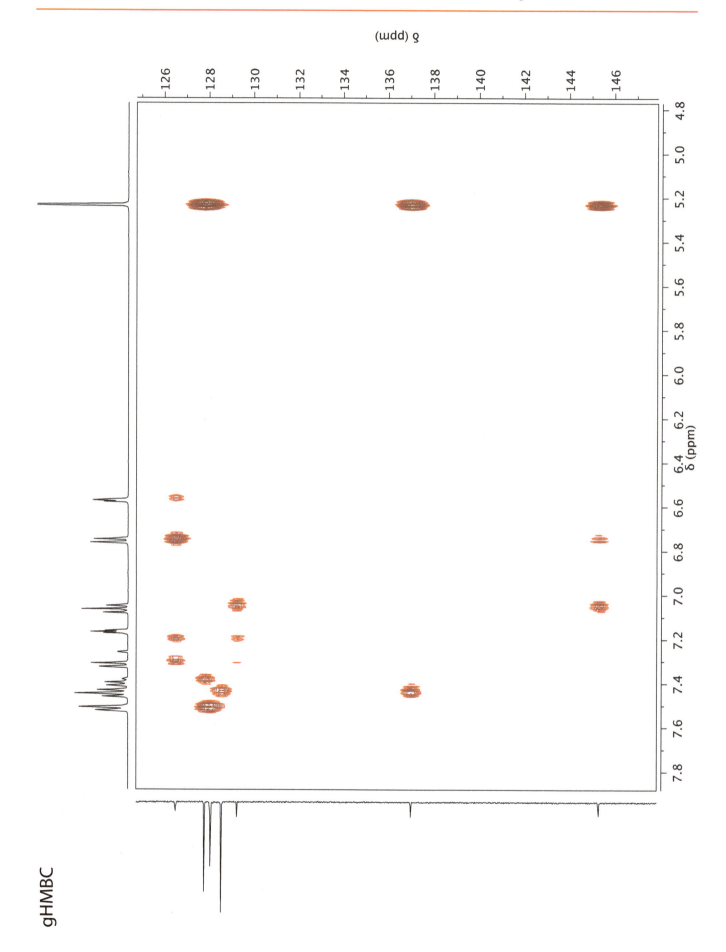

gHMBC

SECTION 2 Problem 35

Using the ¹H, ¹³C, gCOSY, gHSQC and 1D TOCSY spectra, assign all proton and carbon resonances to the structure of the disaccharide below and assign all proton coupling constants.
Spectra acquired in $(CD_3)_2SO$ at 500 MHz (¹H) or 125 MHz (¹³C).

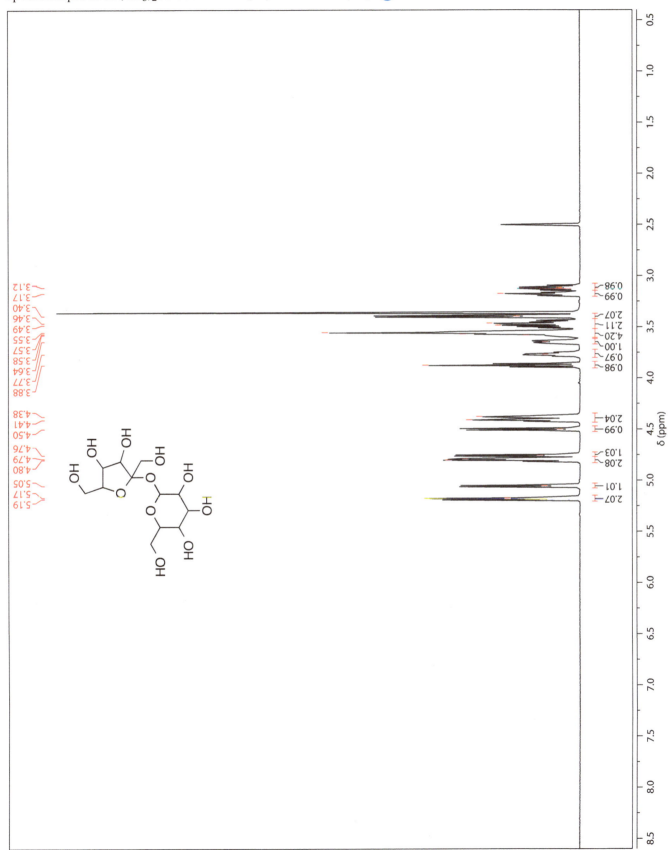

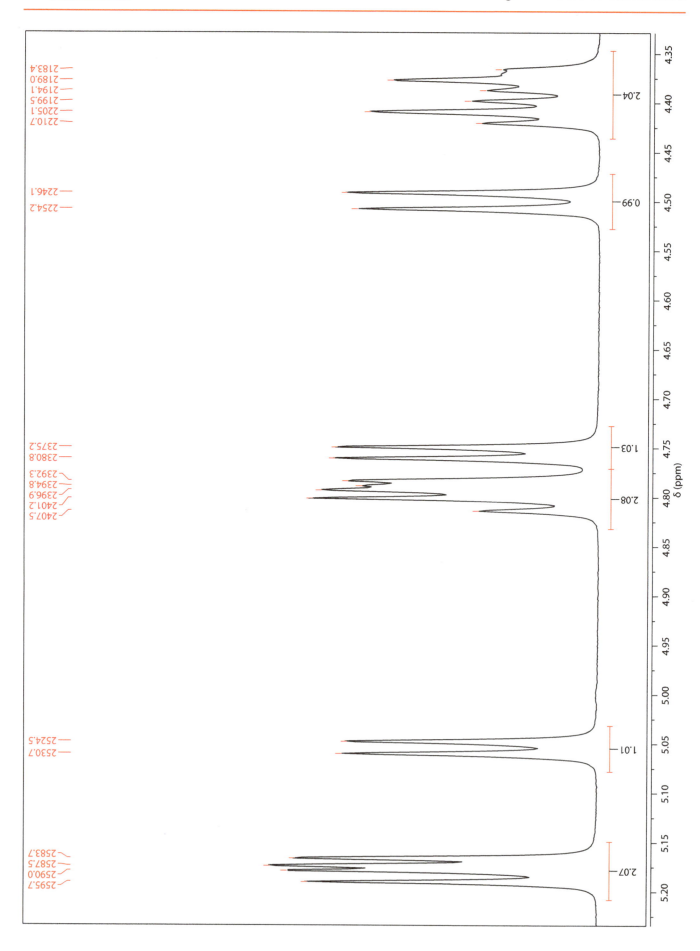

SECTION 2   Problem 35   Problems in Organic Structure Determination

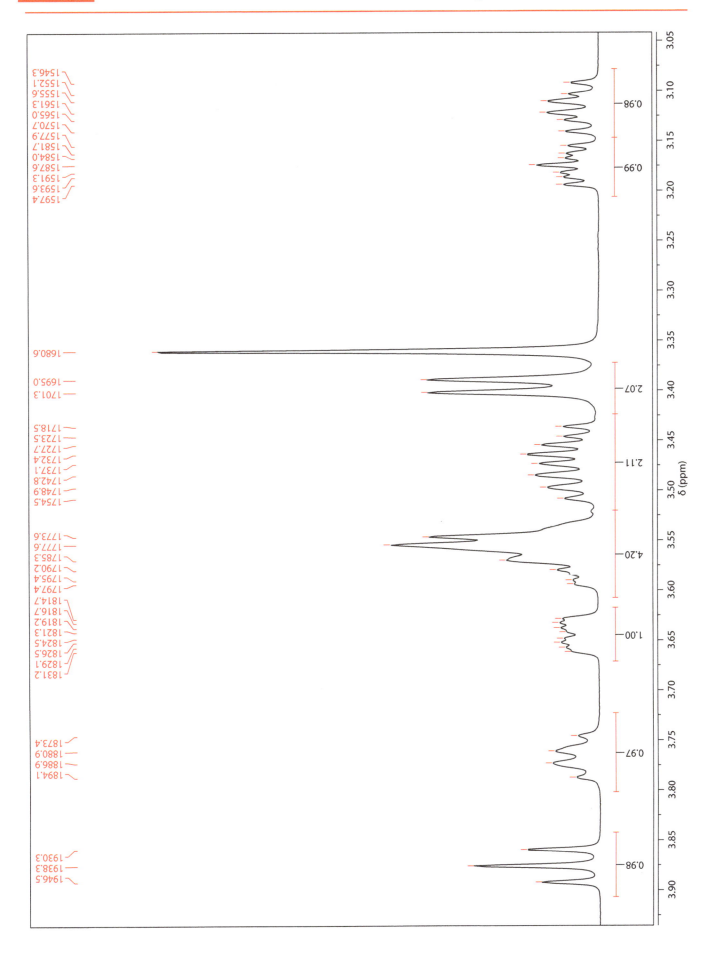

SECTION 2 Problem 35

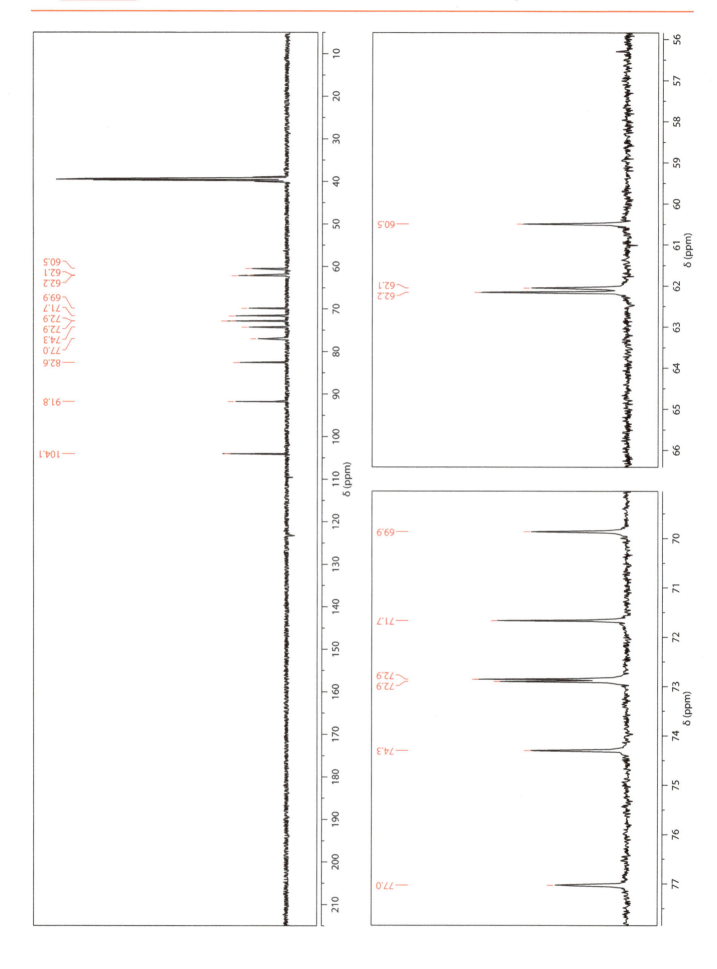

SECTION 2 Problem 35

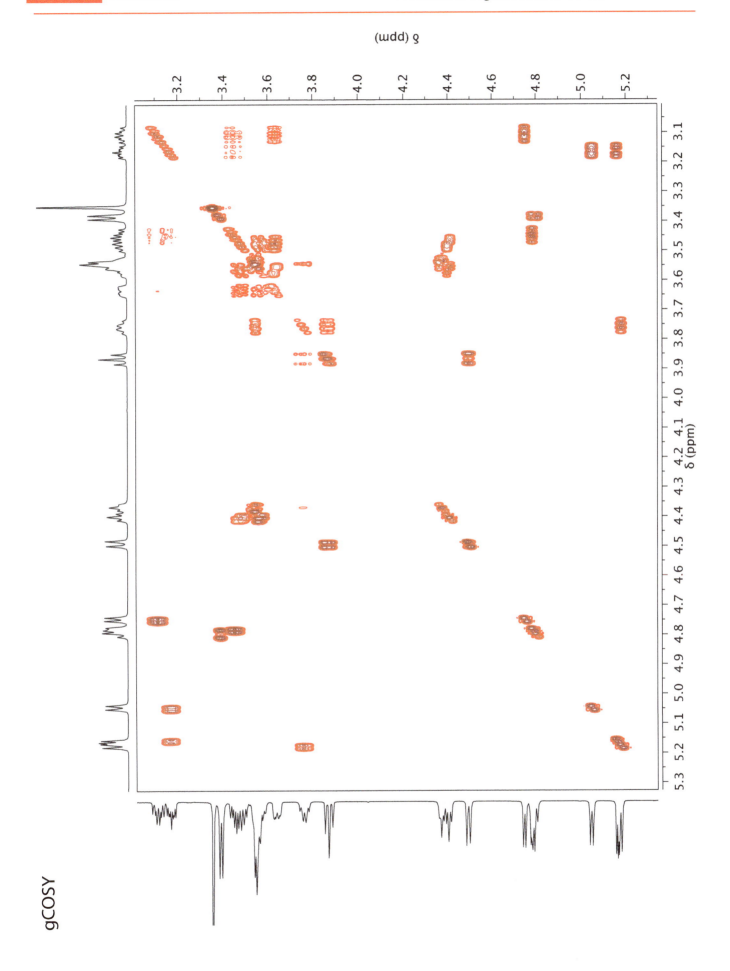

gCOSY

Problem 35

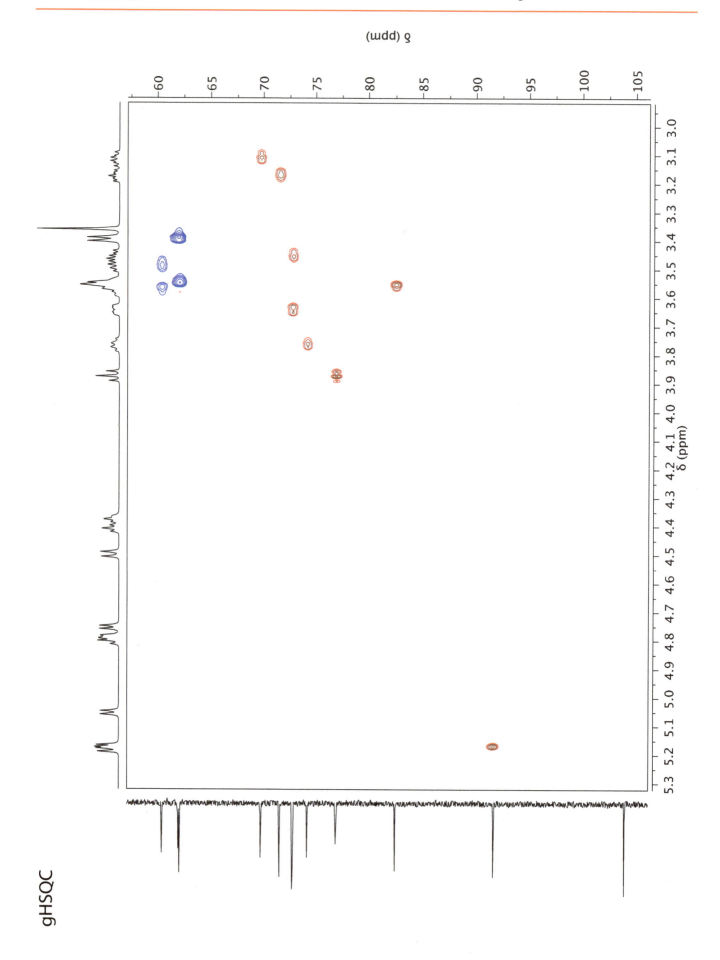

gHSQC

Problem 35

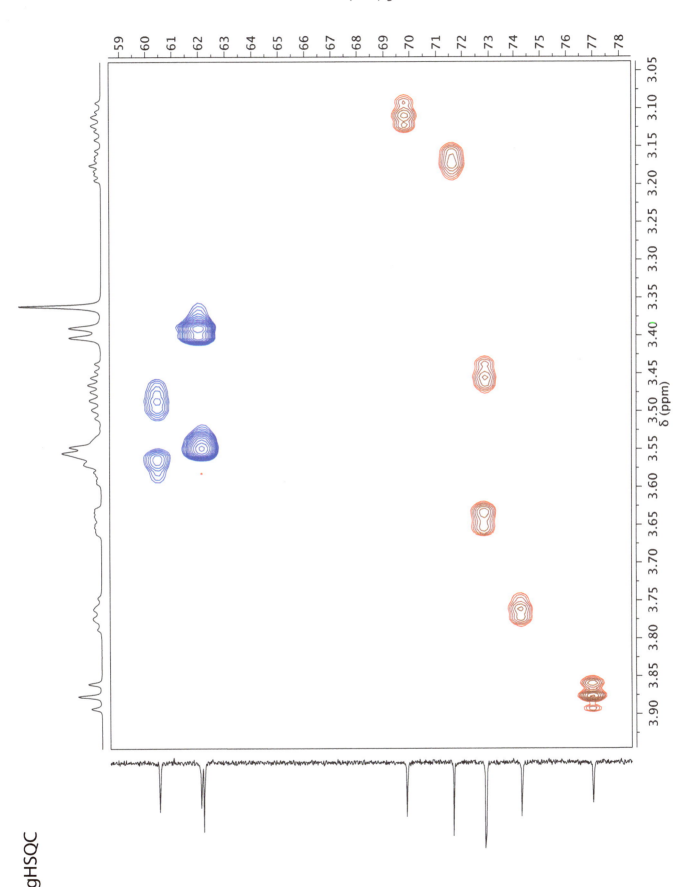

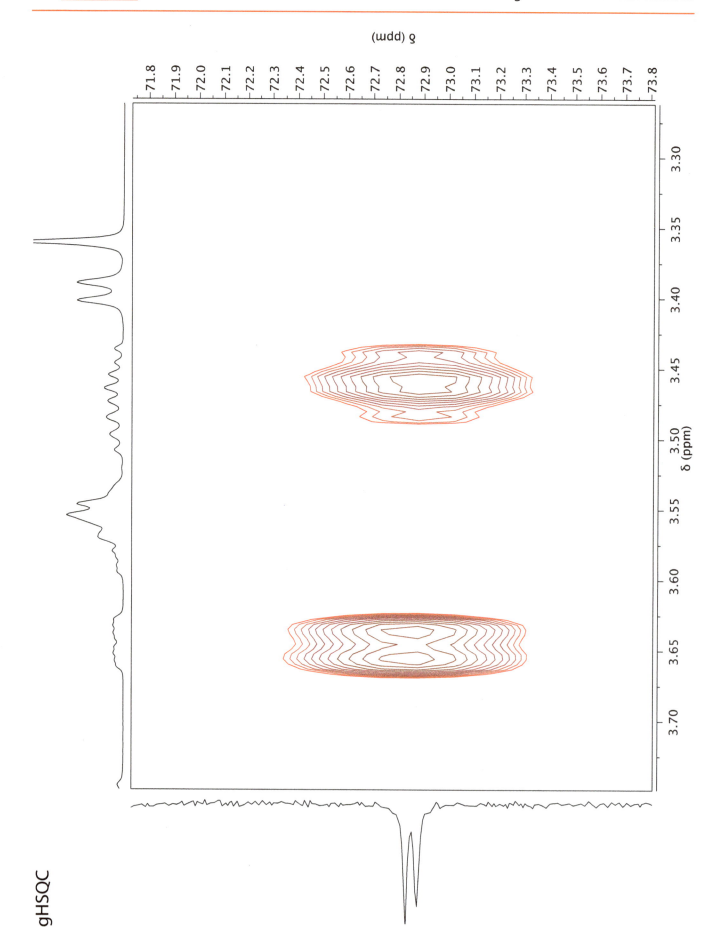

SECTION 2 Problem 35
Problems in Organic Structure Determination

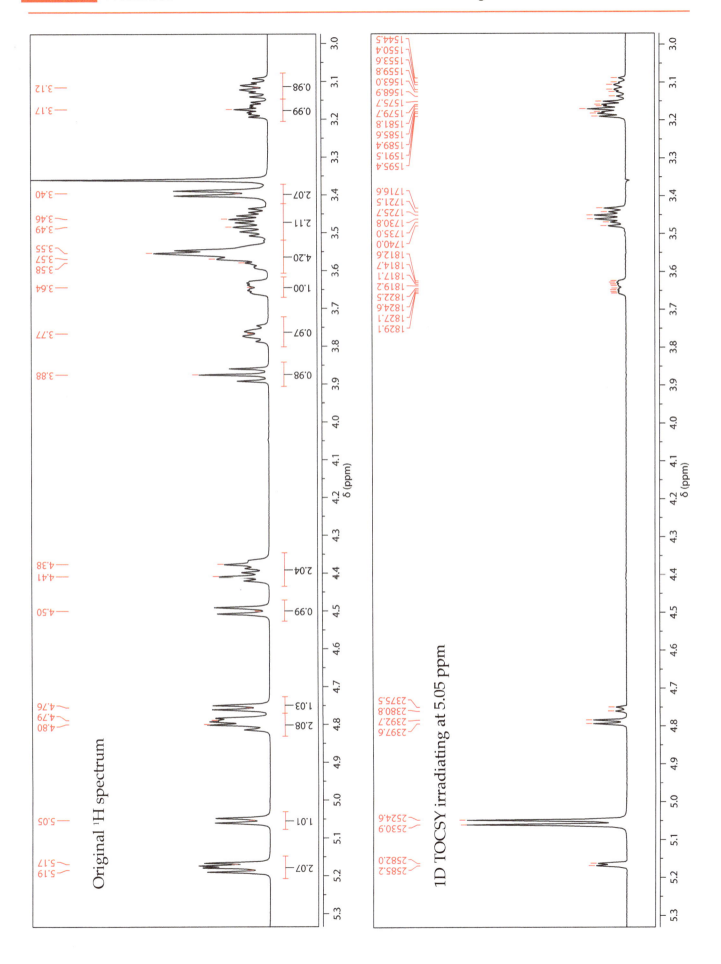

Problem 35

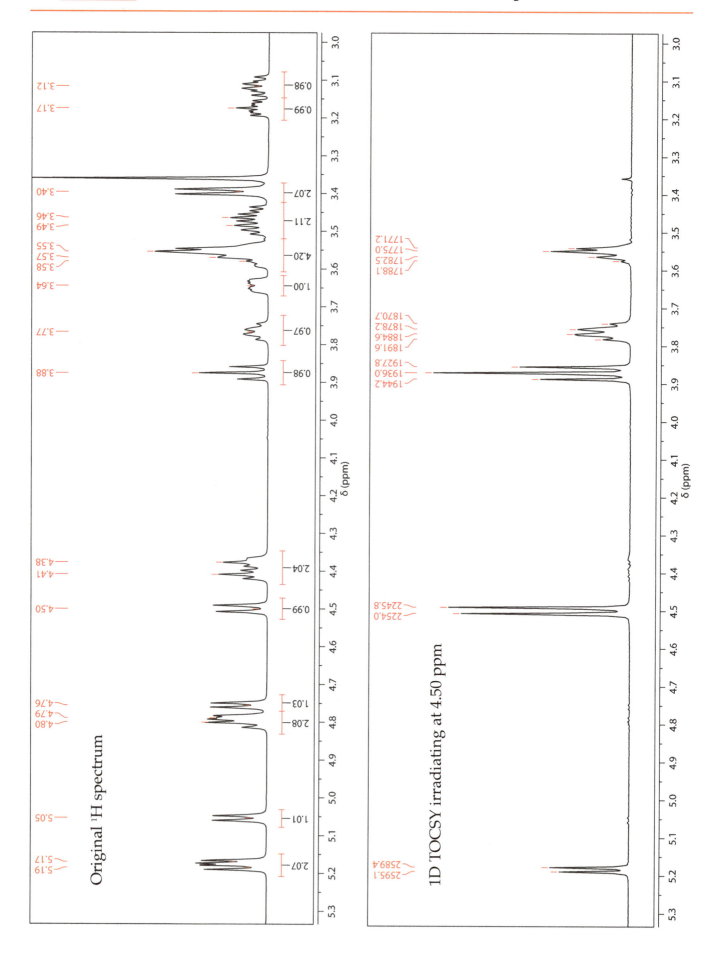

SECTION 2   Problem 35

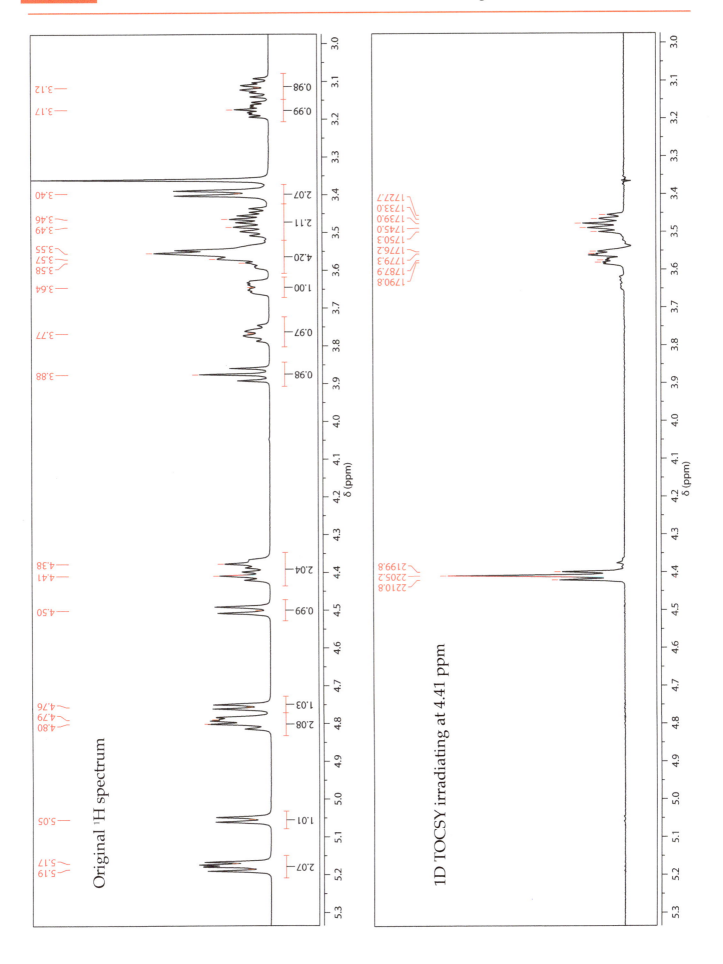

Problem 35

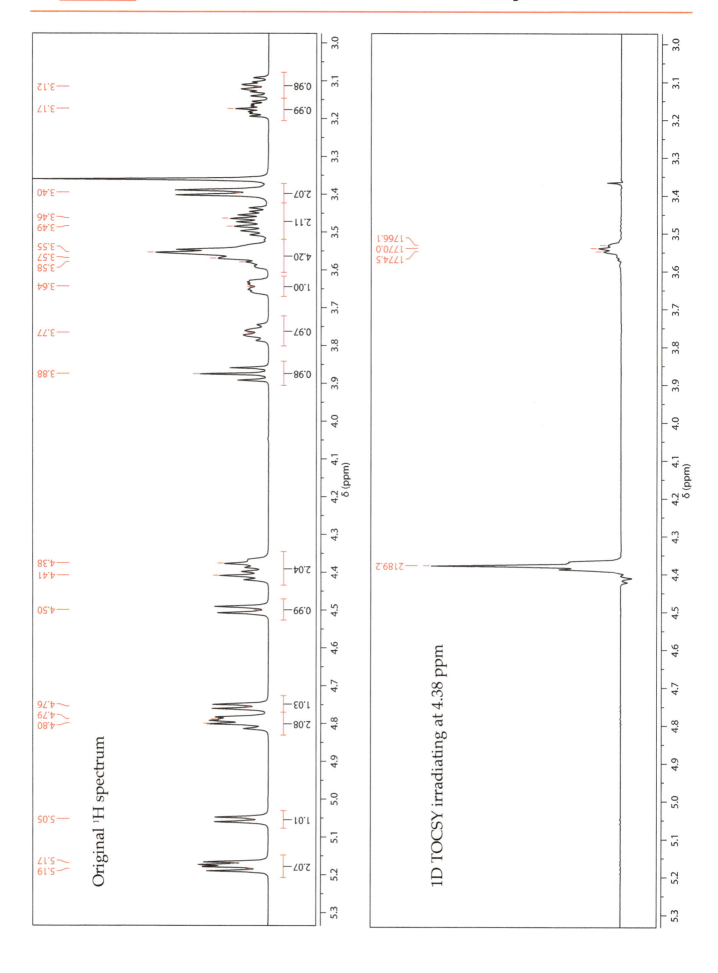

SECTION 2 Problem 35

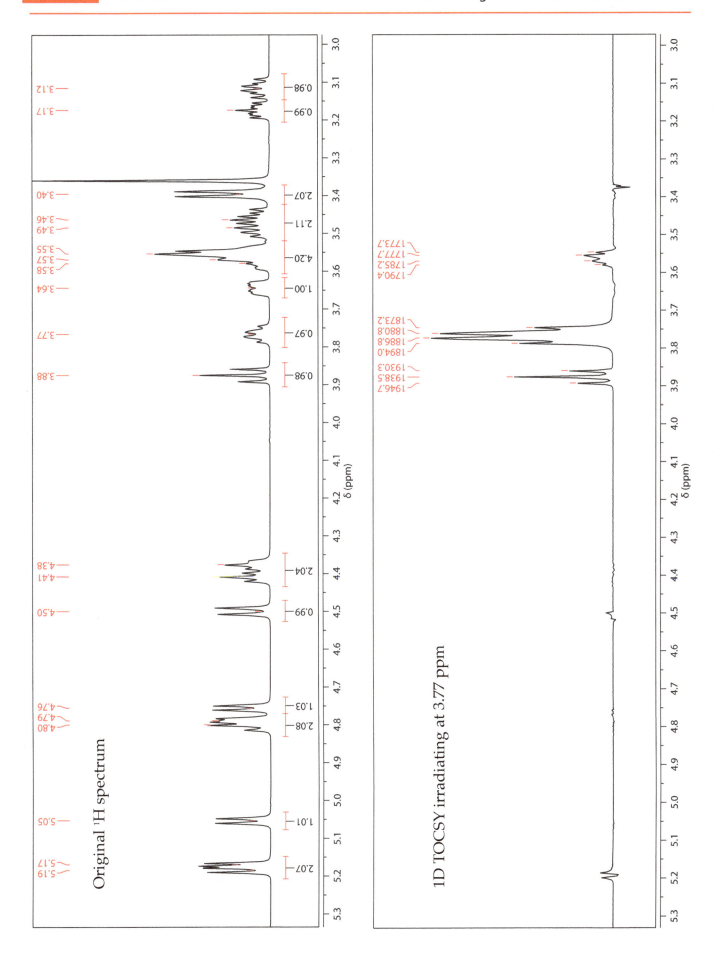

Using the ¹H, ¹³C, gCOSY, gHSQC and selective decoupling spectra, assign all proton and carbon resonances to the structure of (R)-carvone below and assign all proton coupling constants.
Spectra acquired in CDCl₃ at 500 MHz (¹H) or 125 MHz (¹³C).

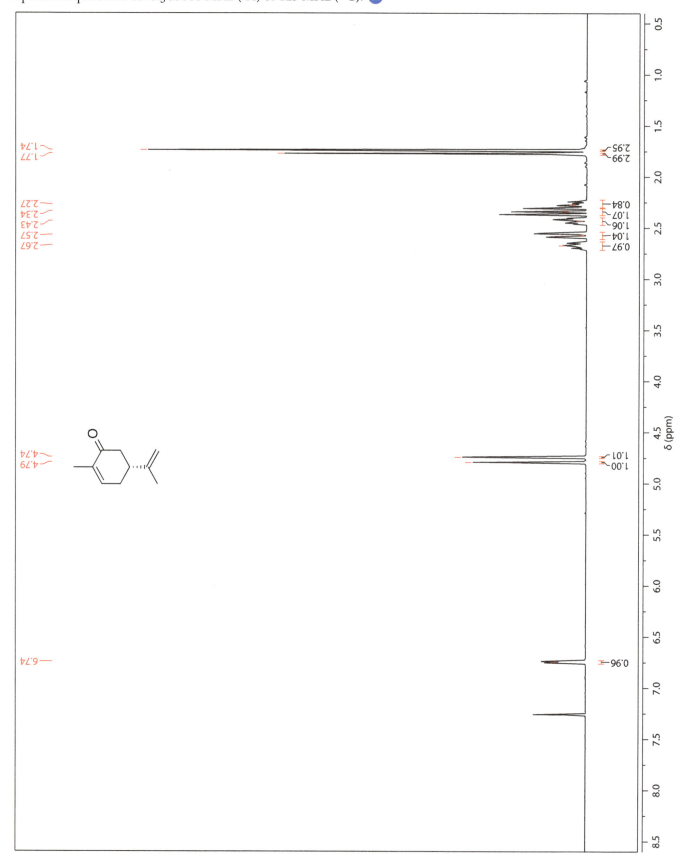

SECTION 2   Problem 36   Problems in Organic Structure Determination

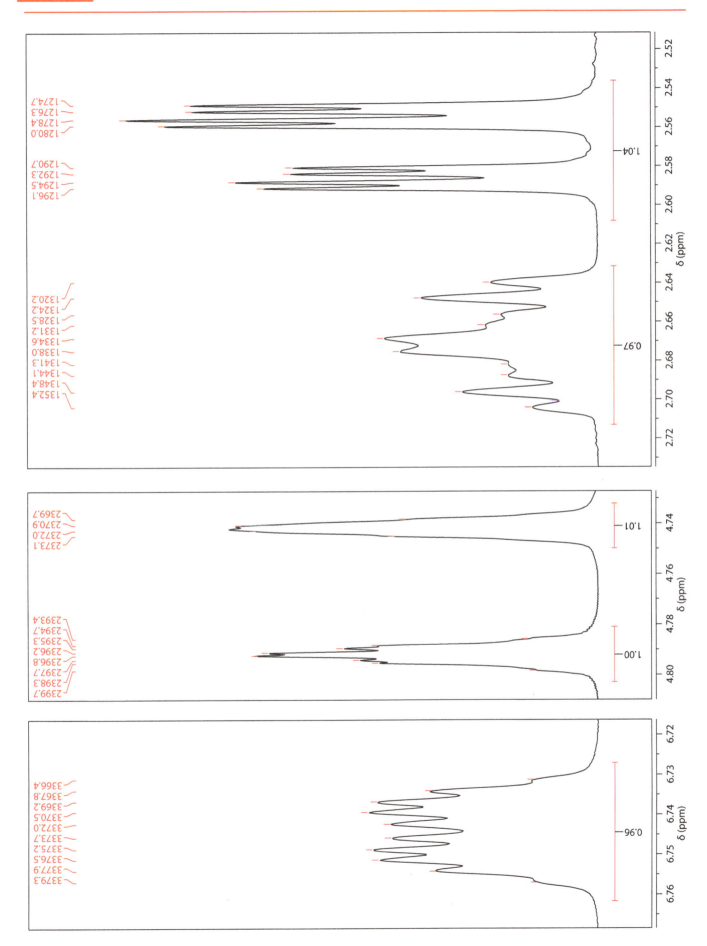

Problem 36

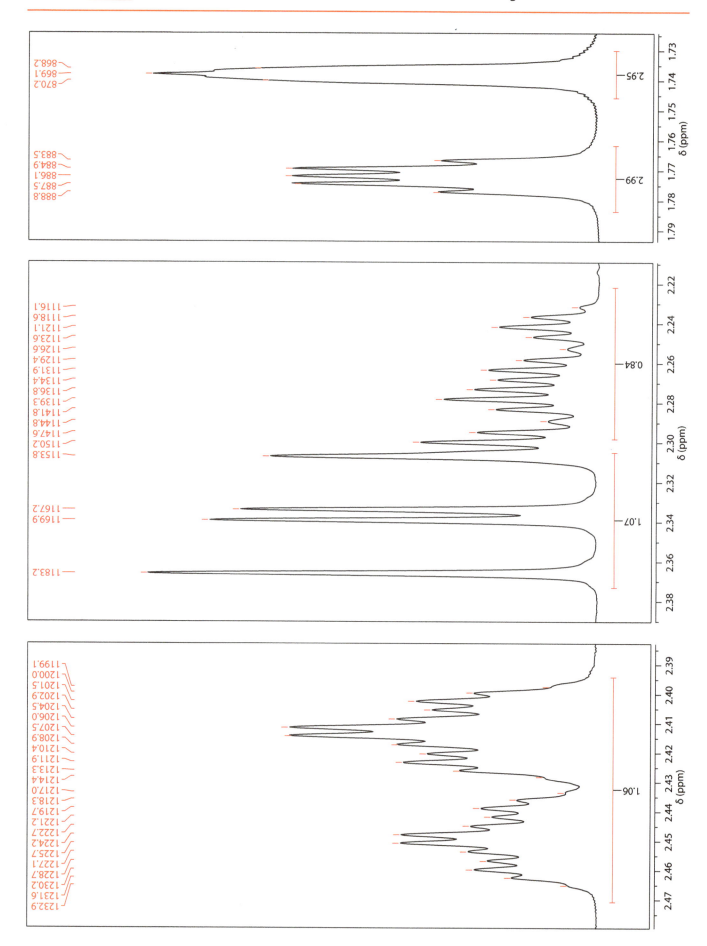

SECTION 2 Problem 36

Problems in Organic Structure Determination

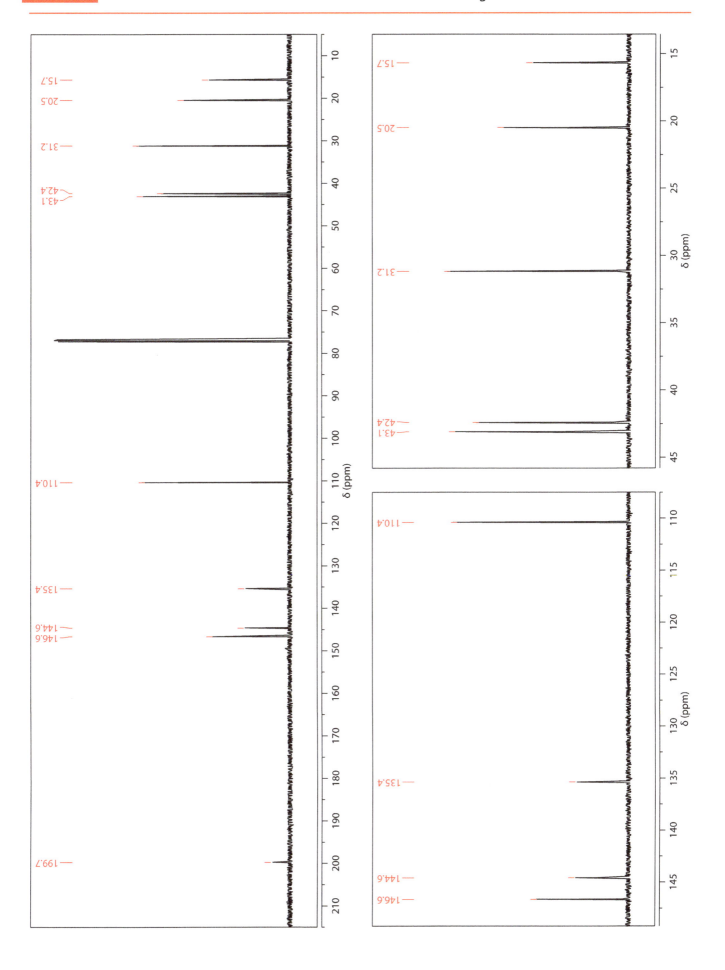

Problem 36

gCOSY

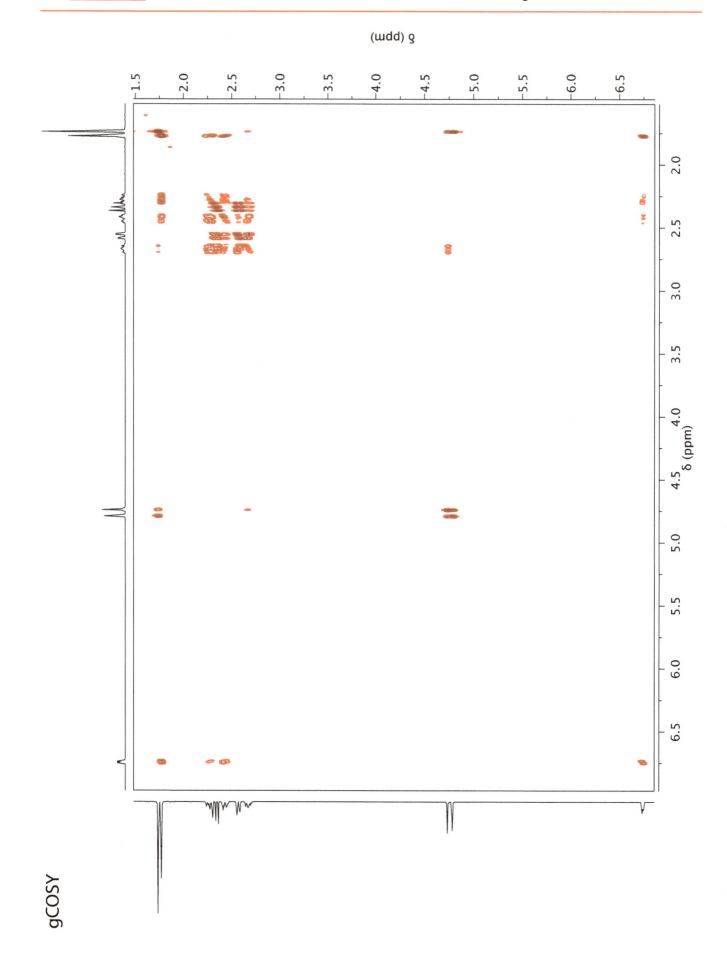

SECTION 2 Problem 36

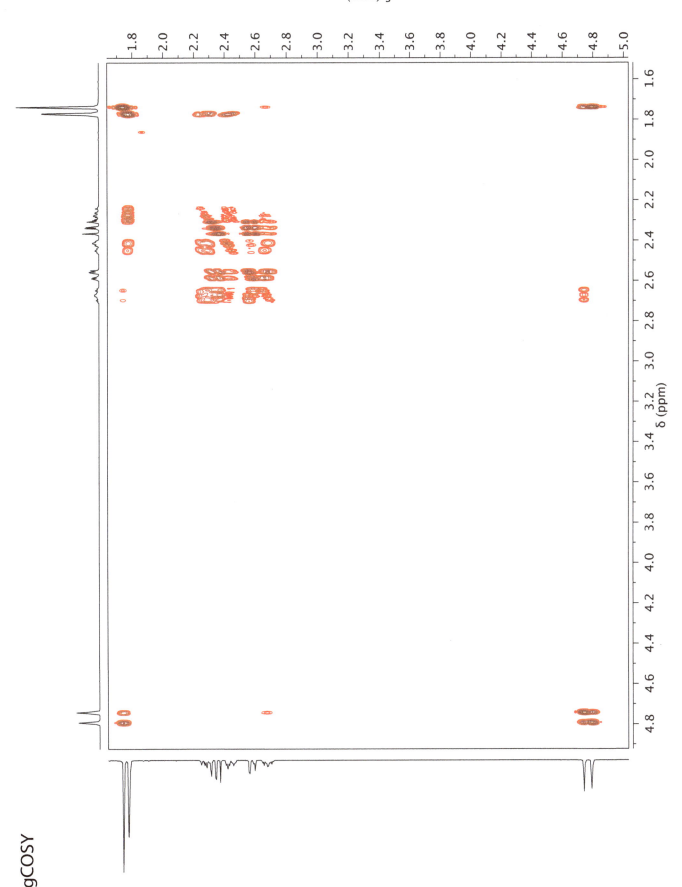

gCOSY

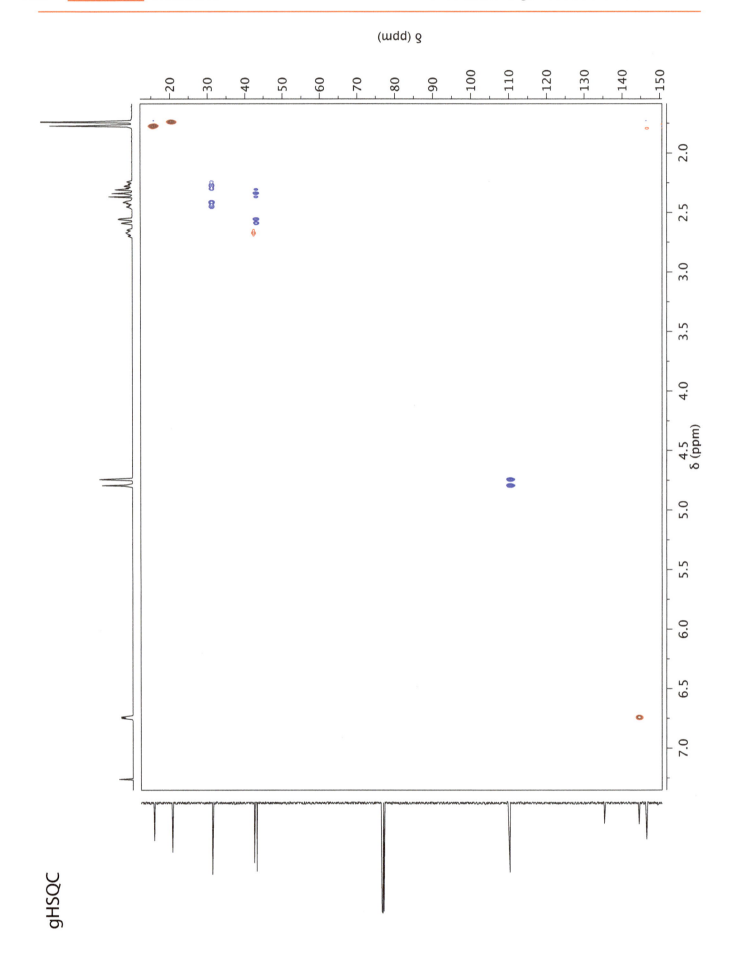

Problem 36

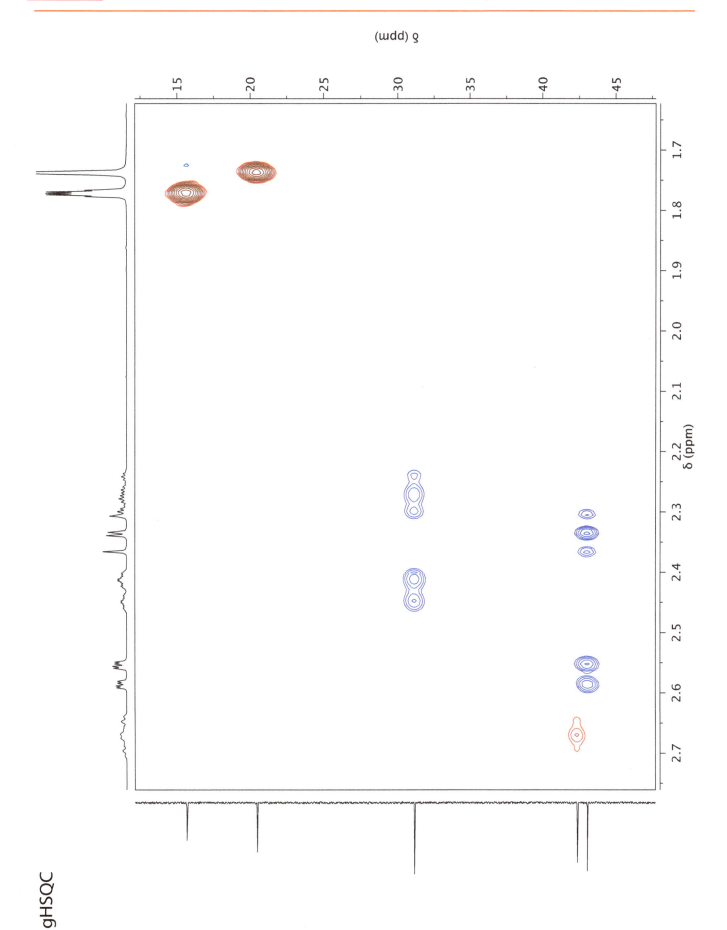

Problem 36

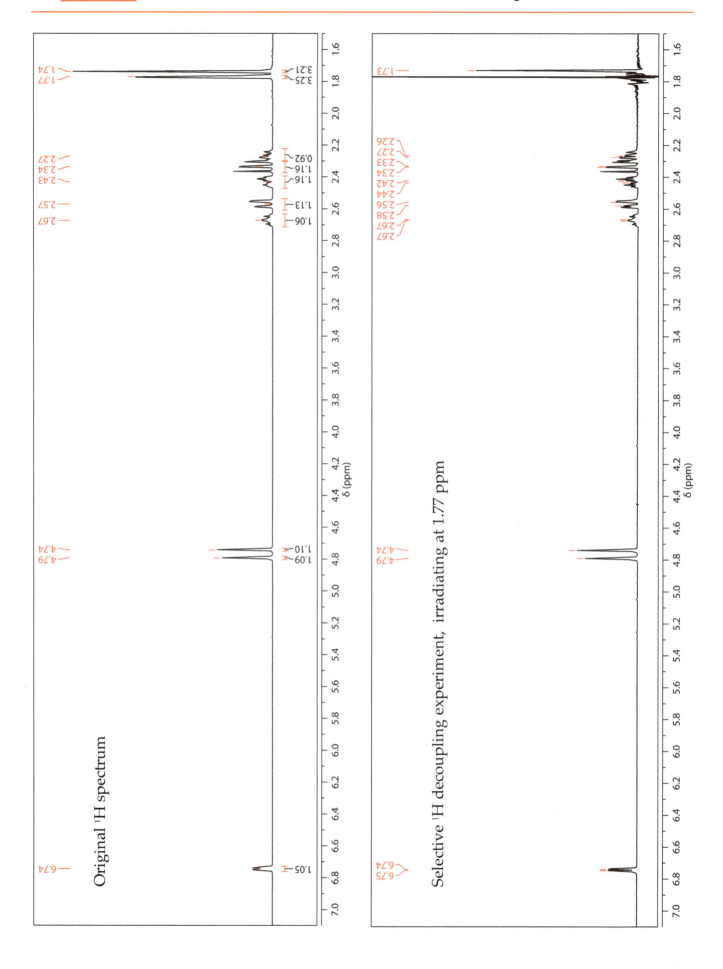

SECTION 2 Problem 36

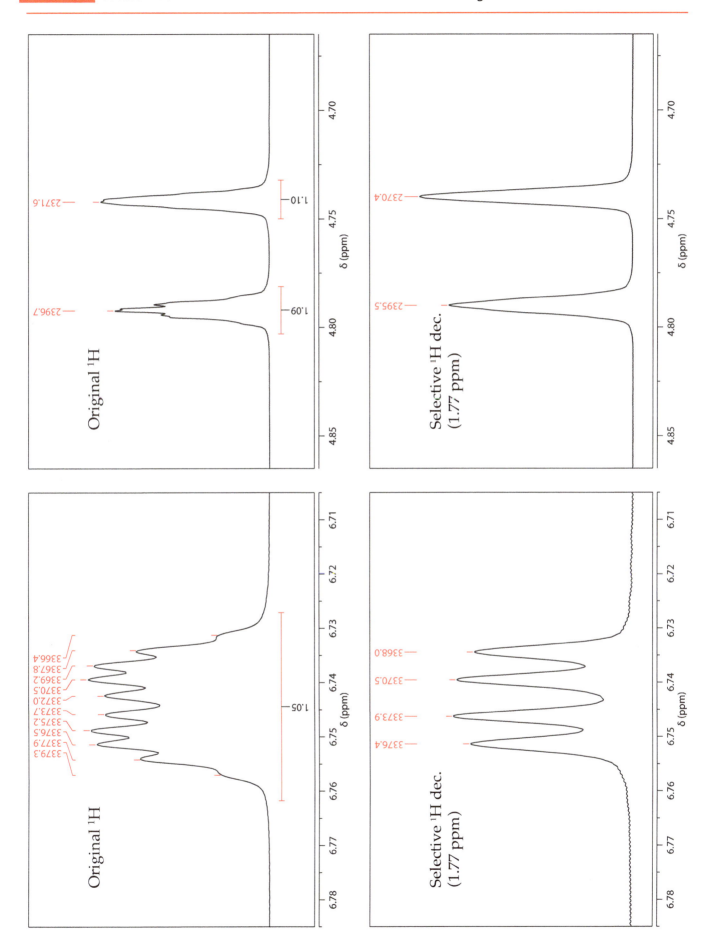

Problem 36

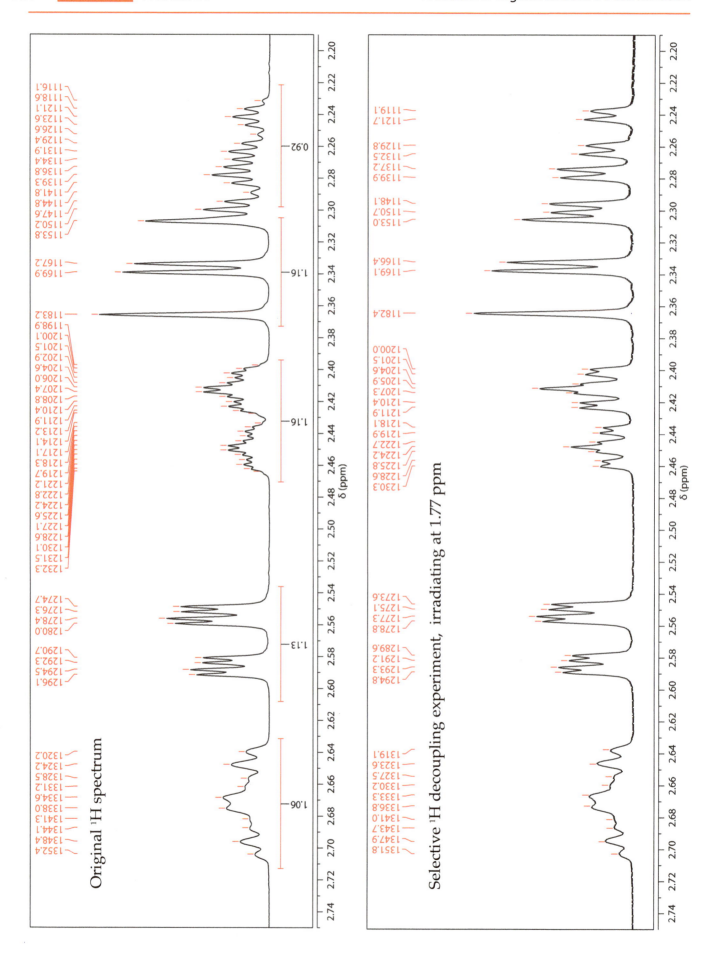

SECTION 2 Problem 37

Using the ¹H, ¹³C, gCOSY, gHSQC, gHMBC and selective 1D-TOCSY spectra, assign all proton and carbon resonances to the structure of 1,8-diazobicyclo[5.4.0]undec-7-ene below.
Spectra acquired in CDCl₃ at 500 MHz (¹H) or 125 MHz (¹³C).

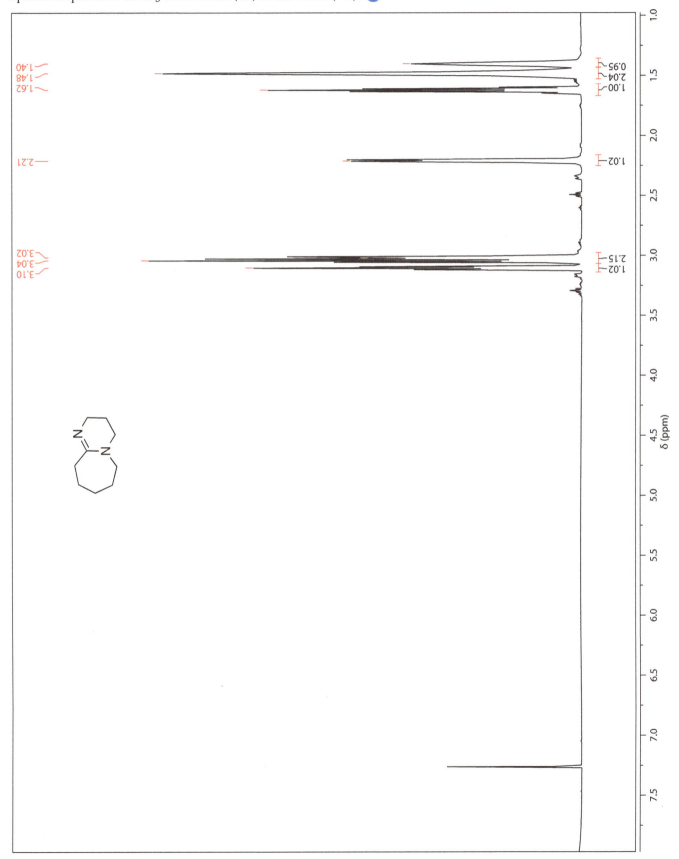

Problem 37

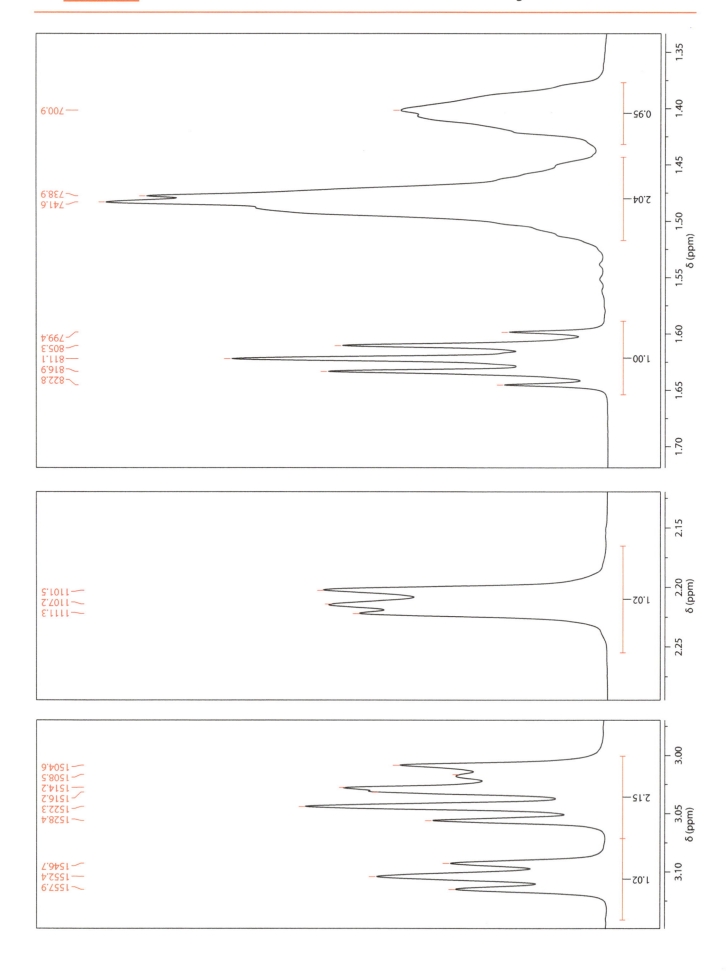

SECTION 2 Problem 37

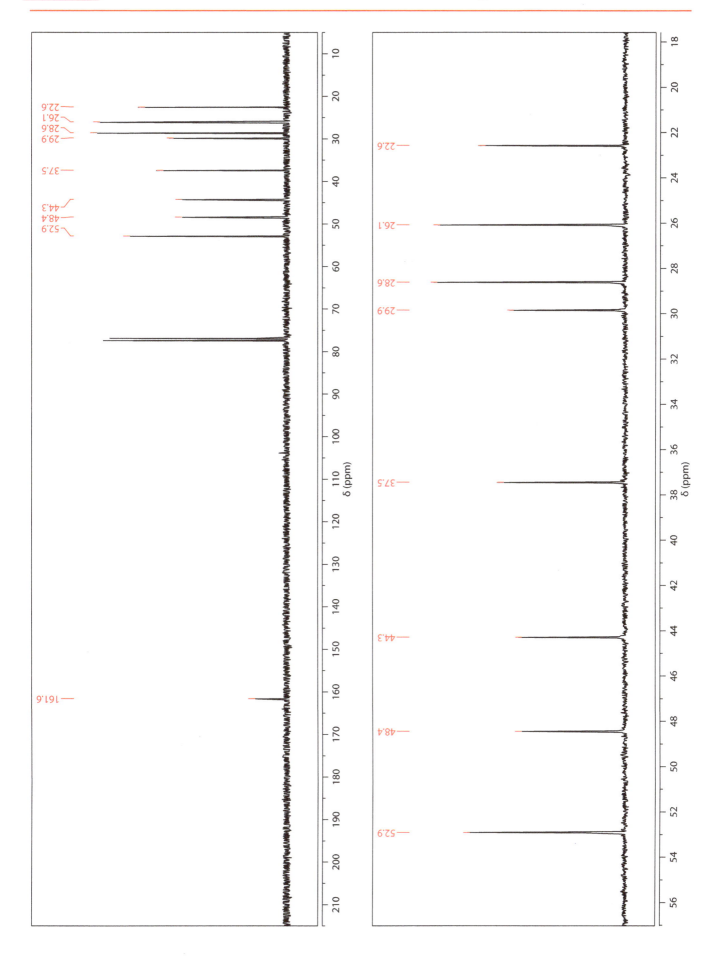

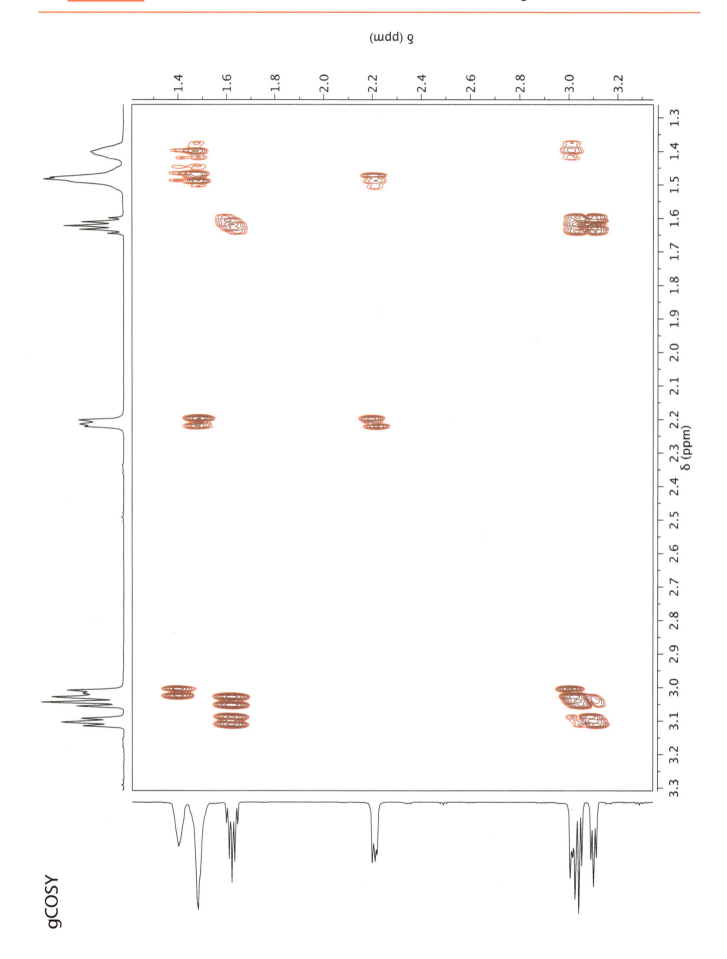

SECTION 2 Problem 37

gHSQC

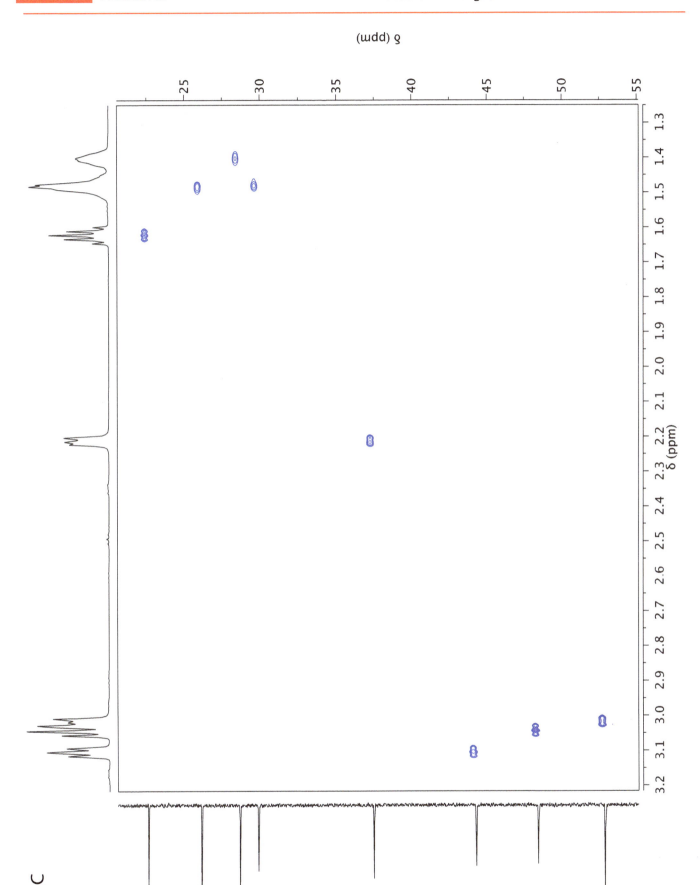

Problem 37

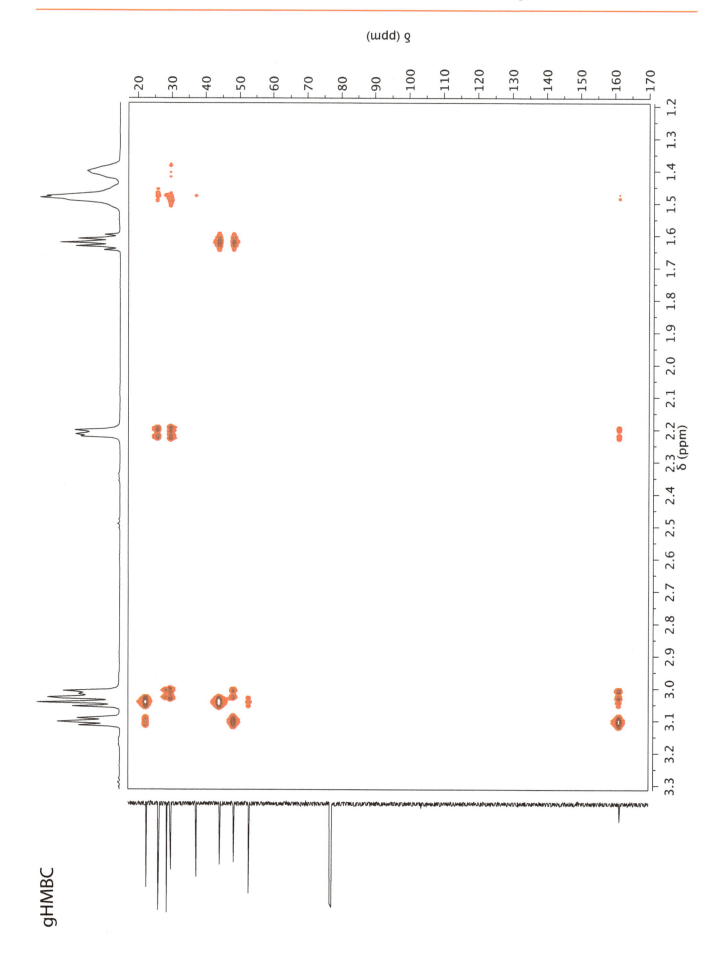

gHMBC

SECTION 2 Problem 37

Problems in Organic Structure Determination

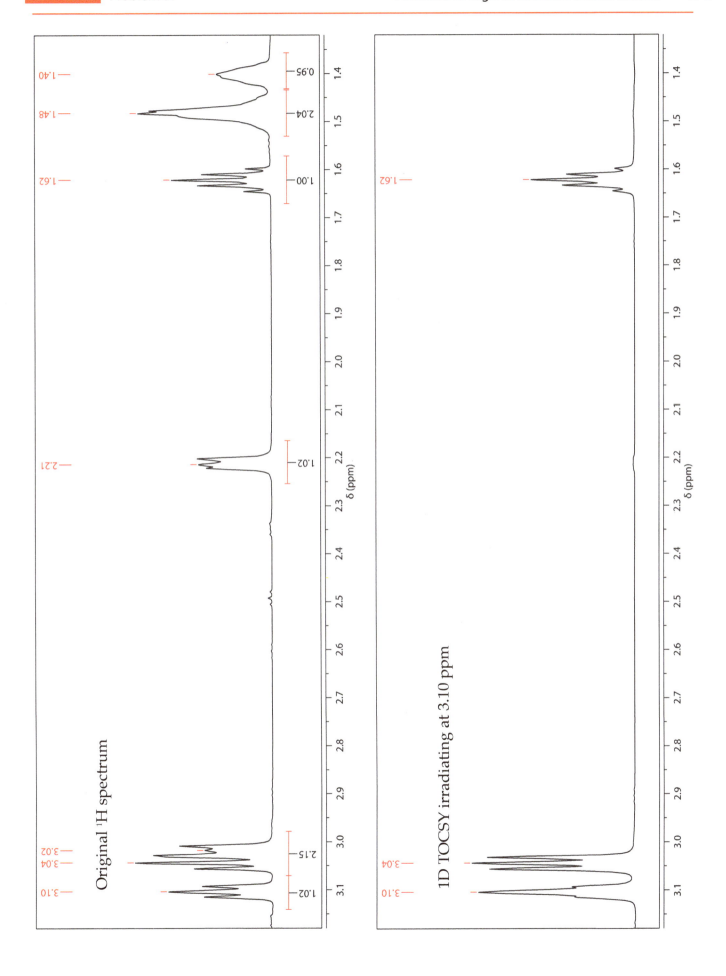

Problem 37

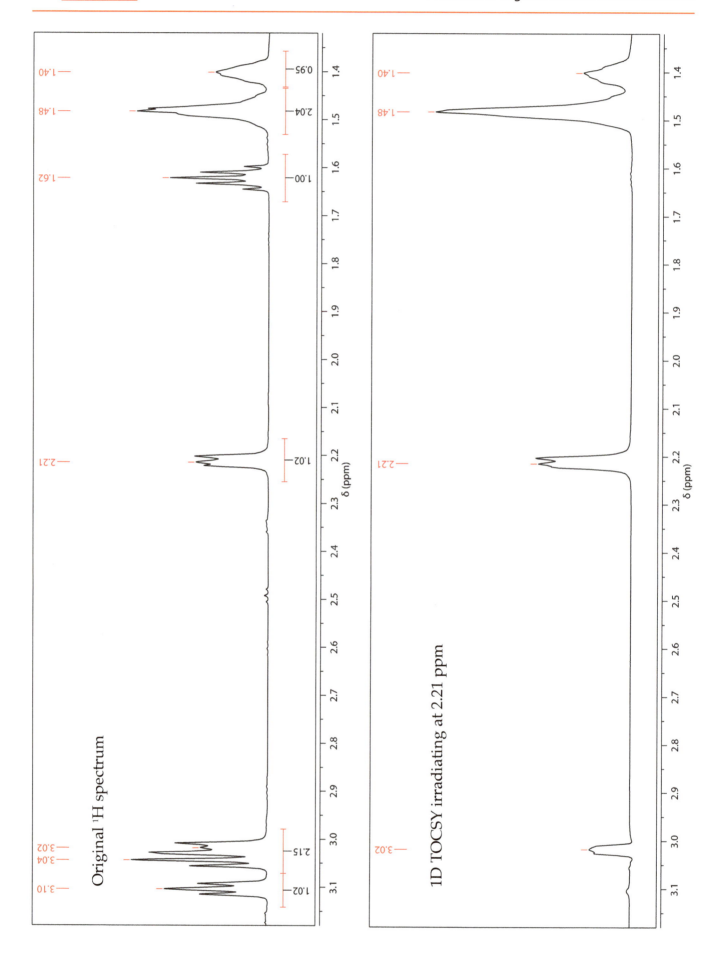

SECTION 2 Problem 38 — Problems in Organic Structure Determination

Using the ^{1}H, ^{13}C, gCOSY and gHSQC spectra, assign all proton and carbon resonances to the structure below and assign all proton coupling constants.
Spectra acquired in $(CD_3)_2SO$ at 500 MHz (^{1}H) or 125 MHz (^{13}C).

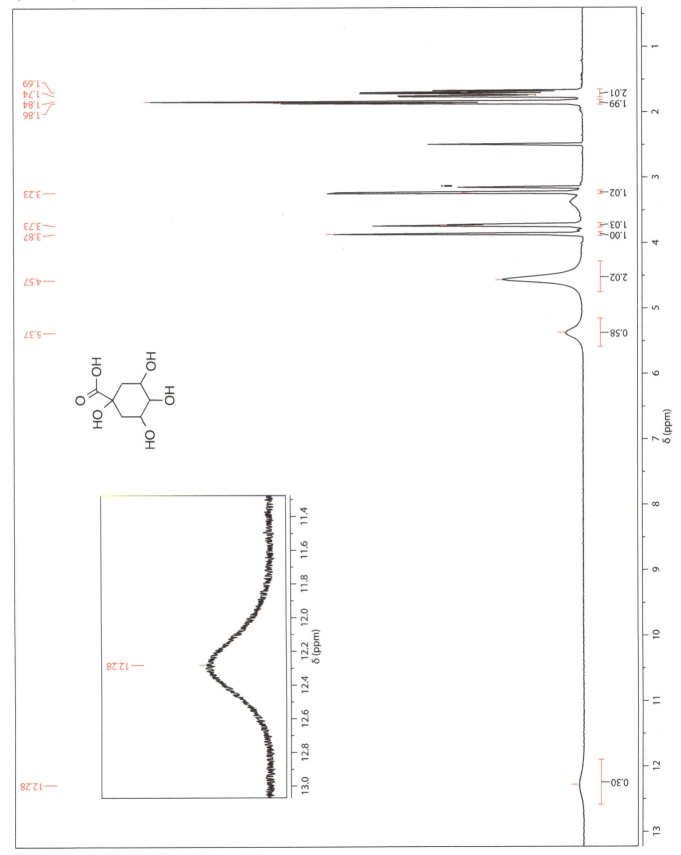

Problem 38

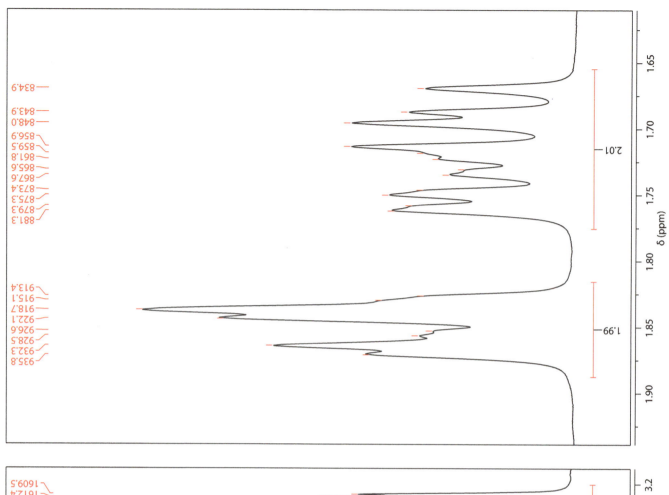

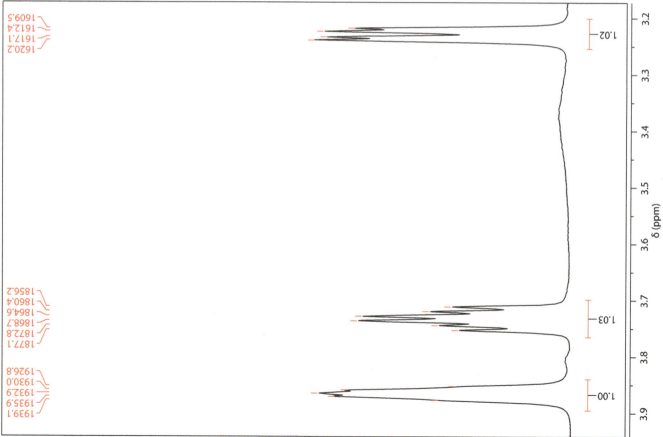

SECTION 2 Problem 38

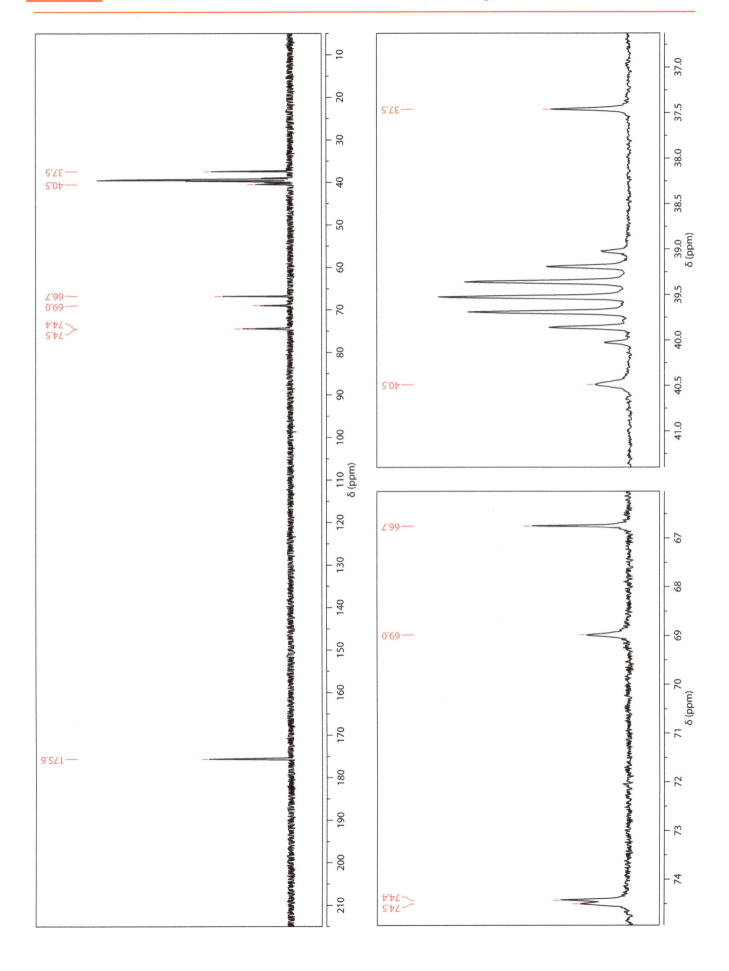

Problem 38

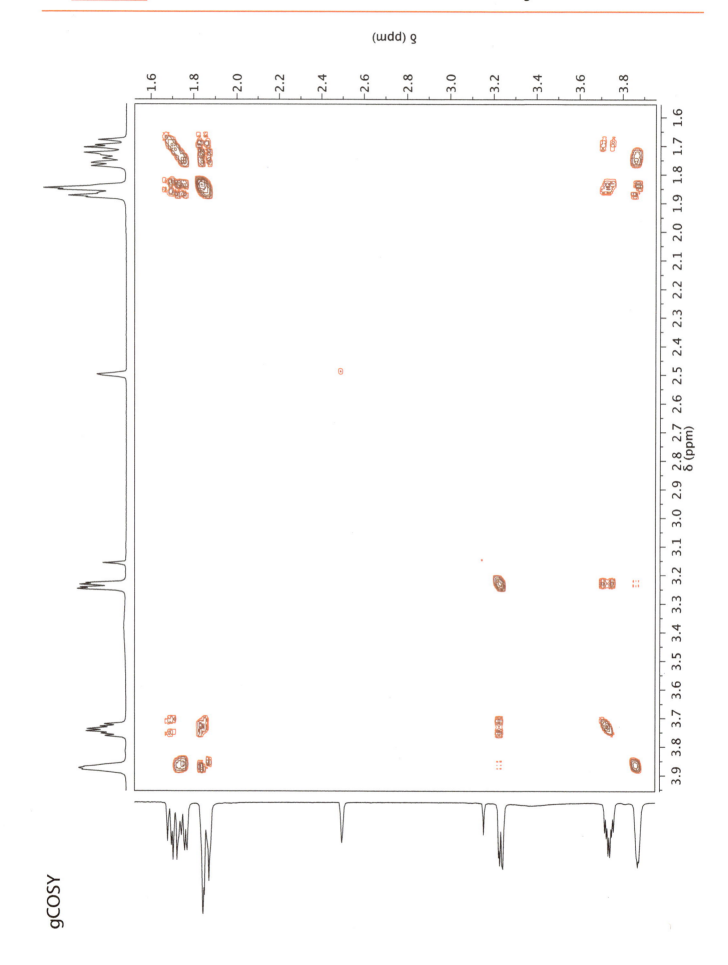

gCOSY

Problem 38

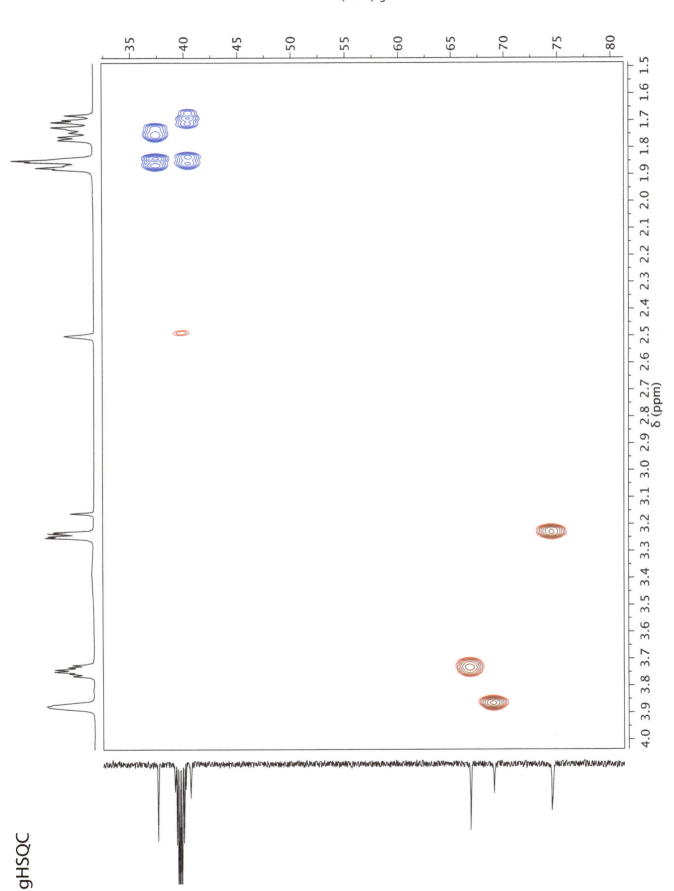

Using the ^1H, ^{13}C, gCOSY and gHSQC spectra, assign all proton and carbon resonances to the structure of 2'-(trifluoromethyl)acetophenone below.
NOTE: Some ^{13}C peak picking provided in Hz, to allow extraction of C-F coupling constants.
Spectra acquired in $CDCl_3$ at 500 MHz (^1H) or 125 MHz (^{13}C).

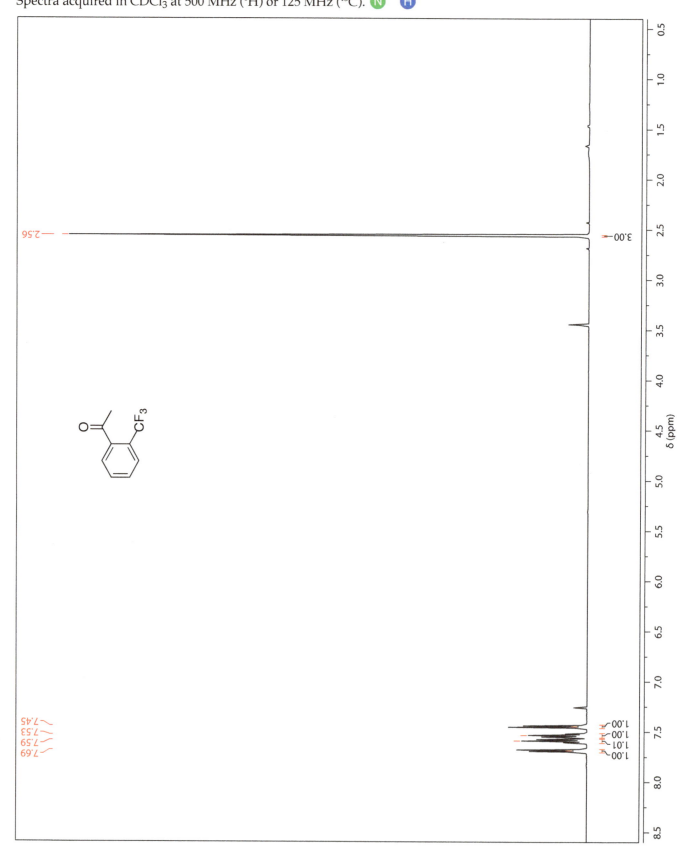

SECTION 2 Problem 39

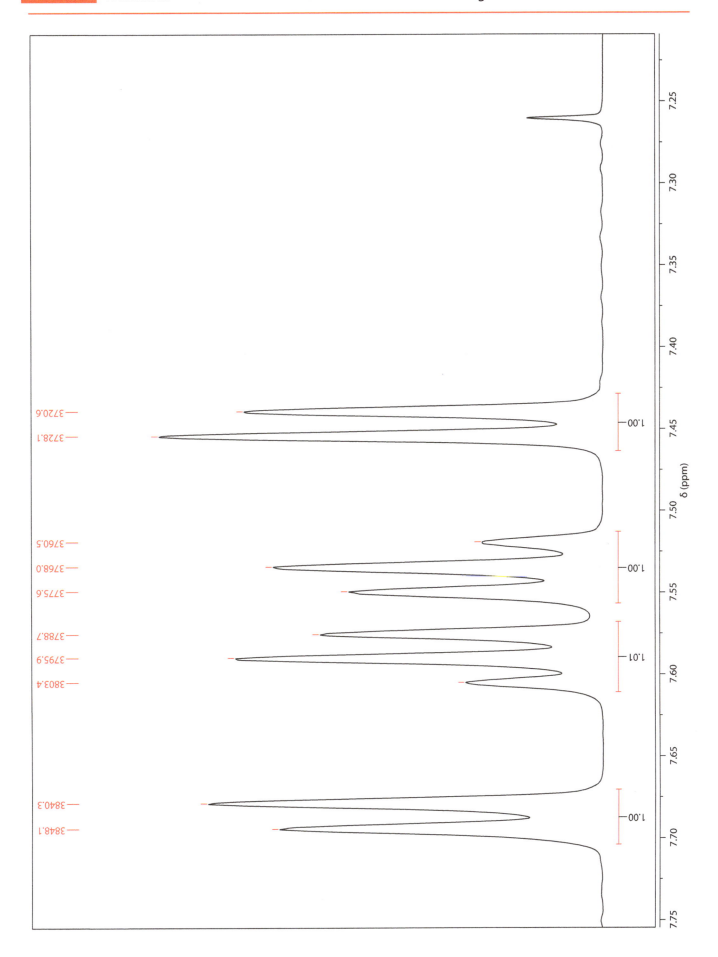

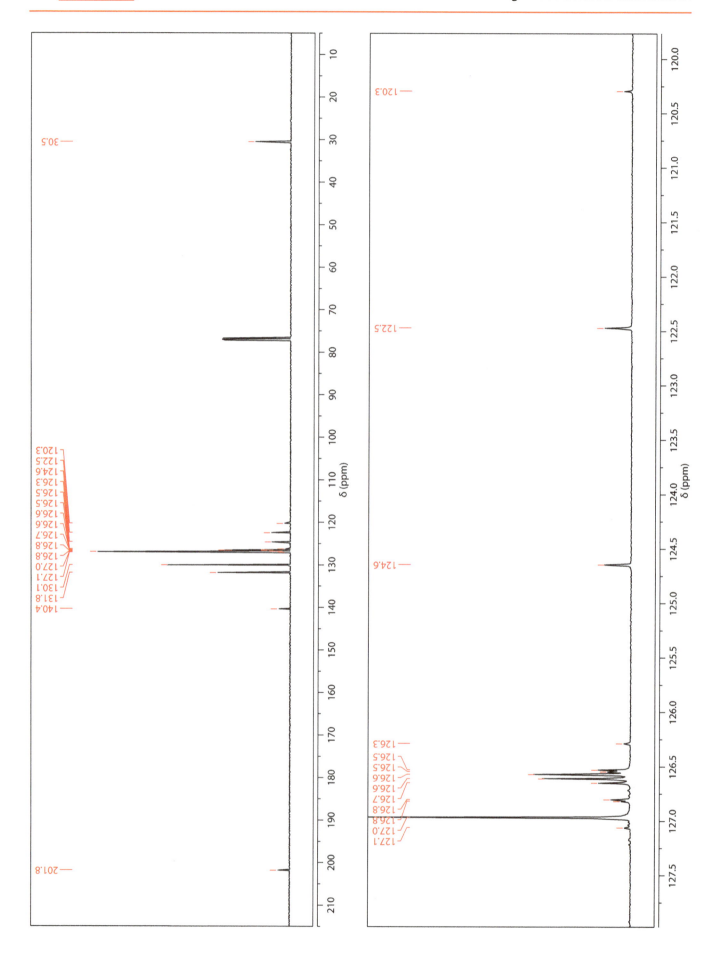

SECTION 2 Problem 39

Problems in Organic Structure Determination 183

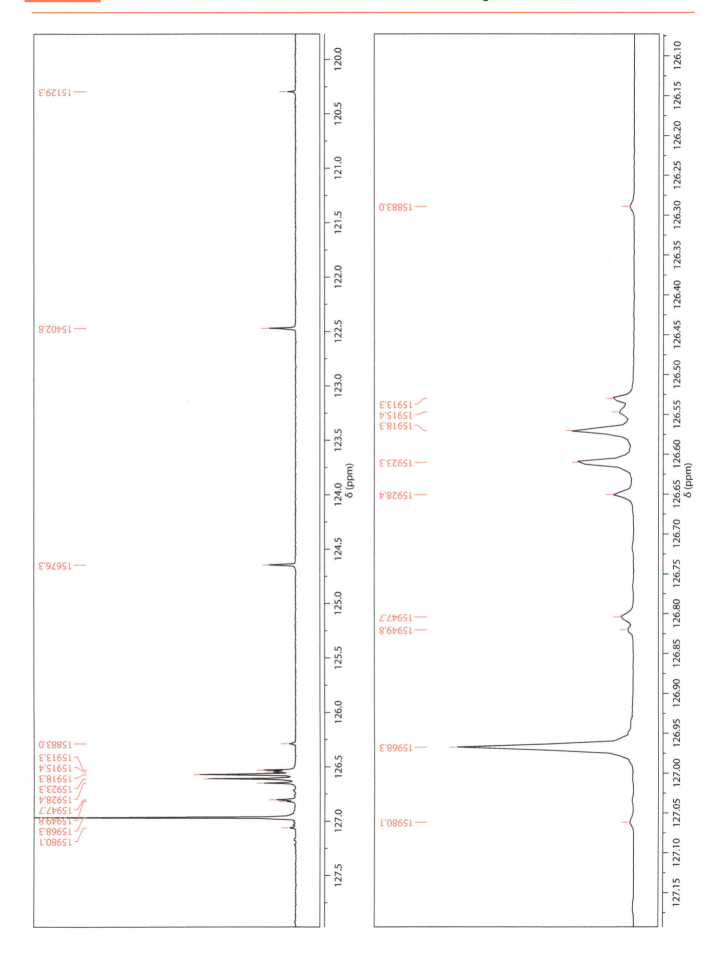

Problem 39

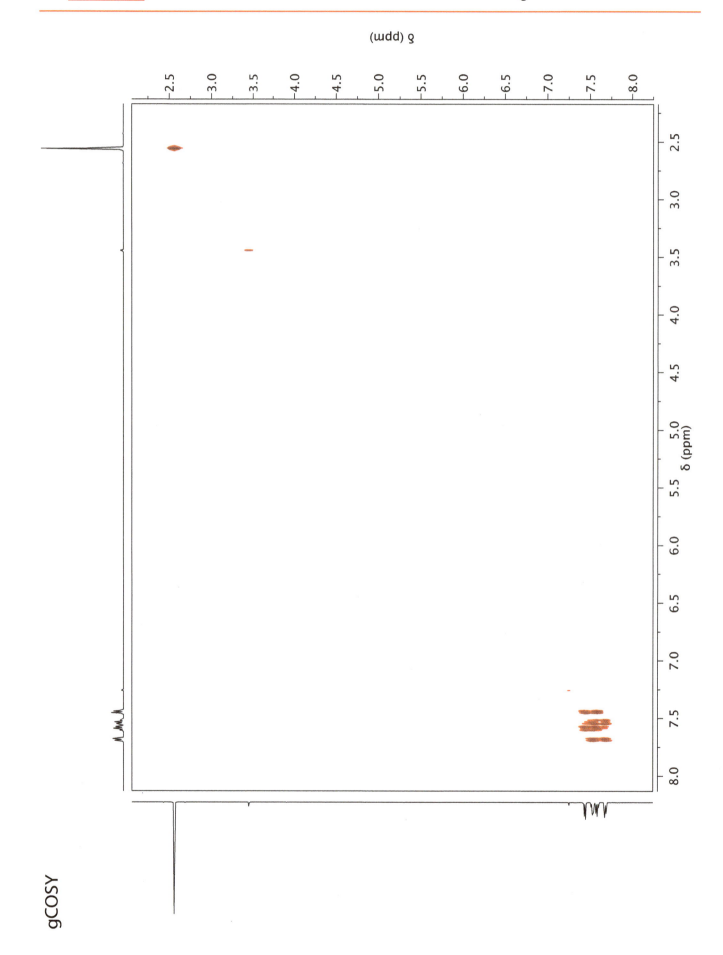

gCOSY

SECTION 2 Problem 39

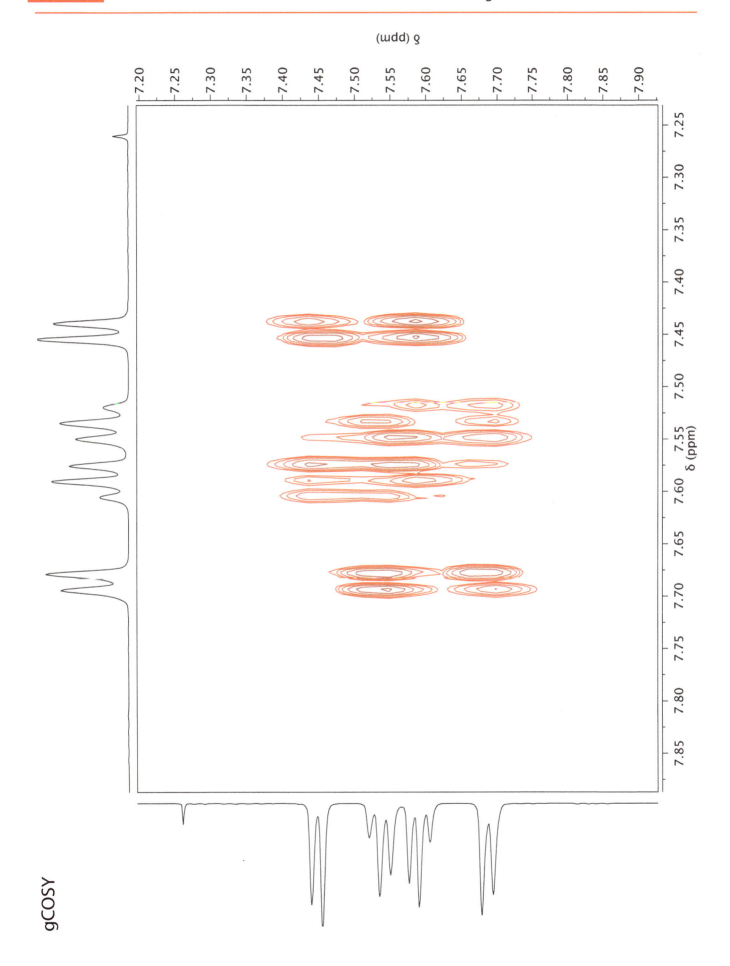

gCOSY

Problem 39

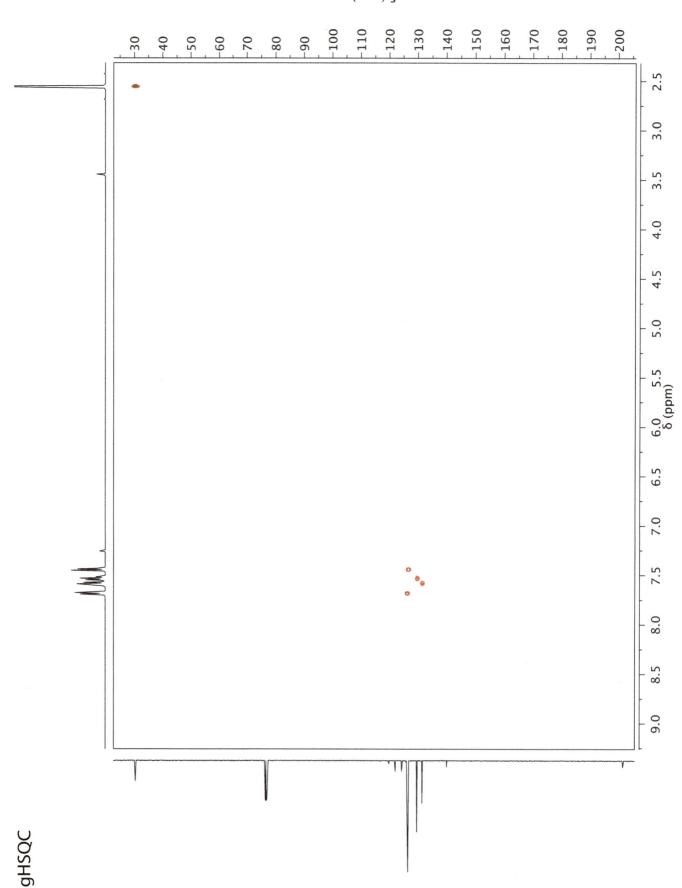

gHSQC

Problem 39

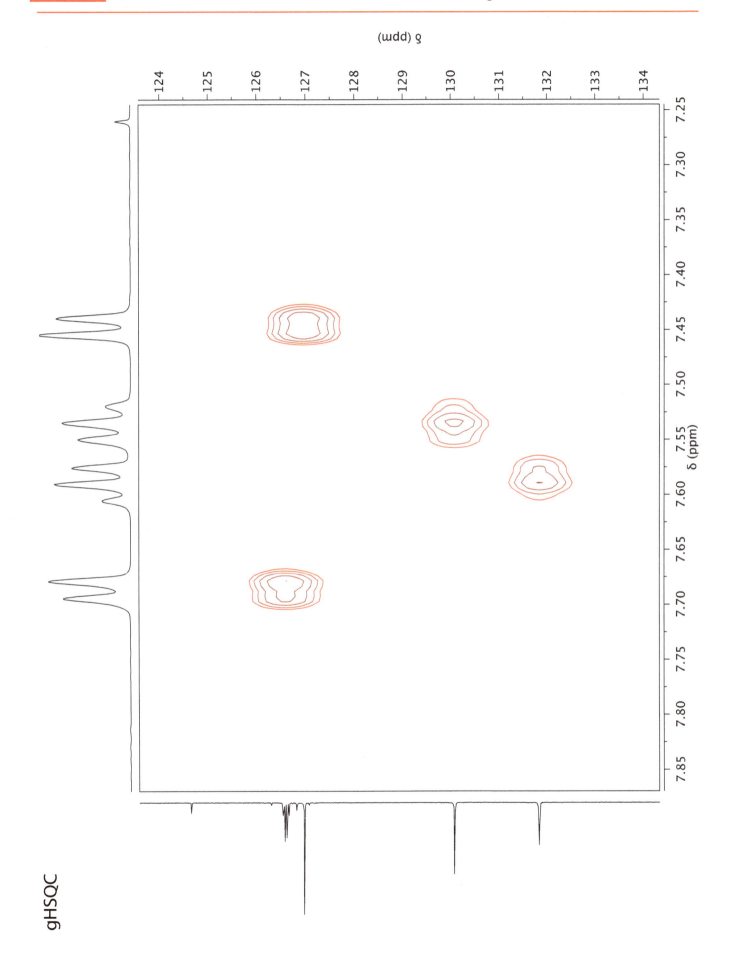

gHSQC

Problem 40

Using the ^1H, ^{13}C, gCOSY, gHSQC and gHMBC spectra, assign all proton and carbon resonances to the structure of imipramine hydrochloride below.
Spectra acquired in $(CD_3)_2SO$ at 500 MHz (^1H) or 125 MHz (^{13}C).

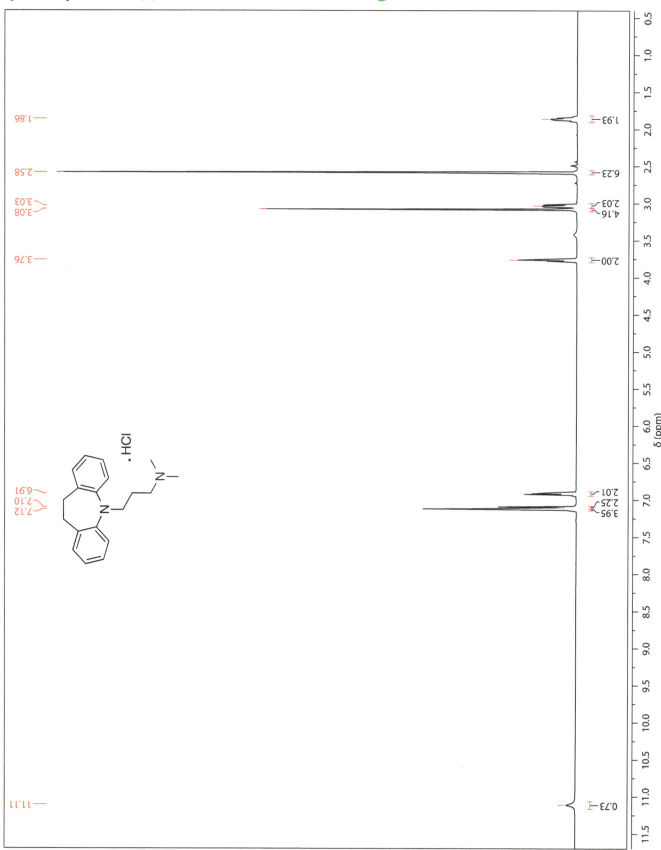

Problem 40

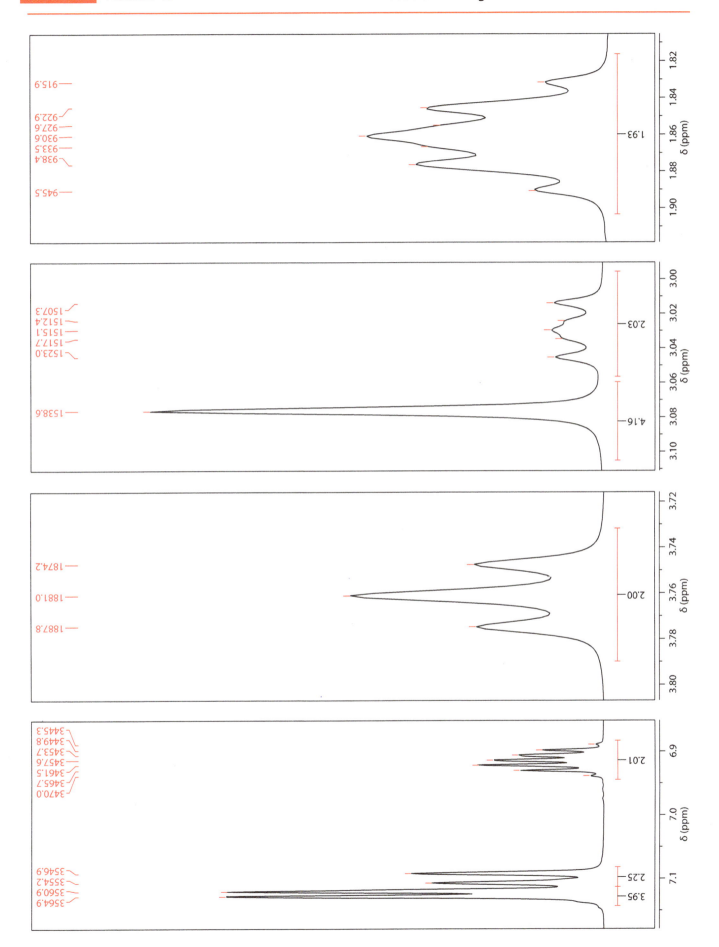

Problem 40

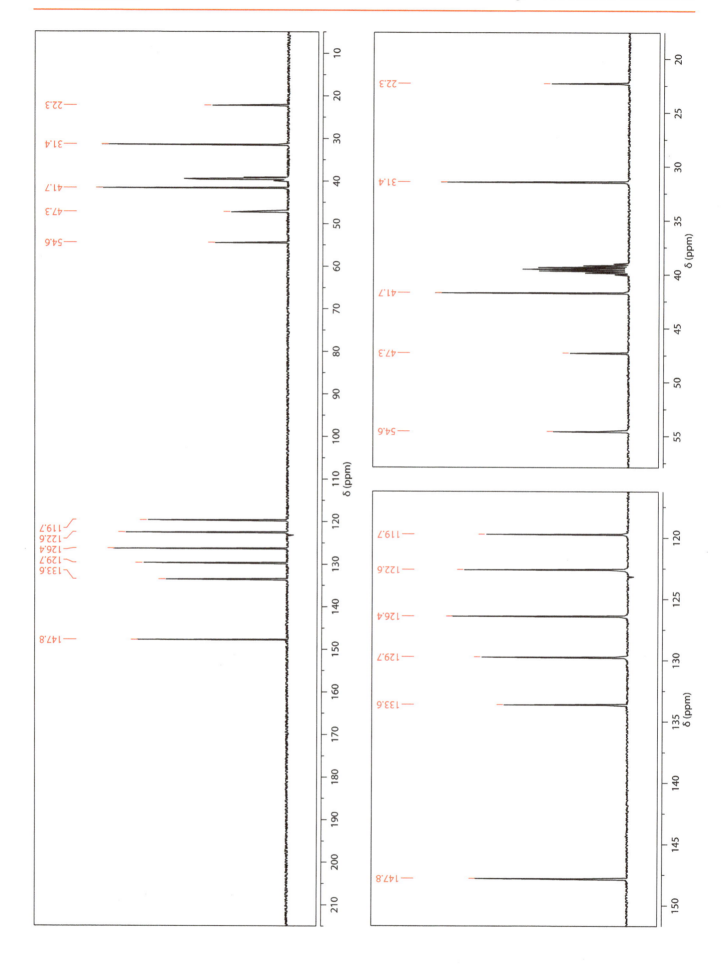

SECTION 2 Problem 40

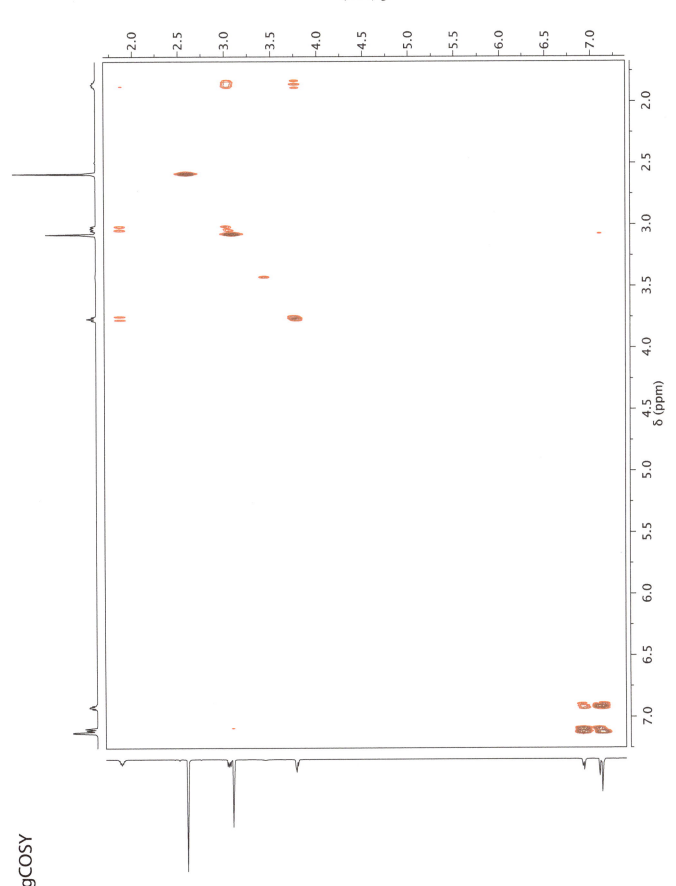

gCOSY

Problem 40

gCOSY

SECTION 2 Problem 40

gCOSY

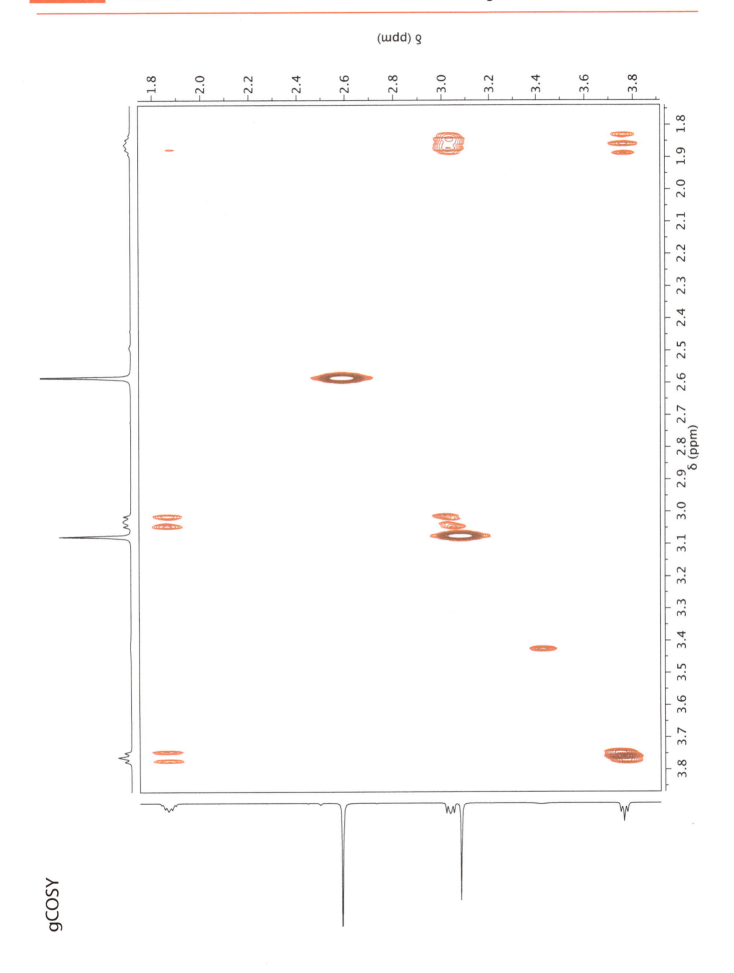

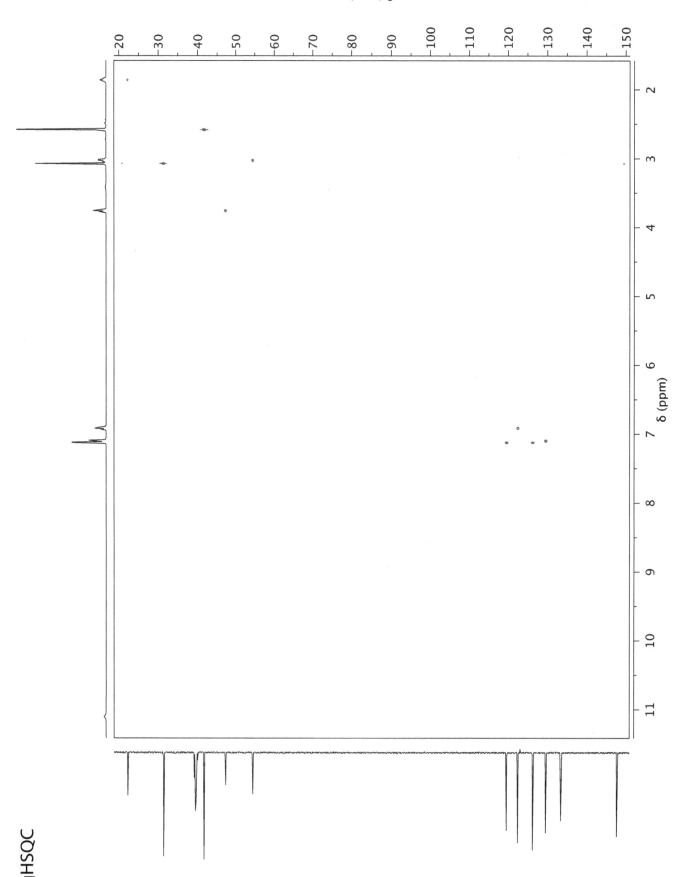

SECTION 2 Problem 40

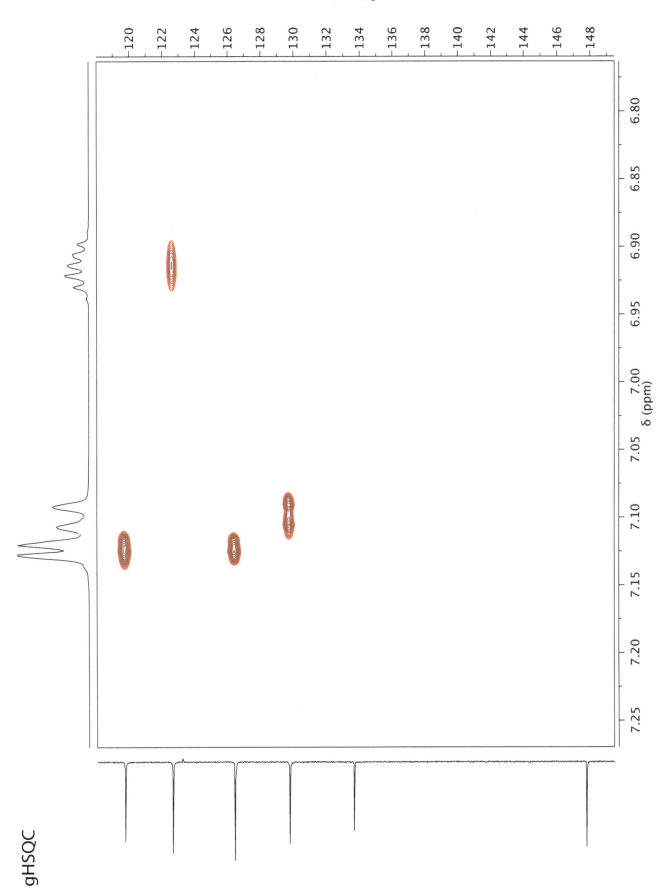

gHSQC

Problem 40

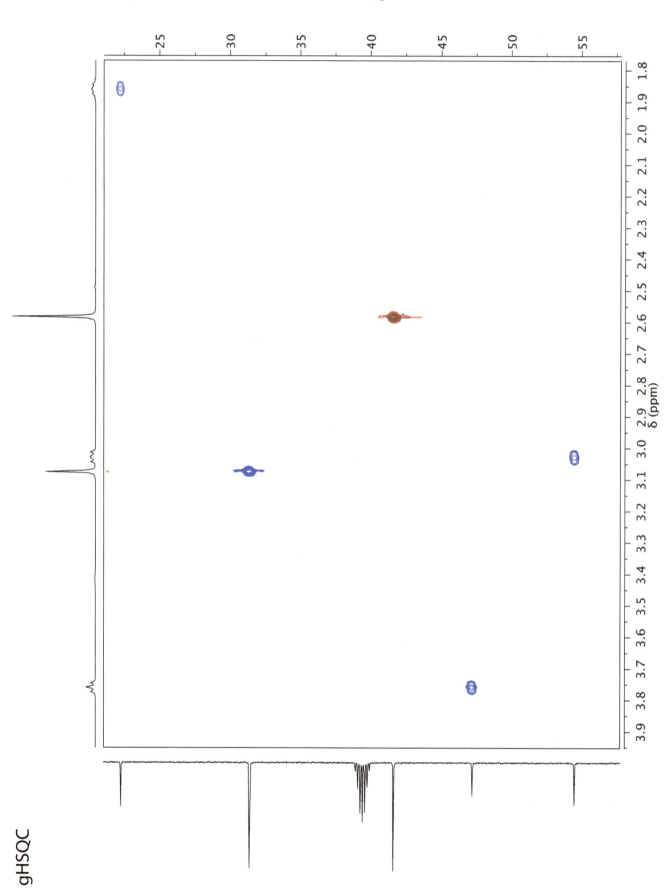

gHSQC

SECTION 2 Problem 40

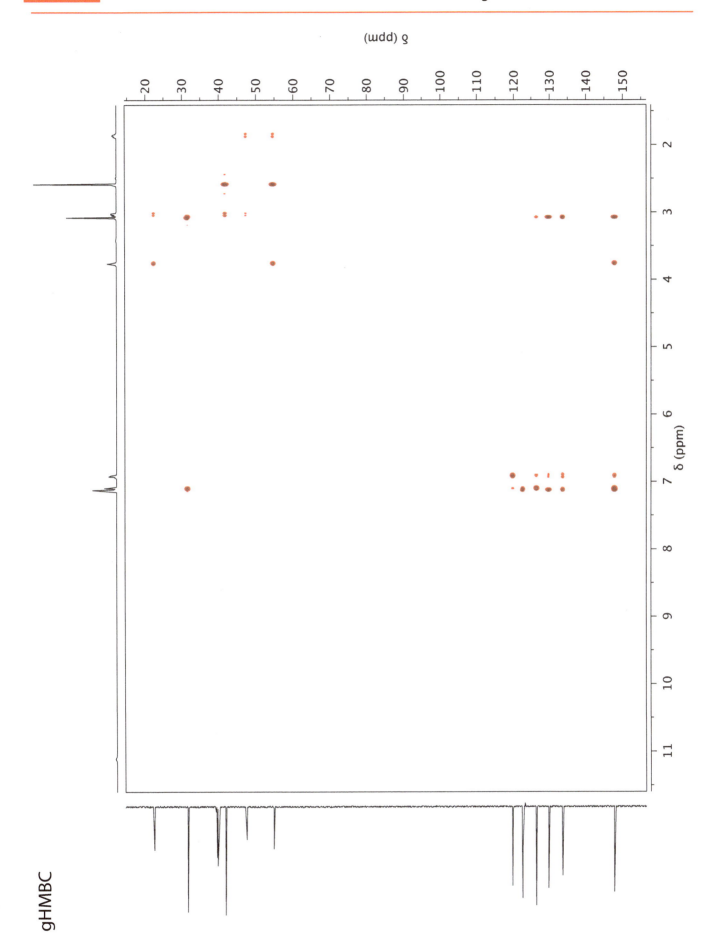

Problem 40

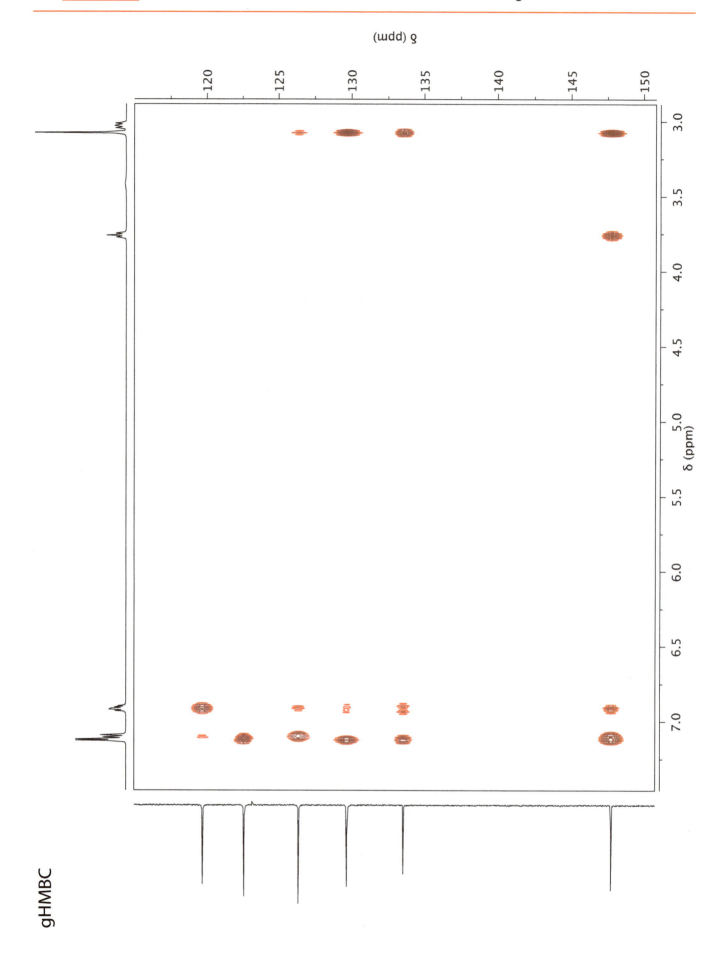

gHMBC

SECTION 2 Problem 40

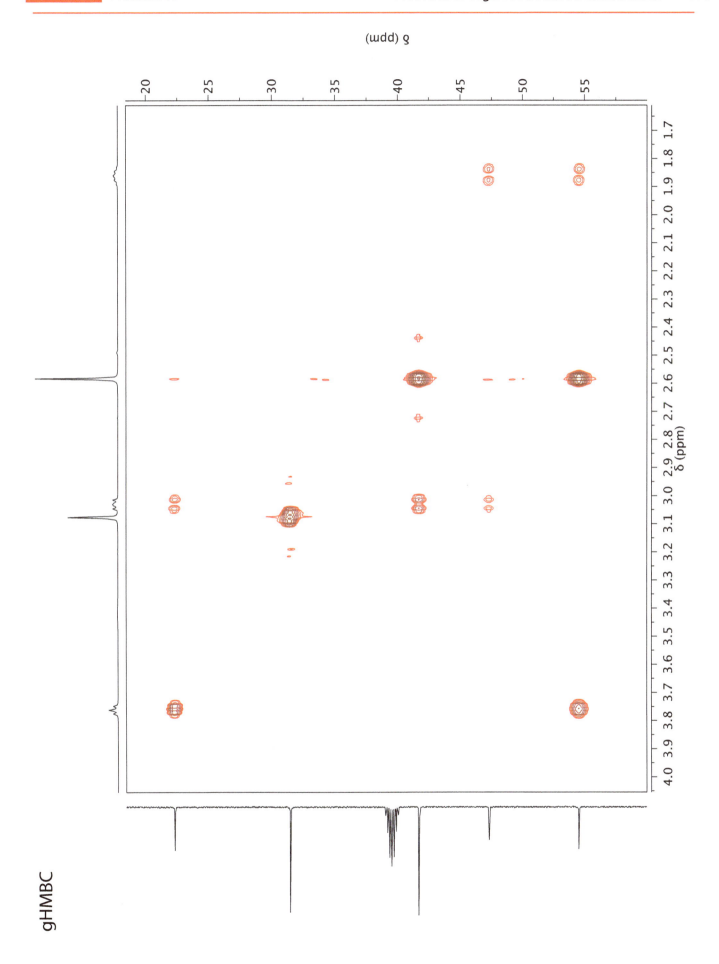

gHMBC

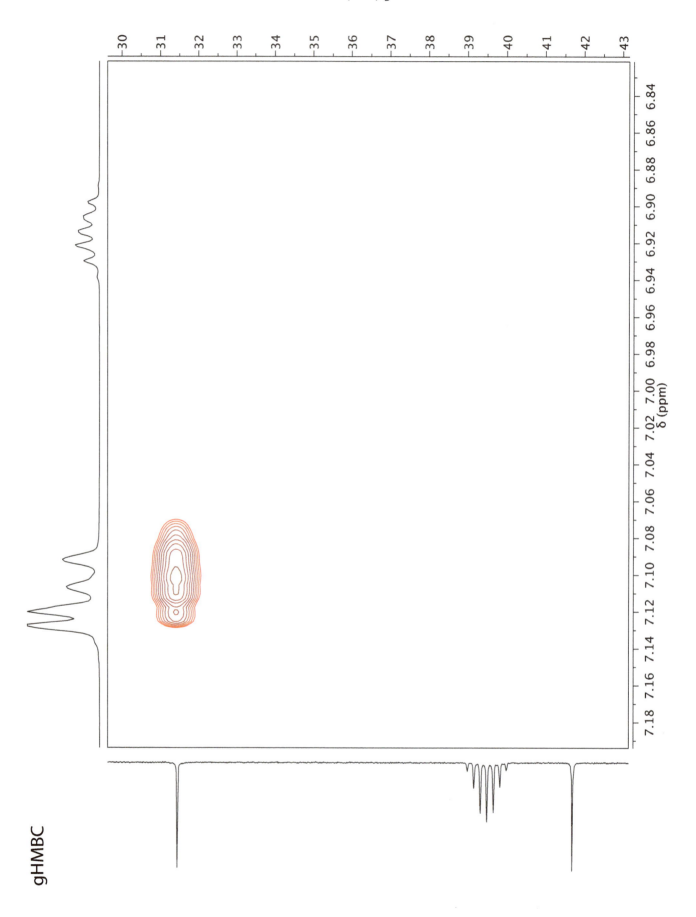

SECTION 2 Problem 41

Using the ^{1}H, ^{13}C, gCOSY, gHSQC and gHMBC spectra, assign all proton and carbon resonances to the structure of gemfibrozil below and assign all proton coupling constants.
Spectra acquired in CDCl$_3$ at 500 MHz (^{1}H) or 125 MHz (^{13}C).

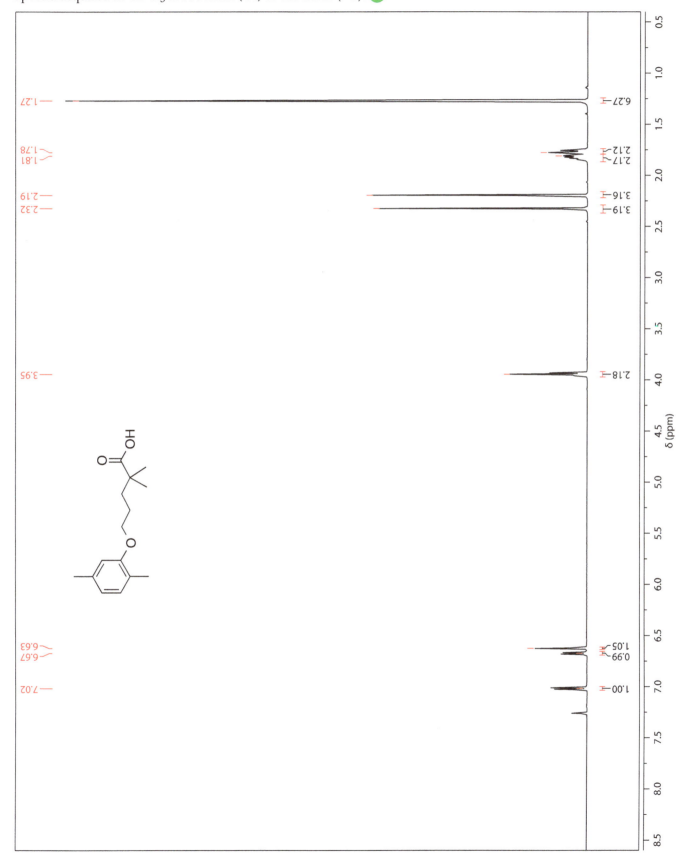

Problem 41

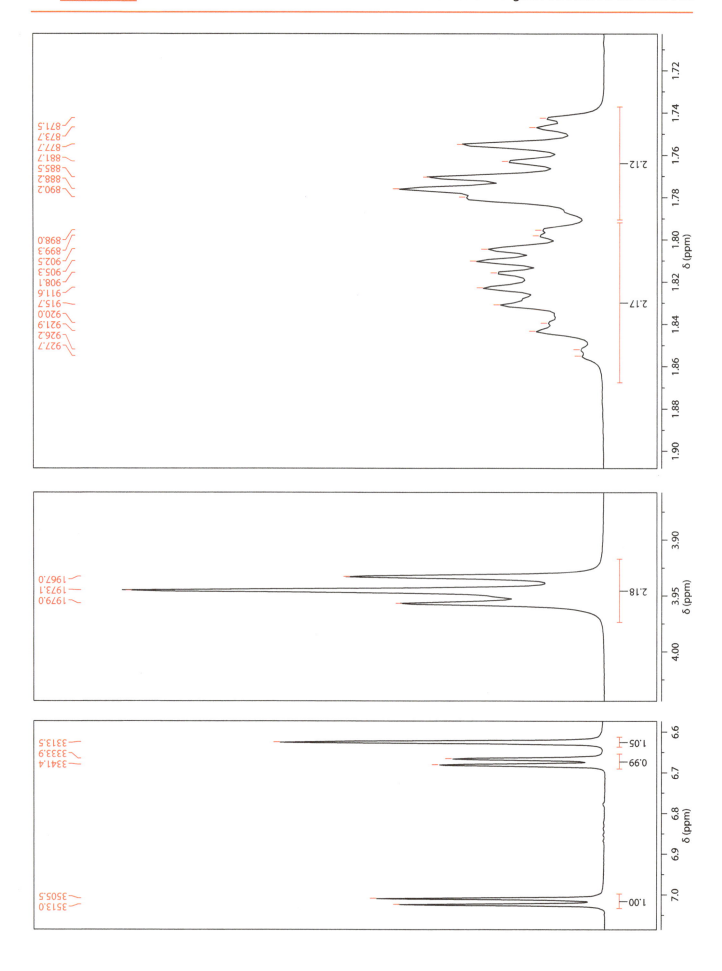

SECTION 2 Problem 41

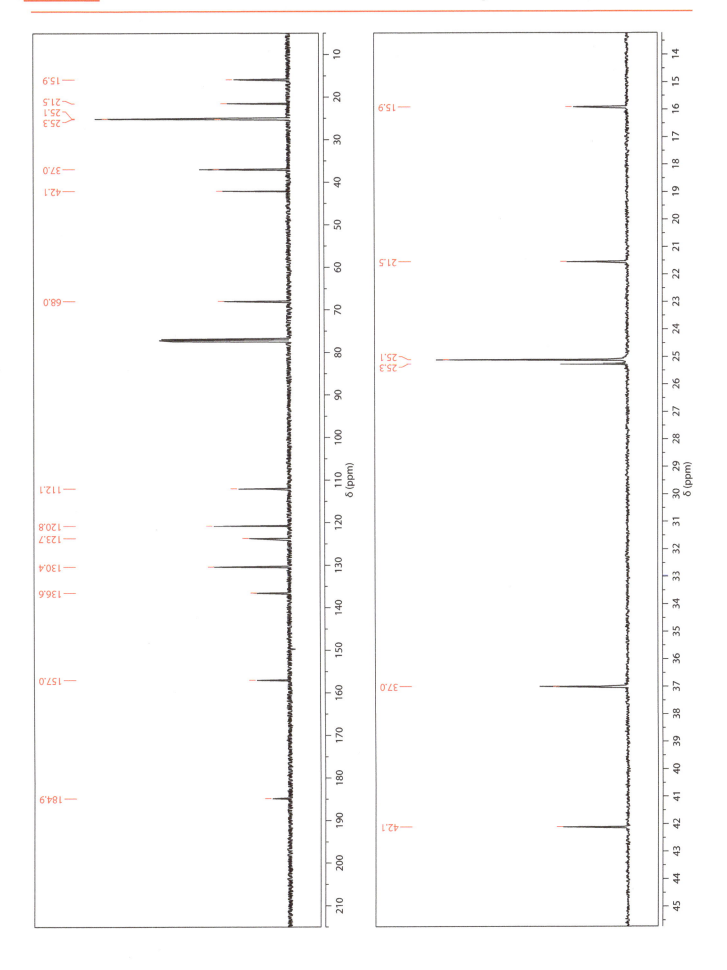

Problem 41

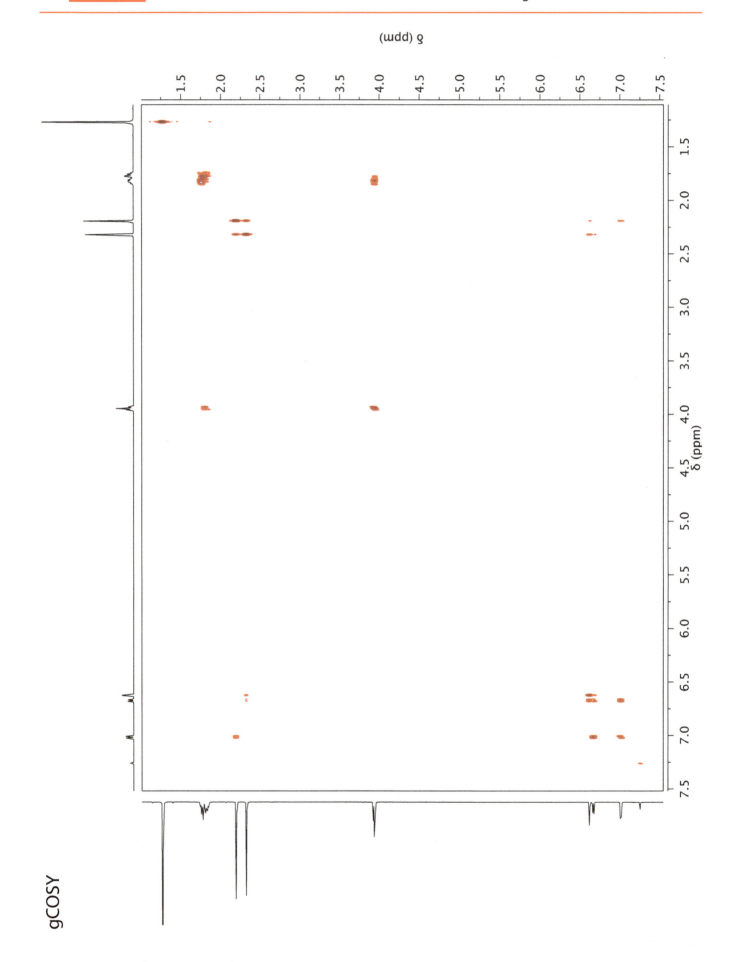

gCOSY

SECTION 2 Problem 41

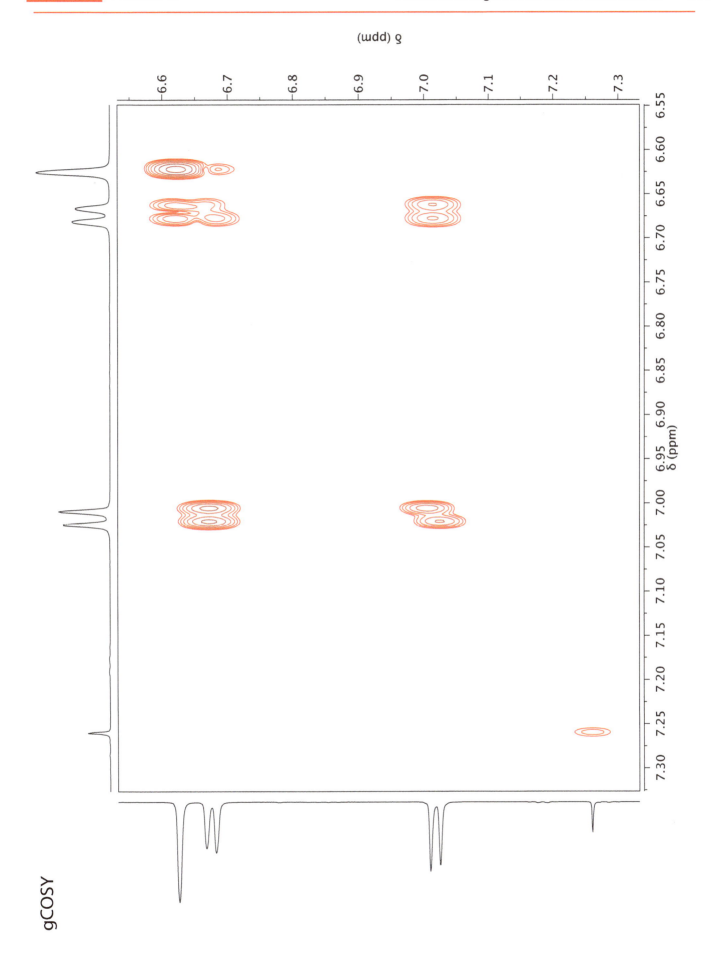

gCOSY

Problem 41

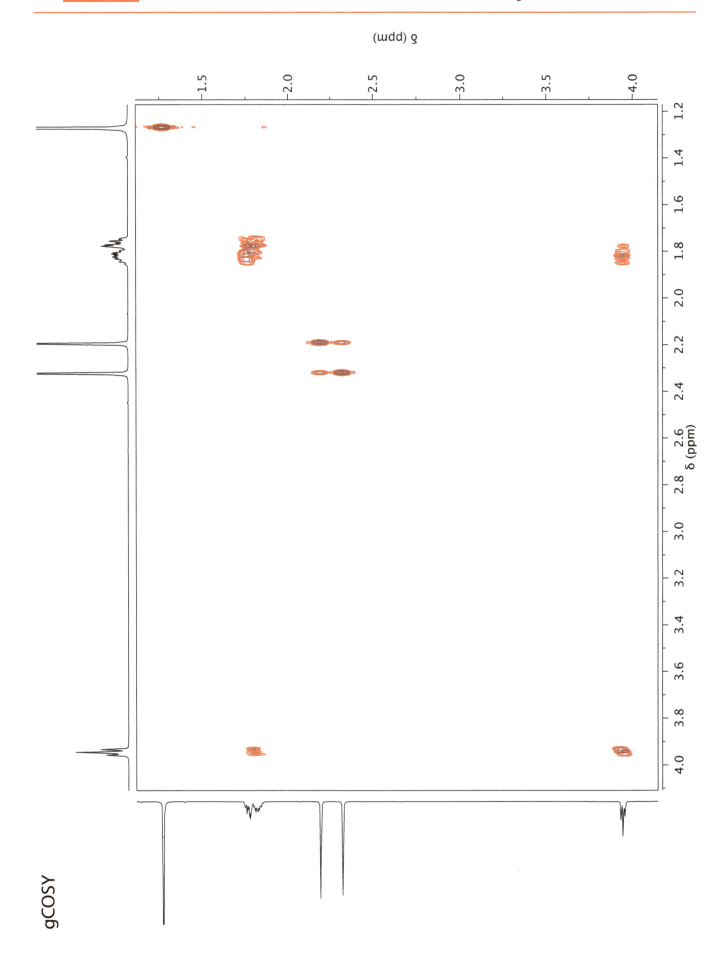

gCOSY

SECTION 2 Problem 41

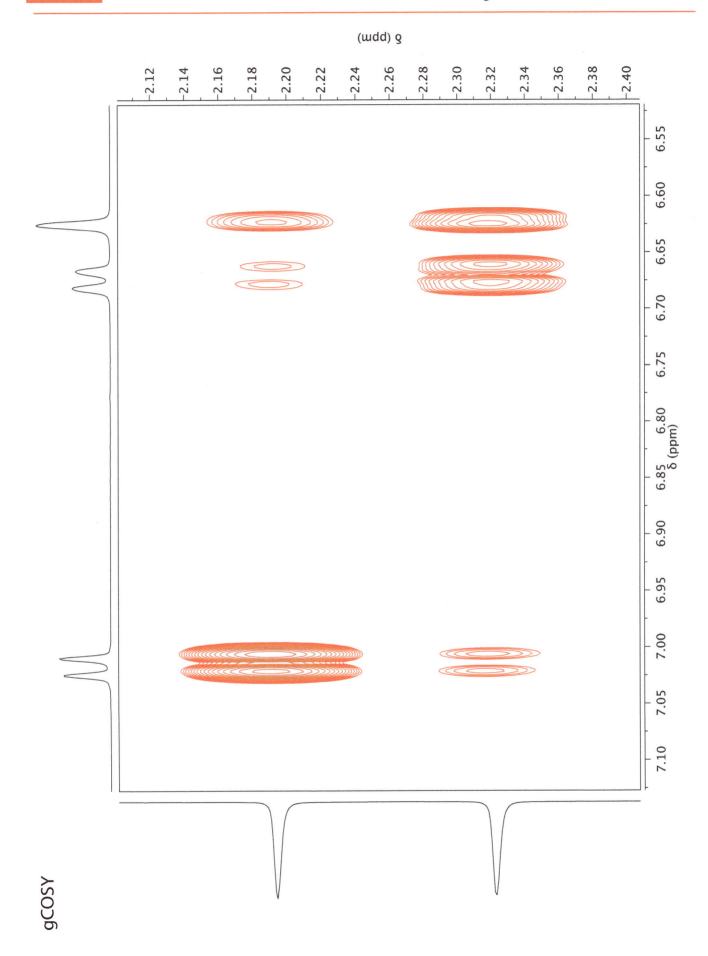

gCOSY

Problem 41

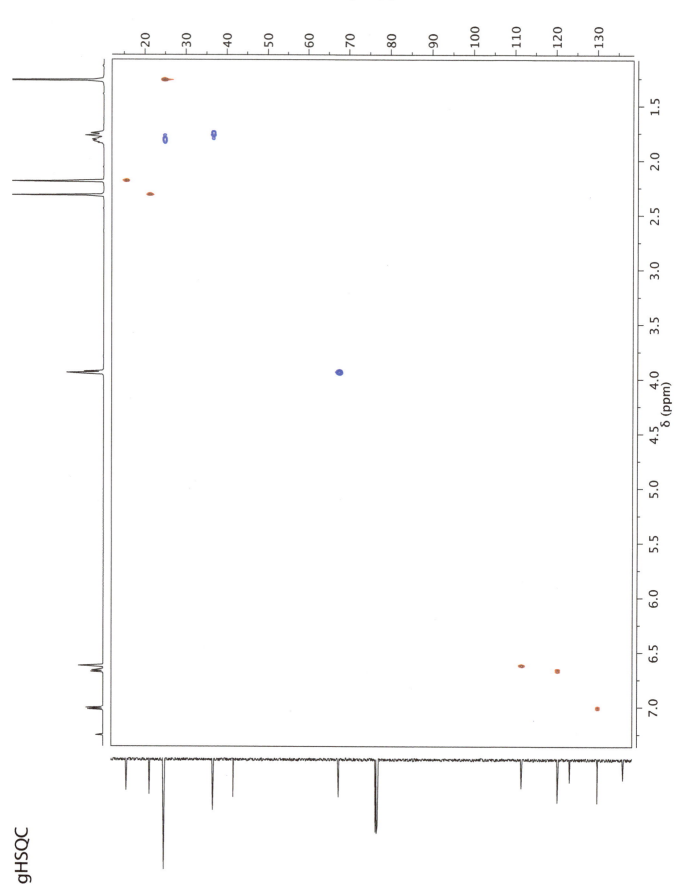

gHSQC

SECTION 2 Problem 41

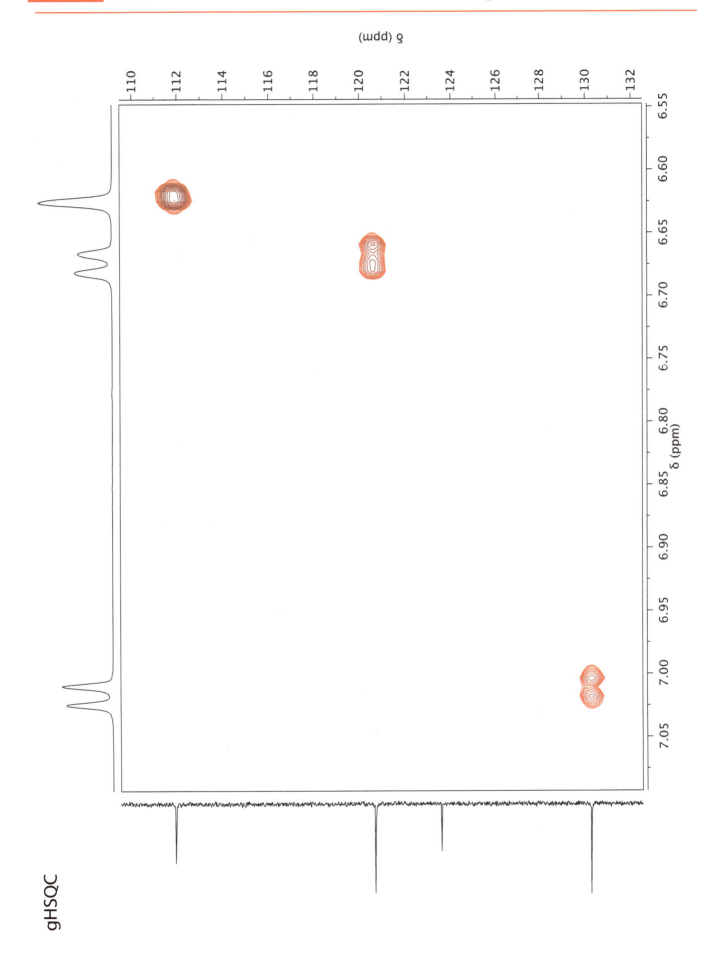

gHSQC

Problem 41

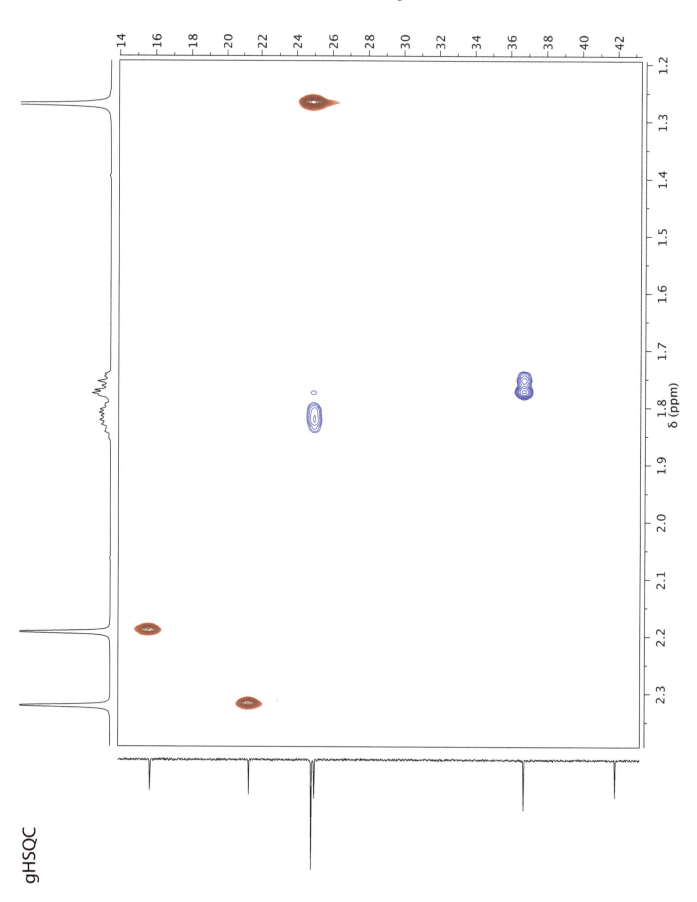

gHSQC

SECTION 2 Problem 41

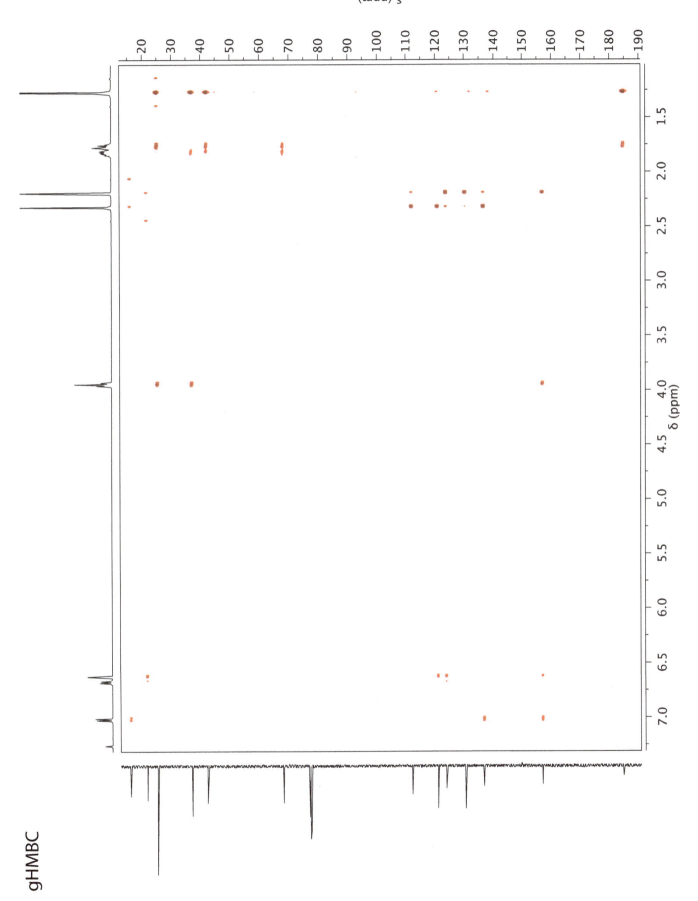

gHMBC

Problem 41

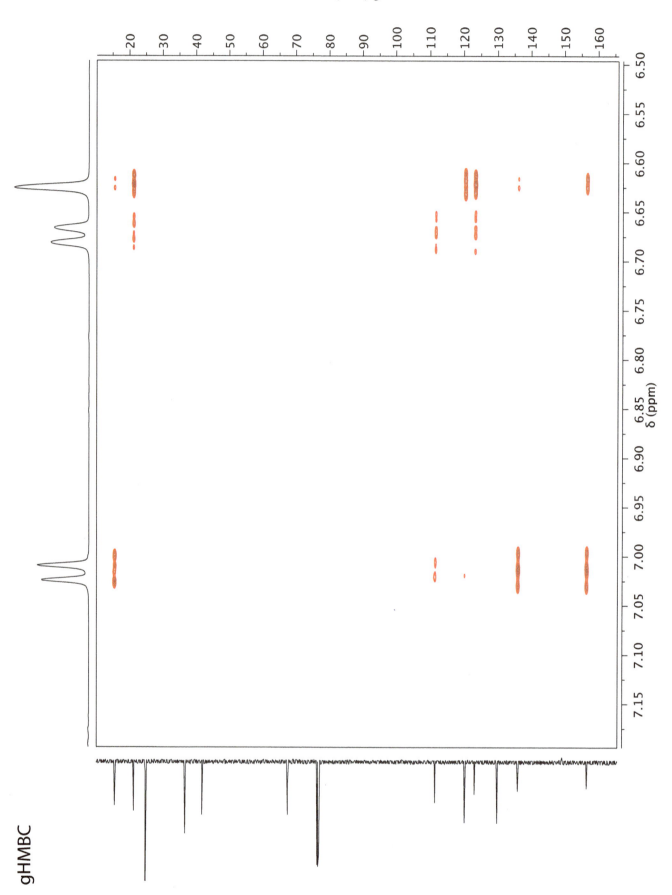

SECTION 2 Problem 41

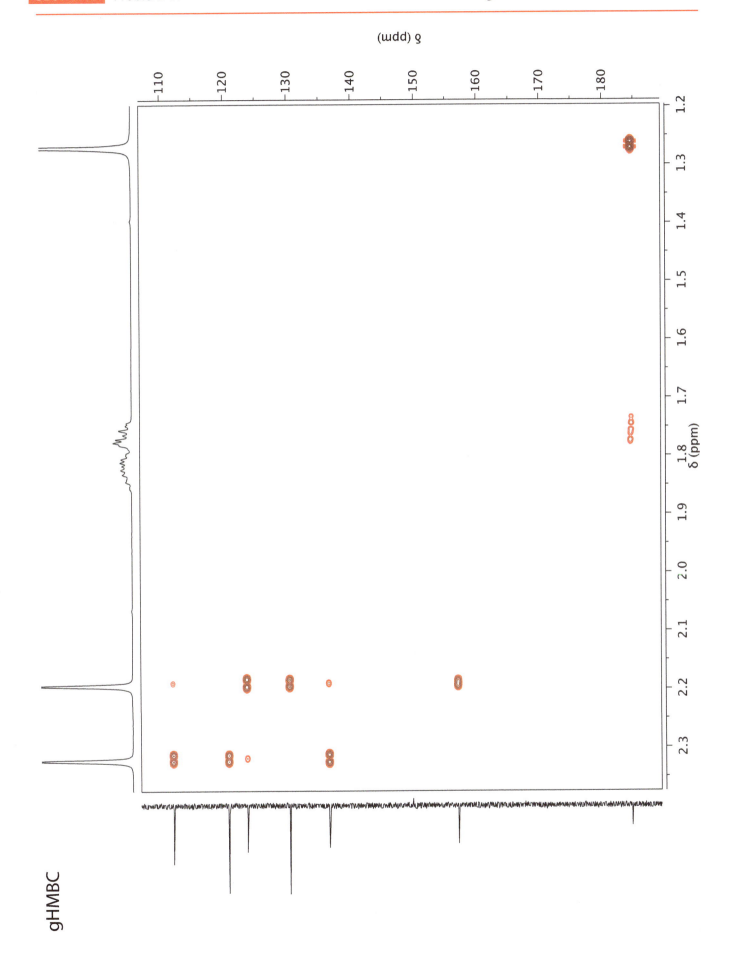

gHMBC

Problem 41

gHMBC

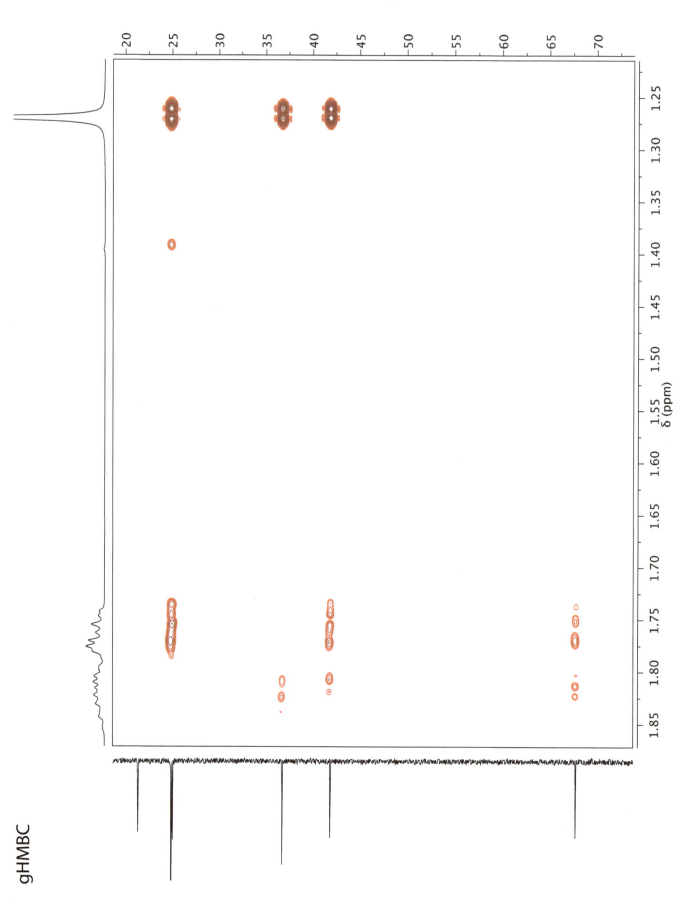

SECTION 2 Problem 42

Using the ¹H, ¹³C, gHSQC, gHMBC and selective 1D-TOCSY spectra, assign all proton and carbon resonances to the structure below.
Spectra acquired in CDCl$_3$ at 500 MHz (¹H) or 125 MHz (¹³C).

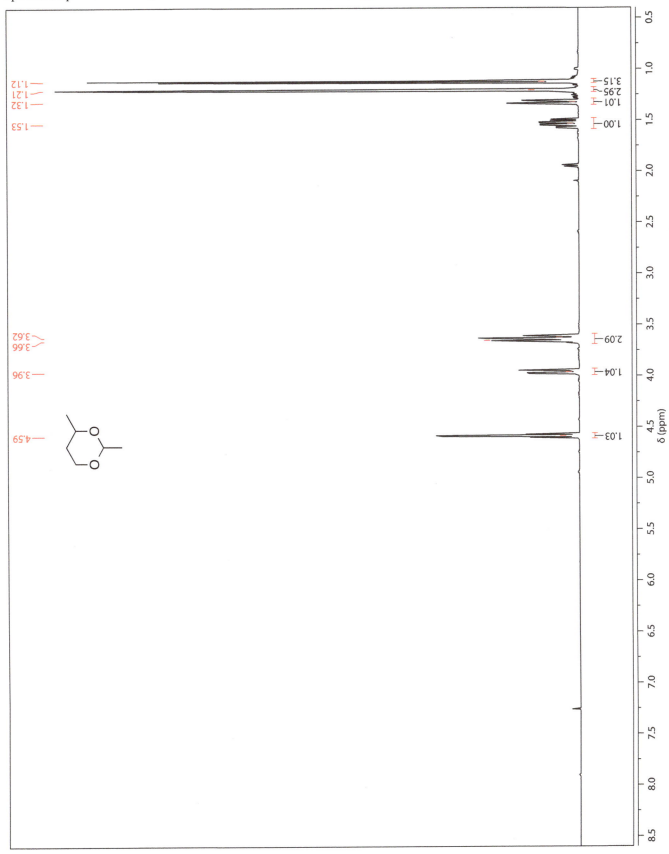

Problem 42

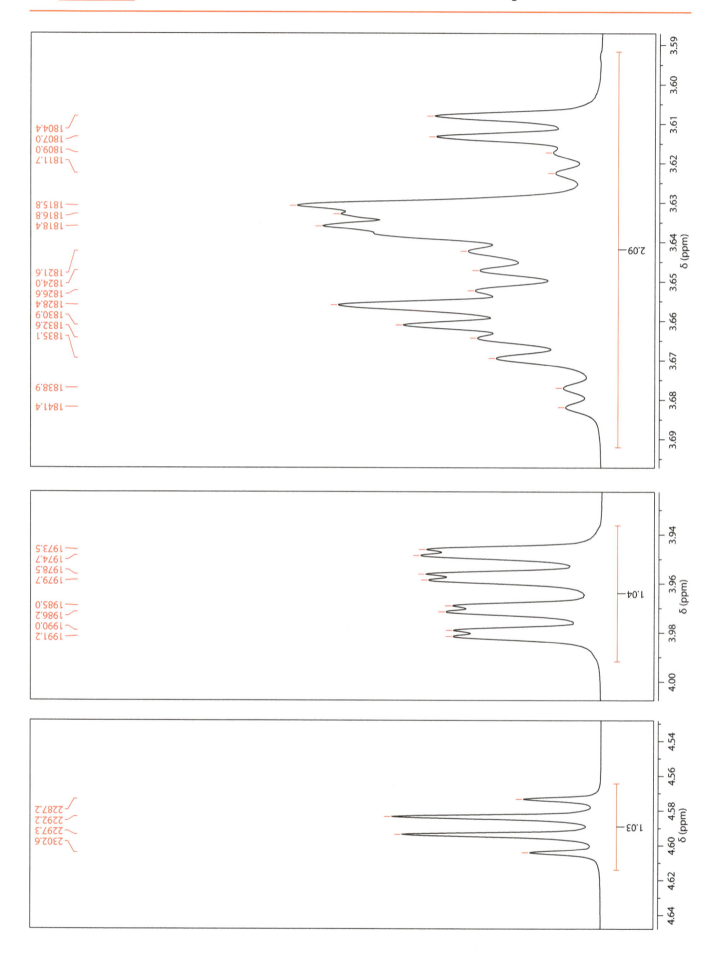

Problem 42

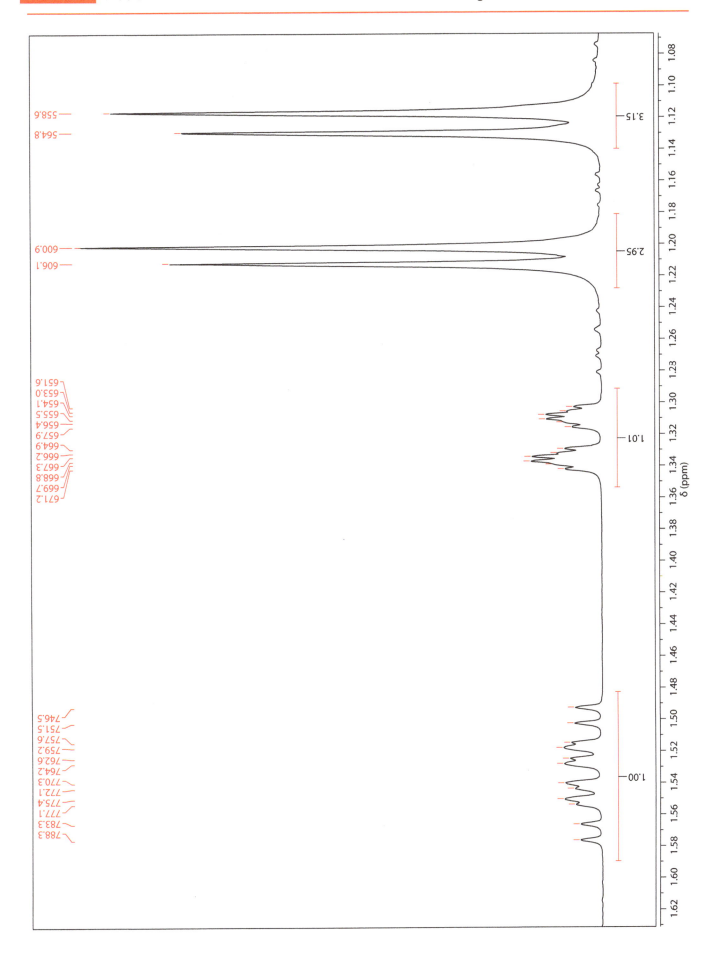

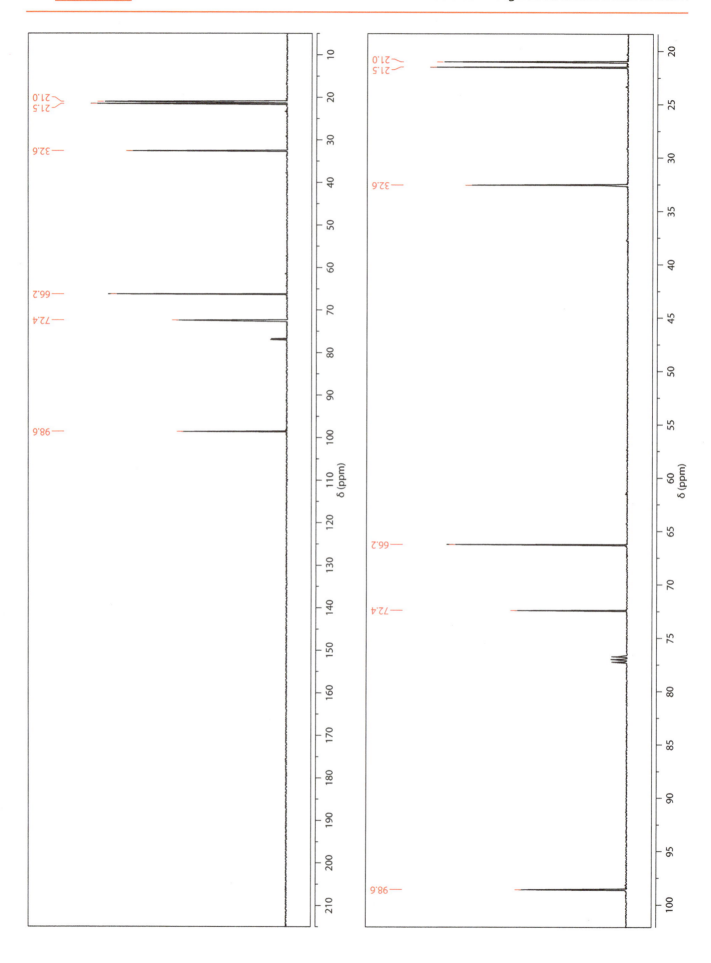

SECTION 2 Problem 42

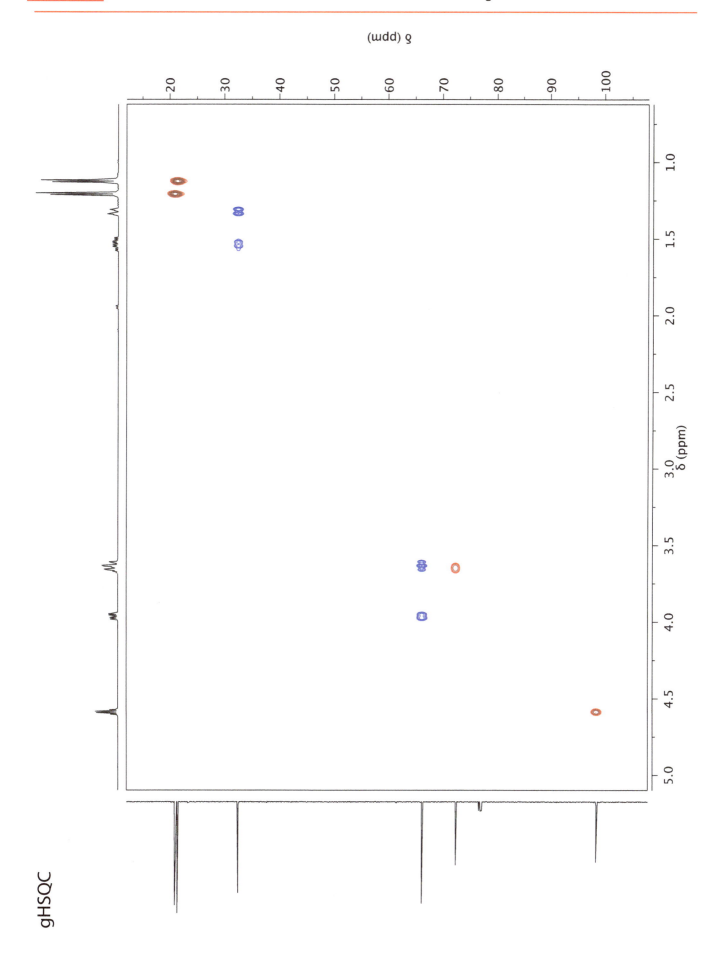

gHSQC

Problem 42

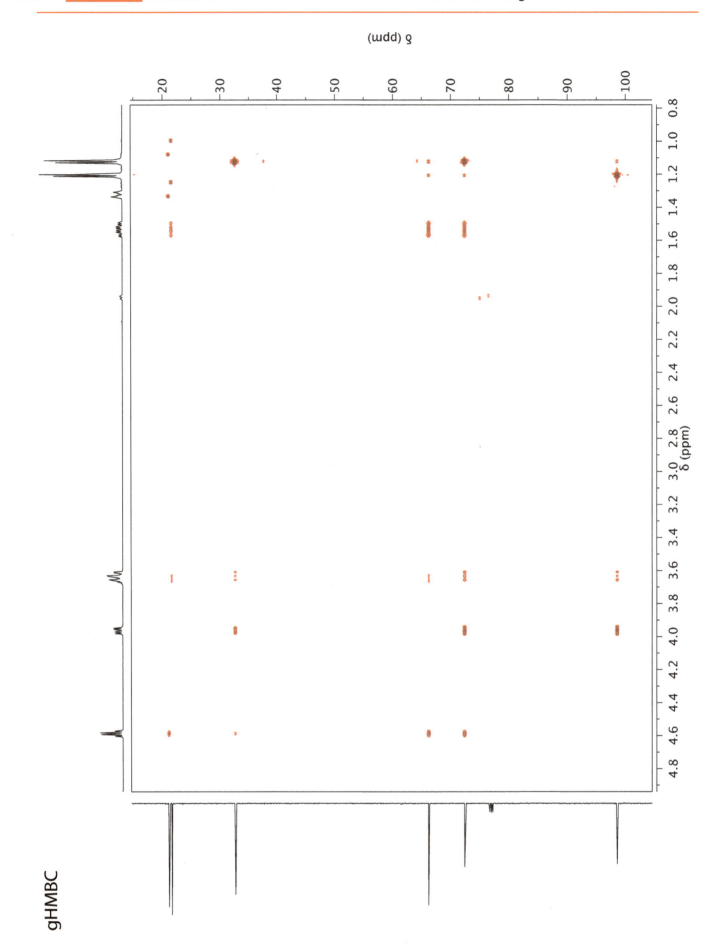

gHMBC

SECTION 2 Problem 42

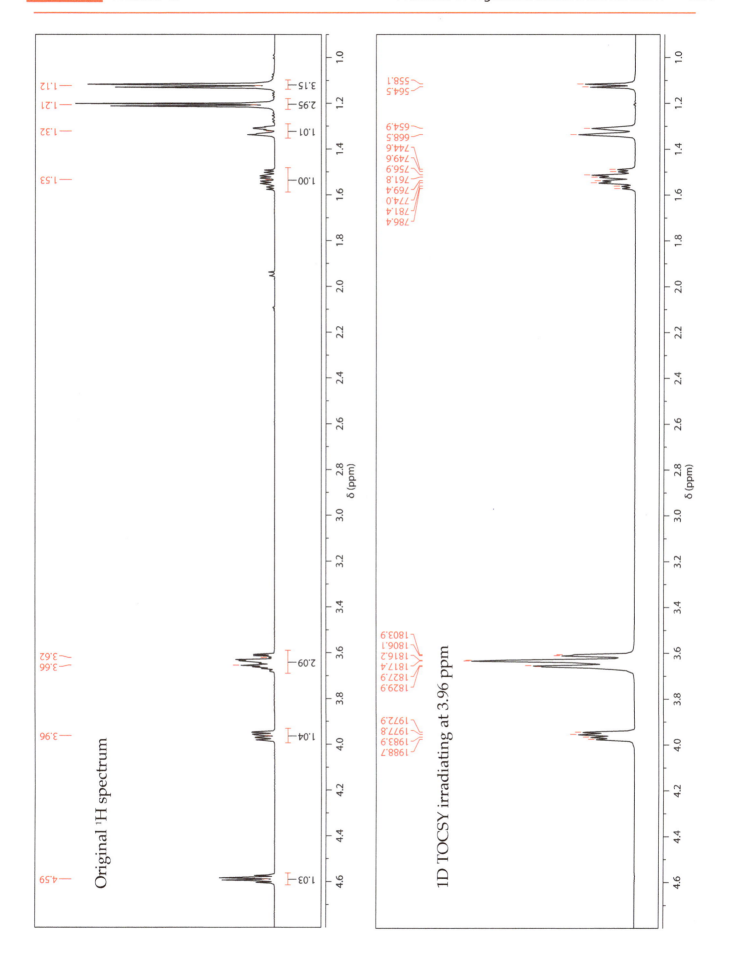

SECTION 2 Problem 43

Which of the following compounds is the correct structure for the spectra below?

Spectra acquired in CDCl₃ at 500 MHz (^{1}H) or 125 MHz (^{13}C).

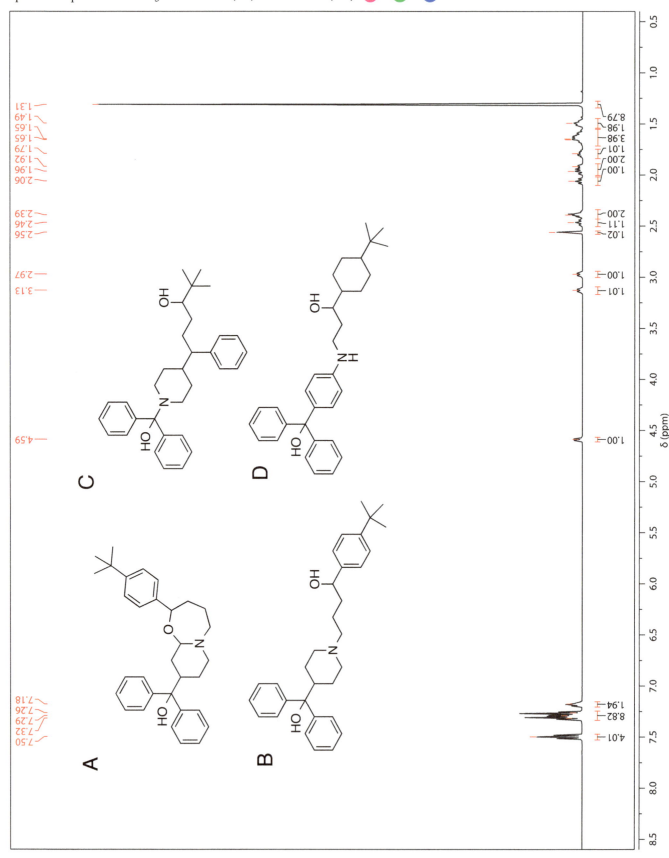

Problem 43

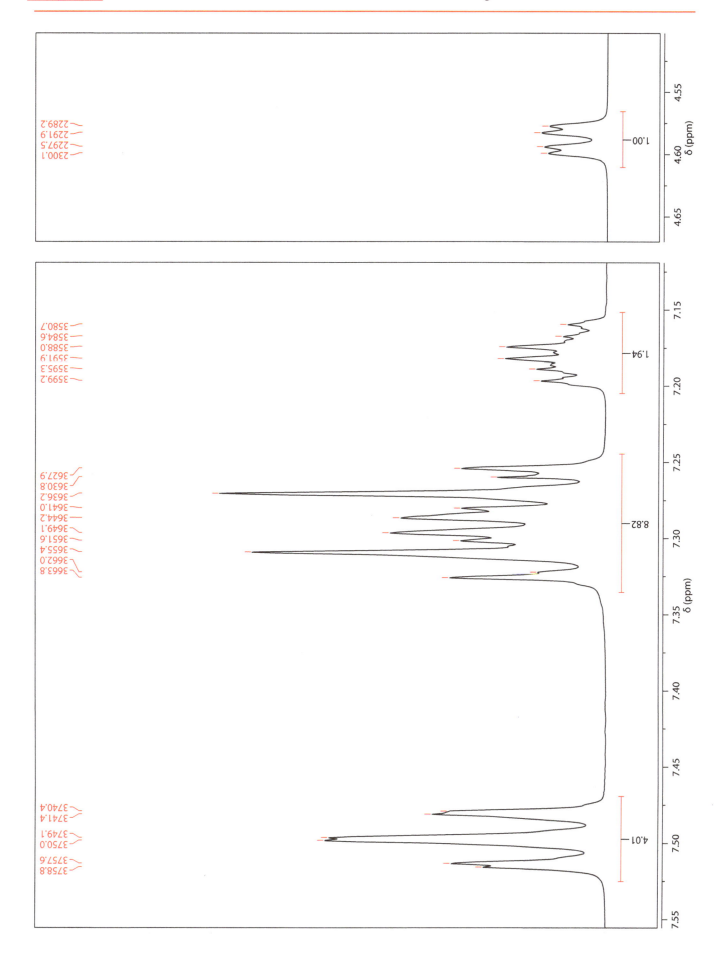

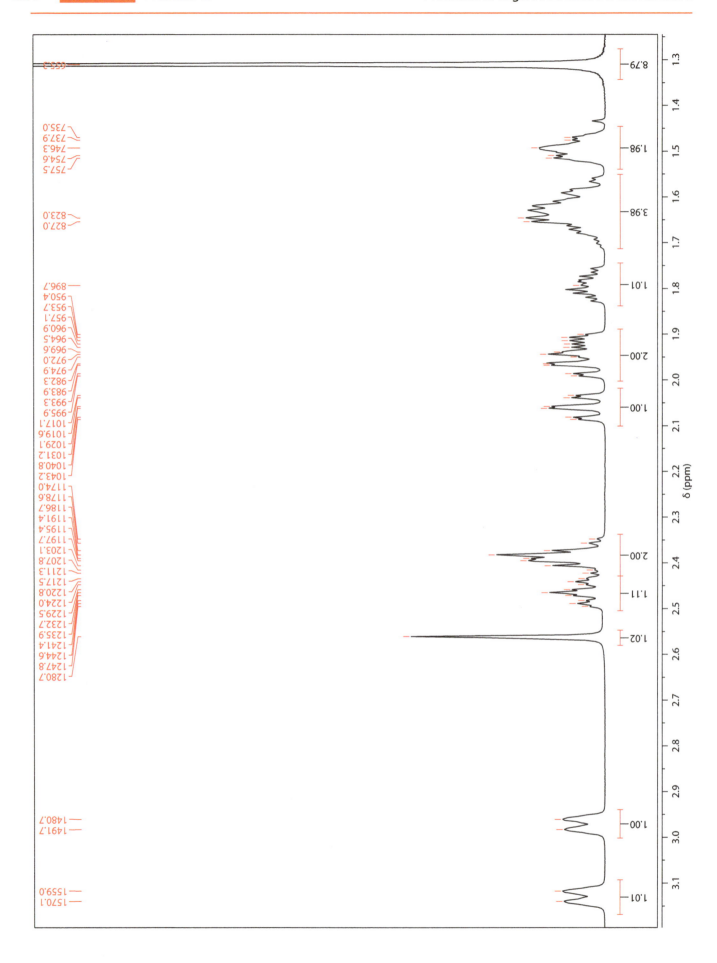

SECTION 2 Problem 43

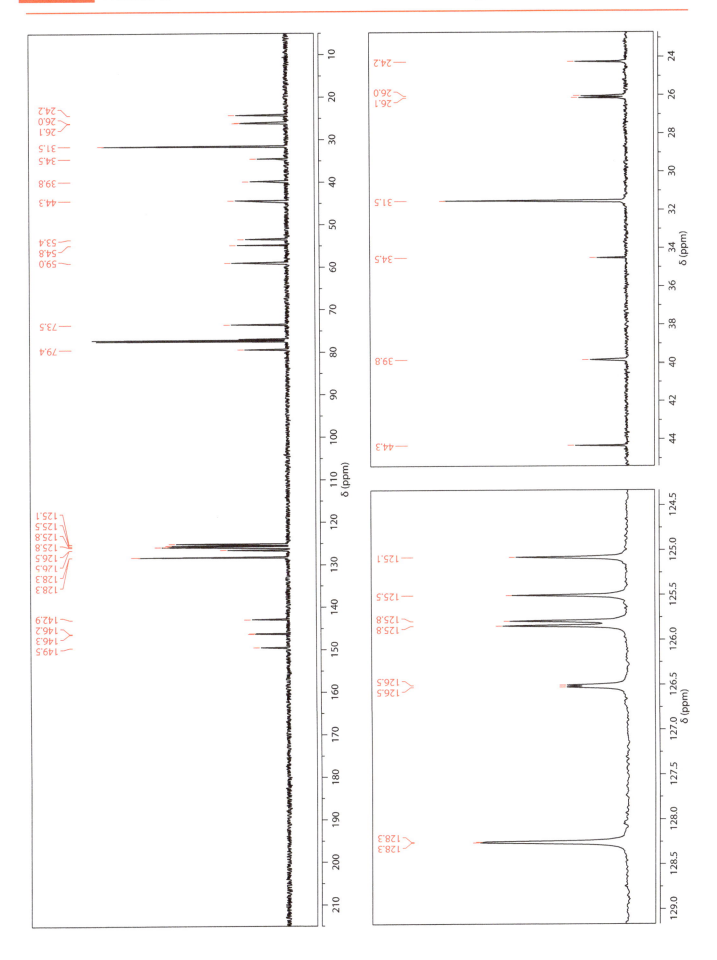

Problem 43

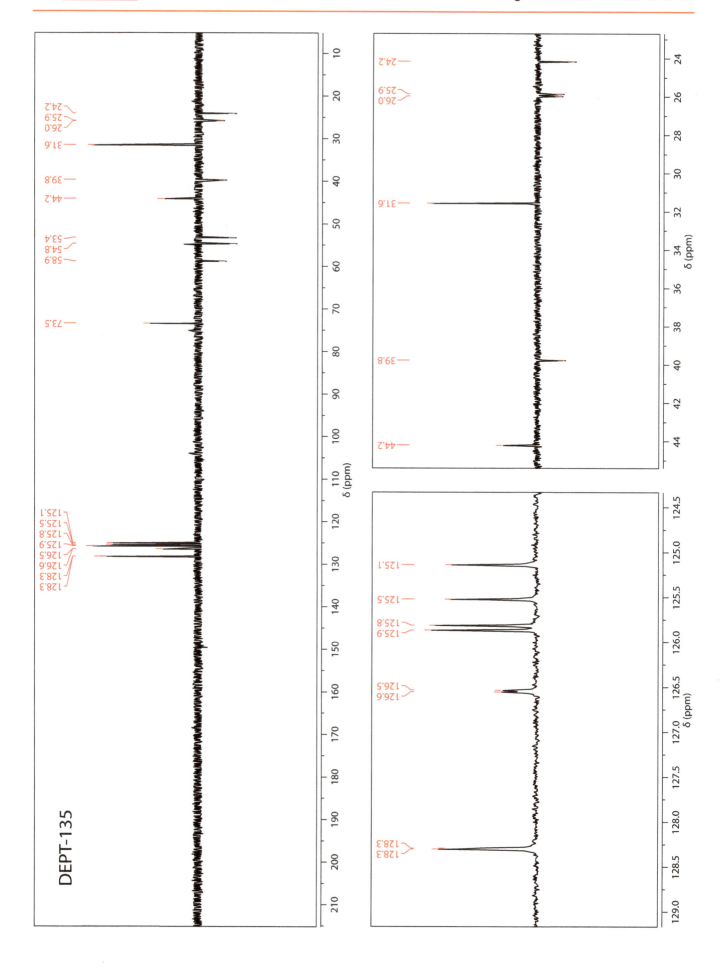

Problem 43

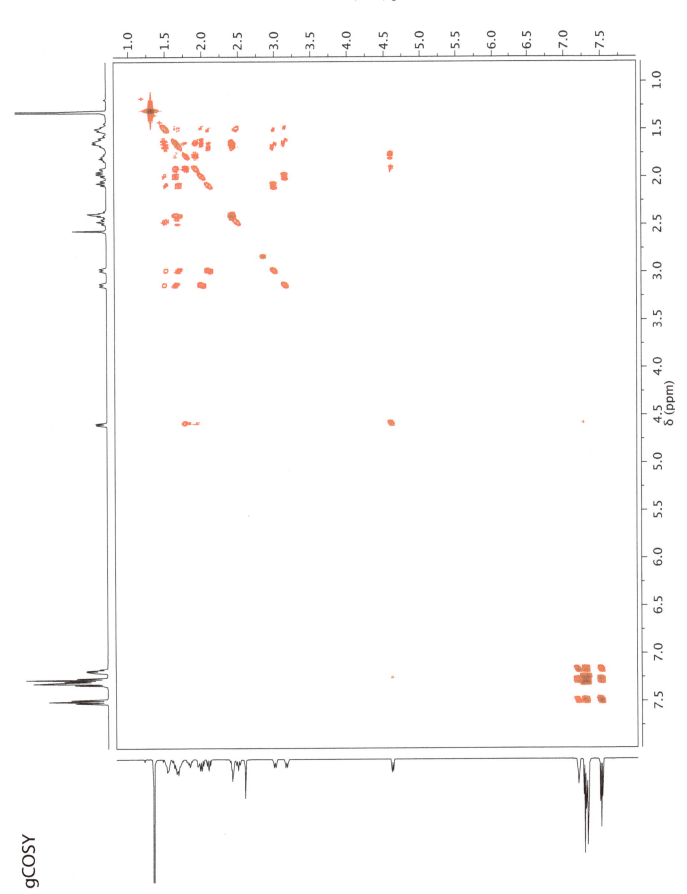

gCOSY

Problem 43

gCOSY

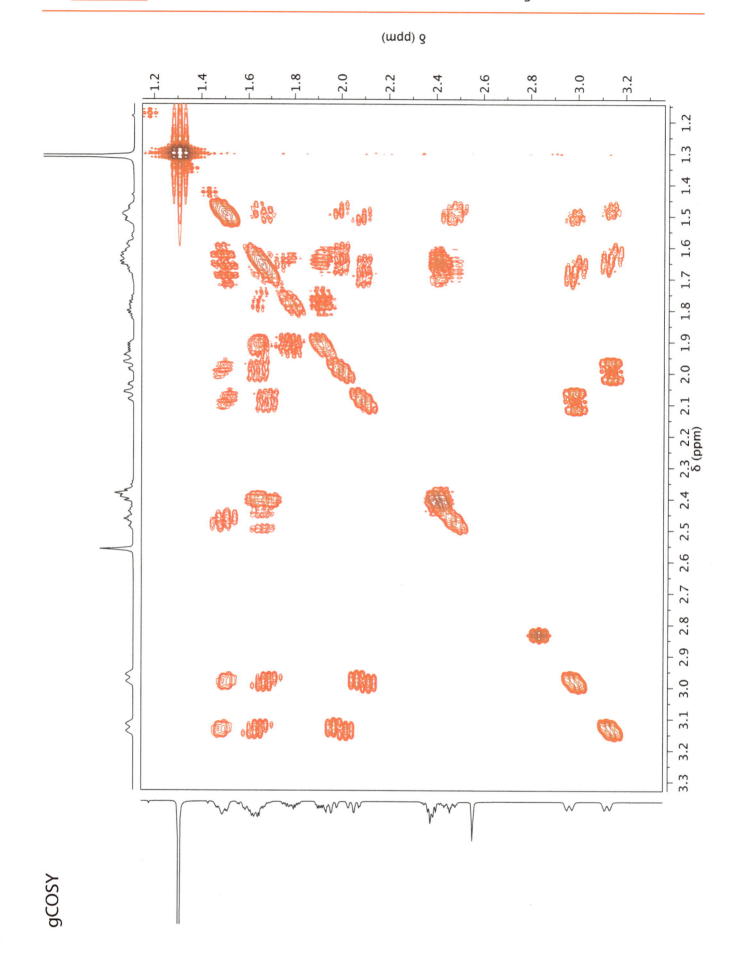

SECTION 2 Problem 43

gHSQC

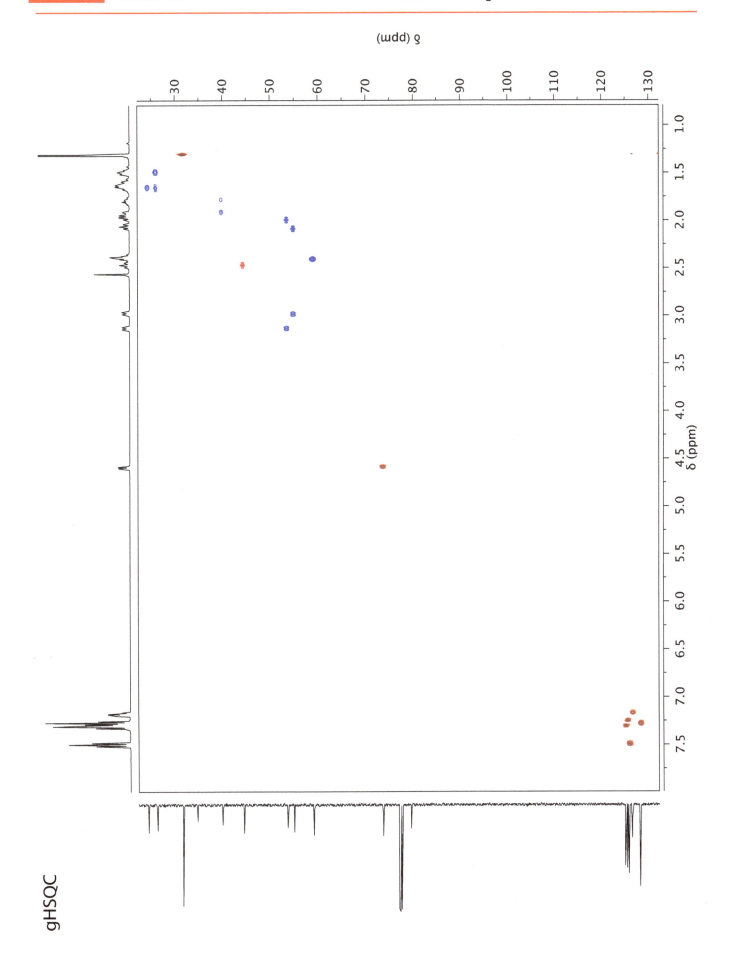

Problem 43

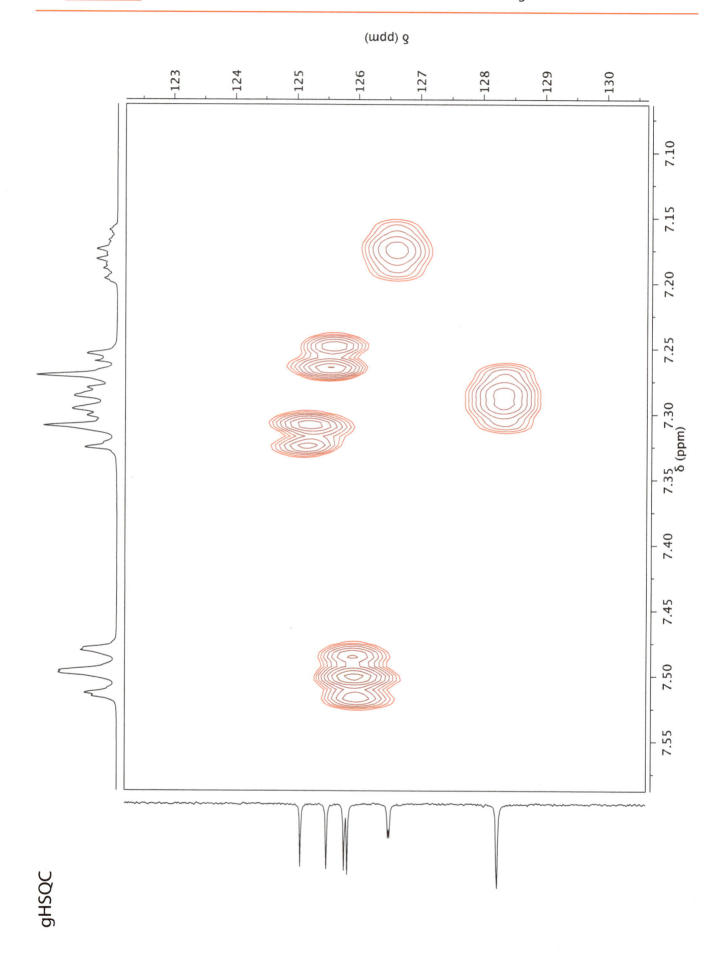

gHSQC

Problem 43

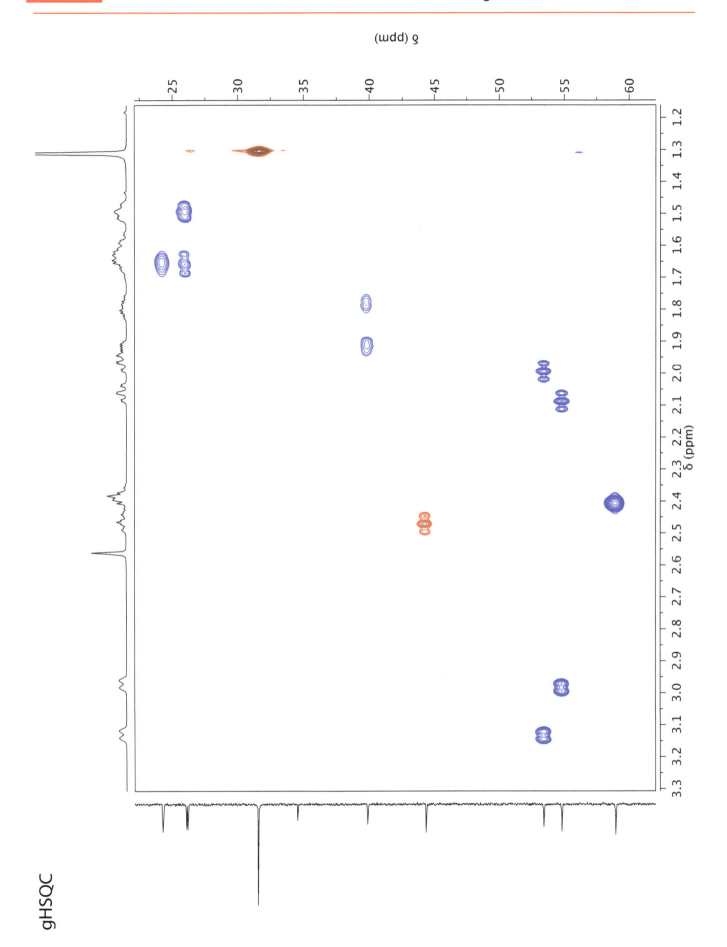

gHSQC

Problem 43

gHMBC

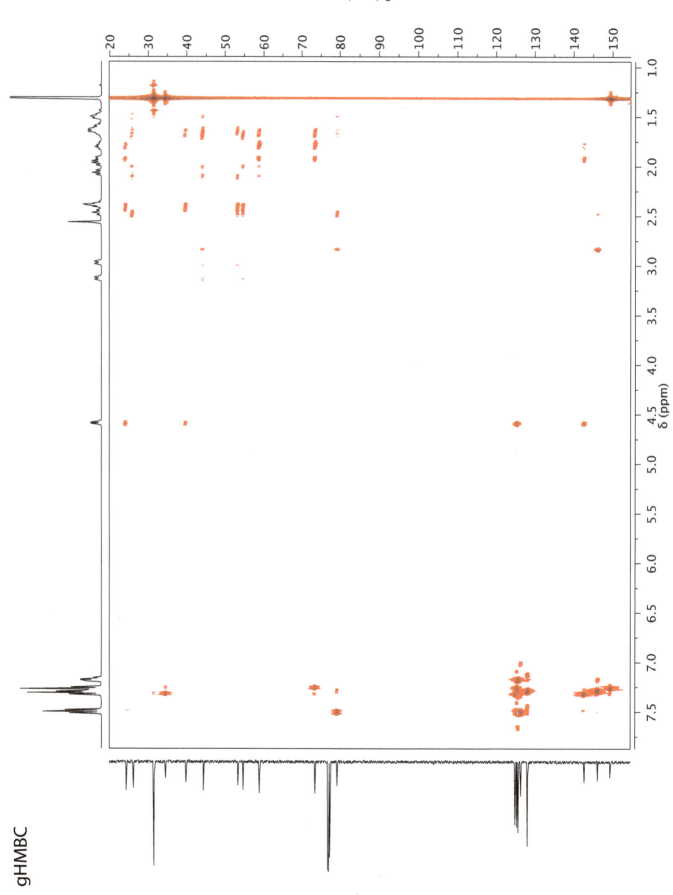

Problem 43

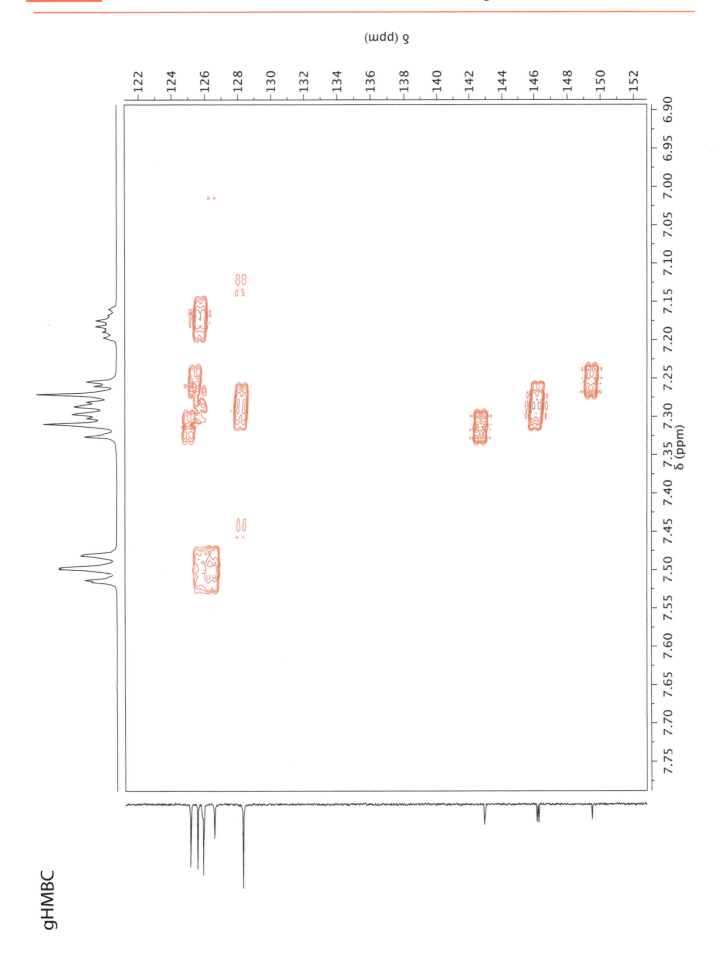

Problem 43

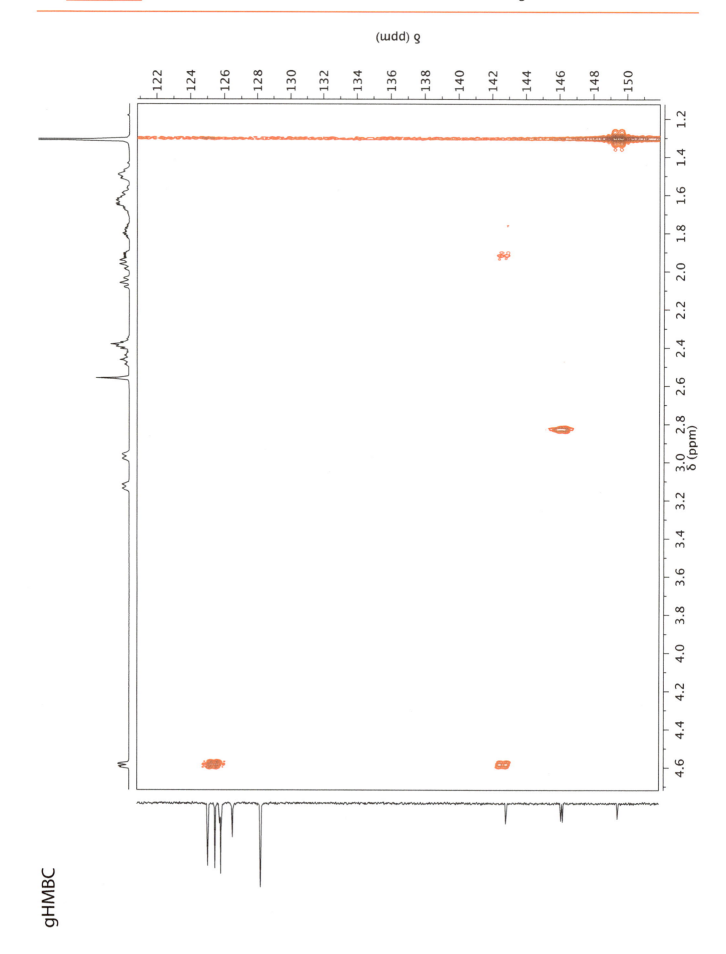

gHMBC

Problem 43

gHMBC

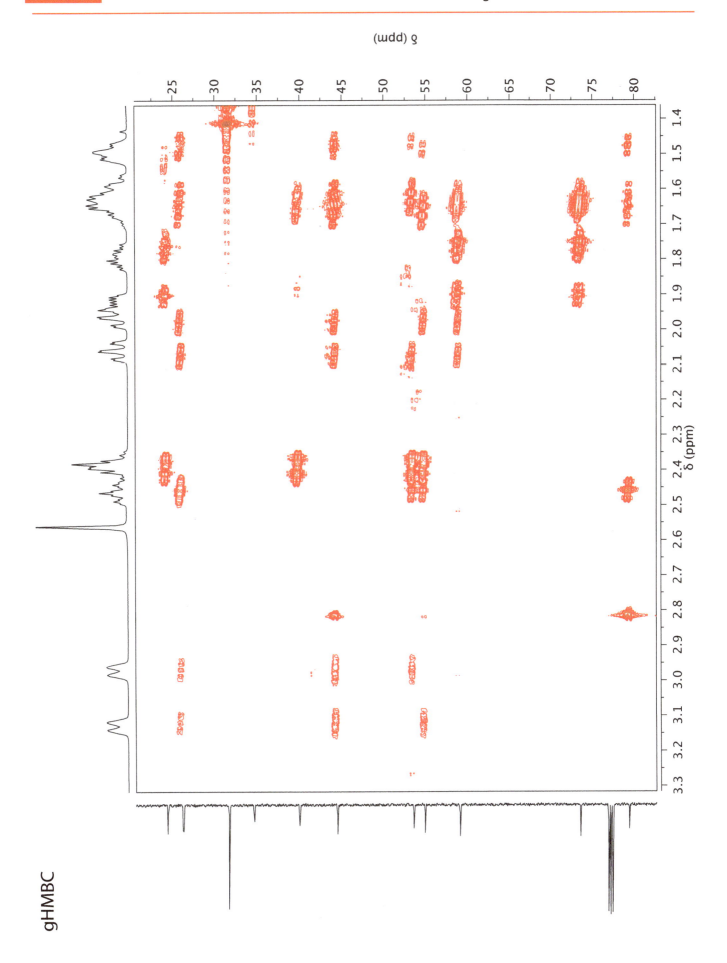

SECTION 2 Problem 44

Problems in Organic Structure Determination

Which of the following compounds is the correct structure for the spectra below?

Spectra acquired in CDCl₃ at 500 MHz (^{1}H) or 125 MHz (^{13}C).

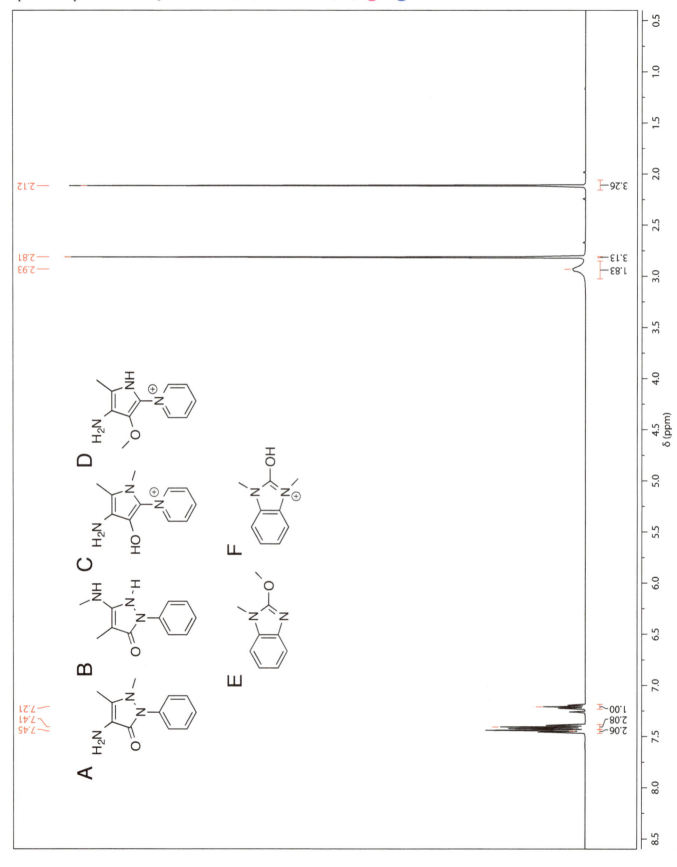

SECTION 2 Problem 44

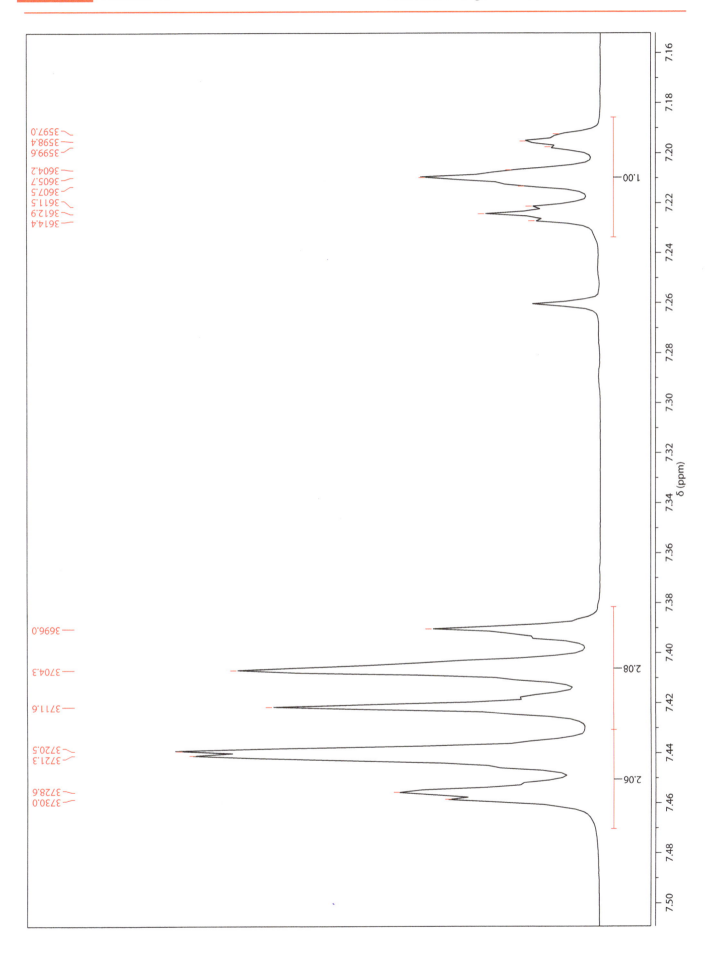

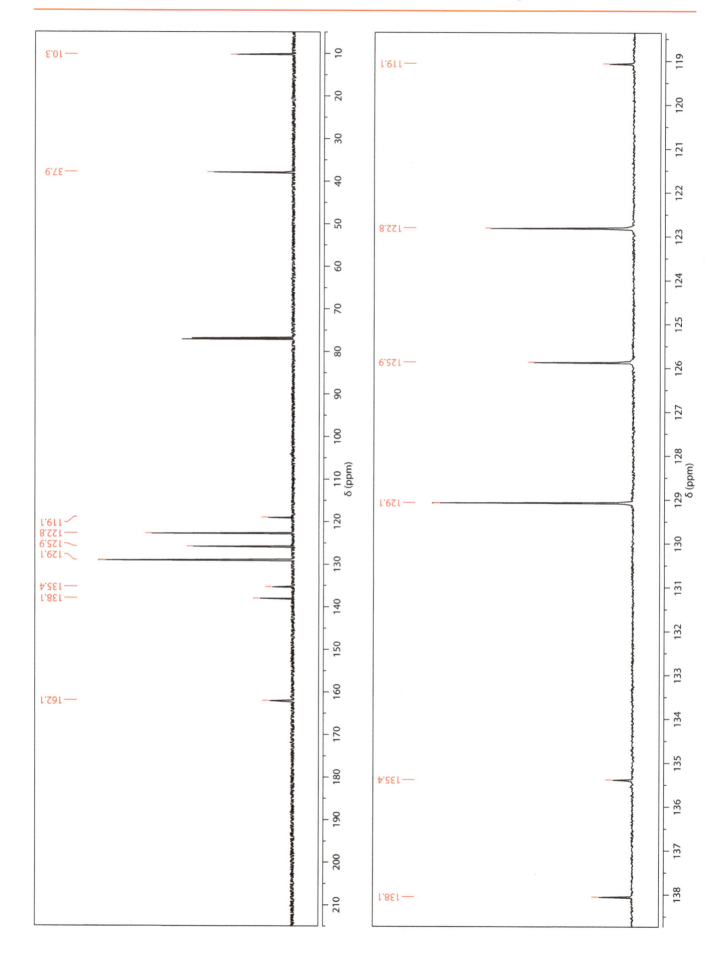

SECTION 2 Problem 44

Problems in Organic Structure Determination

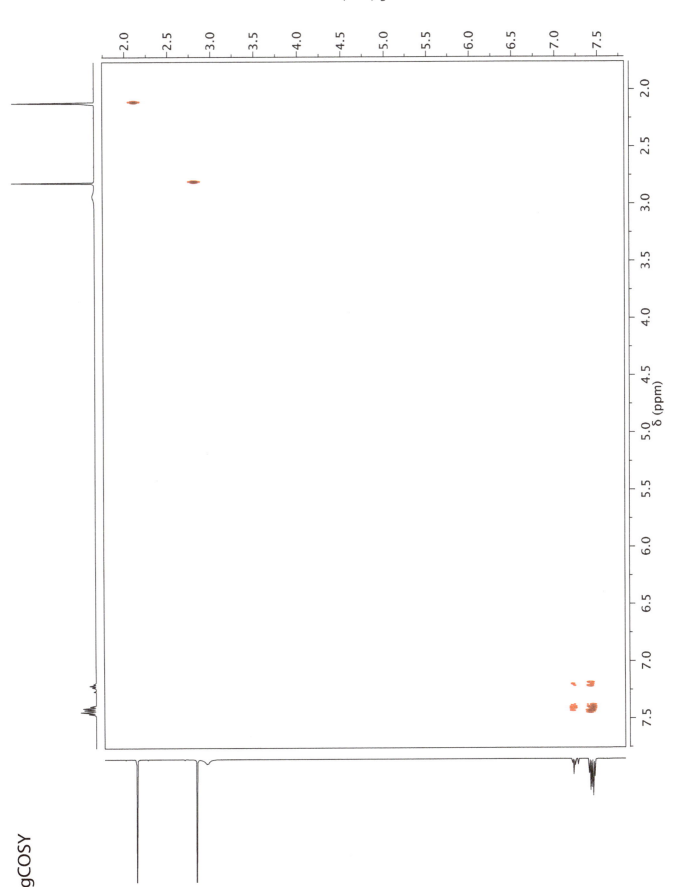

gCOSY

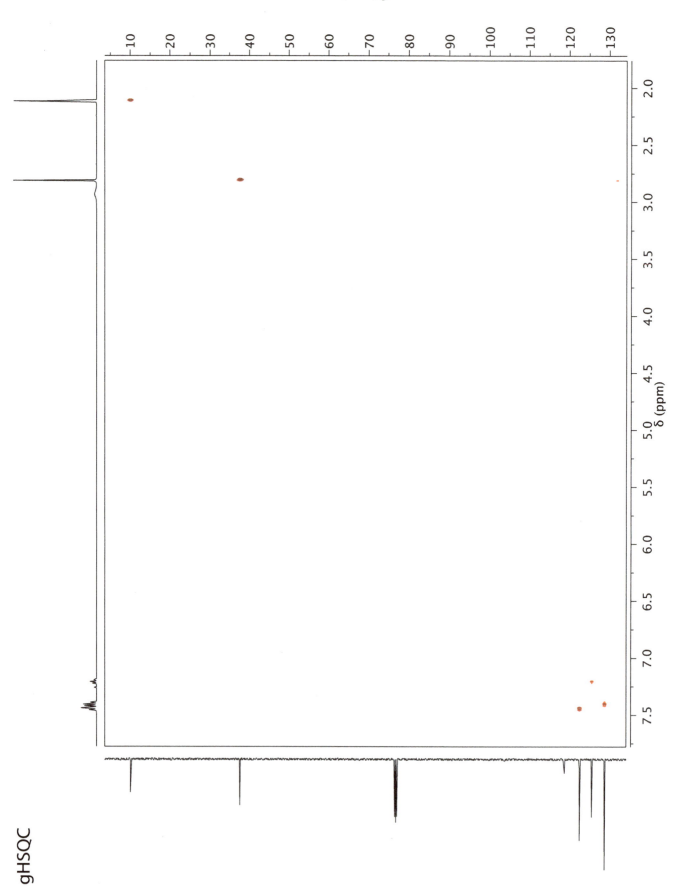

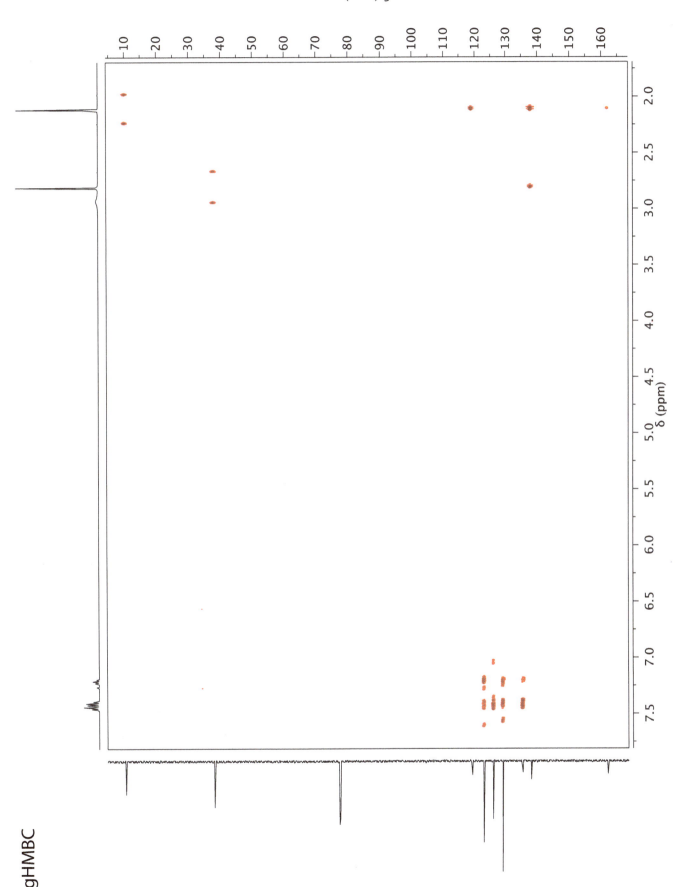

Problem 44

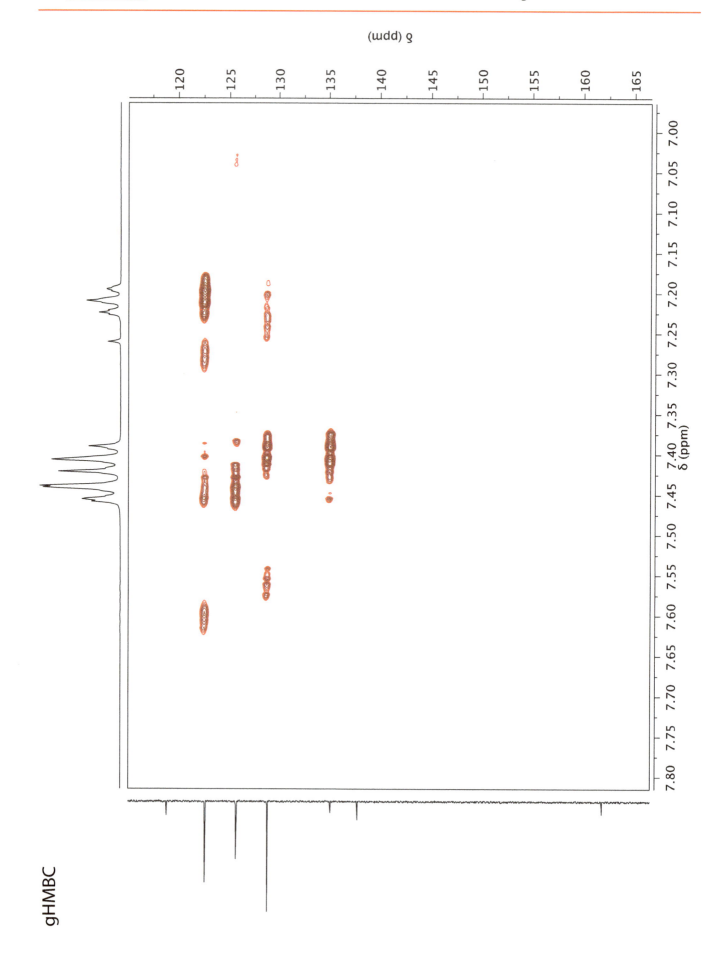

SECTION 2 Problem 45

Using the ¹H, ¹³C, gCOSY, gHSQC and gHMBC spectra, assign all proton and carbon resonances to the structure of 4-(1,3-dioxo-1H,3H-benzo[de]isoquinolin-2-yl)-butanoic acid below.
Spectra acquired in (CD₃)₂SO at 500 MHz (¹H) or 125 MHz (¹³C).

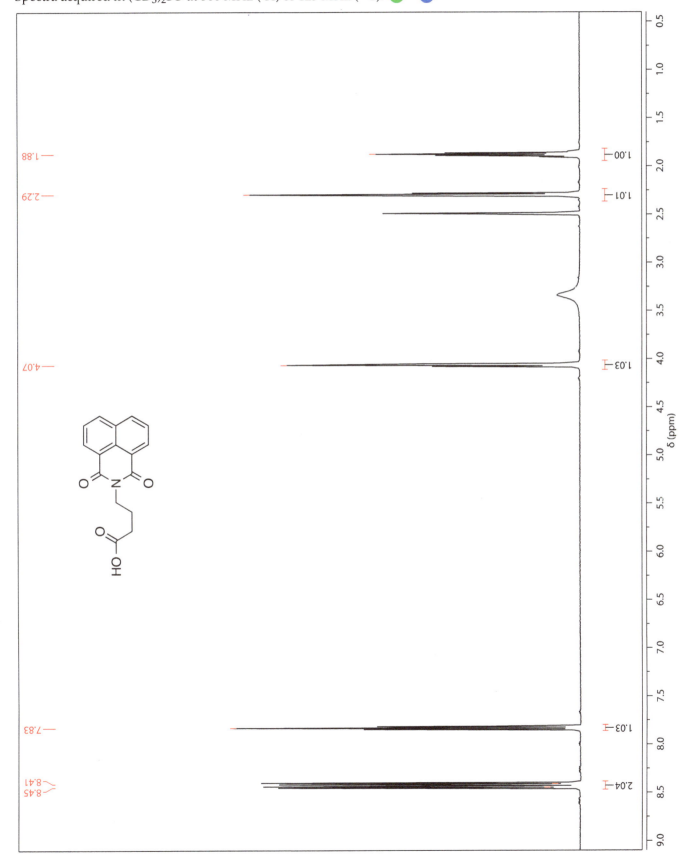

SECTION 2 Problem 45

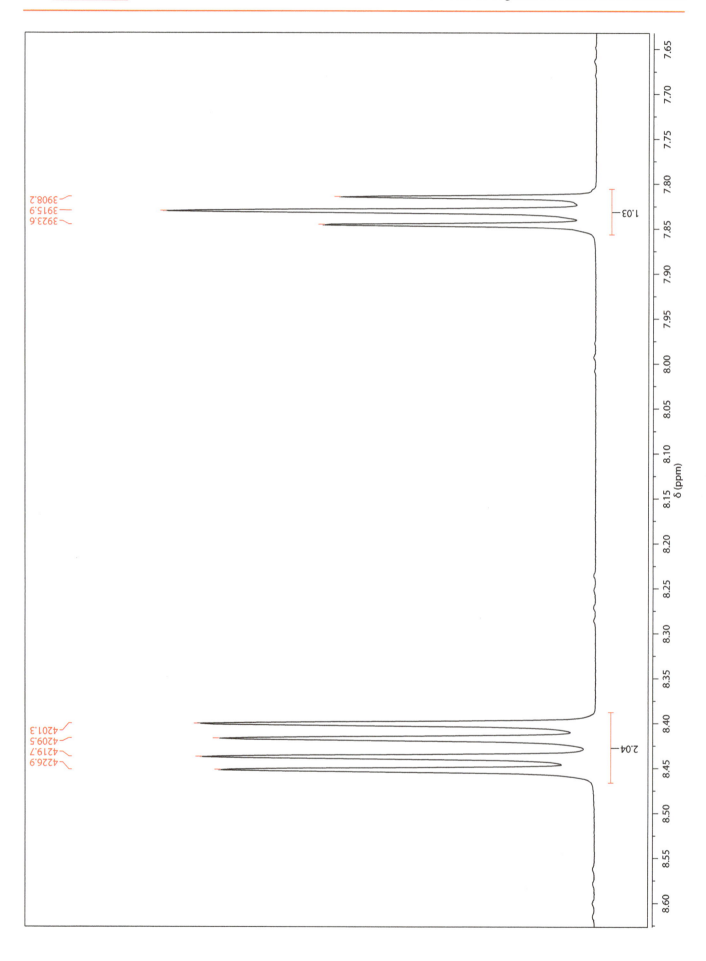

SECTION 2 Problem 45

Problem 45

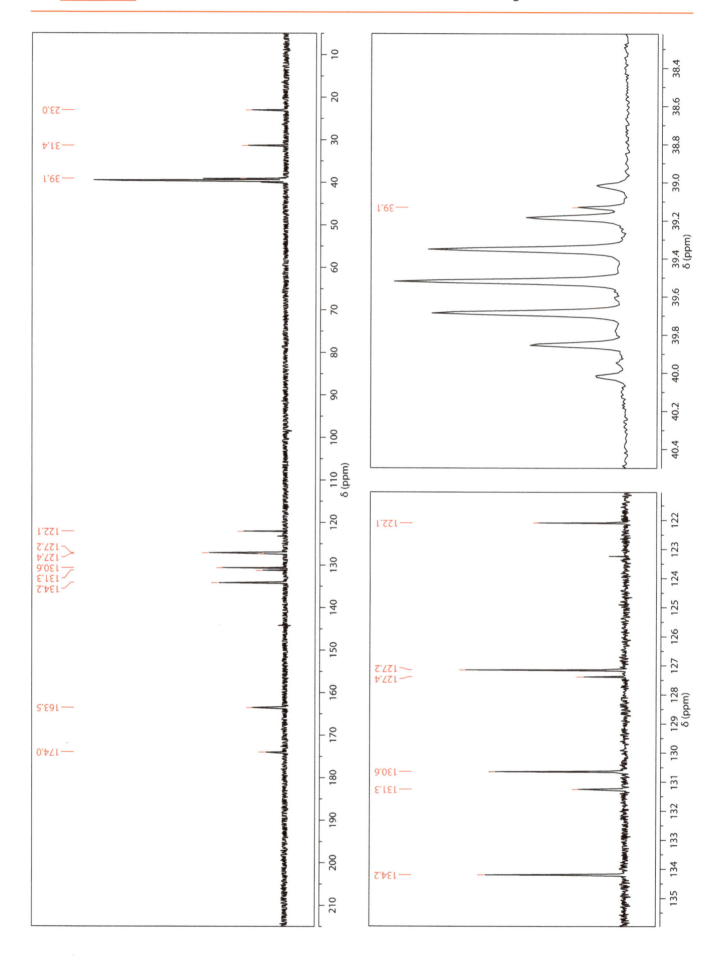

Problem 45

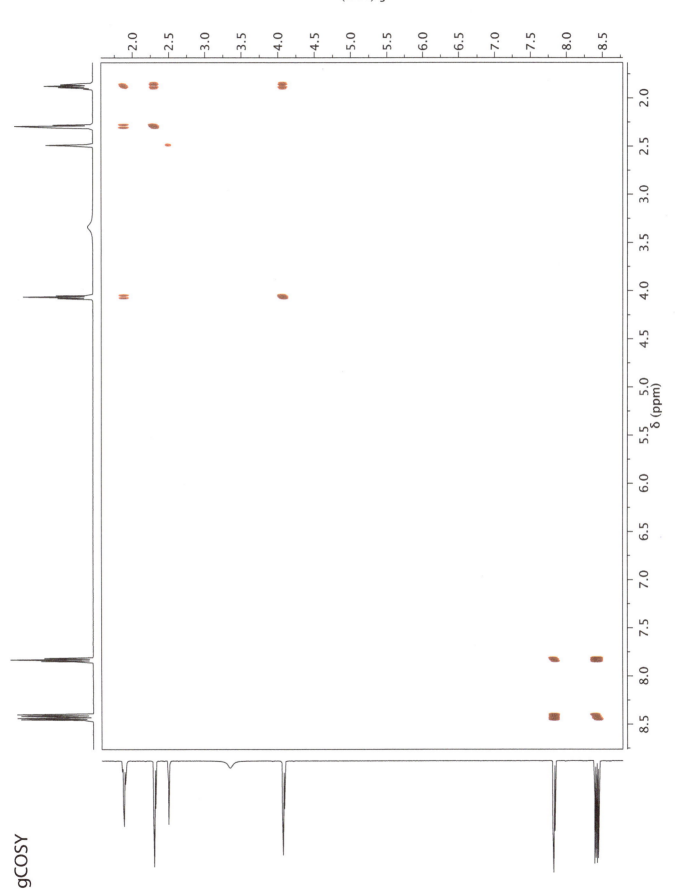

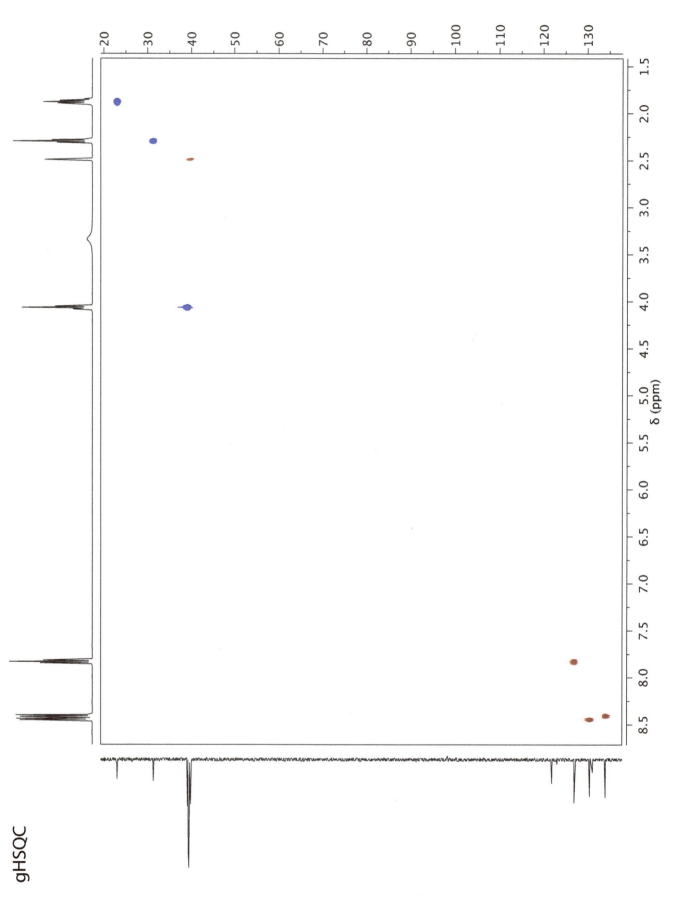

Problem 45

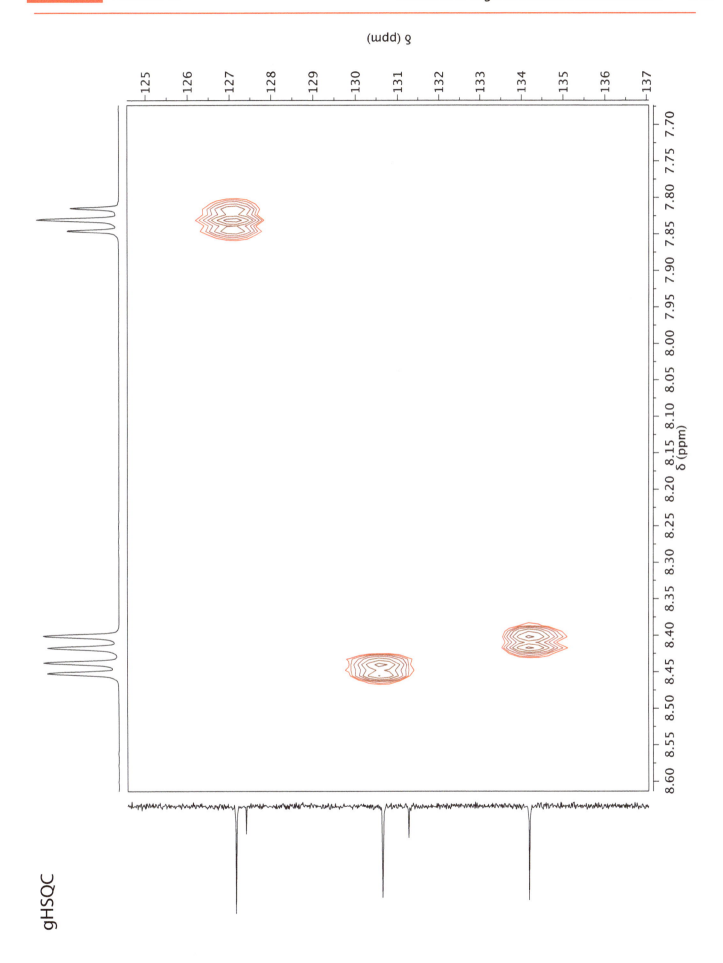

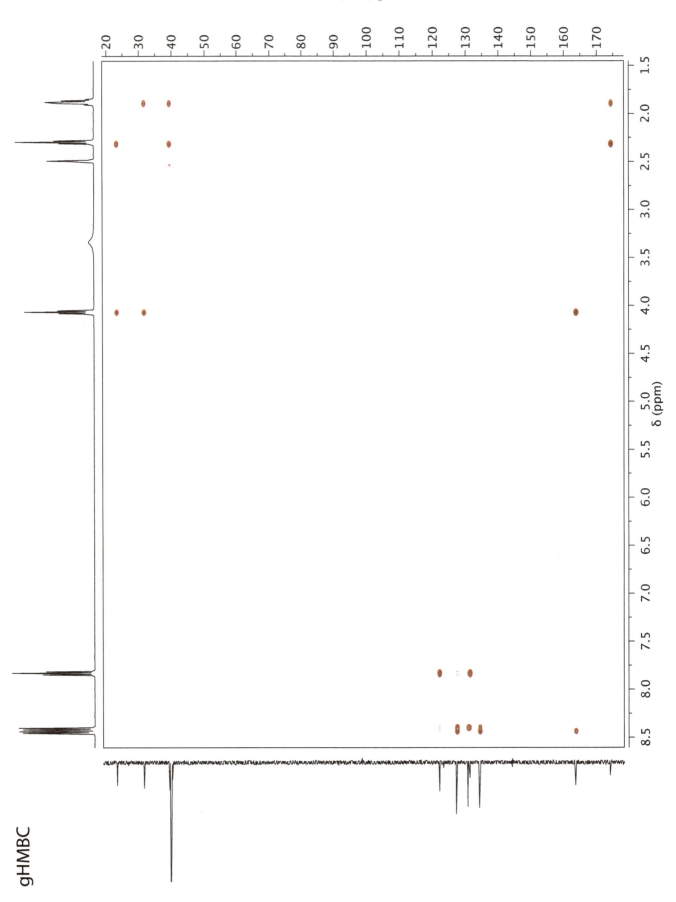

Problem 45

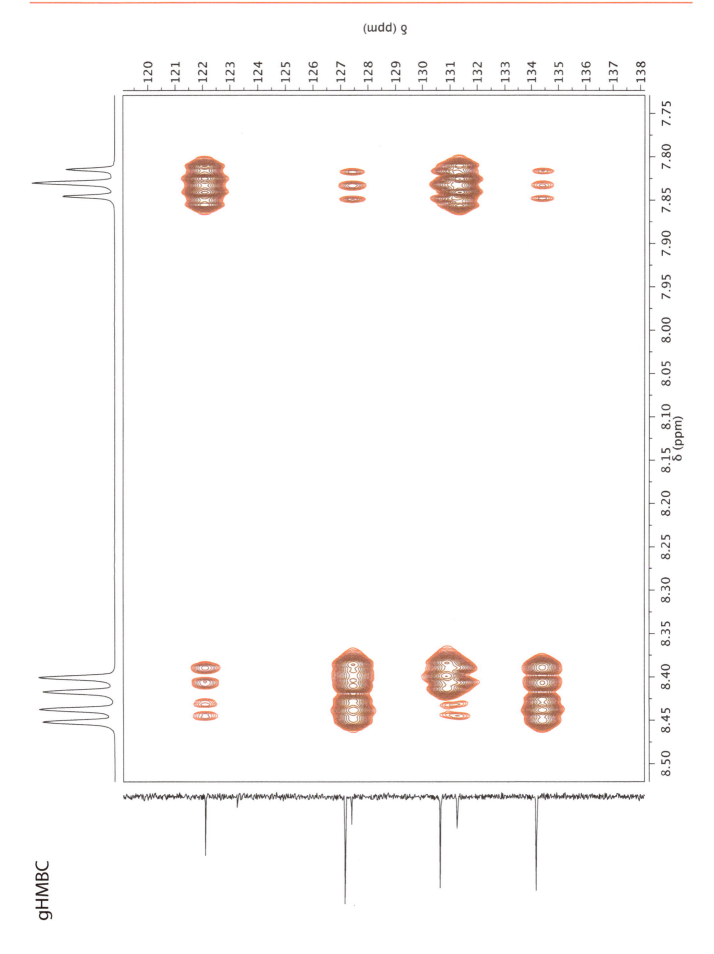

Problem 46

Using the ¹H, ¹³C, gCOSY, gHSQC and gHMBC spectra, assign all proton and carbon resonances to the structure of (S,S)-(+)-N,N'-bis(3,5-di-tert-butylsalicylidene)-1,2-cyclohexanediamine below.
Spectra acquired in CDCl₃ at 500 MHz (¹H) or 125 MHz (¹³C).

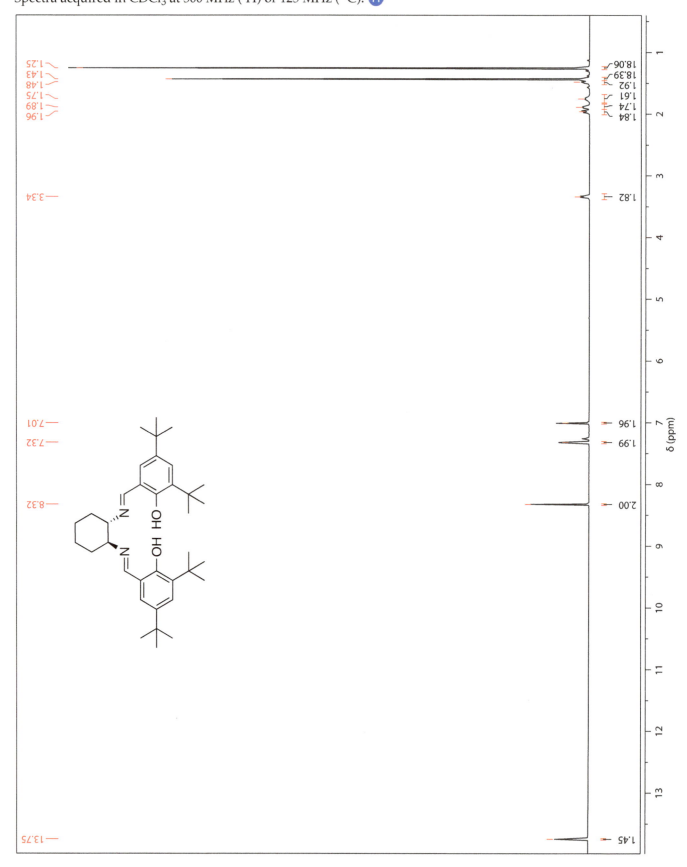

SECTION 2 Problem 46

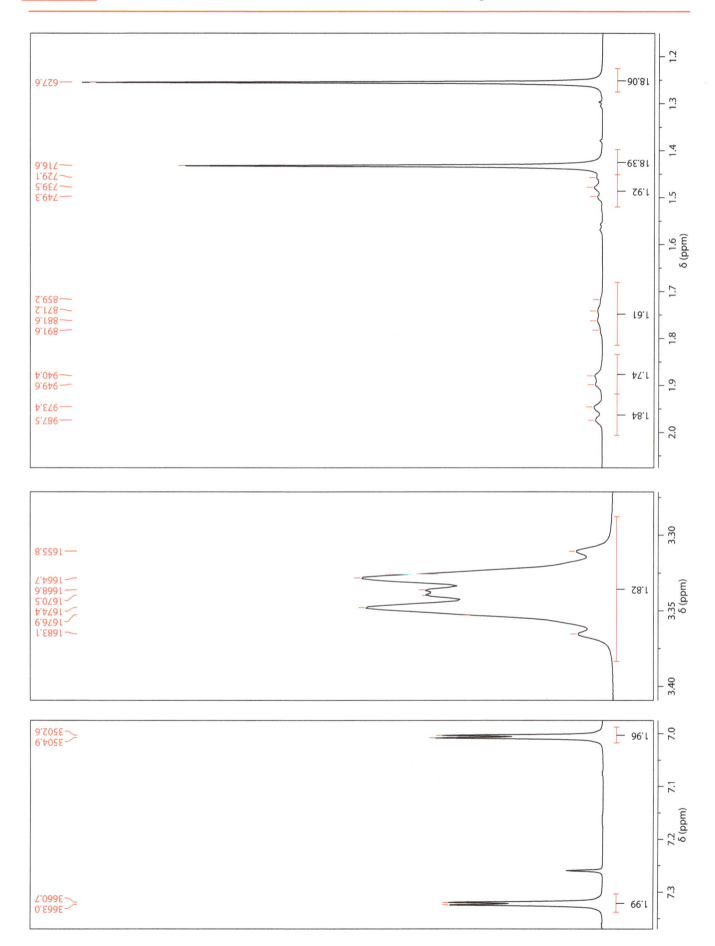

Problem 46

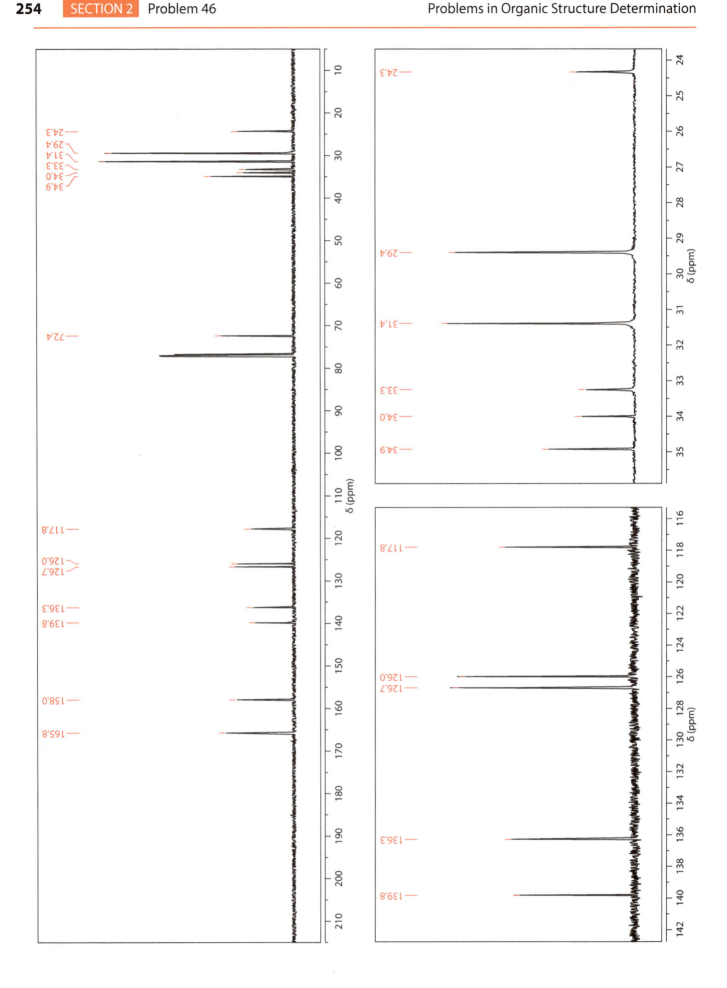

SECTION 2 Problem 46

Problems in Organic Structure Determination

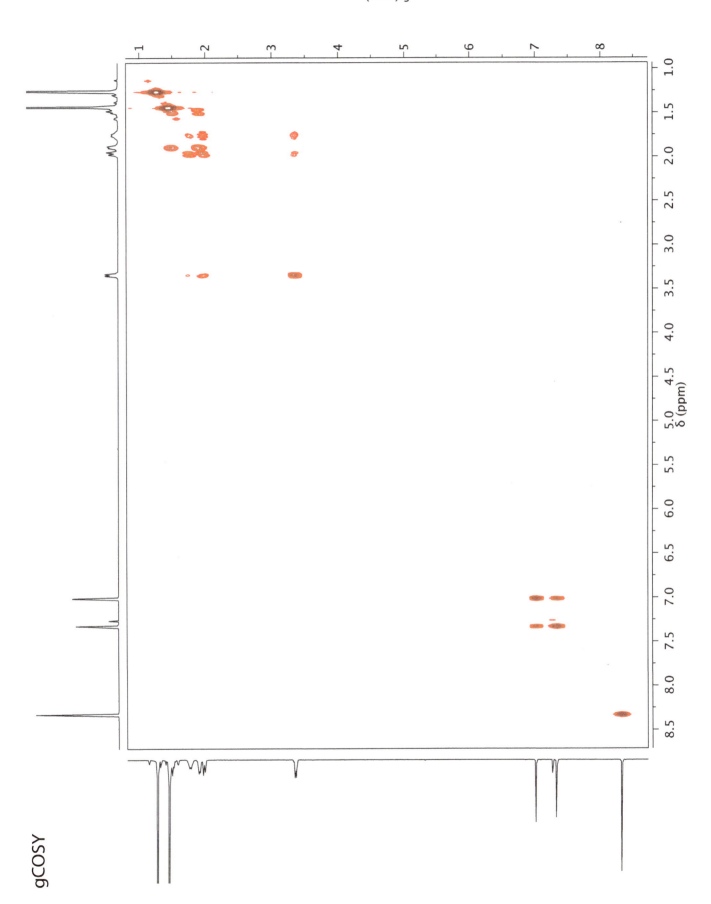

gCOSY

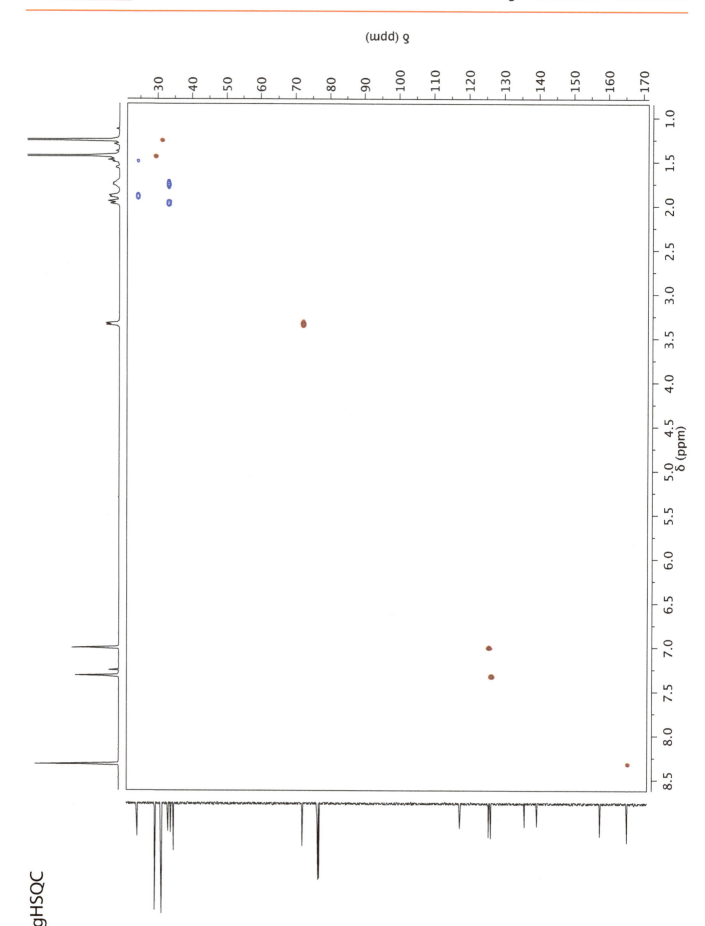

gHSQC

Problem 46

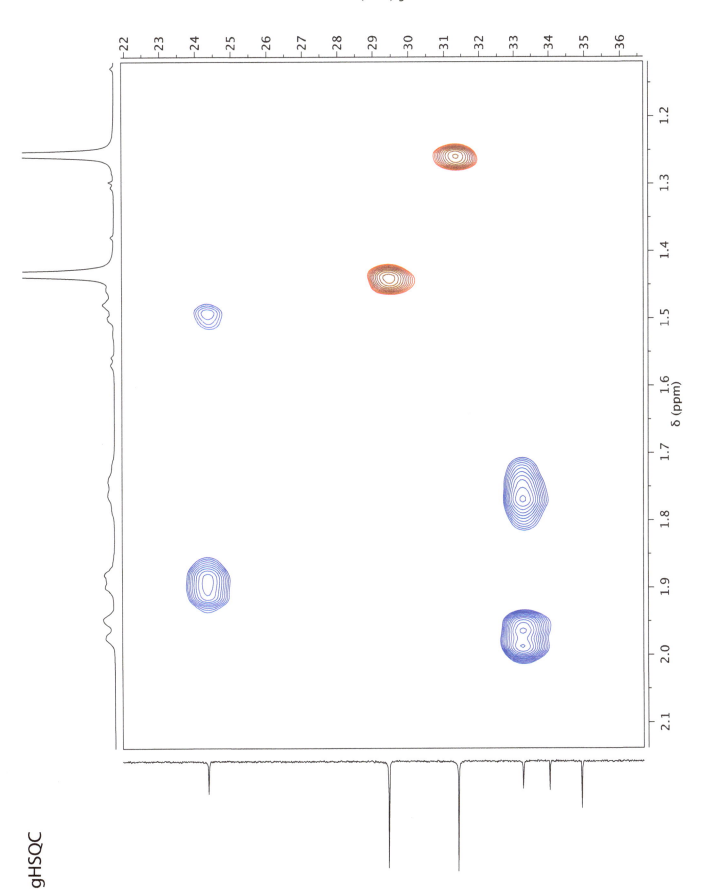

gHSQC

SECTION 2 Problem 46

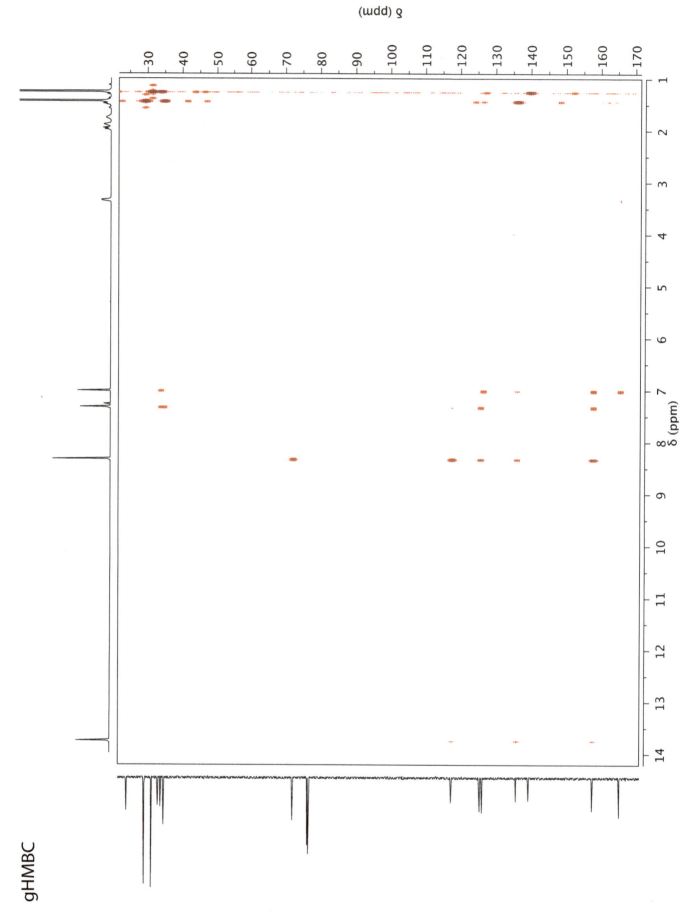

gHMBC

SECTION 2 Problem 46

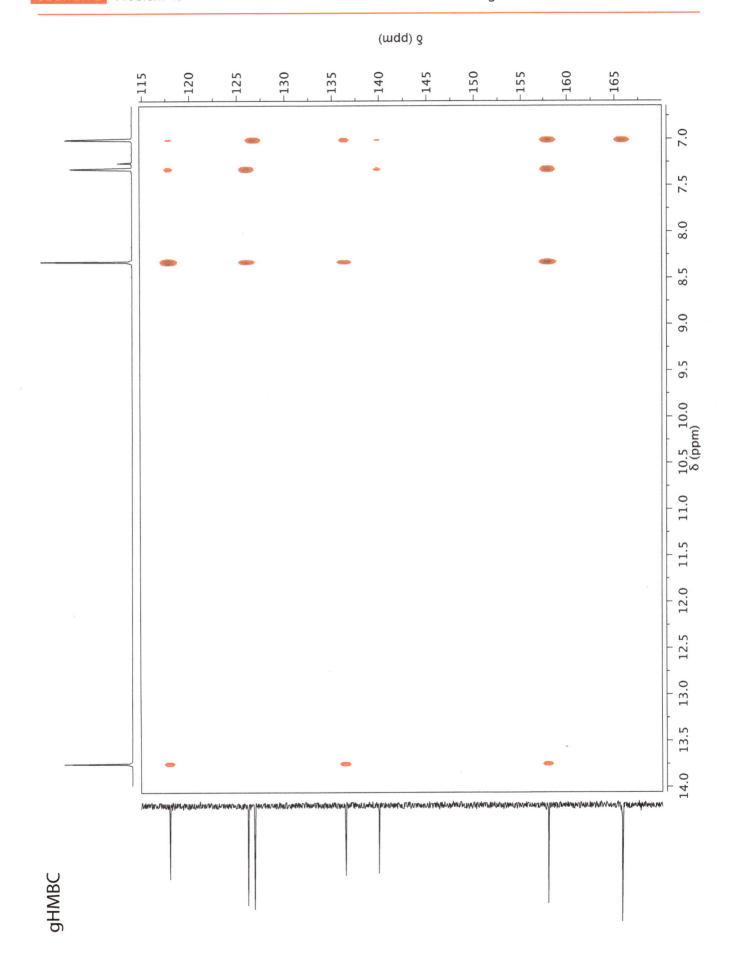

gHMBC

Problem 46

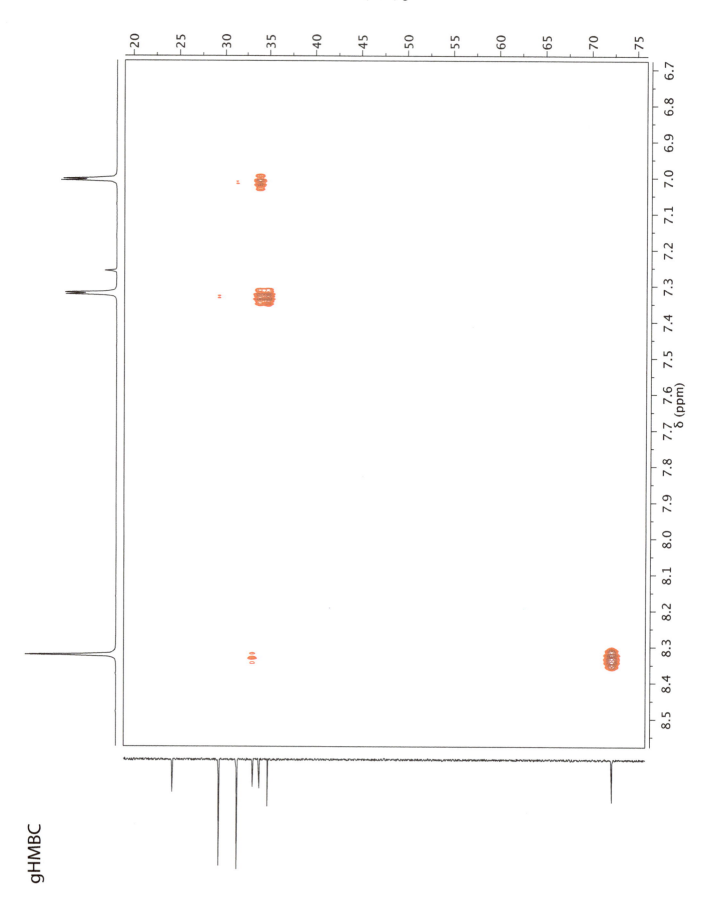

gHMBC

SECTION 2 Problem 46

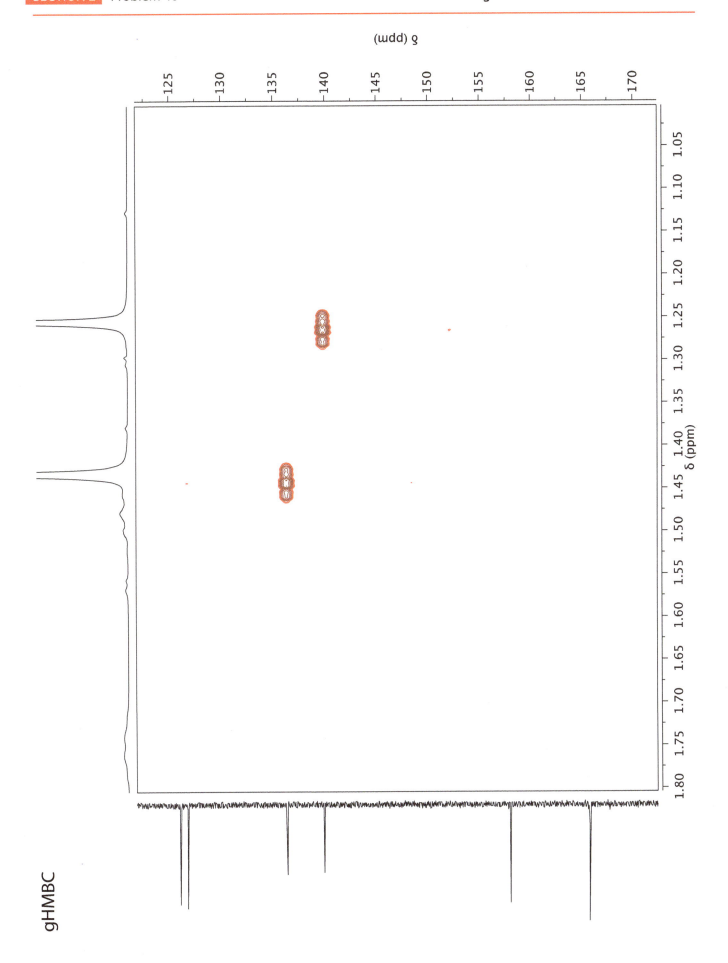

gHMBC

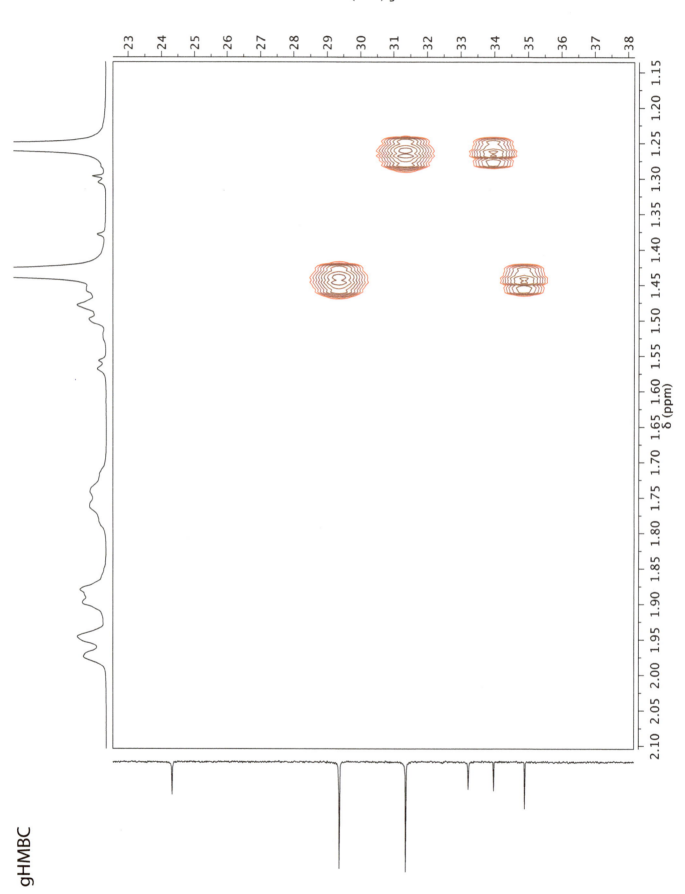

SECTION 2 Problem 47

Problems in Organic Structure Determination

Which of the following compounds is the correct structure for the spectra below?

Spectra acquired in CDCl₃ at 500 MHz (¹H) or 125 MHz (¹³C).

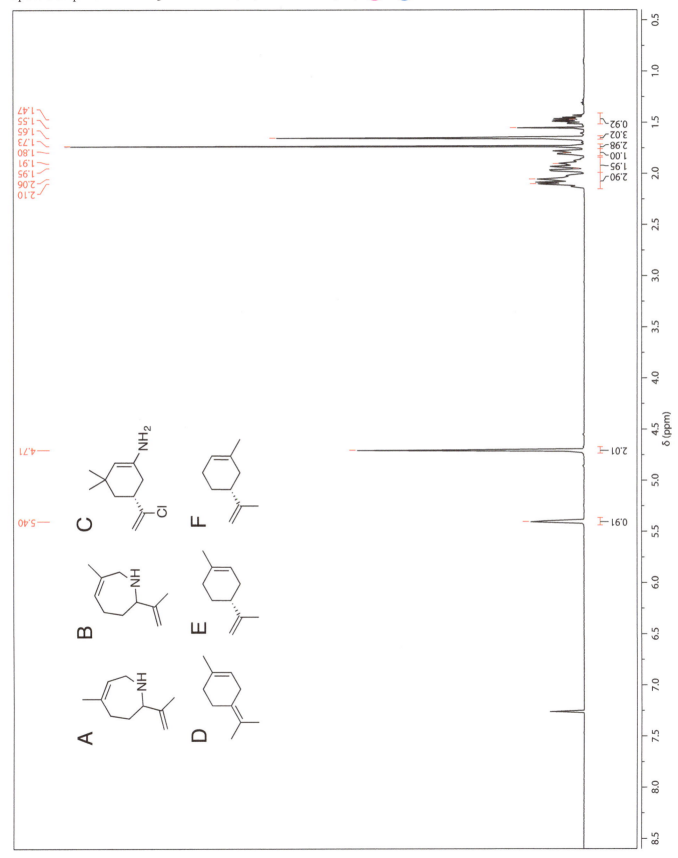

Problem 47

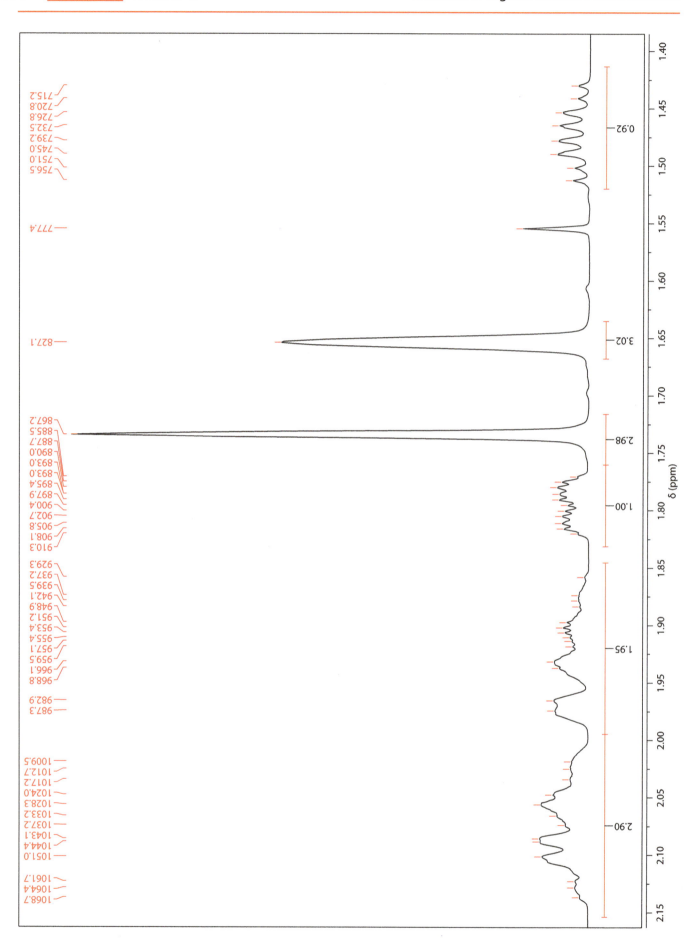

SECTION 2 Problem 47

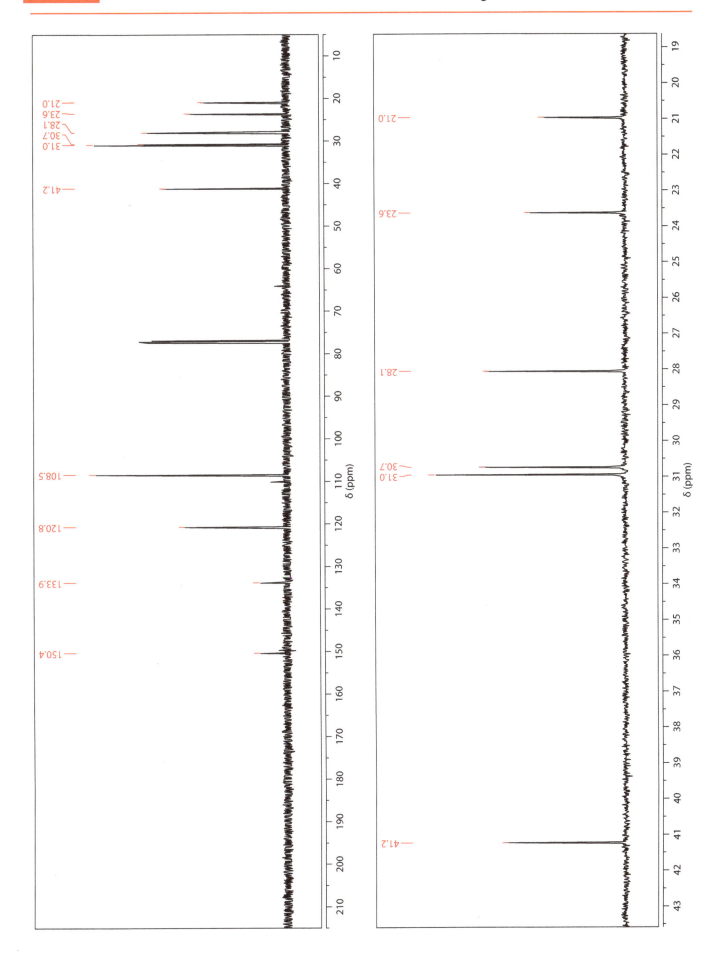

Problem 47

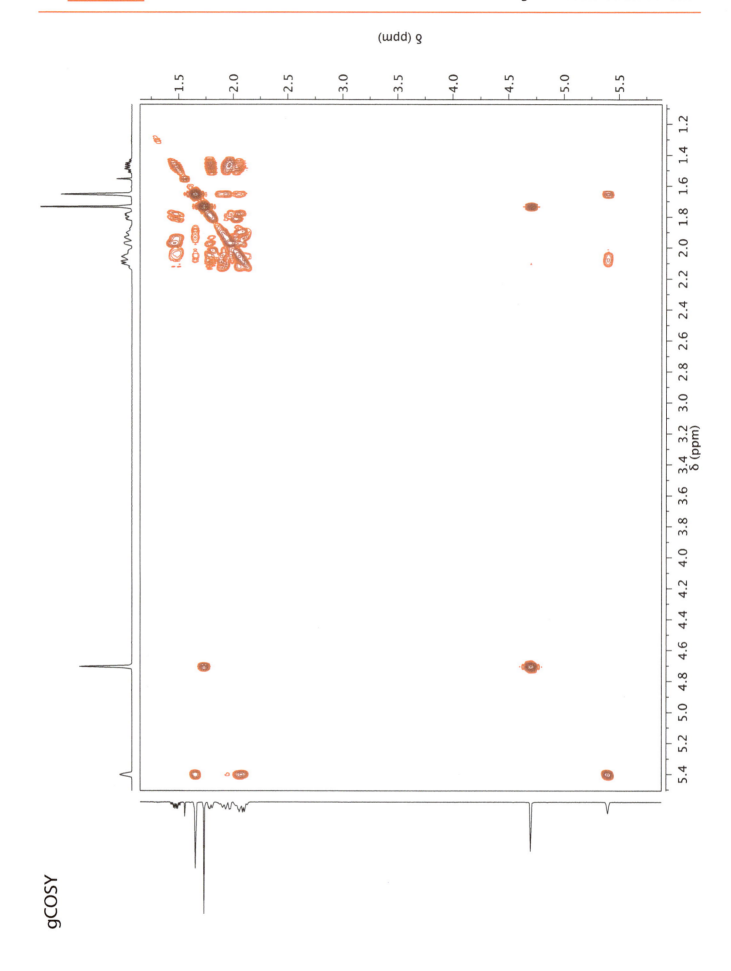

gCOSY

Problem 47

gCOSY

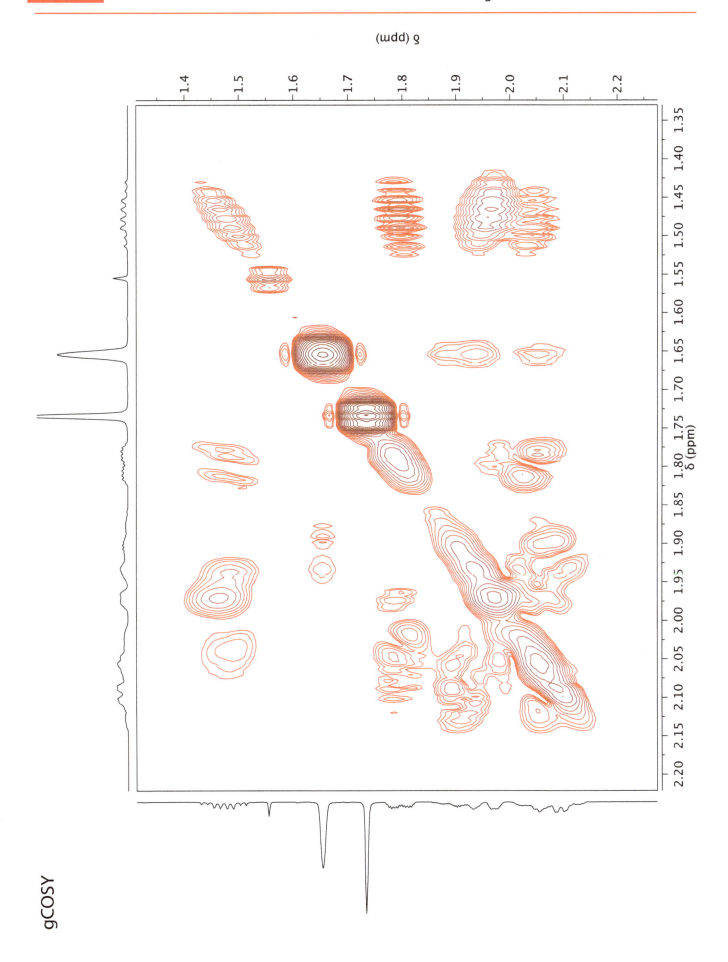

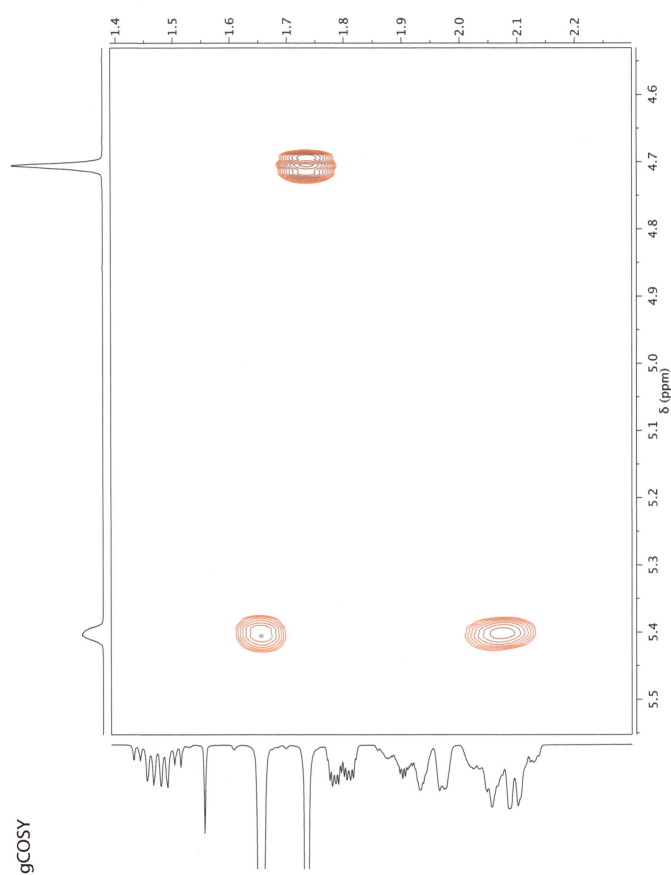

gCOSY

Problem 47

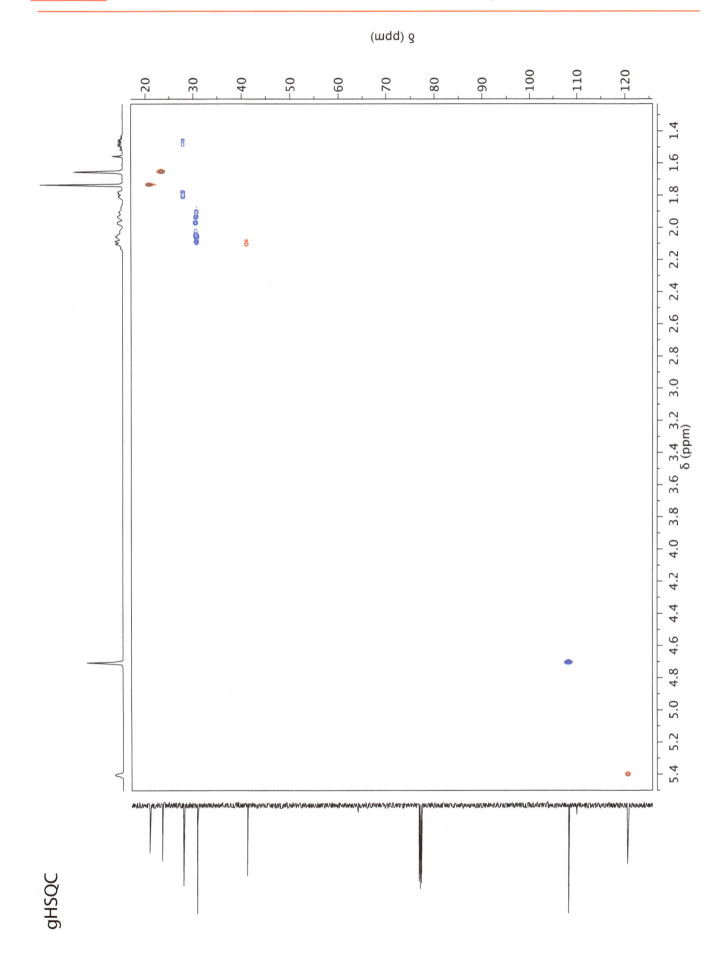

gHSQC

Problem 47

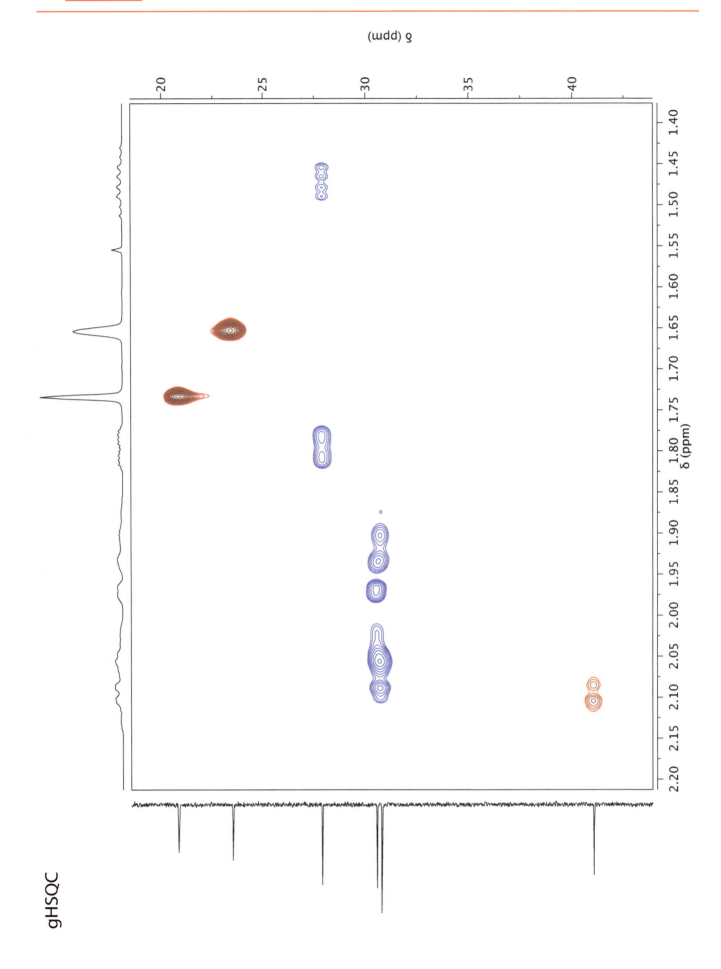

gHSQC

SECTION 2 Problem 47

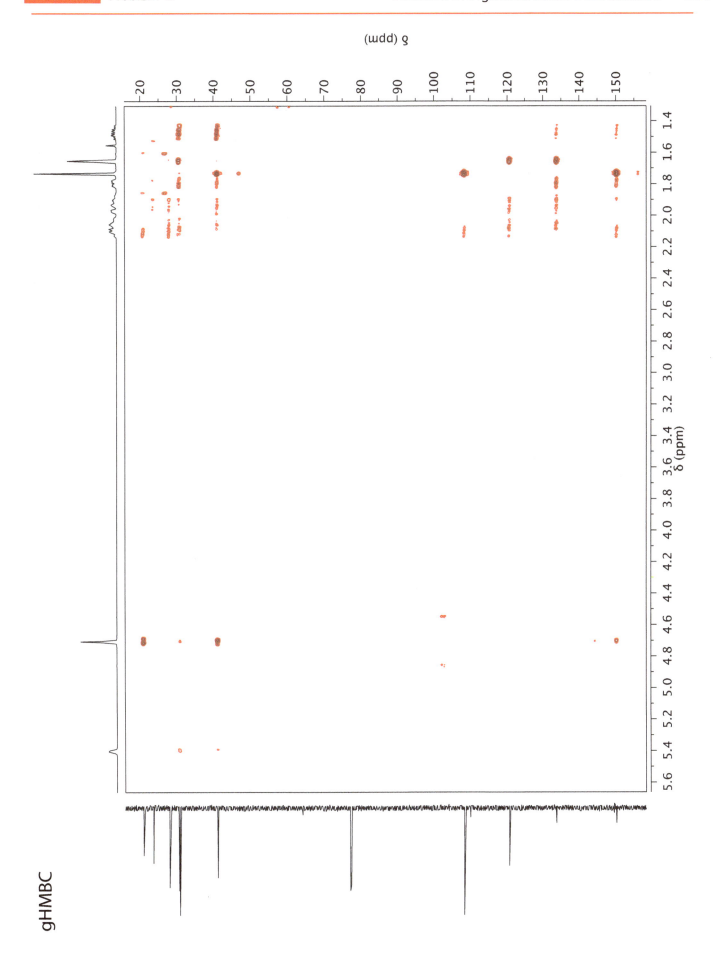

gHMBC

gHMBC

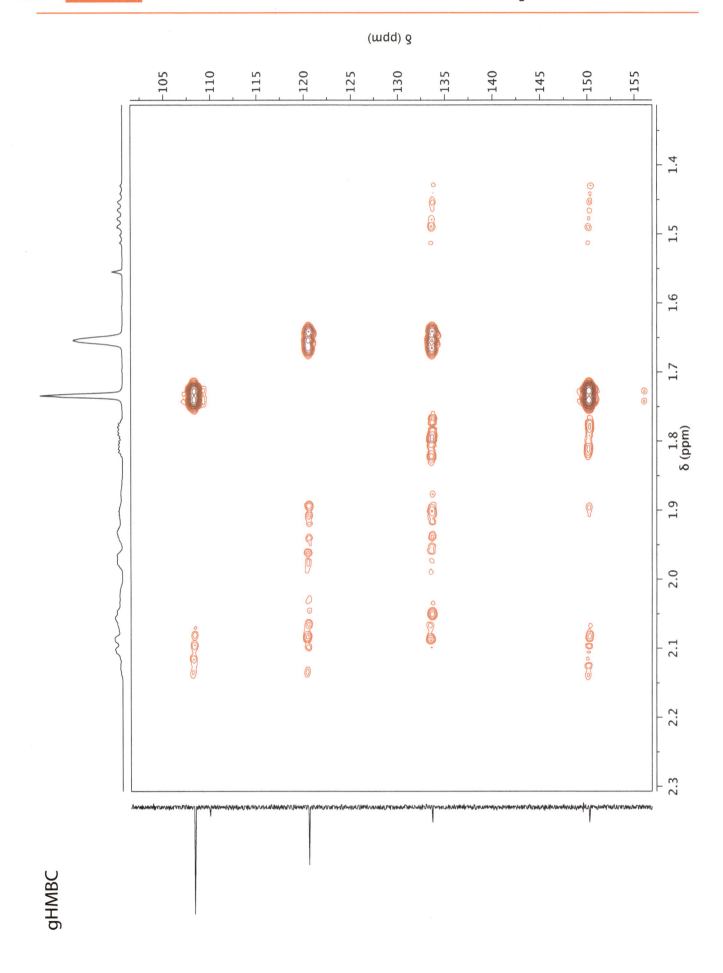

SECTION 2 Problem 47

Problems in Organic Structure Determination

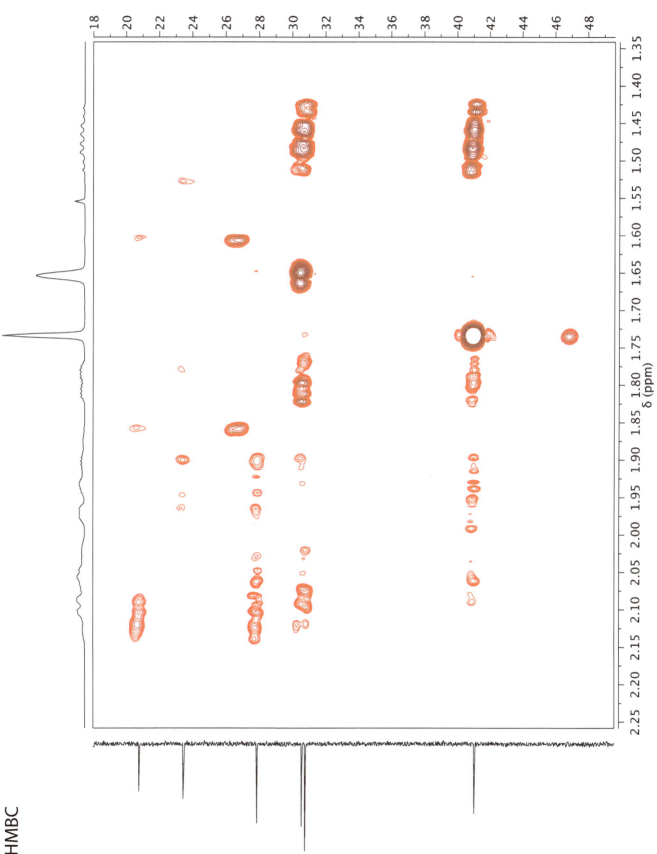

gHMBC

Problem 48

Using the ^{1}H, ^{13}C, gCOSY, gHSQC and gHMBC spectra, assign all proton and carbon resonances to the structure of longifolene below.
Spectra acquired in CDCl$_3$ at 500 MHz (^{1}H) or 125 MHz (^{13}C).

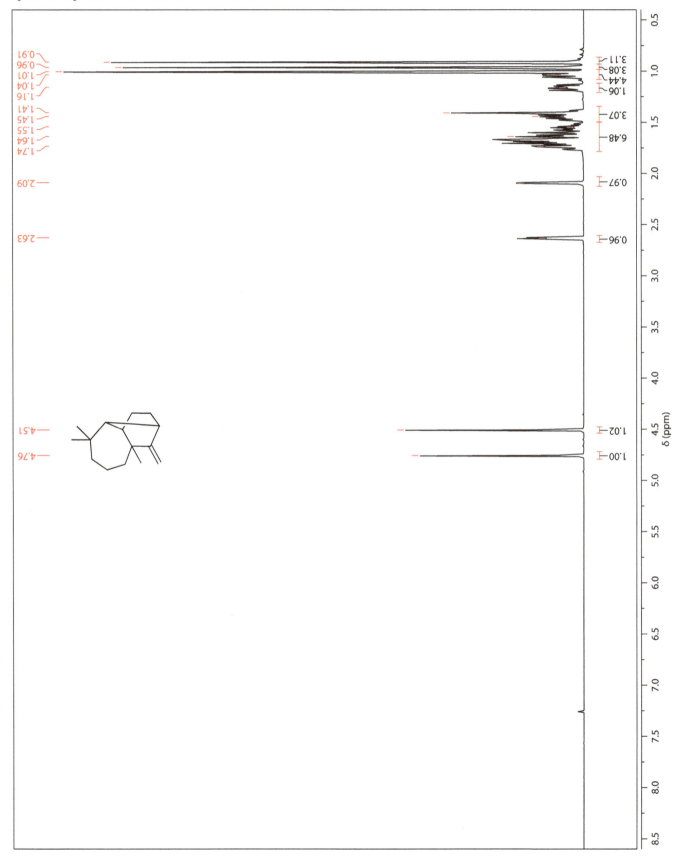

Problem 48

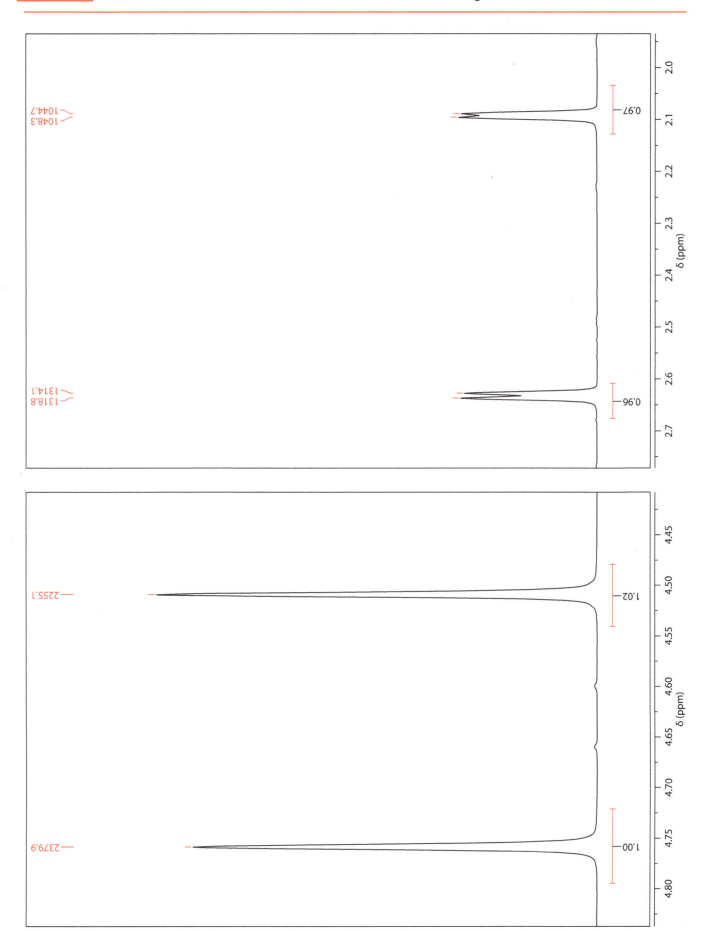

Problem 48

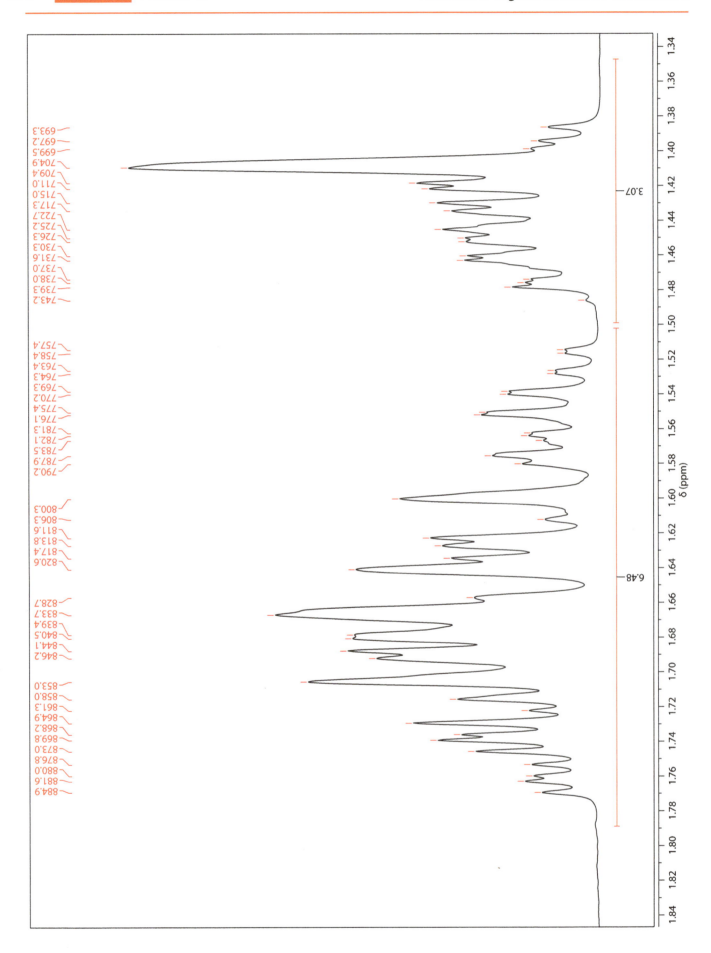

SECTION 2 Problem 48

Problems in Organic Structure Determination

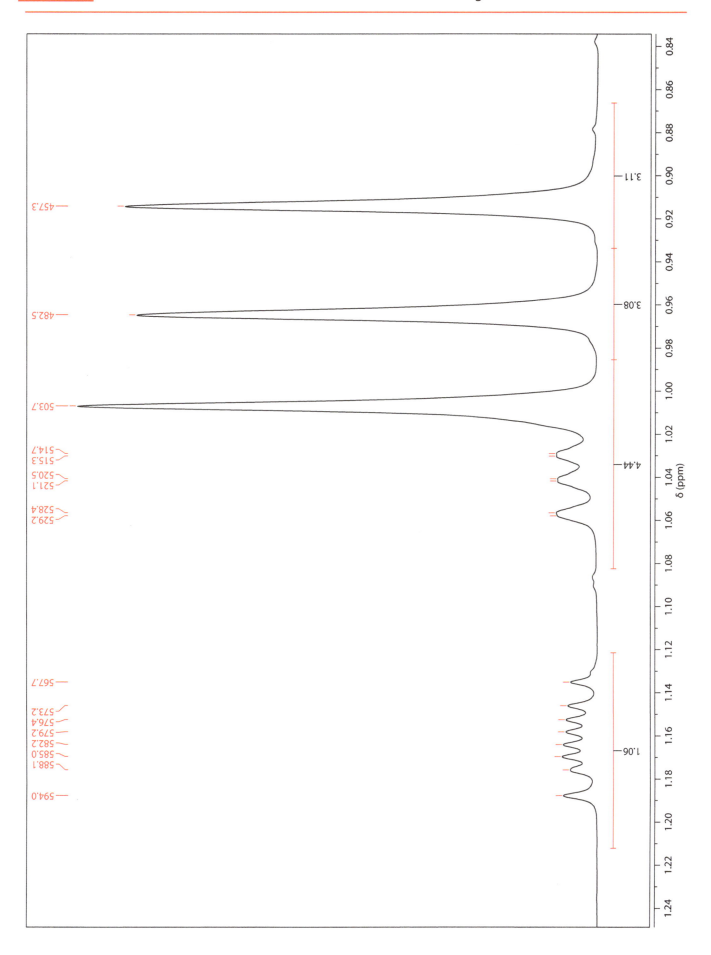

Problem 48

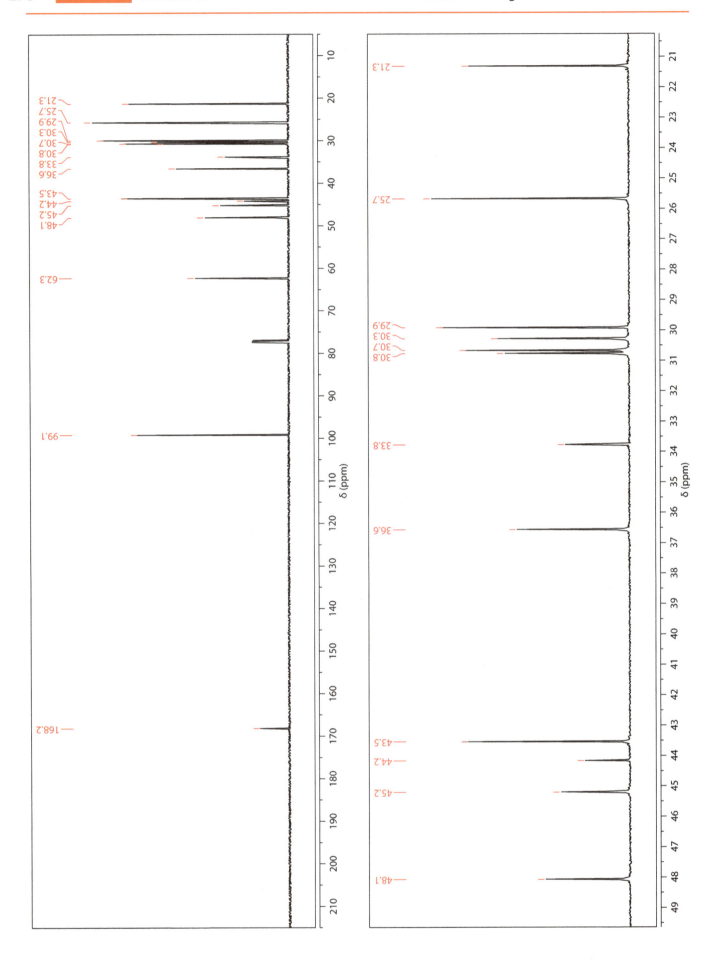

SECTION 2 Problem 48

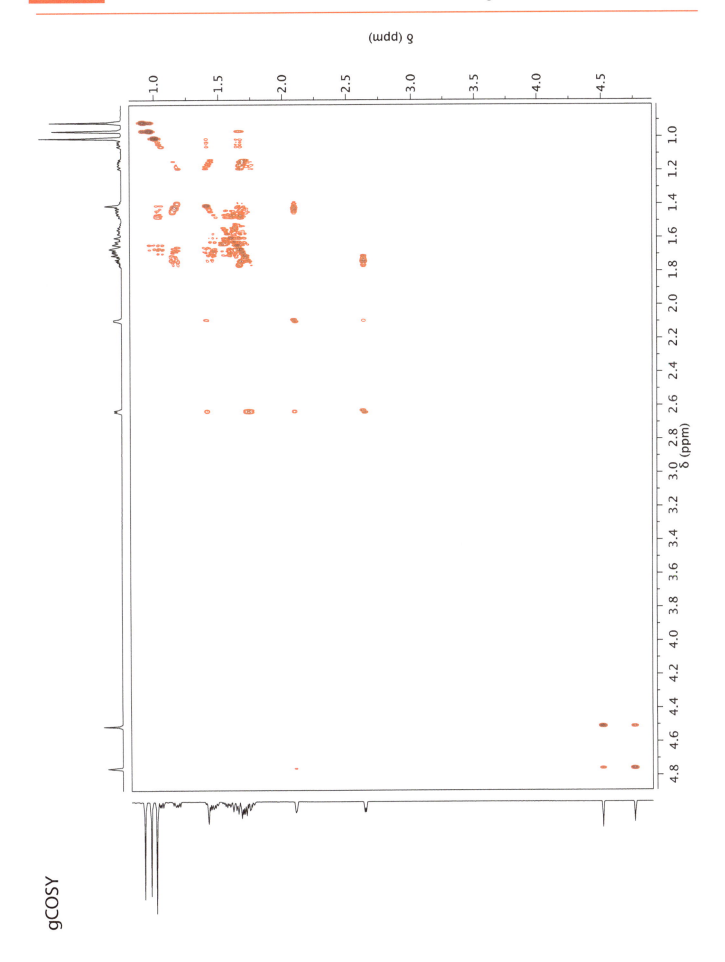

gCOSY

Problem 48

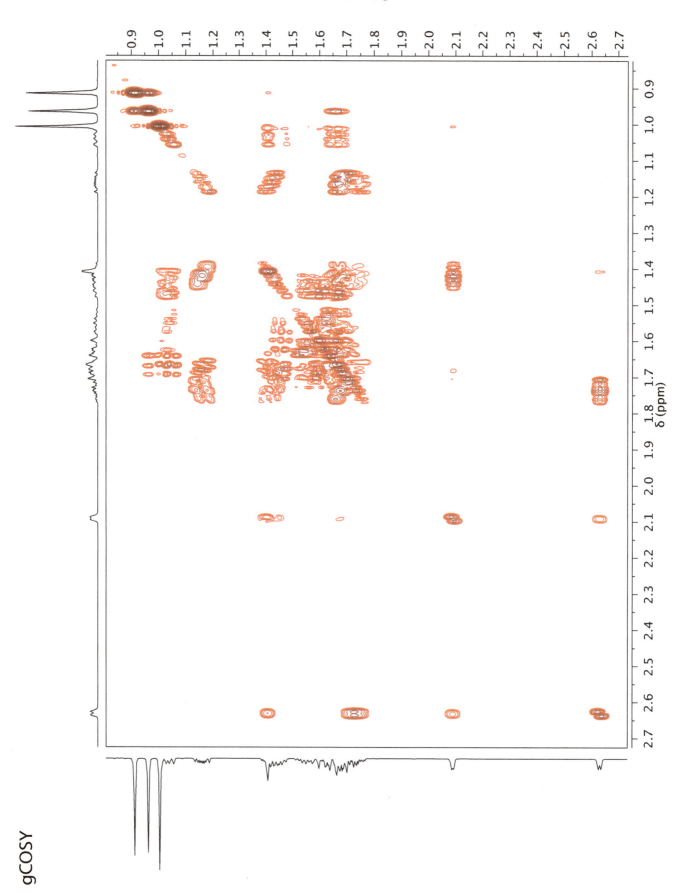

gCOSY

Problem 48

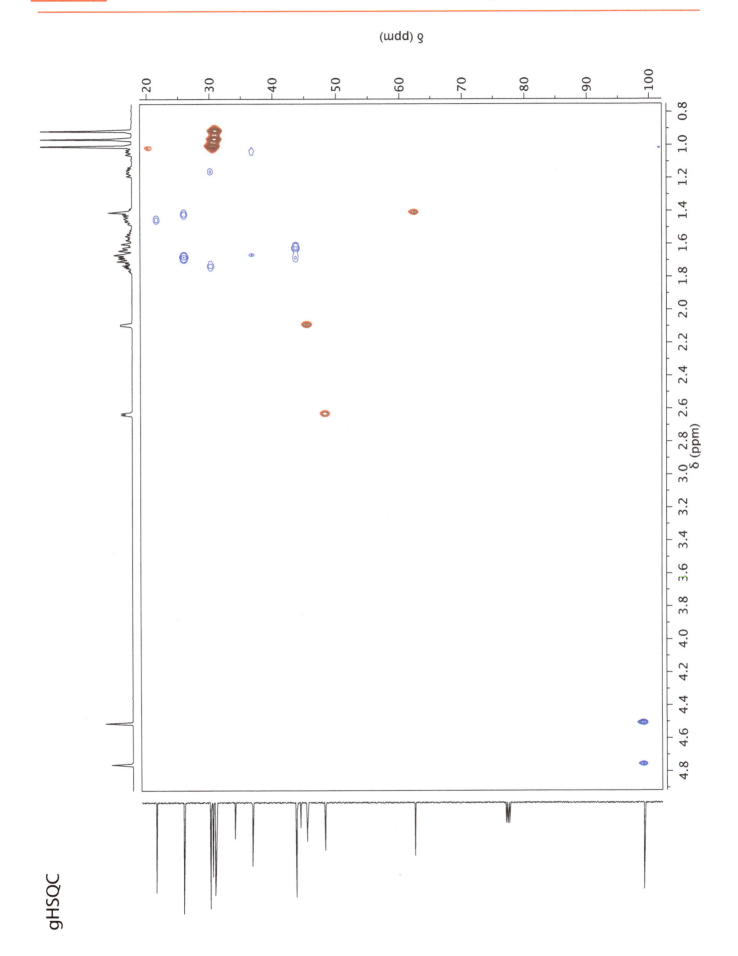

gHSQC

SECTION 2 Problem 48

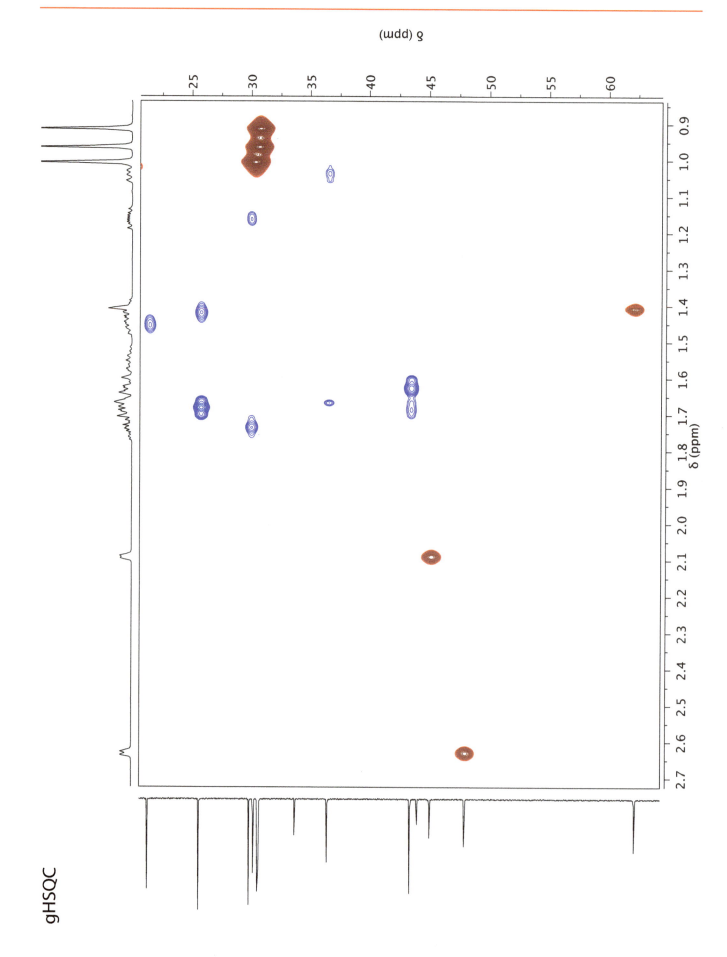

SECTION 2 Problem 48

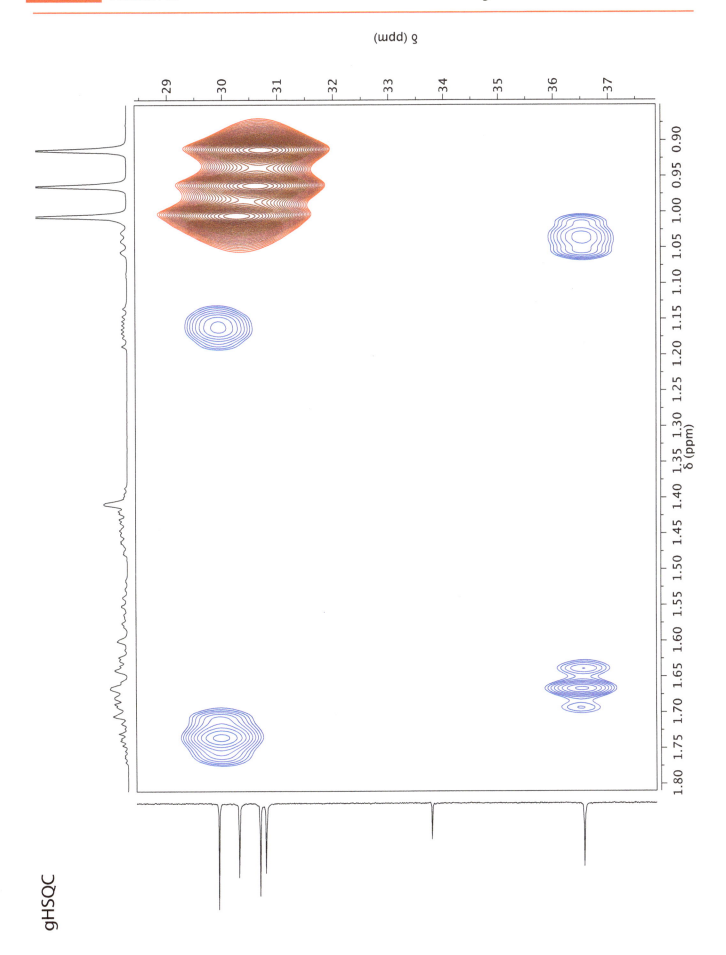

gHSQC

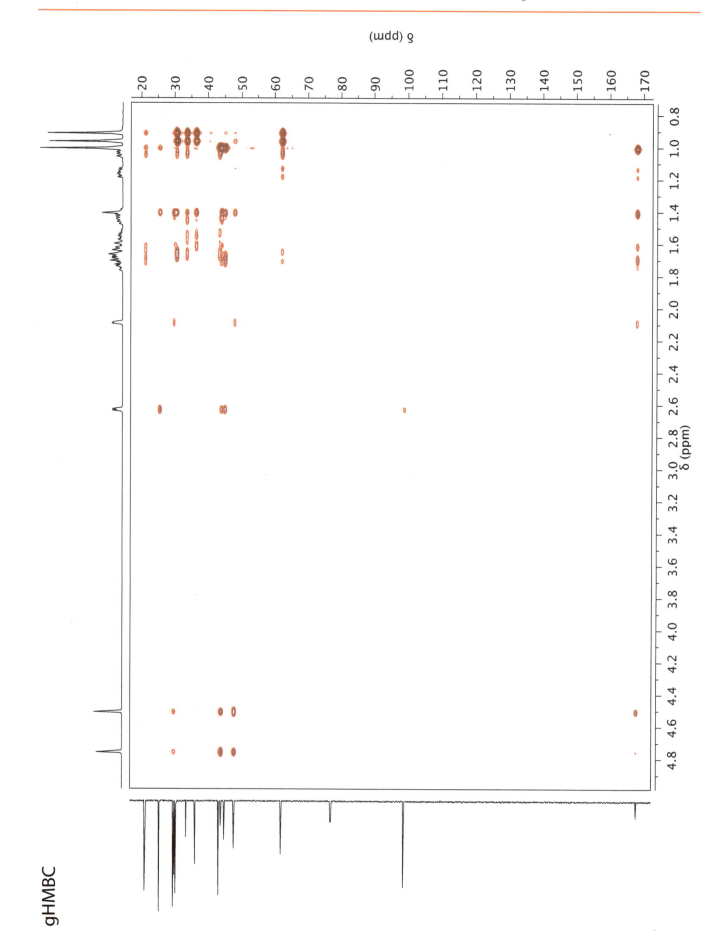

SECTION 2 Problem 48

gHMBC

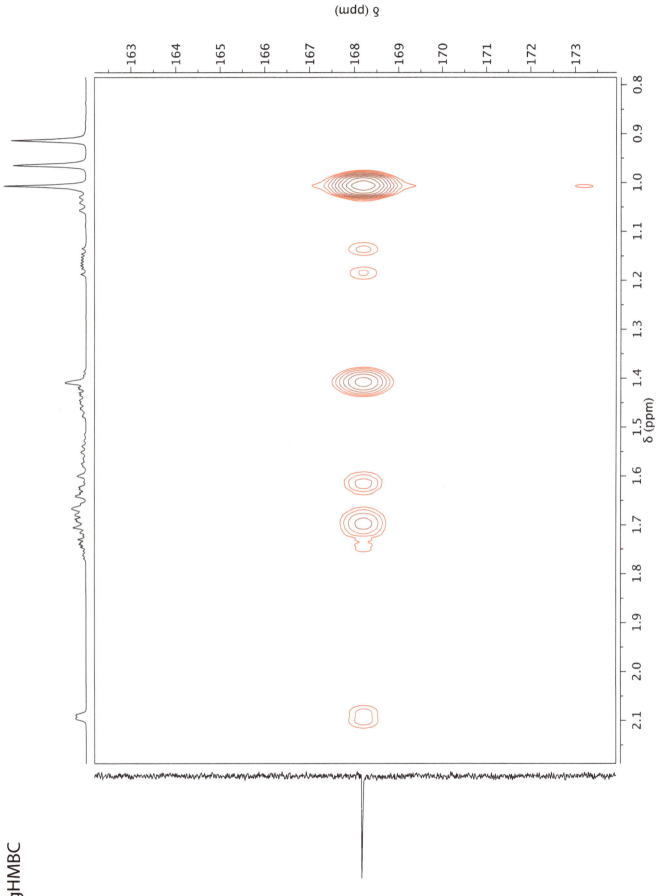

Problem 48

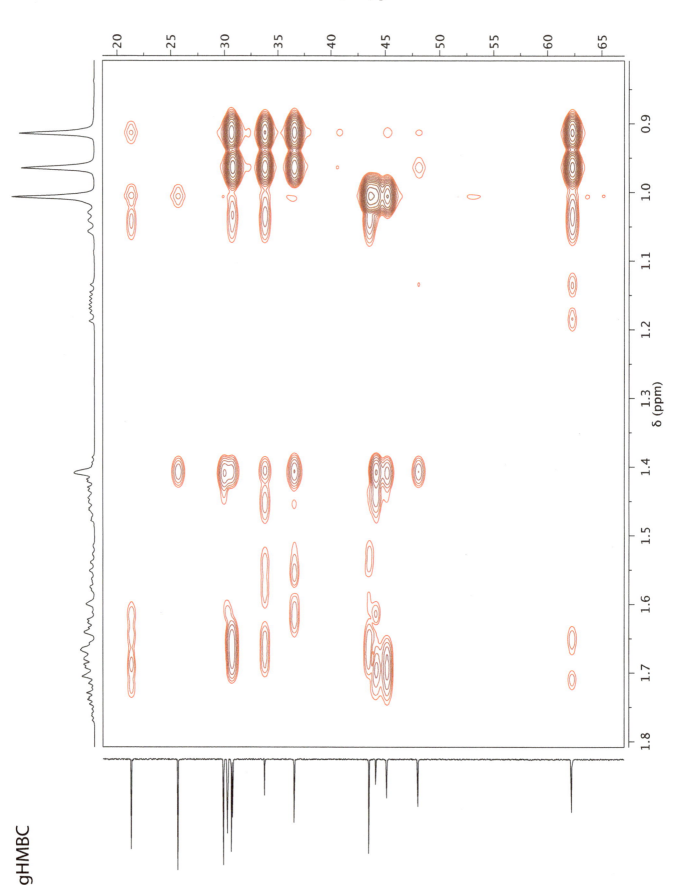

SECTION 2 Problem 48

gHMBC

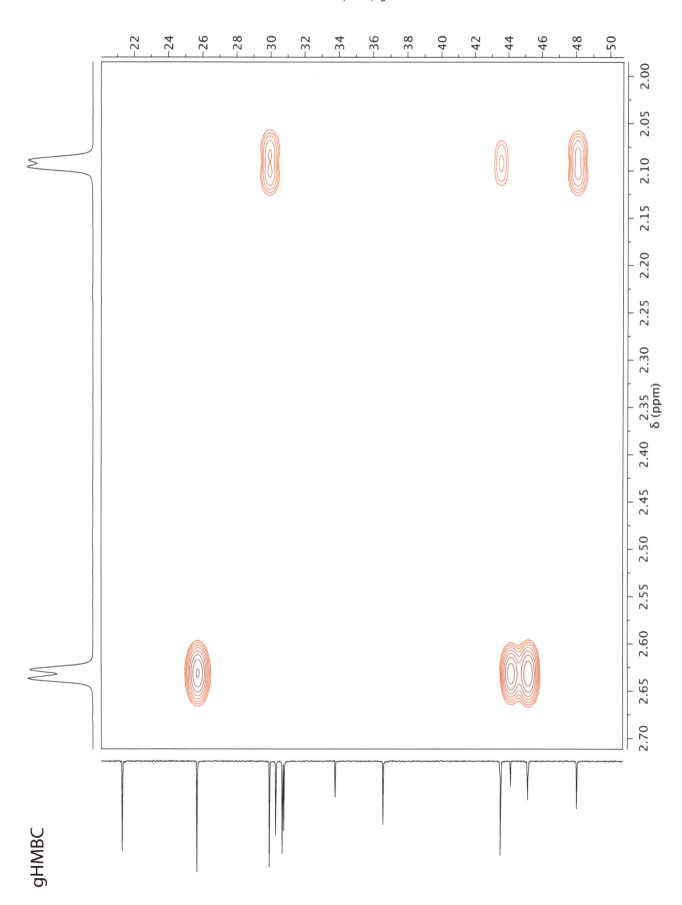

Problem 48

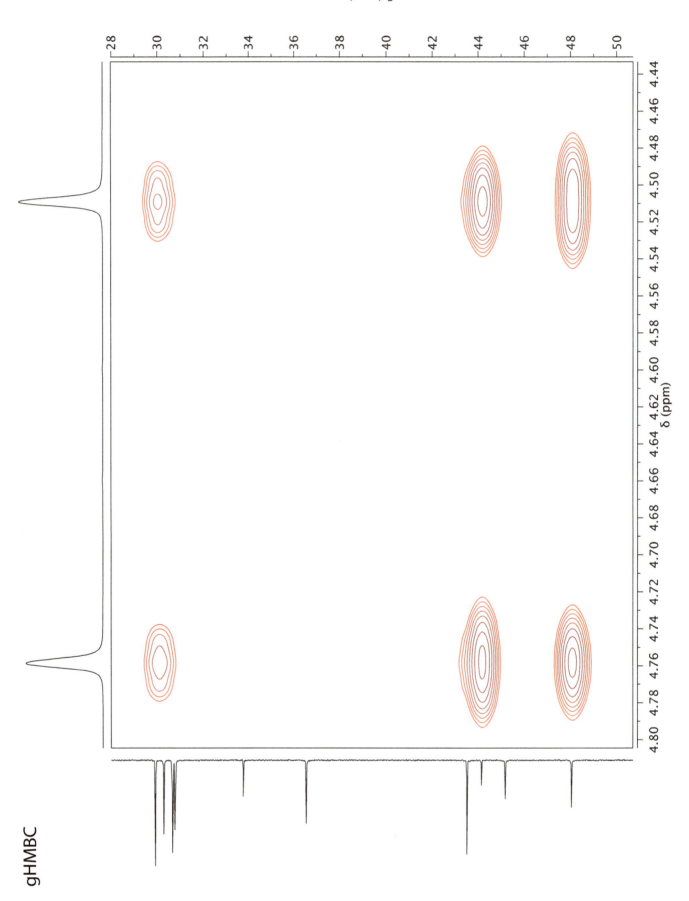

gHMBC

SECTION 2 Problem 49

Using the ¹H, ¹³C, gCOSY, gHSQC, gHMBC and gHSQC-TOCSY spectra, assign all proton and carbon resonances to the structure of D-lactose below.
Spectra acquired in $(CD_3)_2SO$ at 500 MHz (¹H) or 125 MHz (¹³C).

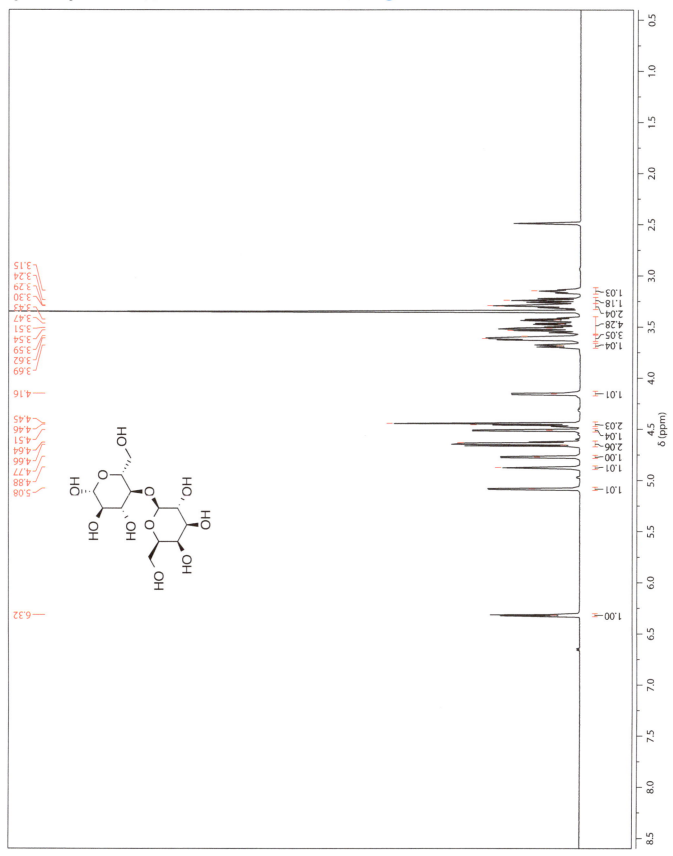

Problem 49

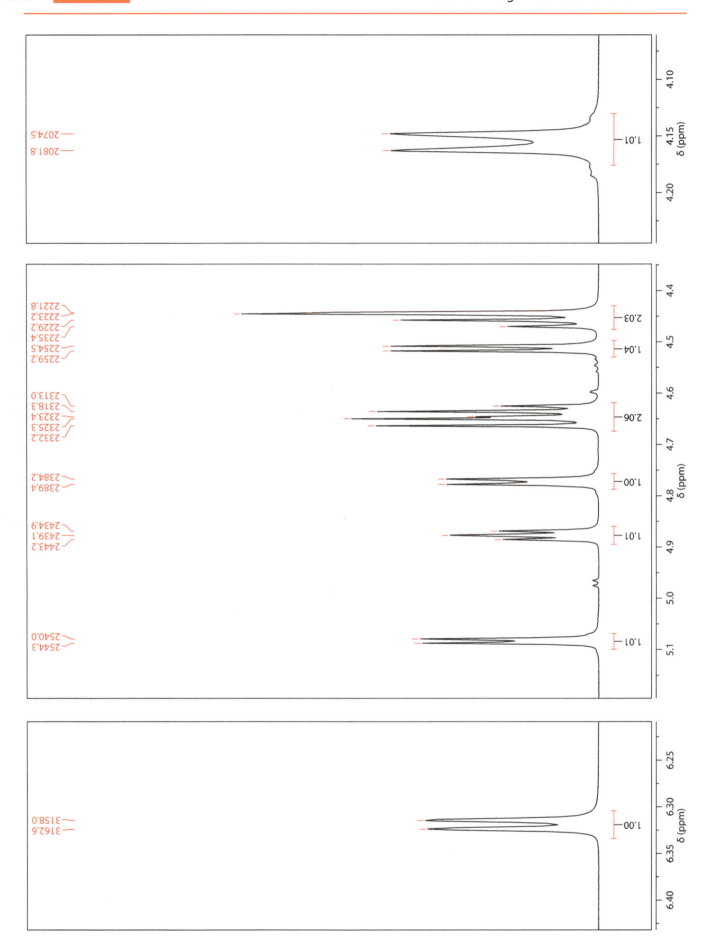

Problem 49

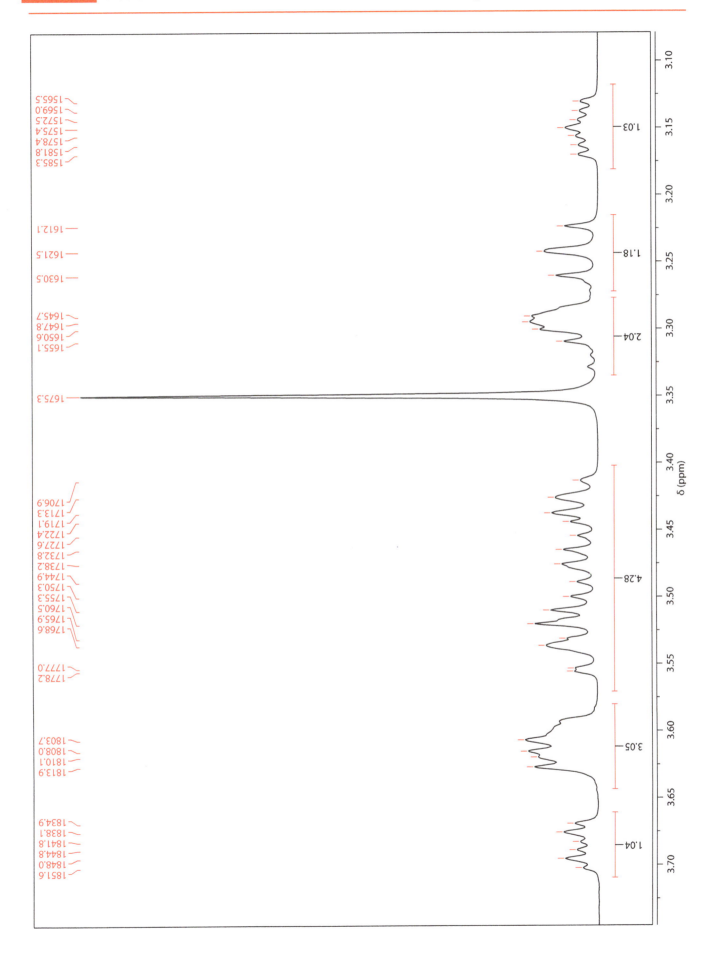

SECTION 2 Problem 49

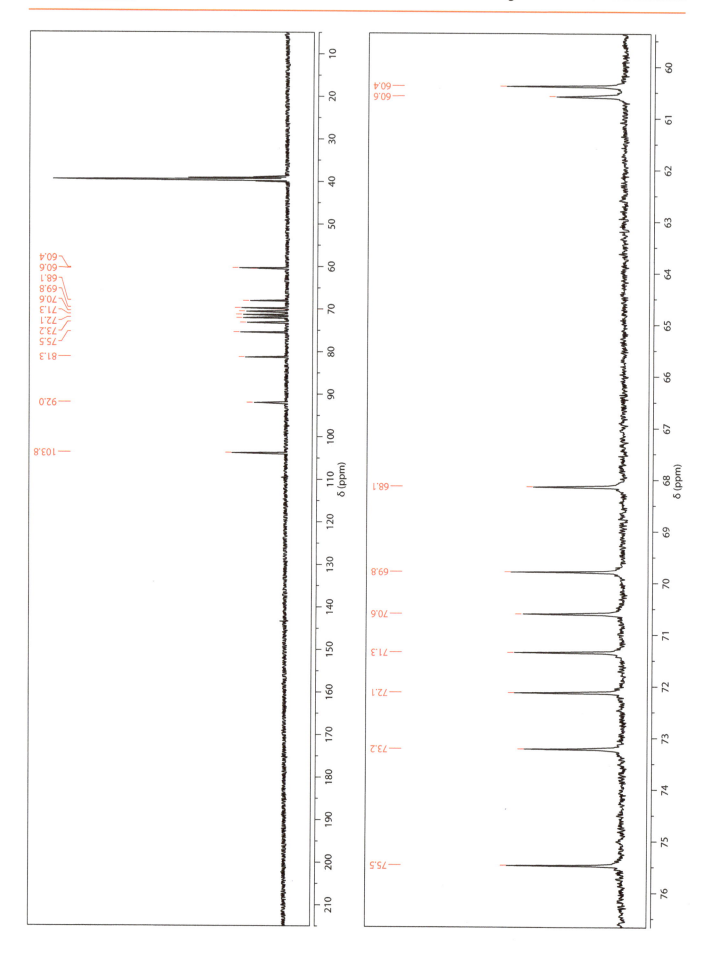

SECTION 2 Problem 49

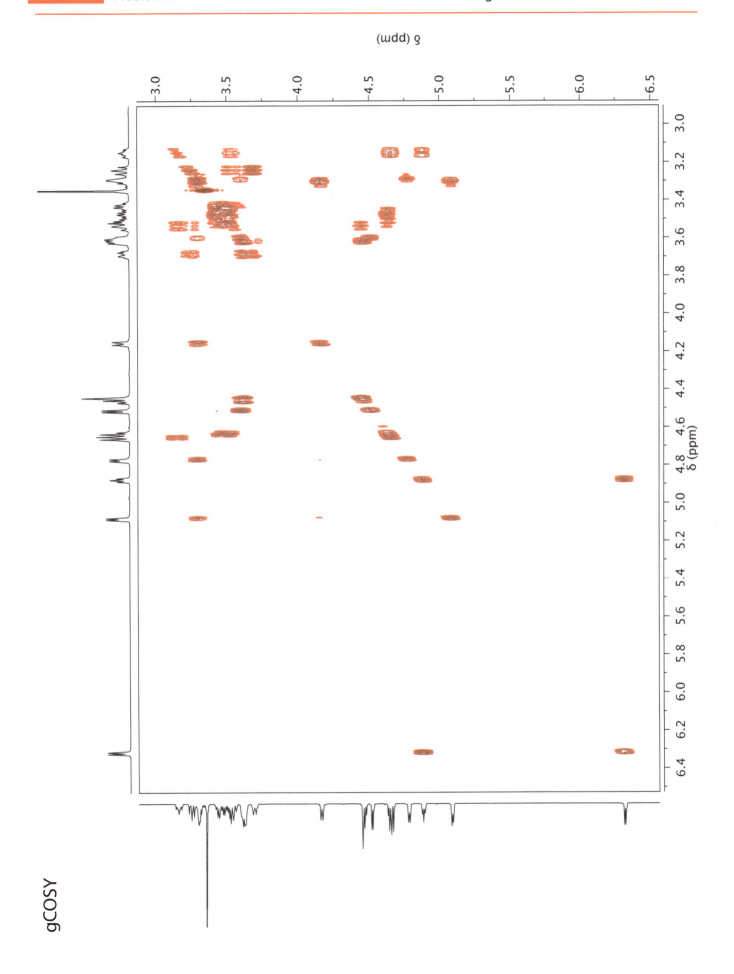

gCOSY

Problem 49

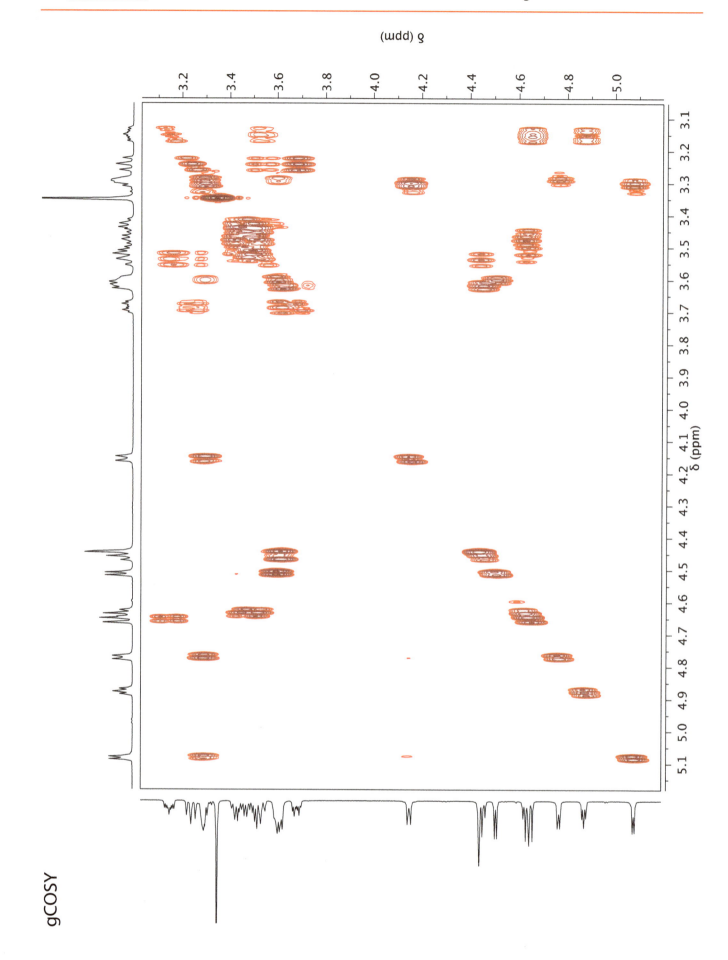

gCOSY

SECTION 2 Problem 49

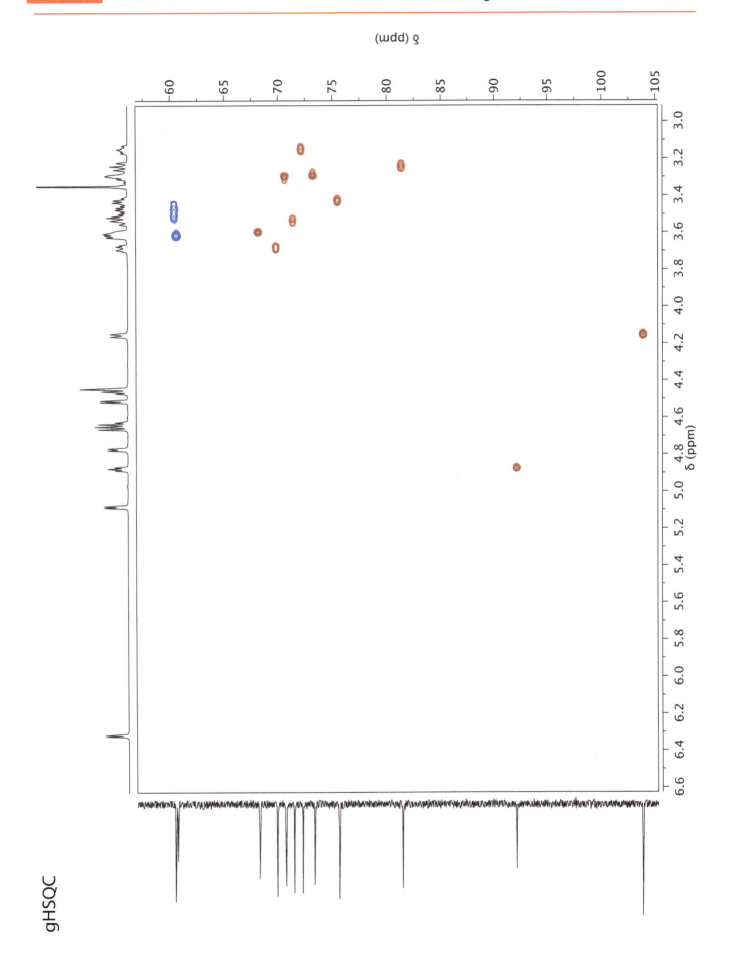

gHSQC

Problem 49

gHMBC

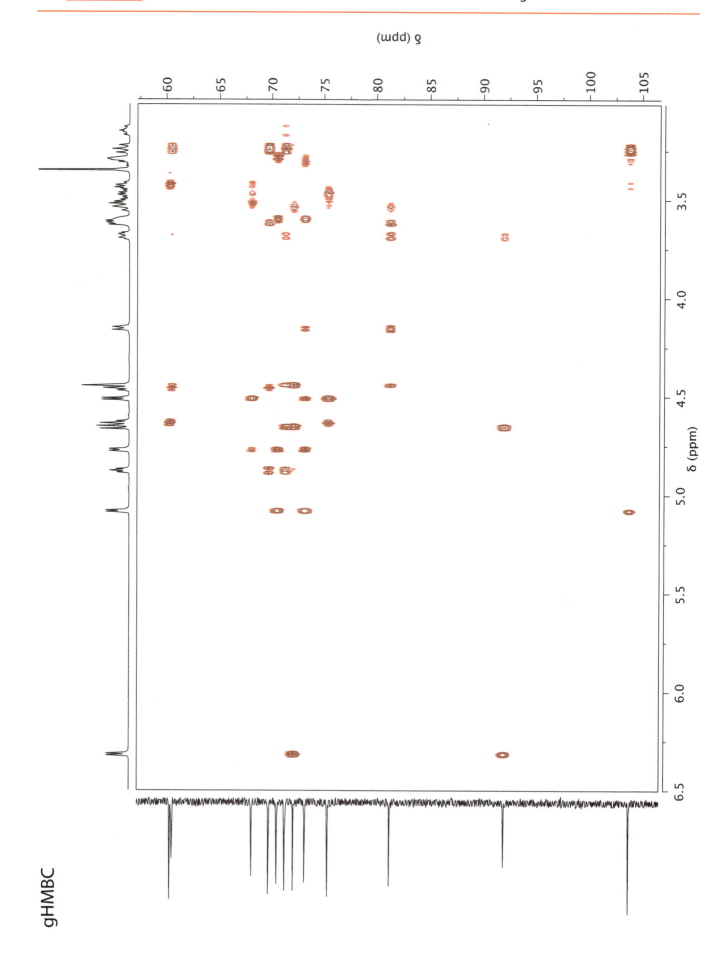

SECTION 2 Problem 49

Problems in Organic Structure Determination

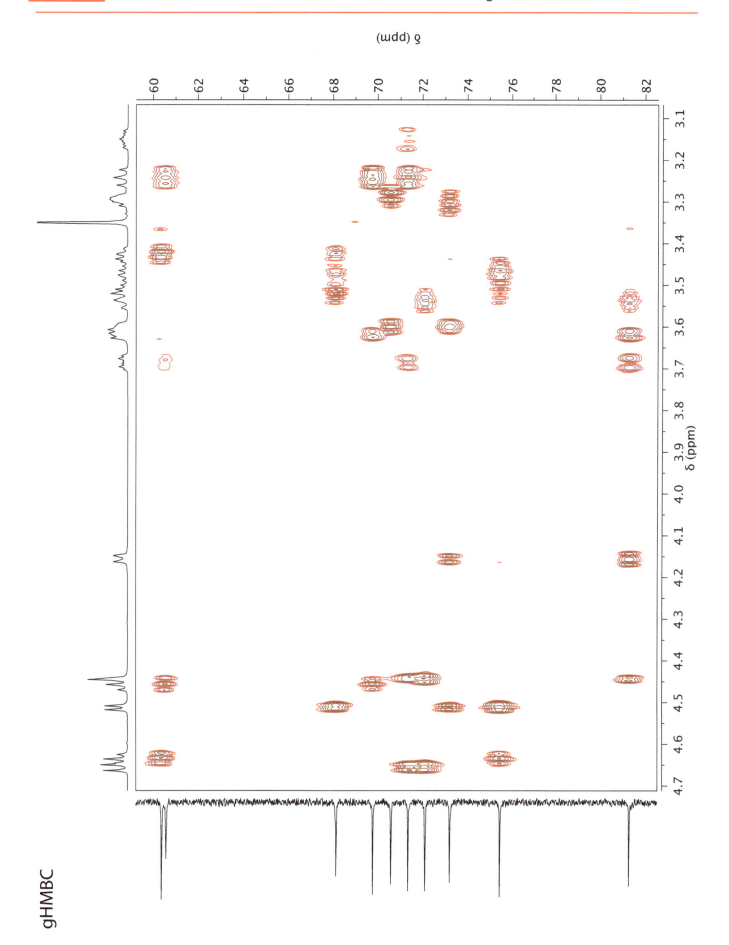

gHMBC

Problem 49

gHSQC-TOCSY

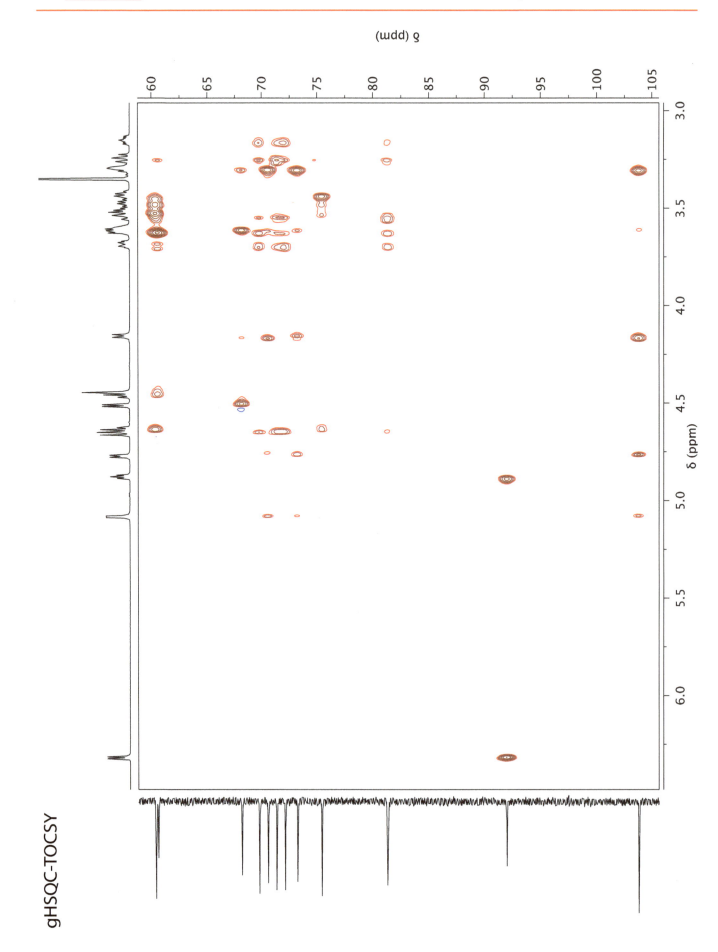

SECTION 2 Problem 50 — Problems in Organic Structure Determination

Which of the following compounds is the correct structure for the spectra below?
NOTE: Some ¹³C peak picking provided in Hz, to allow extraction of C-F coupling constants.
Spectra acquired in $(CD_3)_2SO$ with trifluoroacetic acid (TFA) at 500 MHz (¹H) or 125 MHz (¹³C).

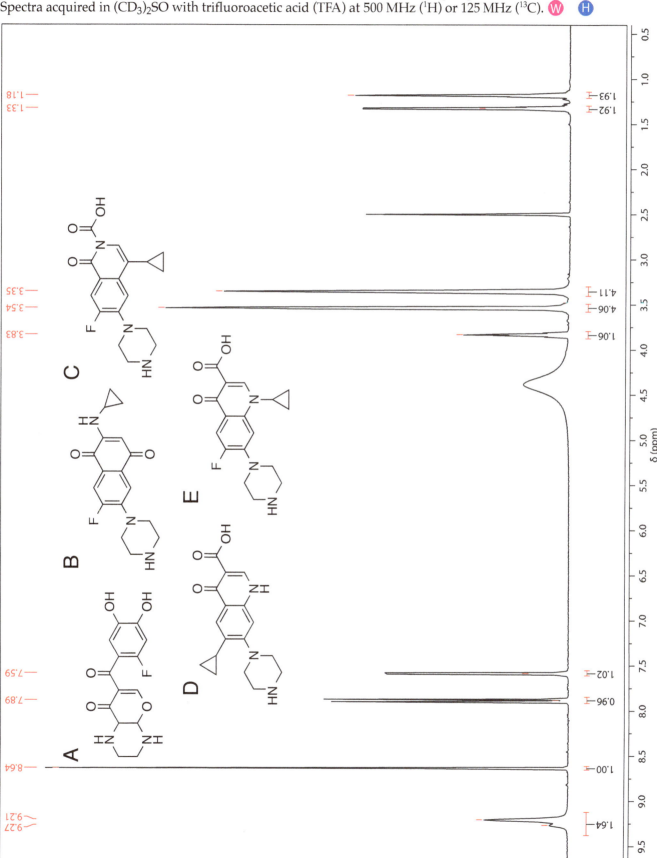

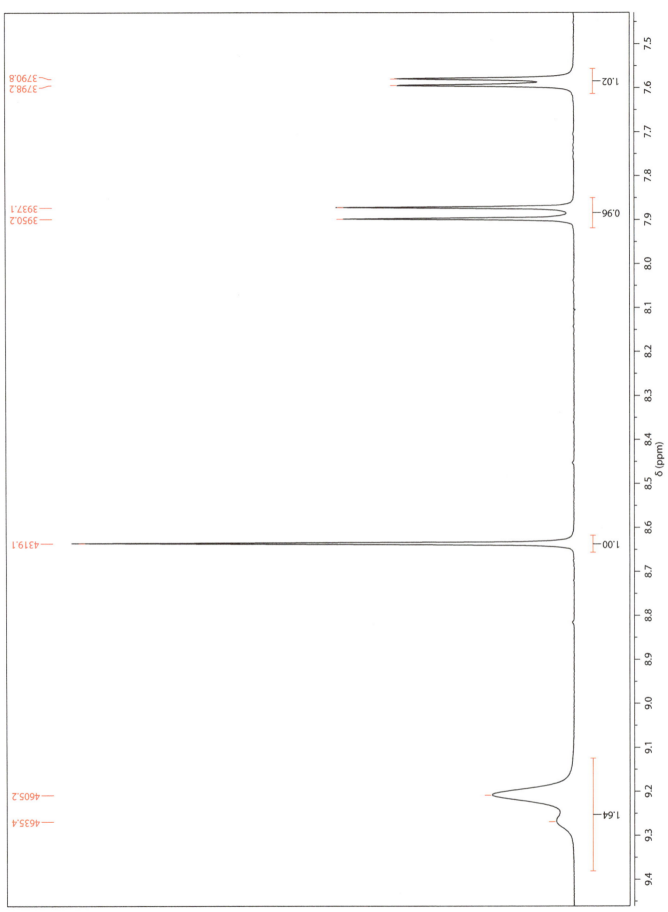

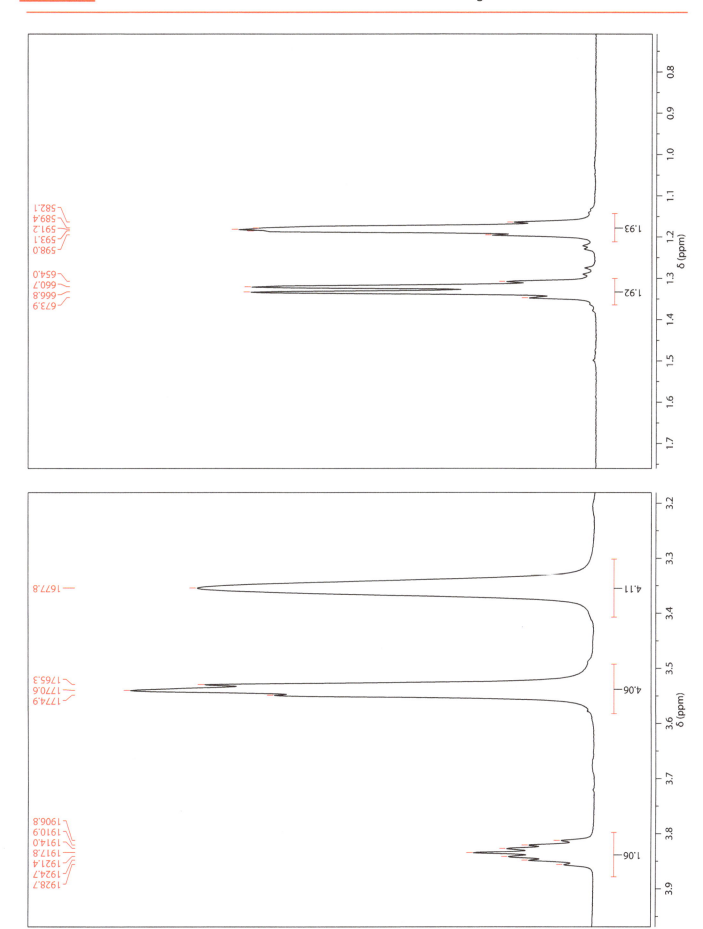

Problem 50

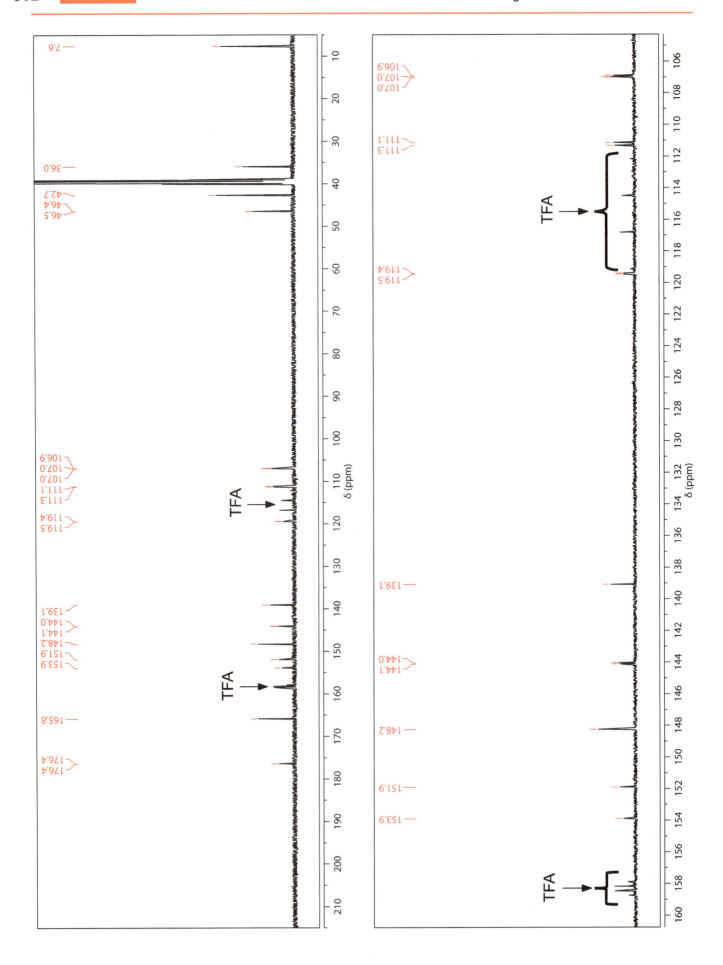

SECTION 2 Problem 50

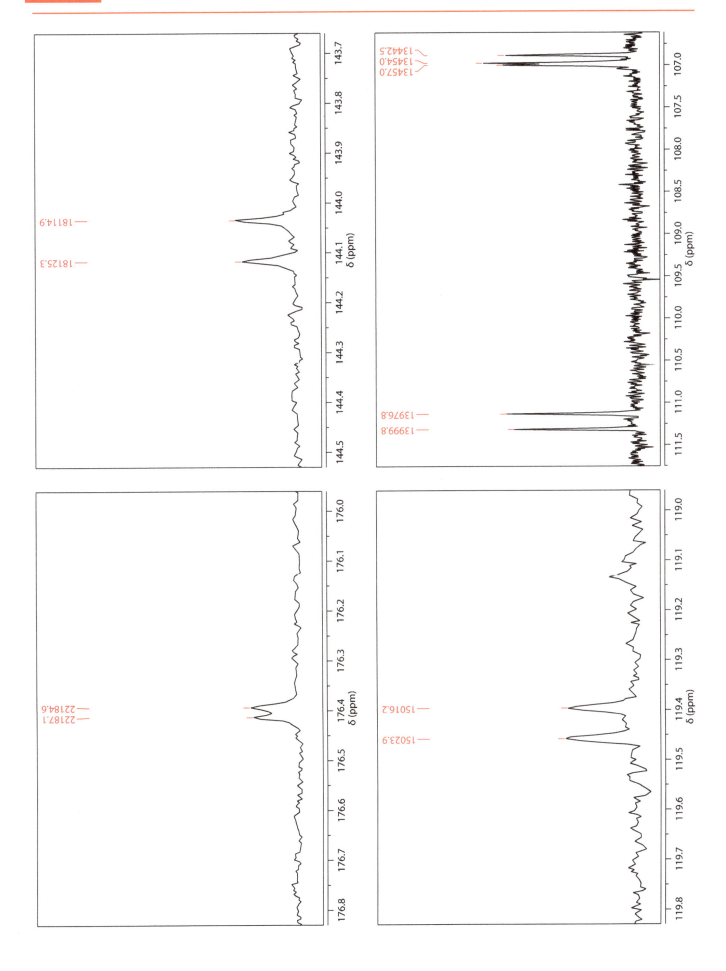

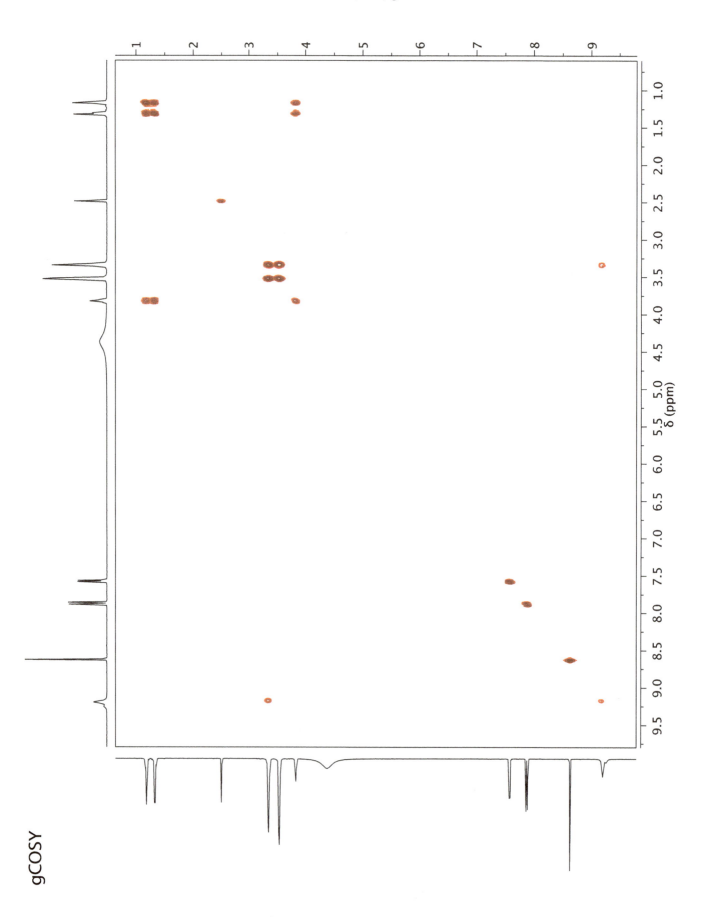

SECTION 2 Problem 50

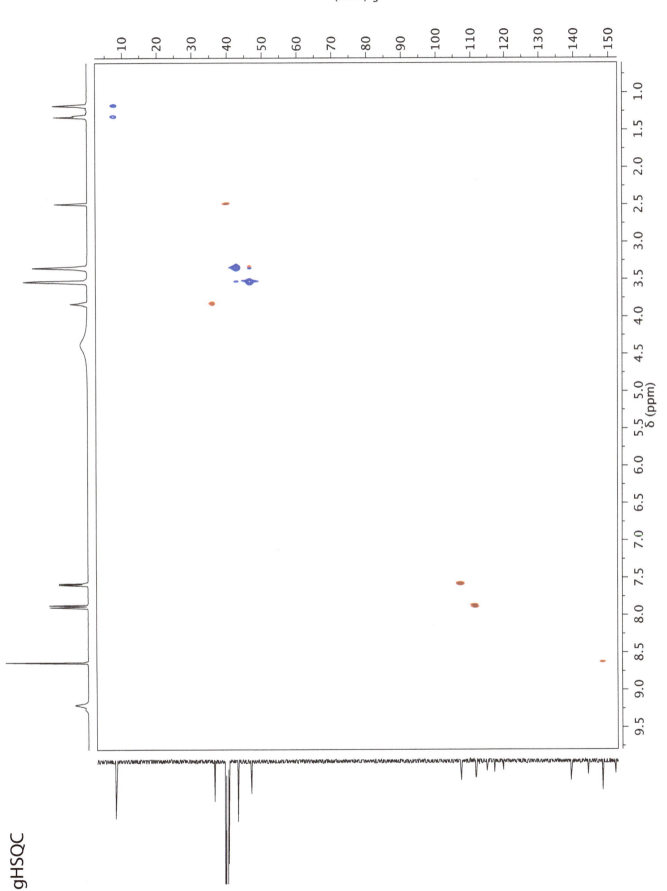

gHSQC

Problem 50

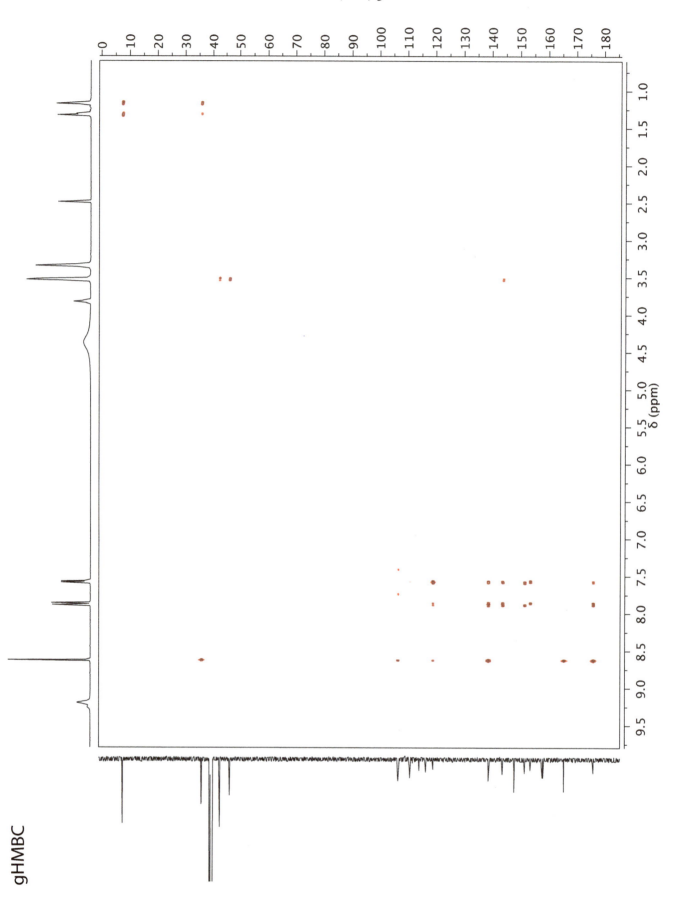

gHMBC

SECTION 2 Problem 50

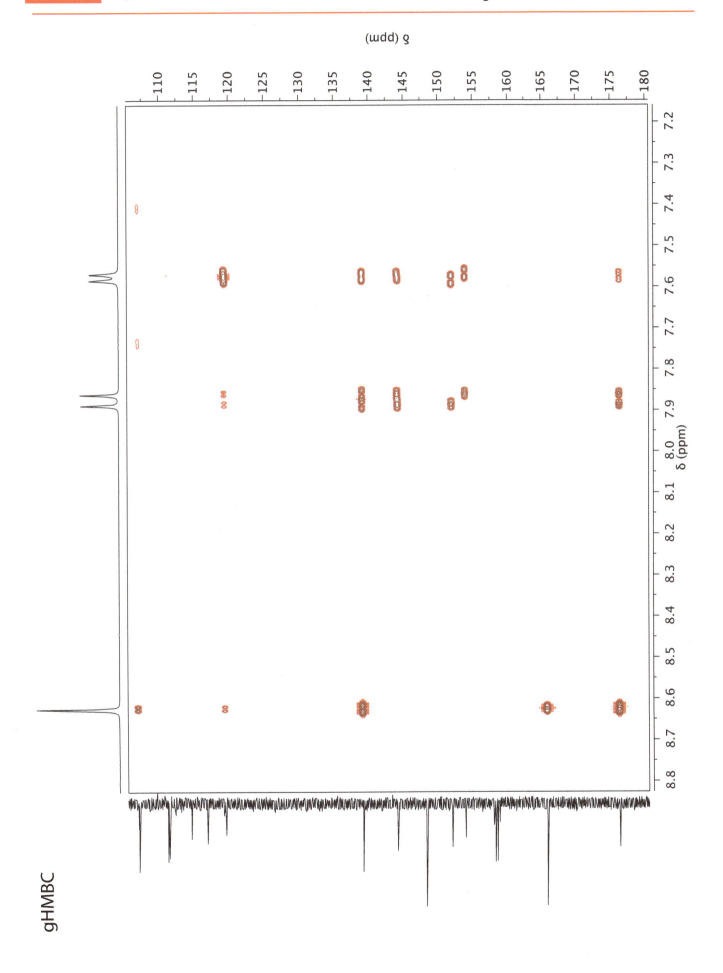

gHMBC

Section 3

Section 3 is the first introduction to *de novo* elucidation of structures using the provided molecular formula and the given NMR spectral data. The complexity of the small molecules increases throughout the section. There is a difference between full structural assignments and the ability to determine the structure of an unknown. In this section you may come across examples where it is not possible to assign all of the ^{1}H and ^{13}C chemical shifts, but it is still possible to determine the correct structure.

LEARNING OBJECTIVES

- Integrate data from all of the provided NMR experiments and molecular formula to assign chemical structures.
- Use NMR data to define compound sub-structures, and assemble these together into a final solution.

EXPERIMENTS INCLUDED

^{1}H, ^{13}C, DEPT-135, gCOSY, gHSQC, gHMBC and 1D NOESY.

Note: All gHSQC experiments are multiplicity edited with cross peaks deriving from CH$_2$ groups depicted in blue, and cross peaks deriving from CH and CH$_3$ groups depicted in red.

TYPES OF MOLECULES

This section contains most of the basic functional groups found in small molecules; alcohols, amines, halides, olefins, alkynes, aromatics and heteroaromatics. This section also contains a number of molecules that have one or more chiral centers.

STRATEGIES FOR SUCCESS

The molecular formula for each compound is provided. A helpful strategy for most of these problems will be to determine the degrees of unsaturation using the following equation:

$$DOU = \frac{(2 \times \#C + 2) - \#H - \#X + \#N}{2}$$

where X= Cl, Br, F, I. Furthermore, a number of problems in this section contain hints that may help determine functional groups that do not have NMR signatures.

As the complexity of the problems increase, it is critical to maintain rigorous record keeping, and to identify all of the proton and carbon atoms in the spectra before attempting to assemble the molecule. We highly recommend the idea of building a table as you begin this process. It is also valuable to look at all of the data to identify features of a molecule (eg. sp^2 carbons, aromatic protons) and elements of symmetry before trying to build fragments.

LEGEND

Spectrum annotations:

i = impurity.

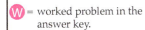 = worked problem in the answer key.

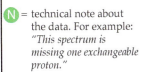 = technical note about the data. For example: *"This spectrum is missing one exchangeable proton."*

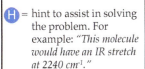 = hint to assist in solving the problem. For example: *"This molecule would have an IR stretch at 2240 cm^{-1}."*

Determine the structure of the unknown compound below with molecular formula C₆H₁₂O.

Spectra acquired in CDCl₃ at 500 MHz (¹H) or 125 MHz (¹³C).

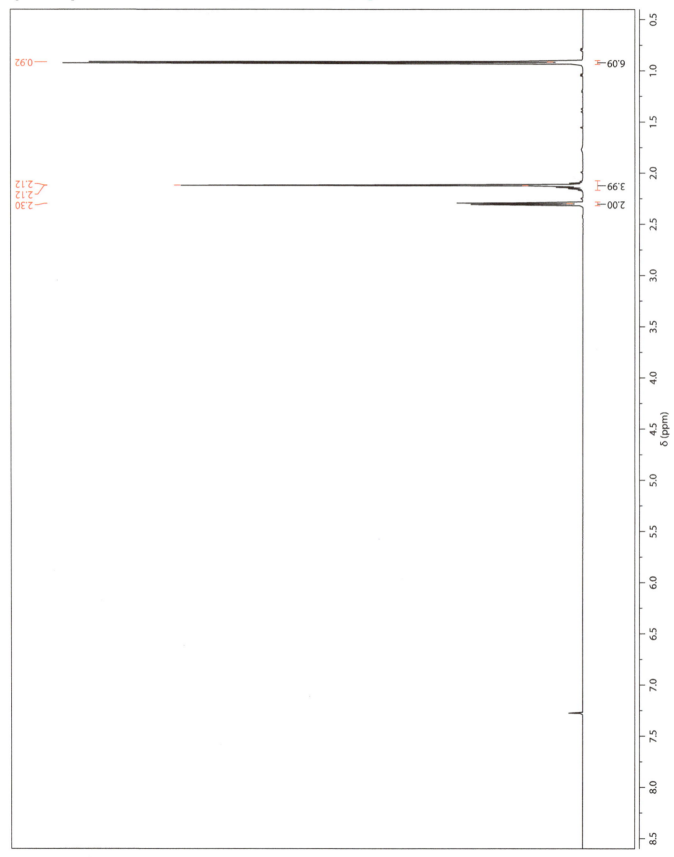

SECTION 3 Problem 51

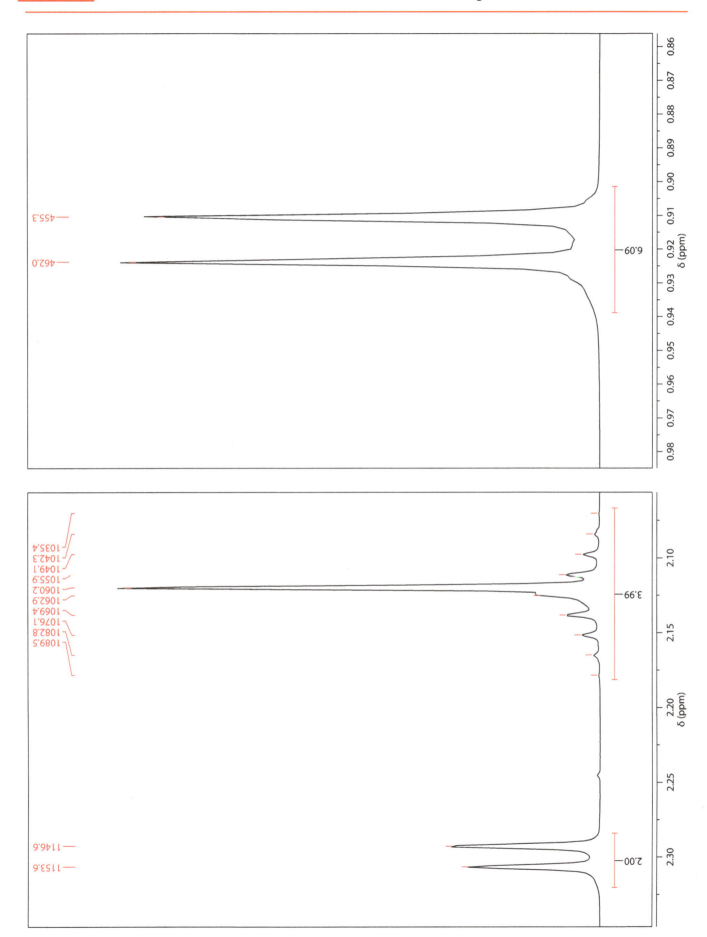

Problem 51

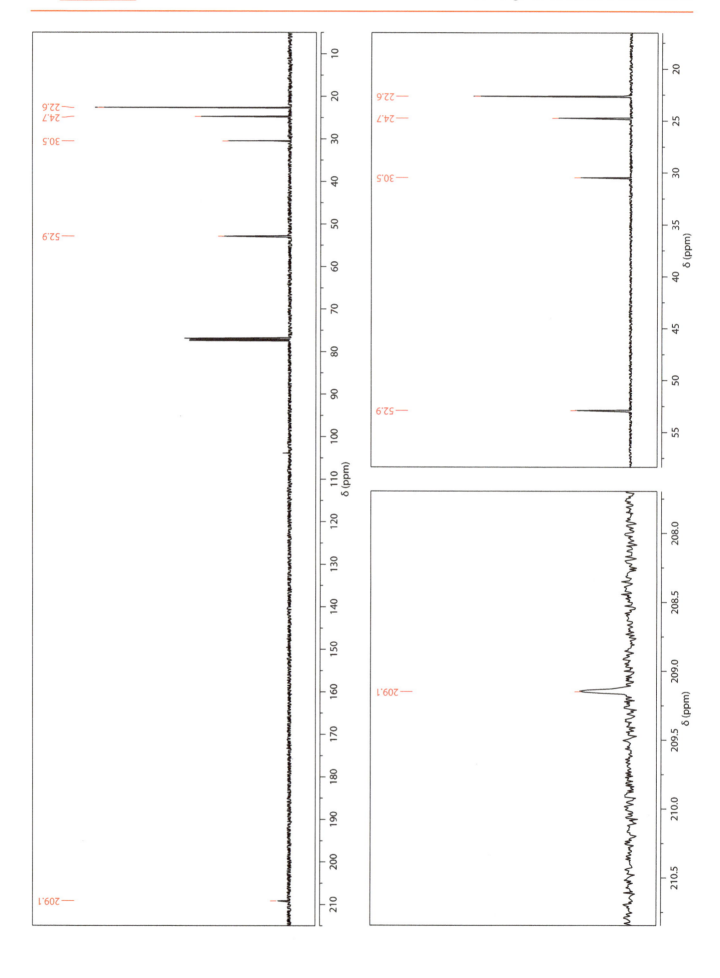

SECTION 3 Problem 52

Determine the structure of the unknown compound below with molecular formula $C_5H_8O_3$.

Spectra acquired in $CDCl_3$ at 500 MHz (^{1}H) or 125 MHz (^{13}C).

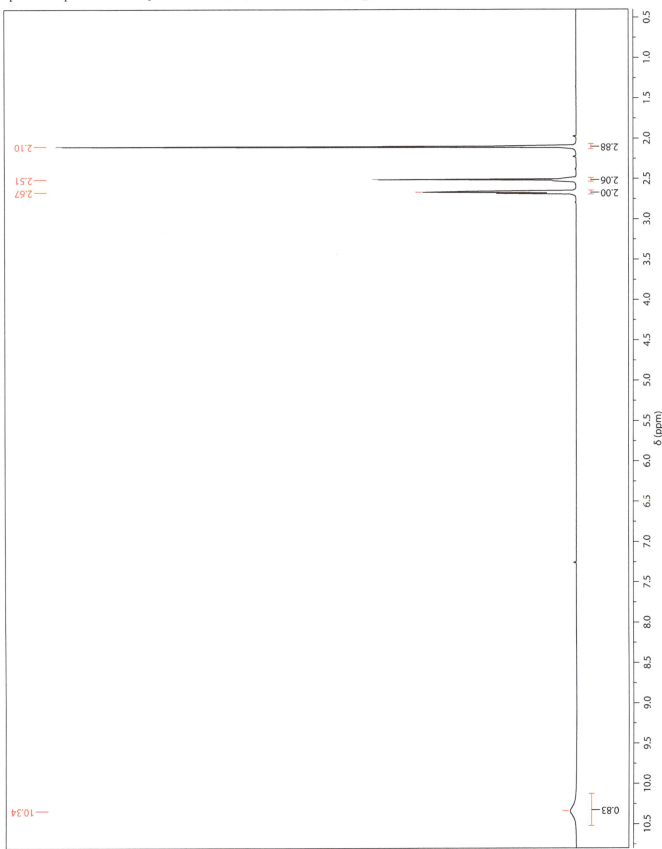

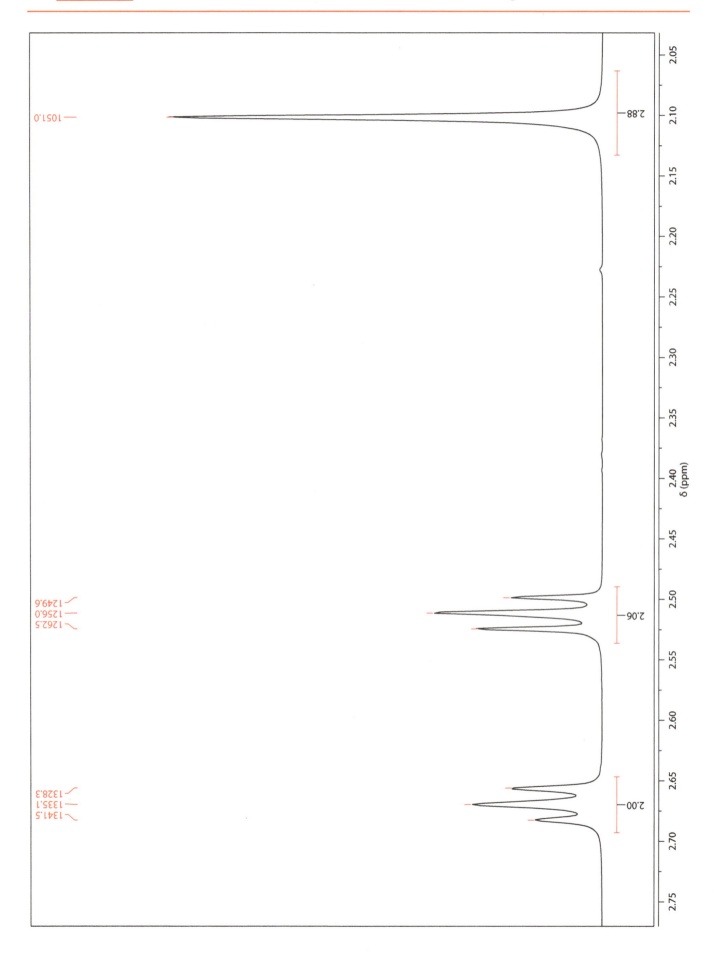

SECTION 3 Problem 52

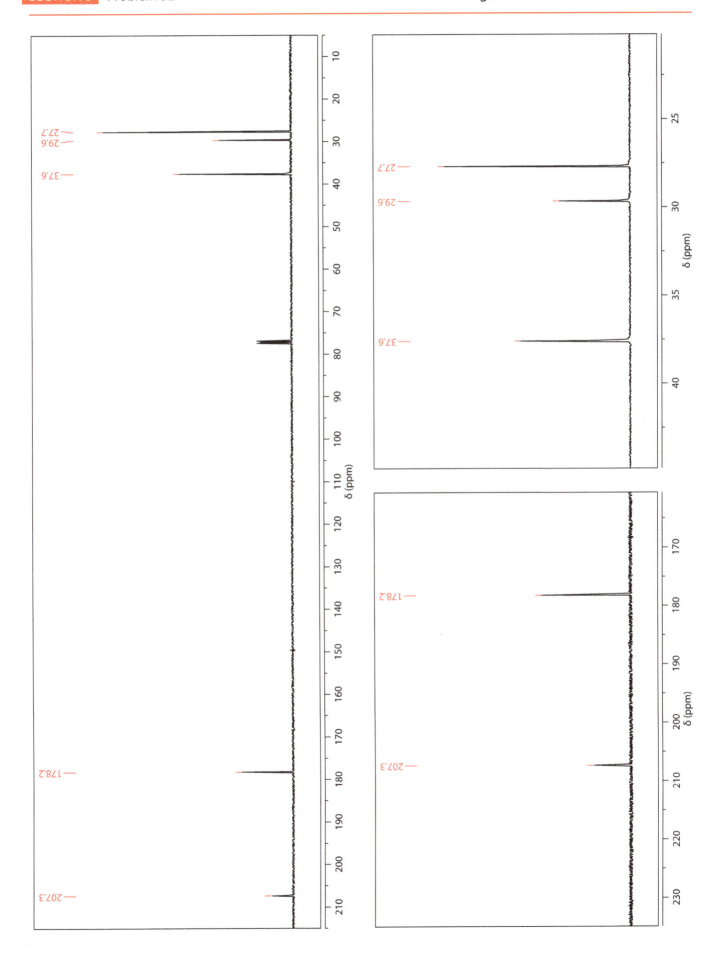

SECTION 3 Problem 53

Determine the structure of the unknown compound below with molecular formula $C_6H_{15}NO$.

Spectra acquired in $CDCl_3$ at 500 MHz (^{1}H) or 125 MHz (^{13}C).

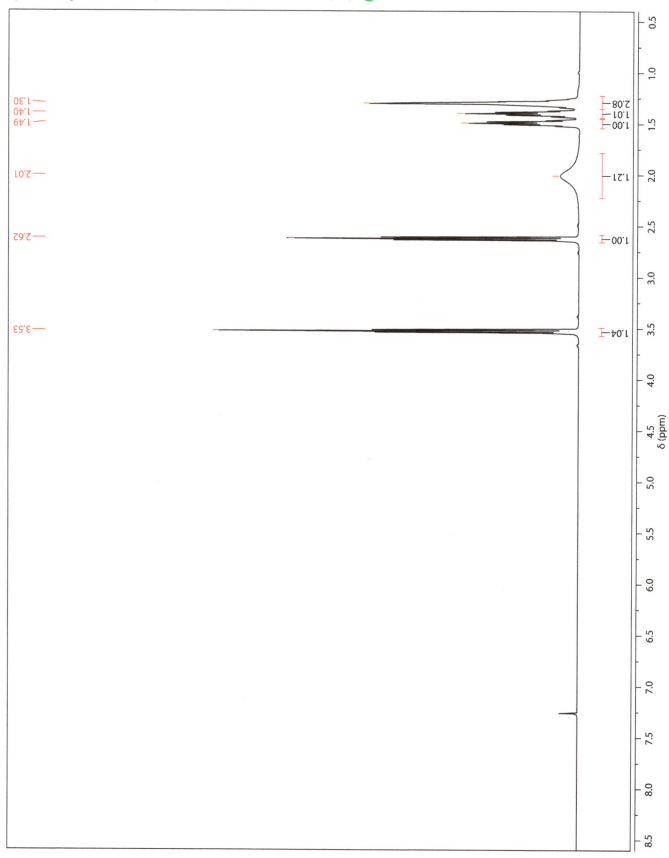

SECTION 3 Problem 53

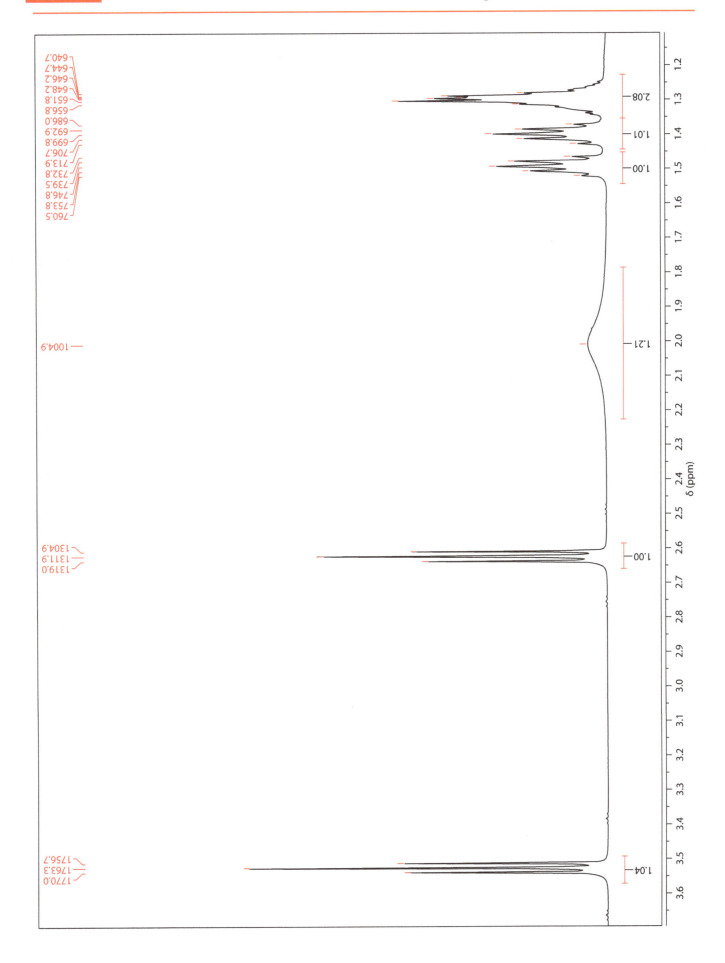

Problem 53

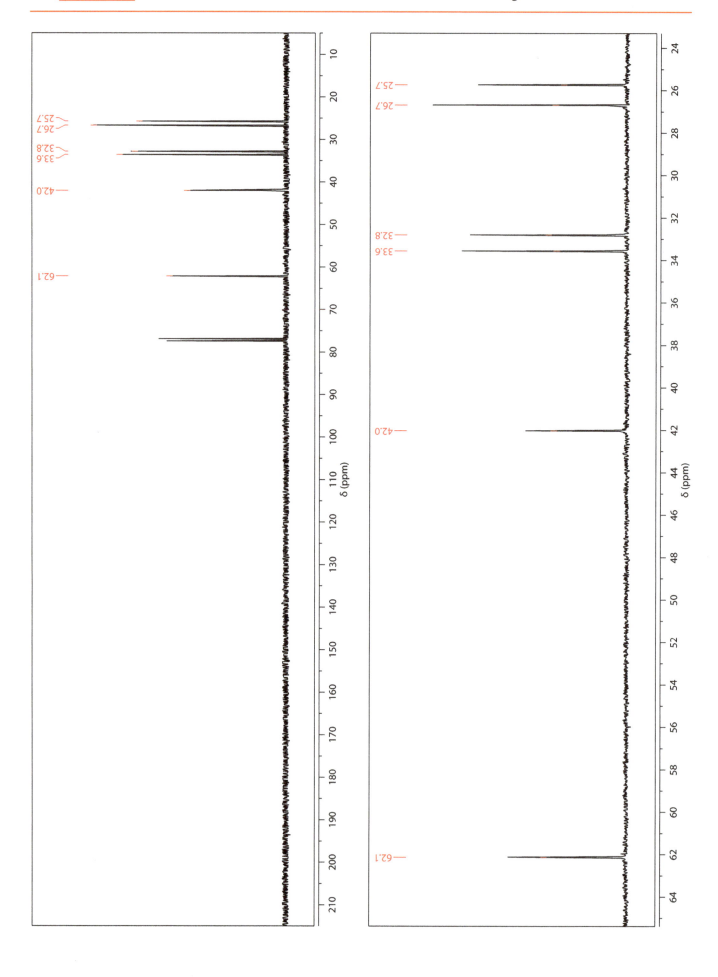

SECTION 3 Problem 54

Determine the structure of the unknown compound below with molecular formula C₅H₉ClO.

Spectra acquired in CDCl₃ at 500 MHz (¹H) or 125 MHz (¹³C).

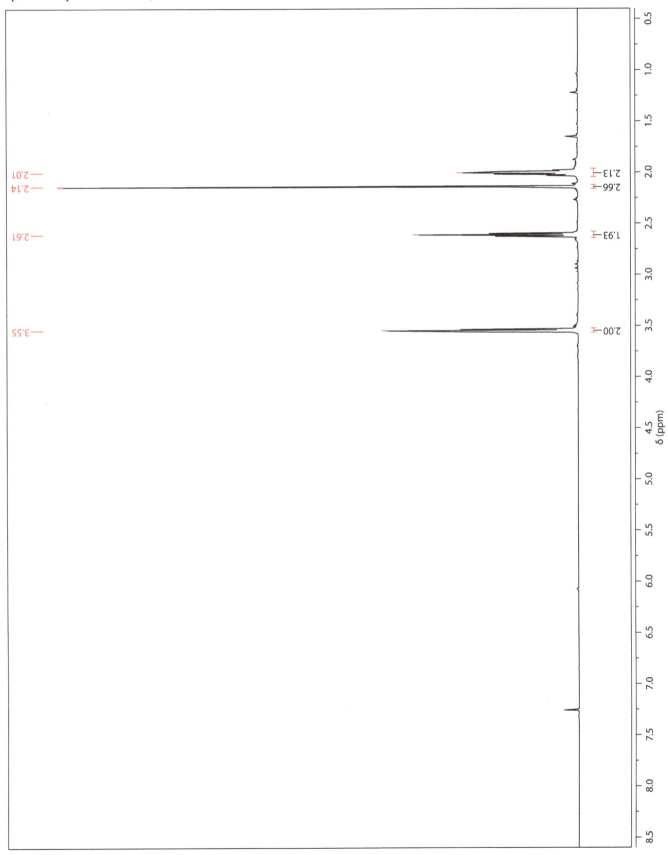

SECTION 3 Problem 54

Peaks (Hz):
- 990.2, 996.9, 1003.5, 1010.1, 1016.8 — 2.13
- 1071.1 — 2.66
- 1299.8, 1306.8, 1313.7 — 1.93
- 1768.0, 1774.3, 1780.6 — 2.00

δ (ppm) axis: 1.9 – 3.6

SECTION 3 Problem 54

Problems in Organic Structure Determination

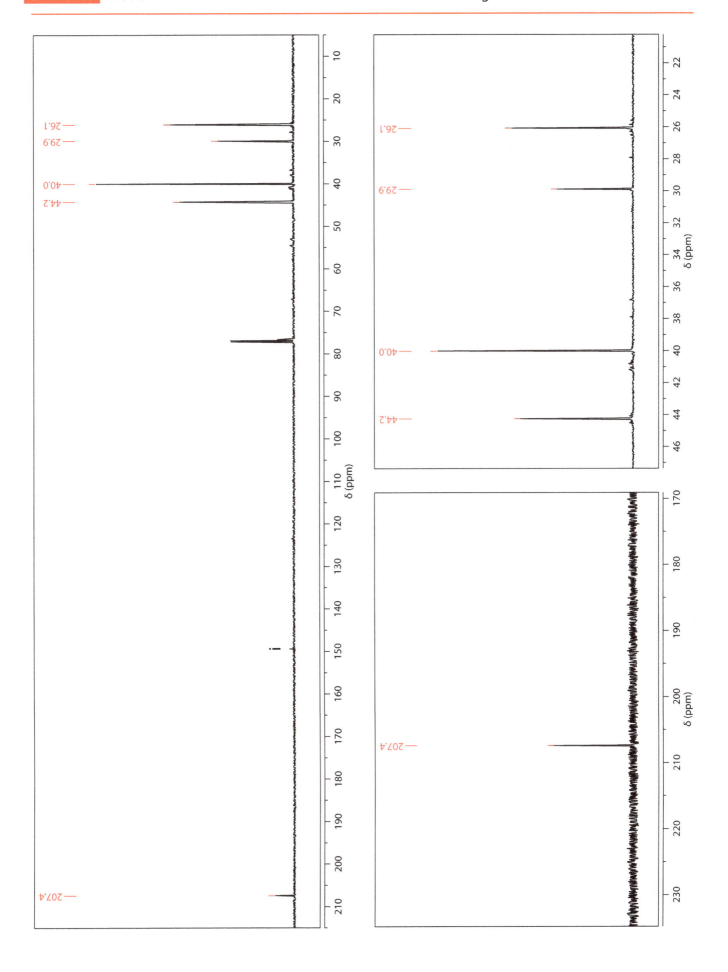

Problem 55

Determine the structure of the unknown compound below with molecular formula $C_5H_8O_2$.

Spectra acquired in $CDCl_3$ at 500 MHz (^1H) or 125 MHz (^{13}C).

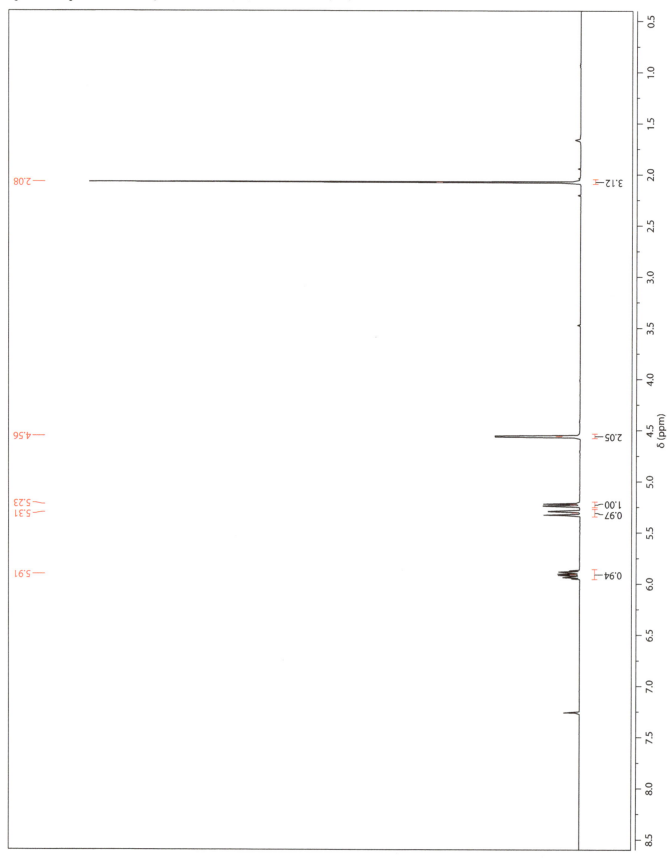

SECTION 3 Problem 55

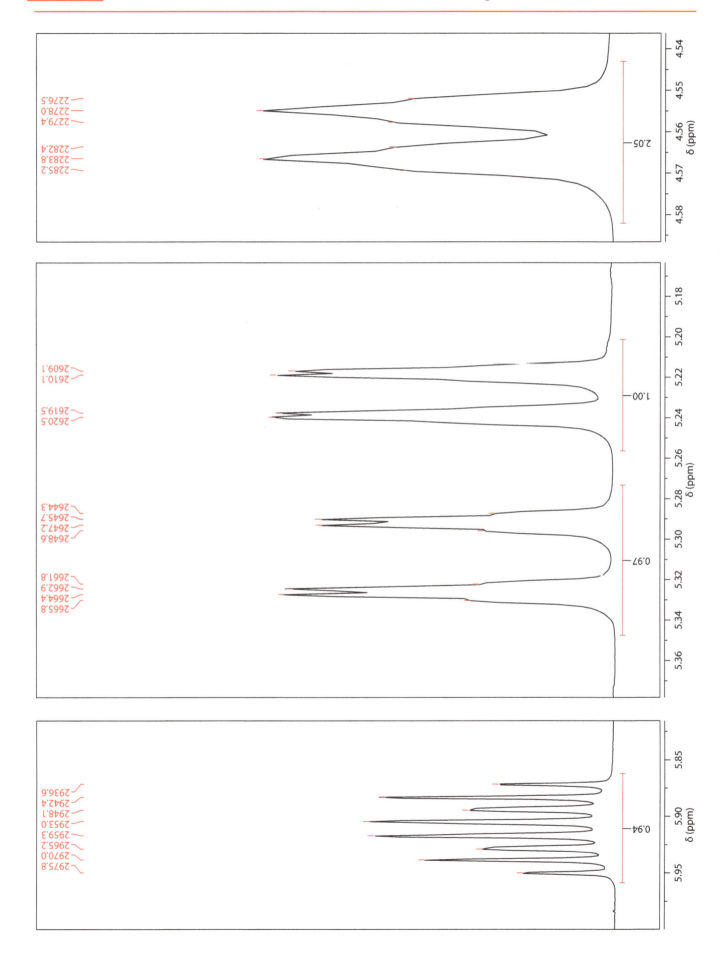

Problem 55

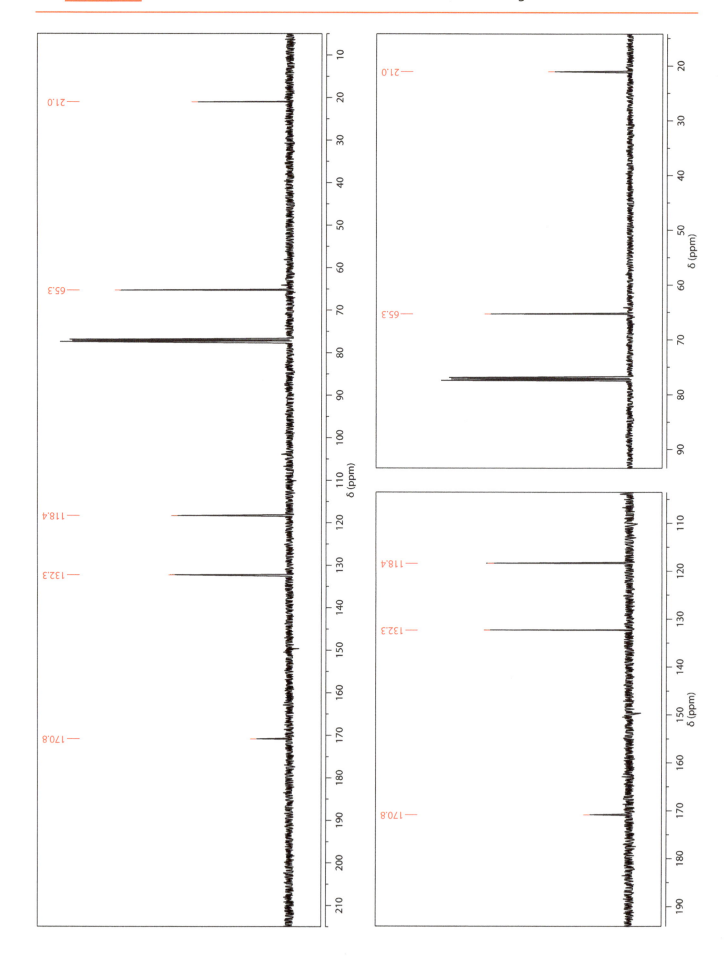

SECTION 3   Problem 56

Determine the structure of the unknown compound below with molecular formula $C_6H_{11}NO$.

Spectra acquired in $CDCl_3$ at 500 MHz (^1H) or 125 MHz (^{13}C).

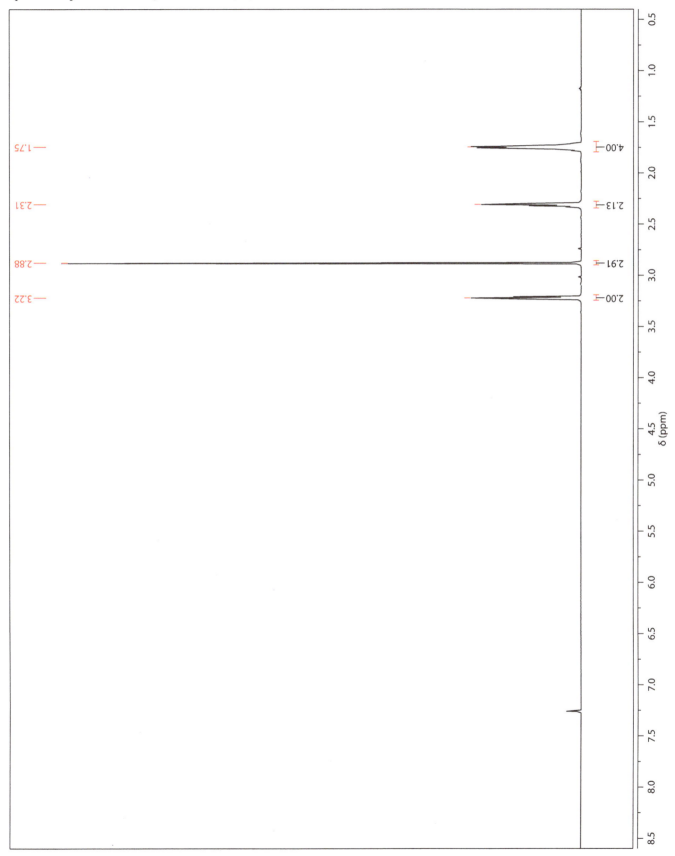

Problem 56

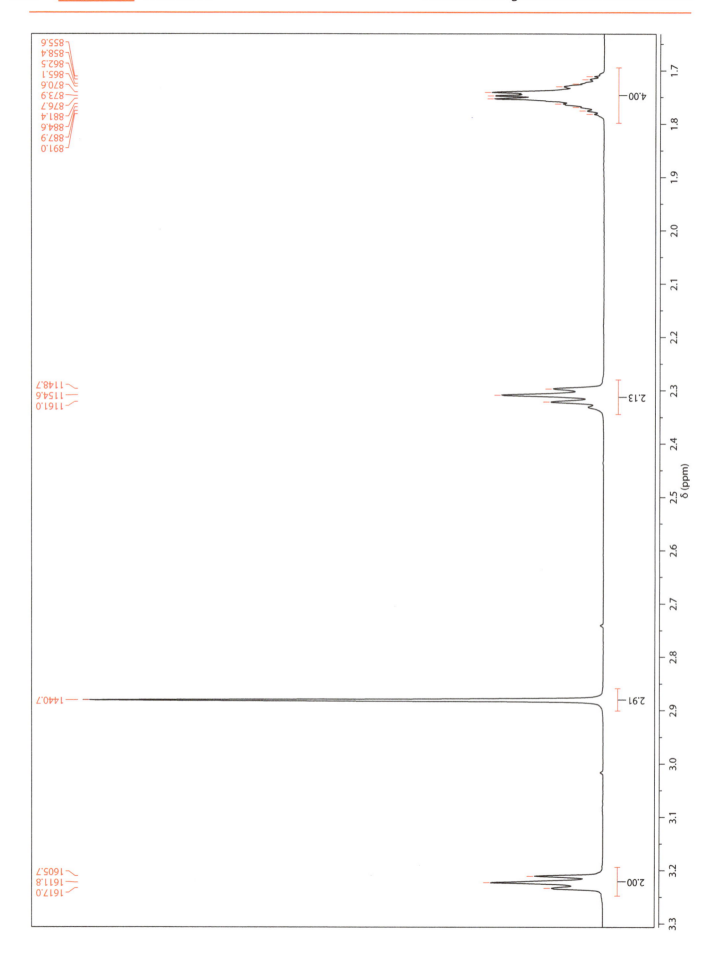

SECTION 3 Problem 56

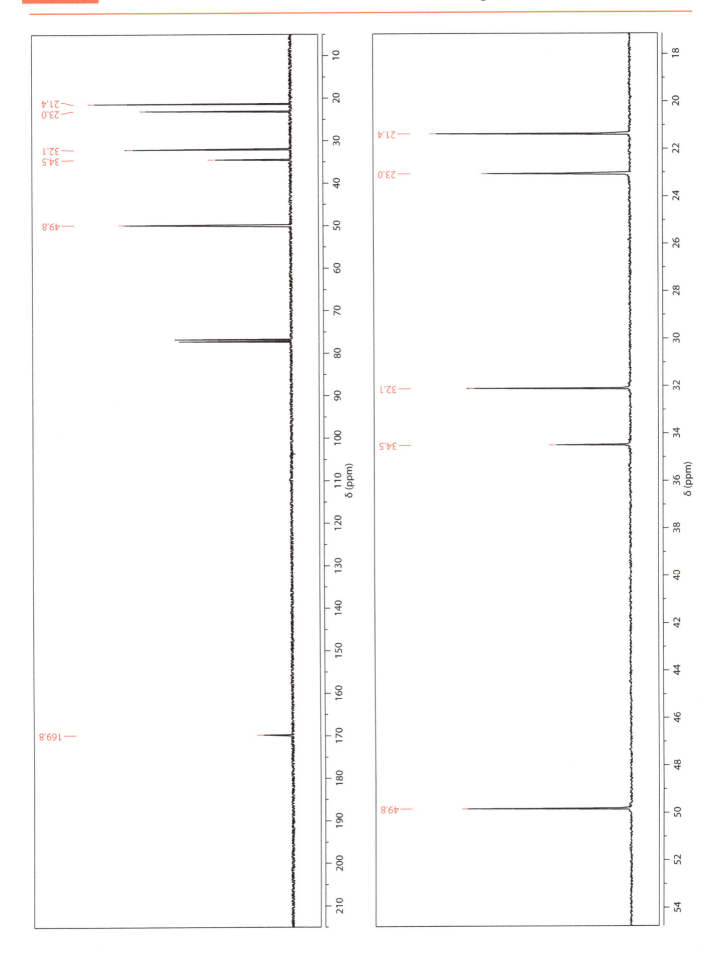

Determine the structure of the unknown compound below with molecular formula C₉H₉BrO.

Spectra acquired in CDCl₃ at 500 MHz (¹H) or 125 MHz (¹³C).

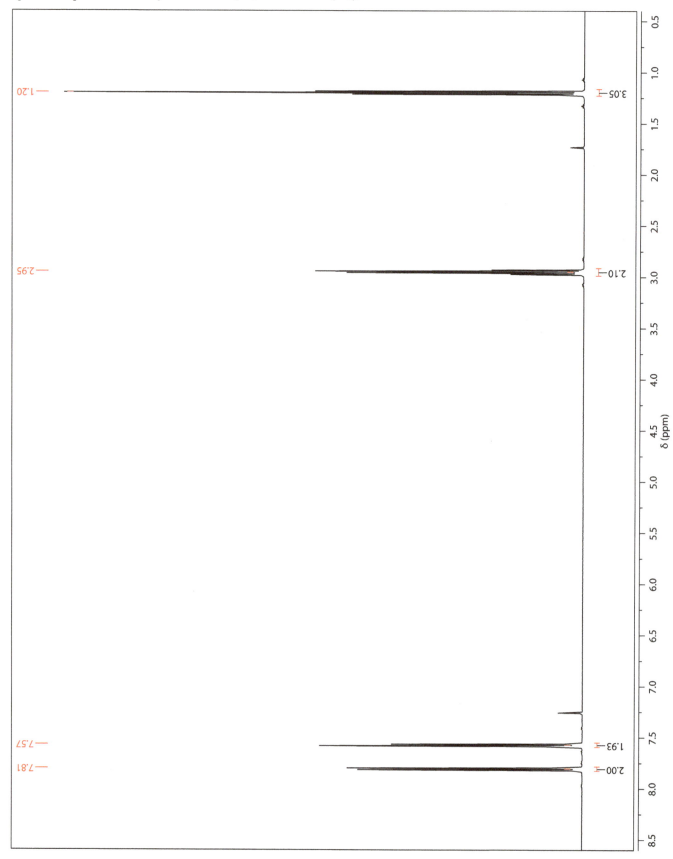

SECTION 3 Problem 57 — Problems in Organic Structure Determination

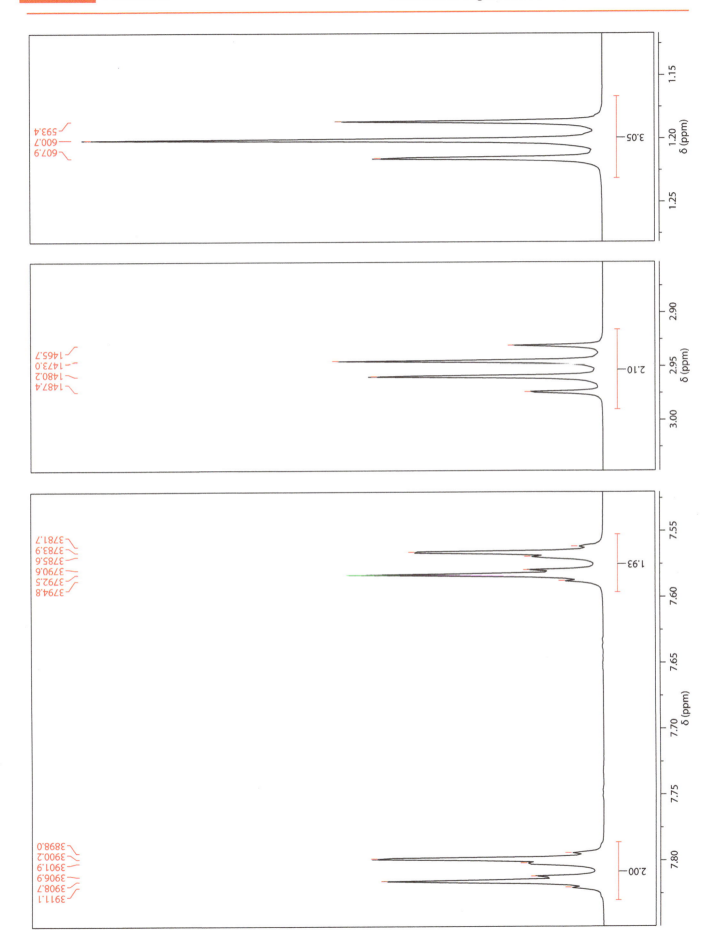

Problem 57

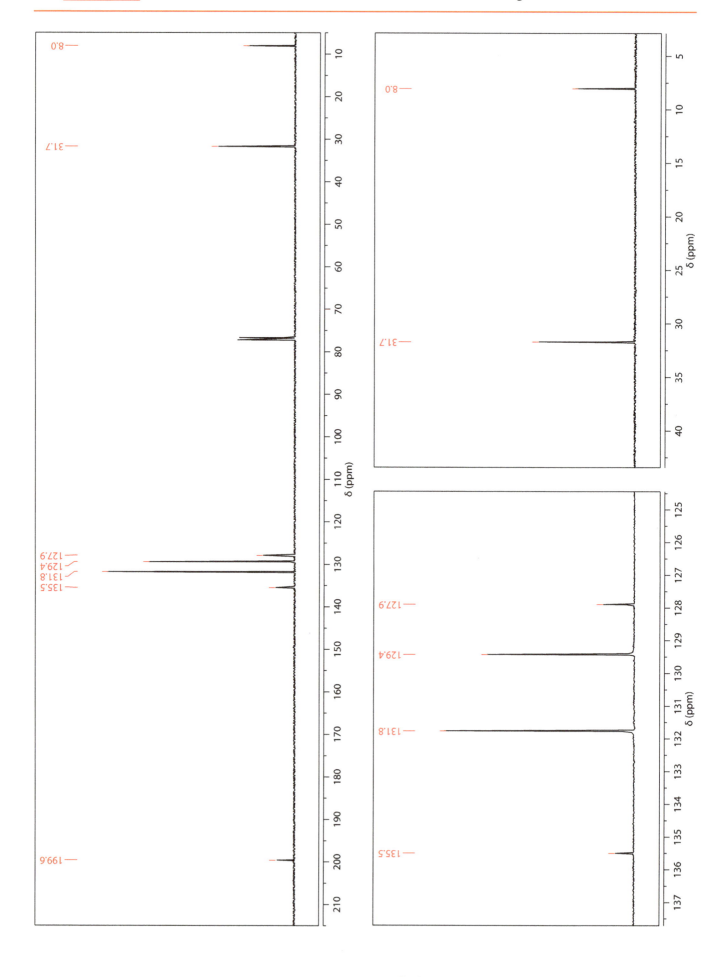

SECTION 3 Problem 58

Determine the structure of the unknown compound below with molecular formula $C_5H_8O_2$.

Spectra acquired in $CDCl_3$ at 500 MHz (^{1}H) or 125 MHz (^{13}C).

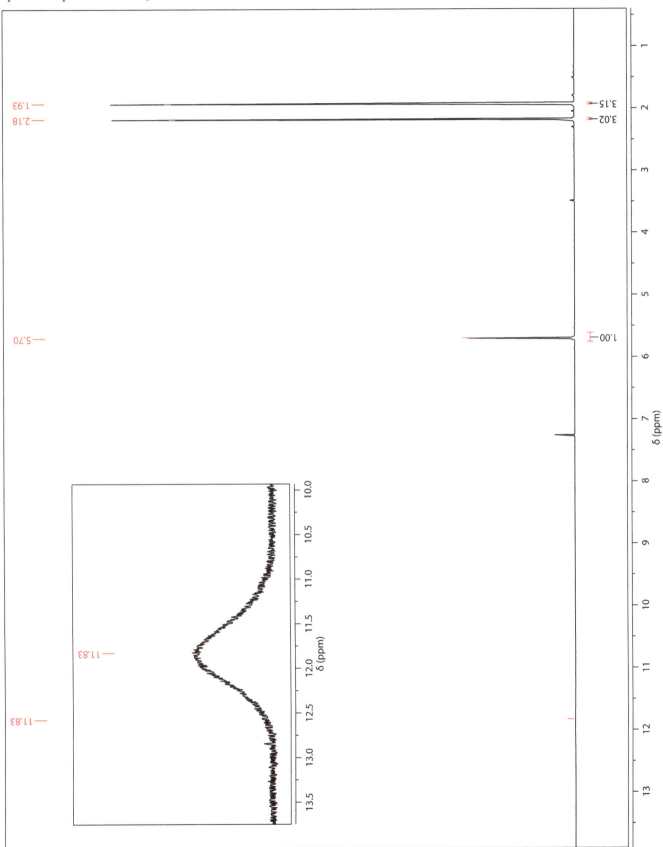

Problem 58

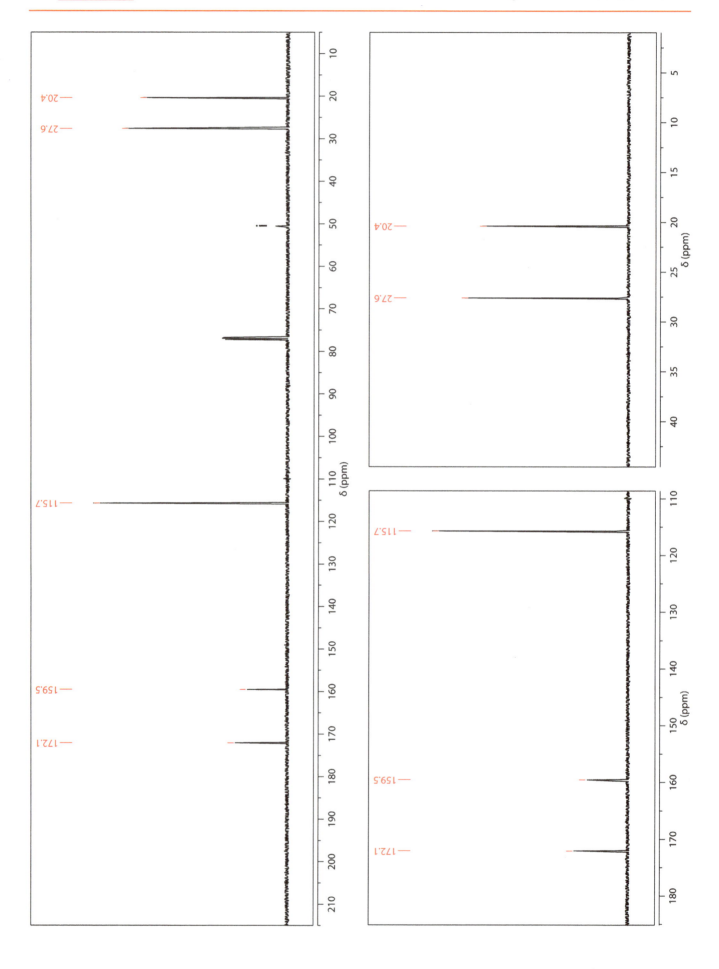

SECTION 3 Problem 59

Determine the structure of the unknown compound below with molecular formula $C_8H_{10}OS$.

Spectra acquired in $CDCl_3$ at 500 MHz (^1H) or 125 MHz (^{13}C).

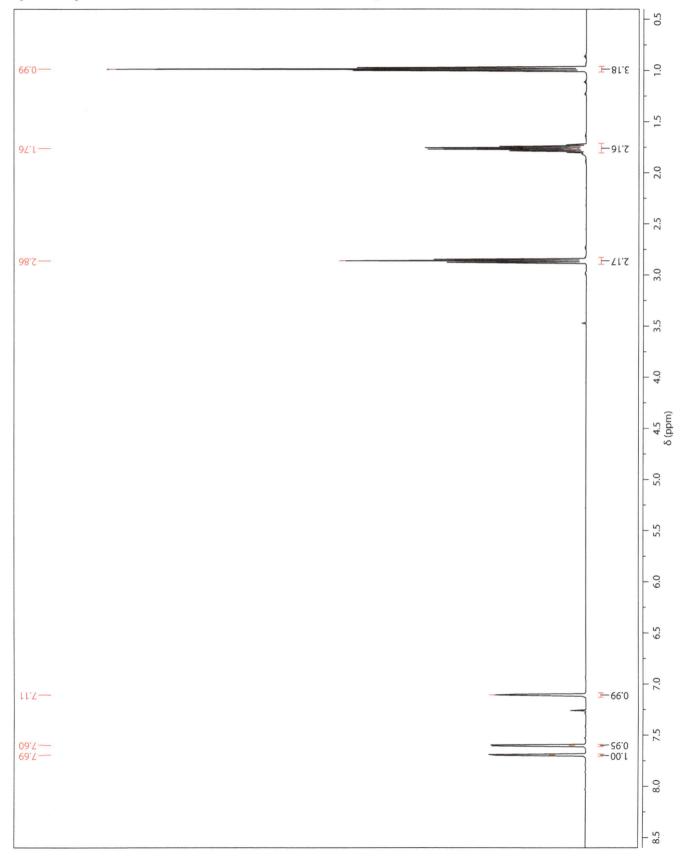

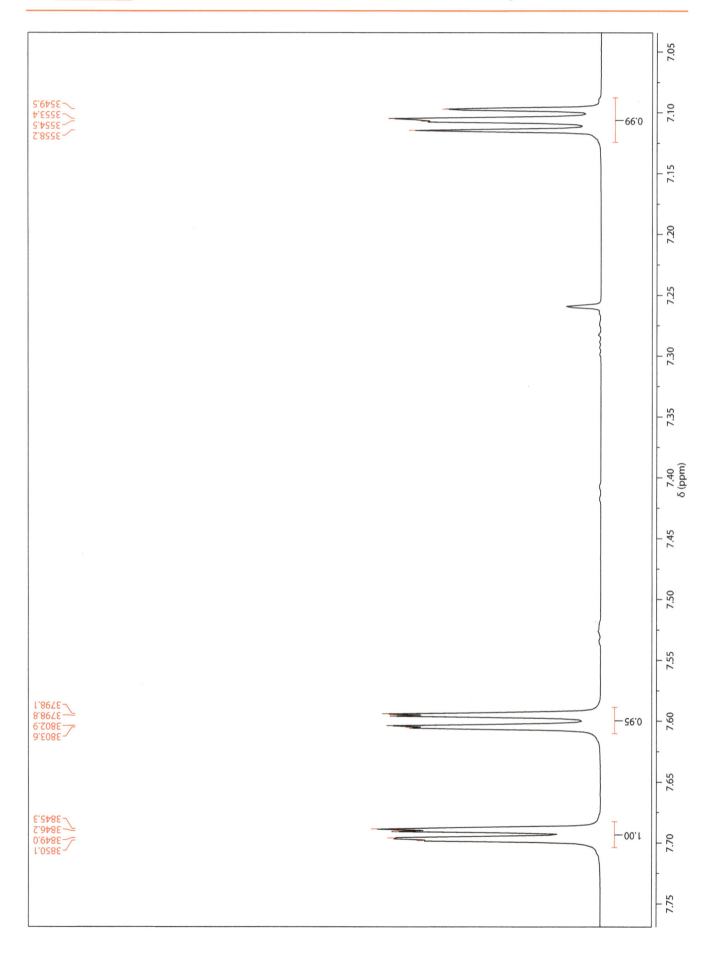

SECTION 3 Problem 59

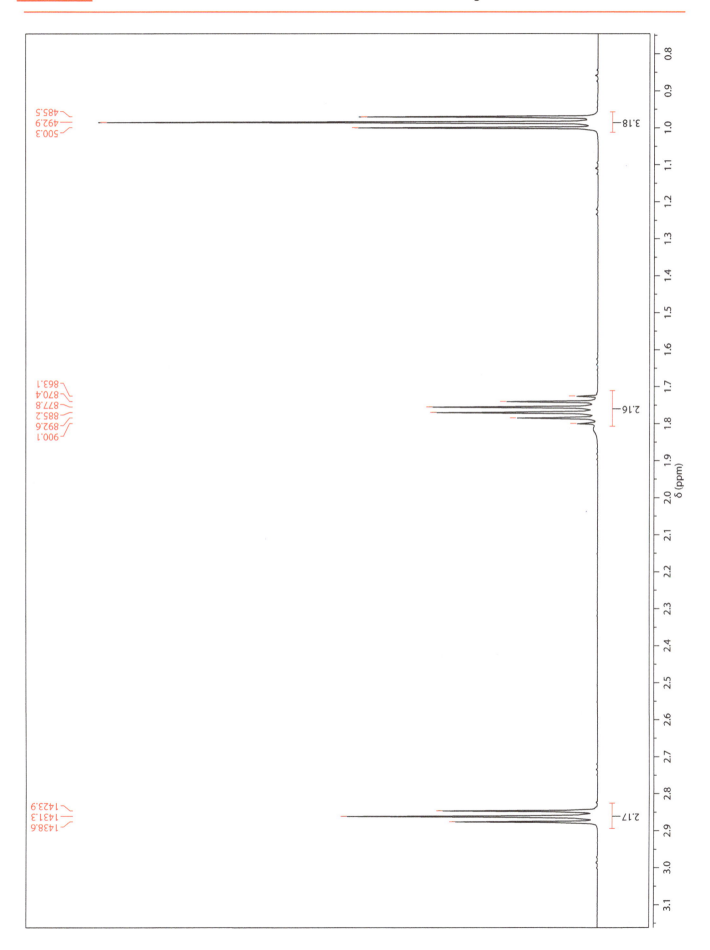

Problem 59

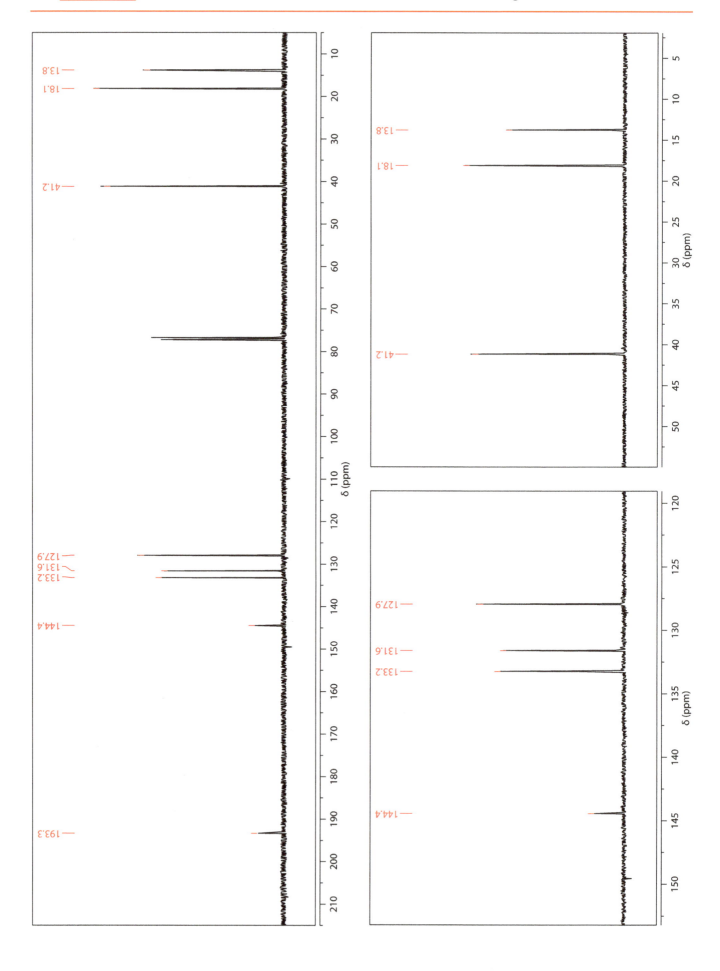

SECTION 3 Problem 60

Problems in Organic Structure Determination 337

Determine the structure of the unknown compound below with molecular formula C_7H_5NO.

Spectra acquired in CD_3OD at 500 MHz (^1H) or 125 MHz (^{13}C).

— 7.52
— 6.87

1.01
1.00

Problem 60

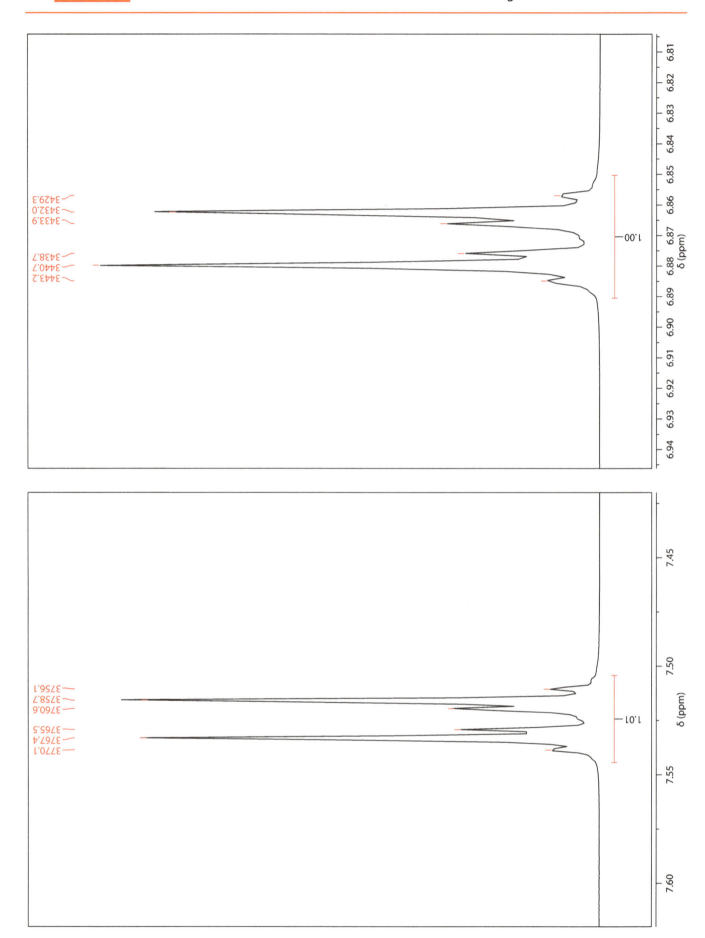

Problem 60

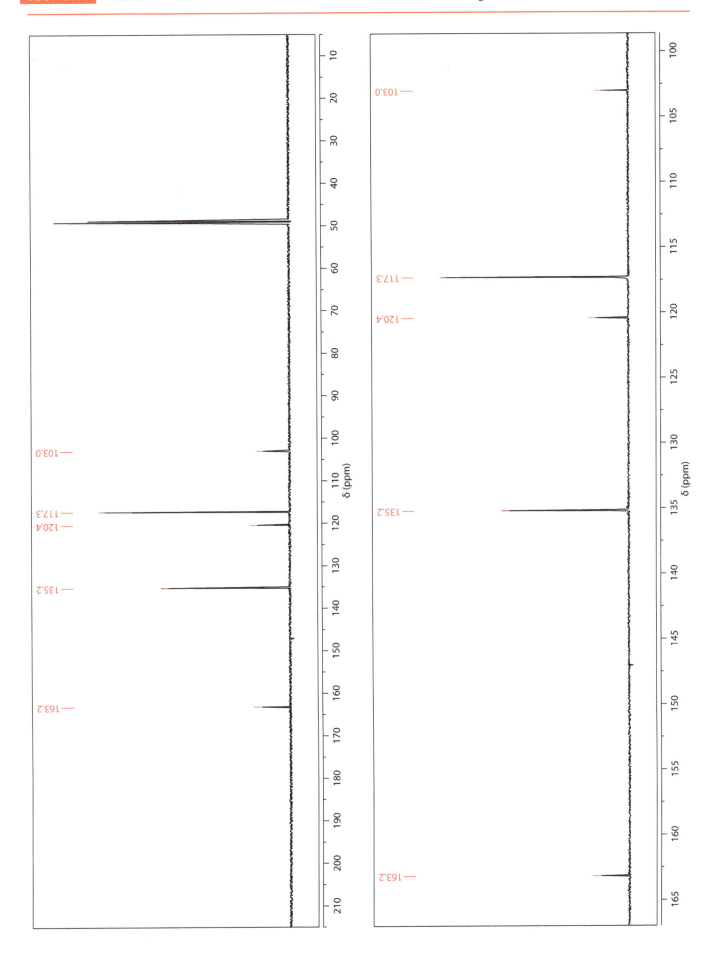

SECTION 3 Problem 61

Problems in Organic Structure Determination

Determine the structure of the unknown compound below with molecular formula $C_9H_8O_3$.

Spectra acquired in $CDCl_3$ at 500 MHz (^1H) or 125 MHz (^{13}C).

Peaks:
- 7.54 (1.02)
- 7.42 (0.96)
- 6.83 (1.00)
- 6.03 (2.18)
- 2.53
- 3.26 (integration)

SECTION 3 Problem 61

Problem 61

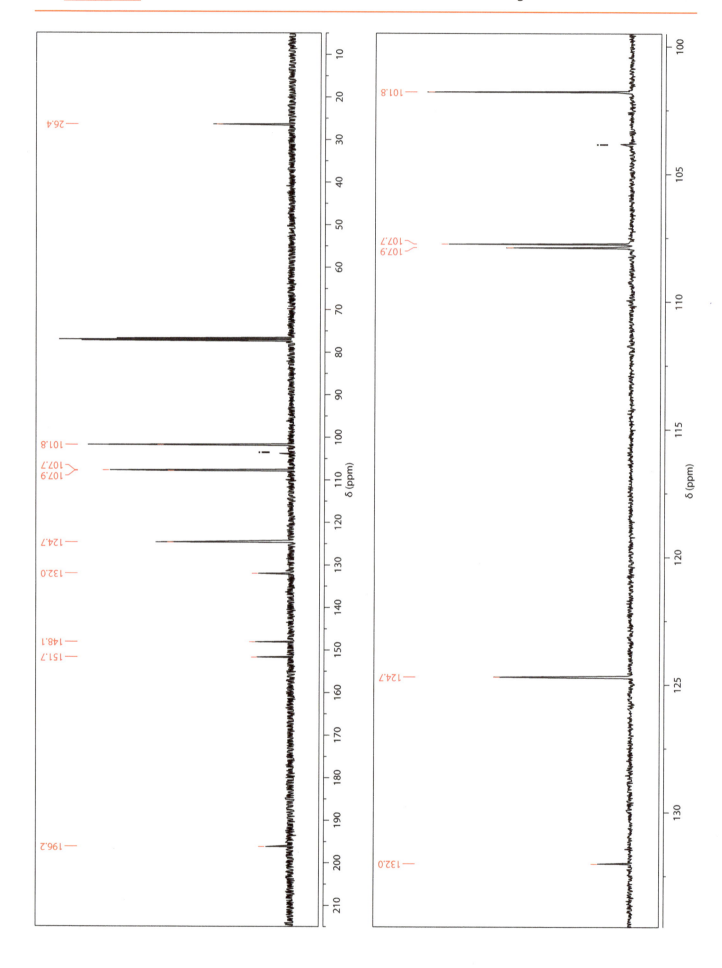

SECTION 3   Problem 62

Problems in Organic Structure Determination

Determine the structure of the unknown compound below with molecular formula $C_9H_8O_3$.

Spectra acquired in CD_3OD at 500 MHz (^1H) or 125 MHz (^{13}C).

Peaks: 7.59, 7.43, 6.79, 6.27

Integrations: 1.00, 2.07, 1.95, 1.00

SECTION 3 Problem 62

SECTION 3 Problem 62

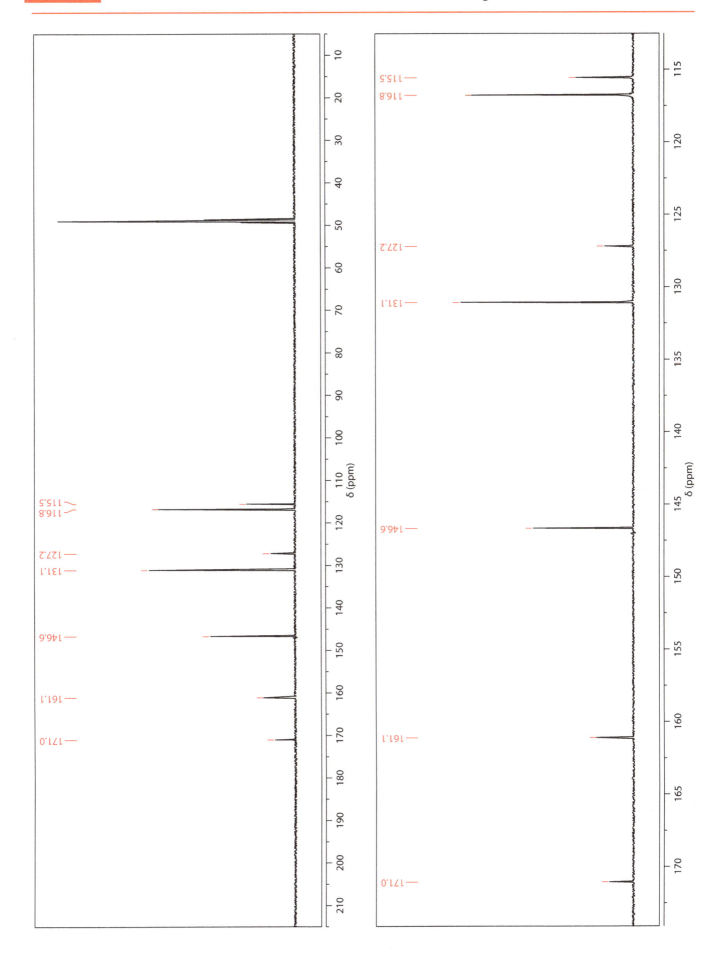

SECTION 3 Problem 63

Determine the structure of the unknown compound below with molecular formula C₉H₉BrO.

Spectra acquired in CDCl₃ at 500 MHz (¹H) or 125 MHz (¹³C).

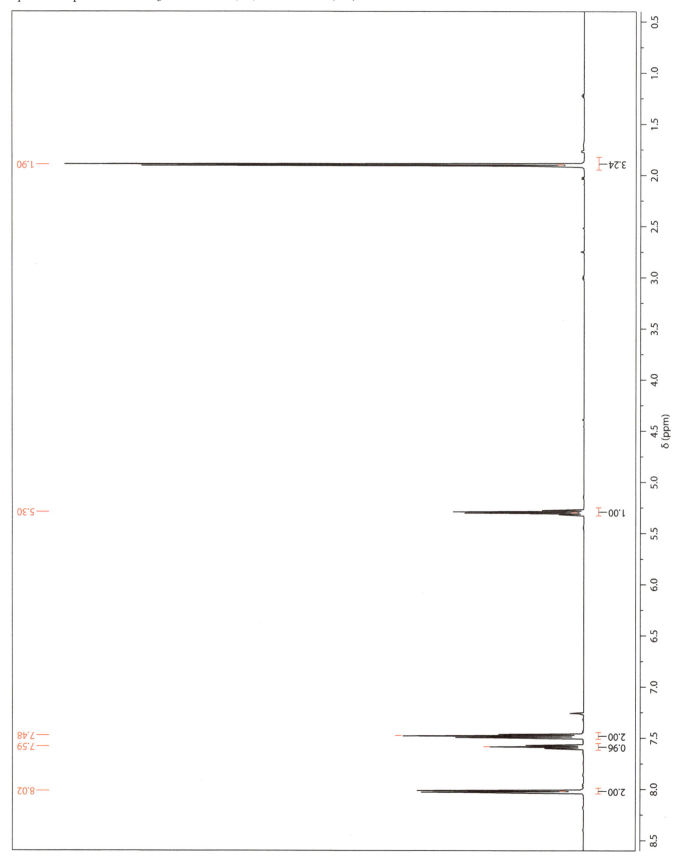

SECTION 3 Problem 63

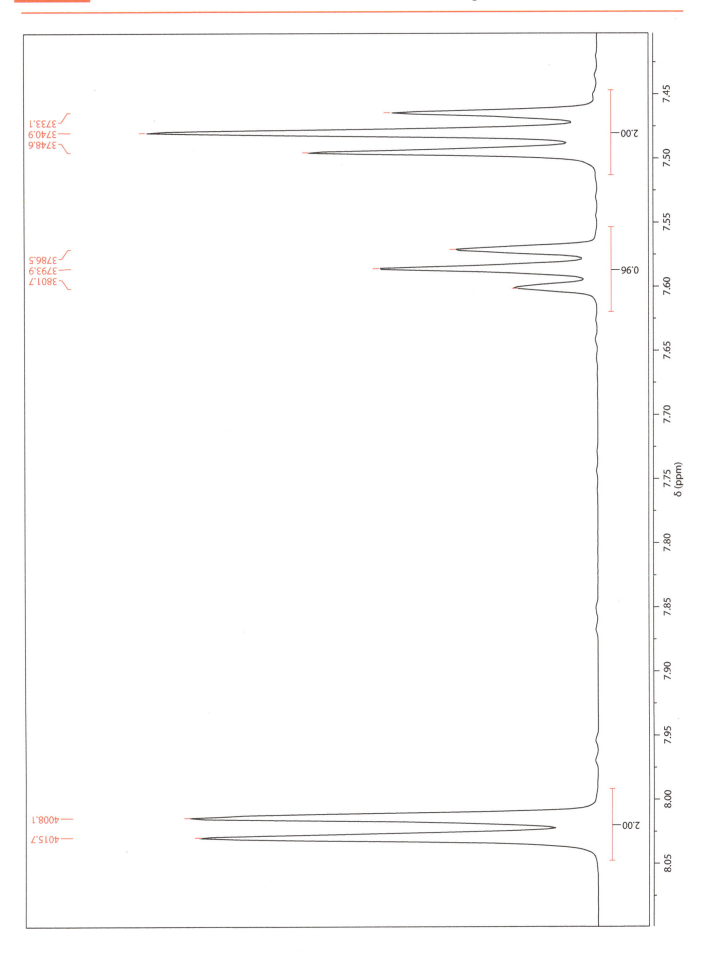

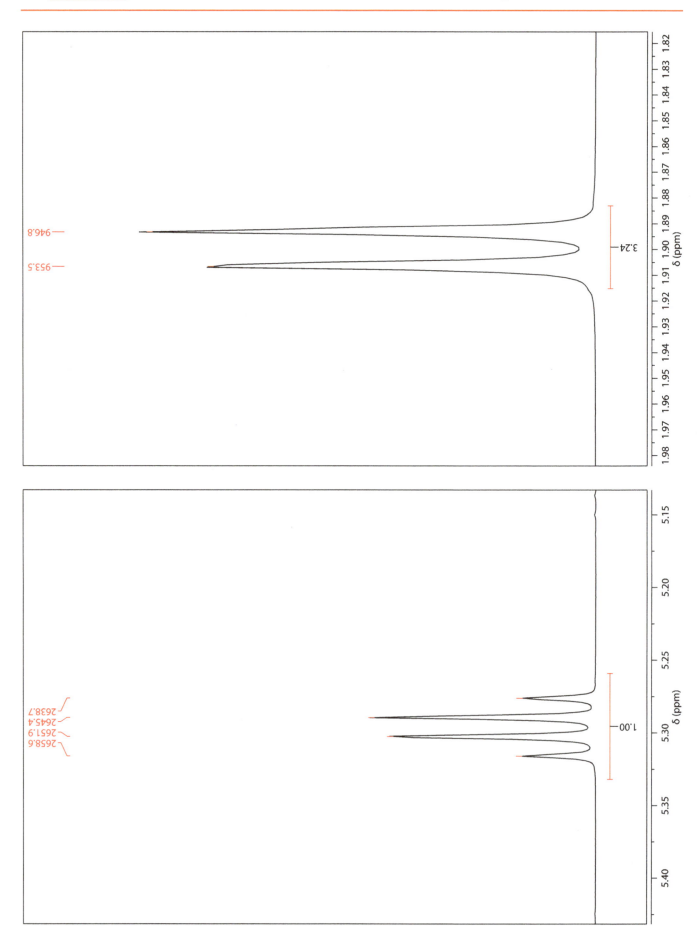

Problem 63

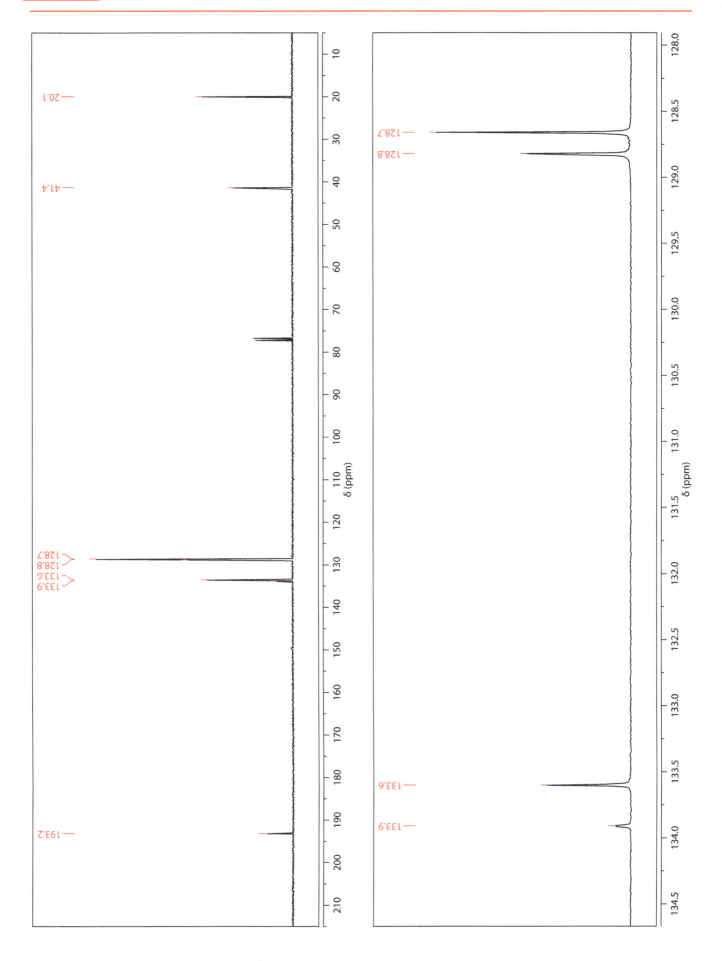

SECTION 3 Problem 64

Determine the structure of the unknown compound below with molecular formula C_5H_3NO.

Spectra acquired in $CDCl_3$ at 500 MHz (^1H) or 125 MHz (^{13}C).

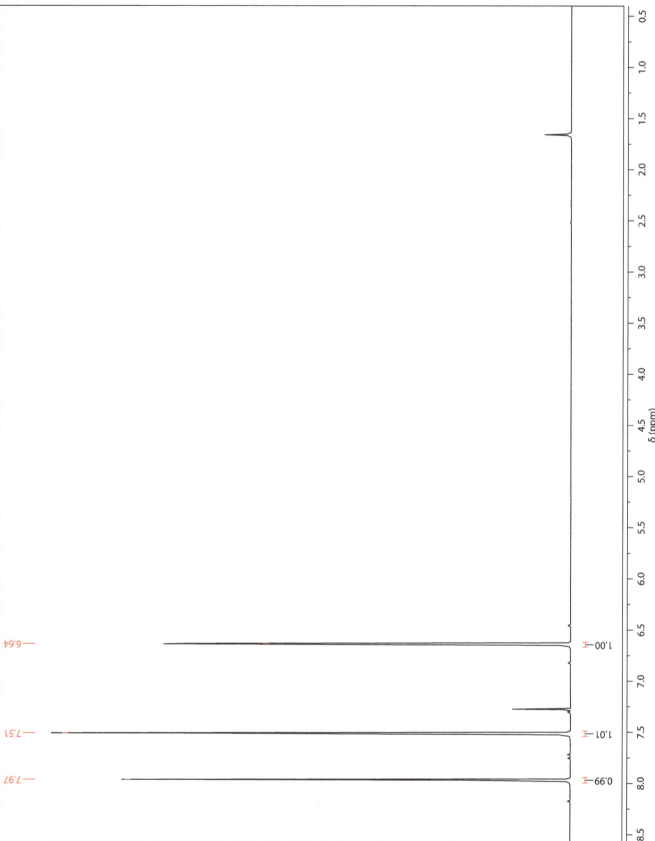

SECTION 3 Problem 64

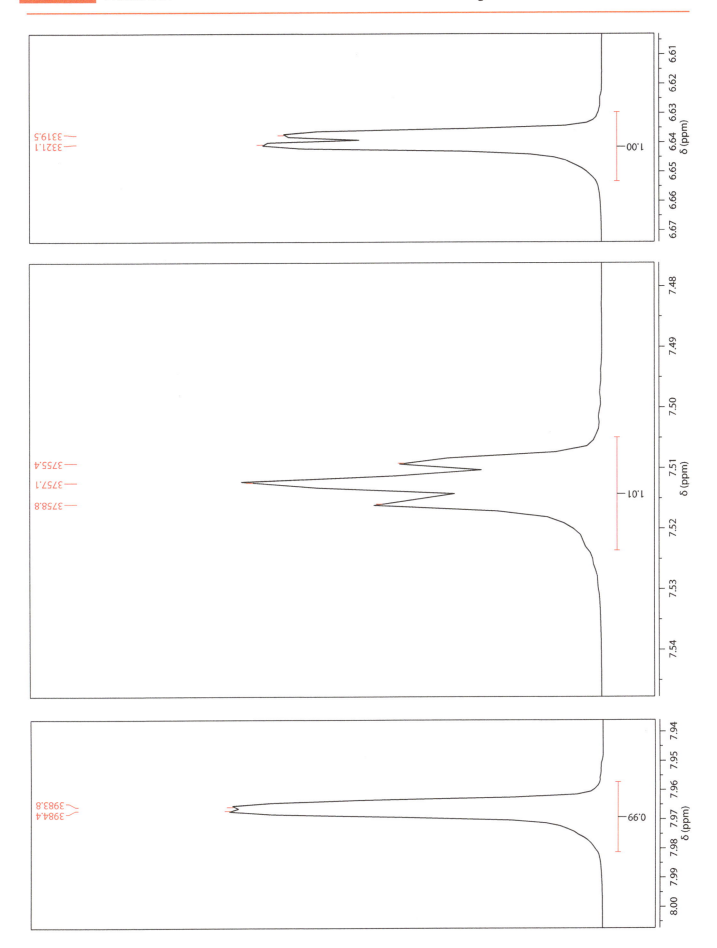

SECTION 3 Problem 64

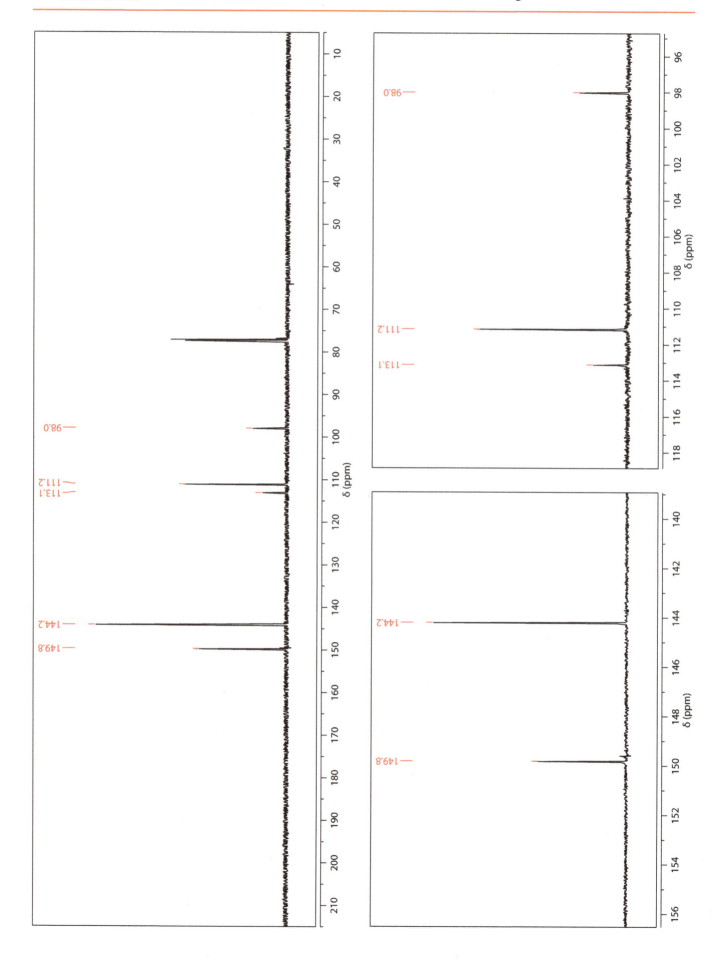

SECTION 3 Problem 65

Determine the structure of the unknown compound below with molecular formula $C_7H_6N_2$.

Spectra acquired in $CDCl_3$ at 500 MHz (^{1}H) or 125 MHz (^{13}C).

Problem 65

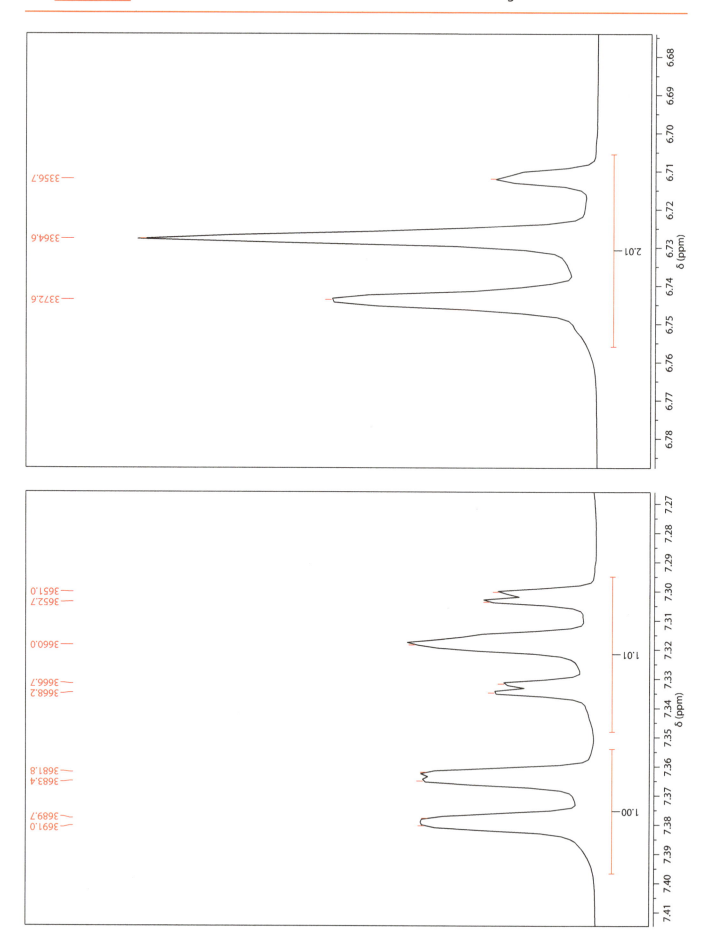

SECTION 3 Problem 65

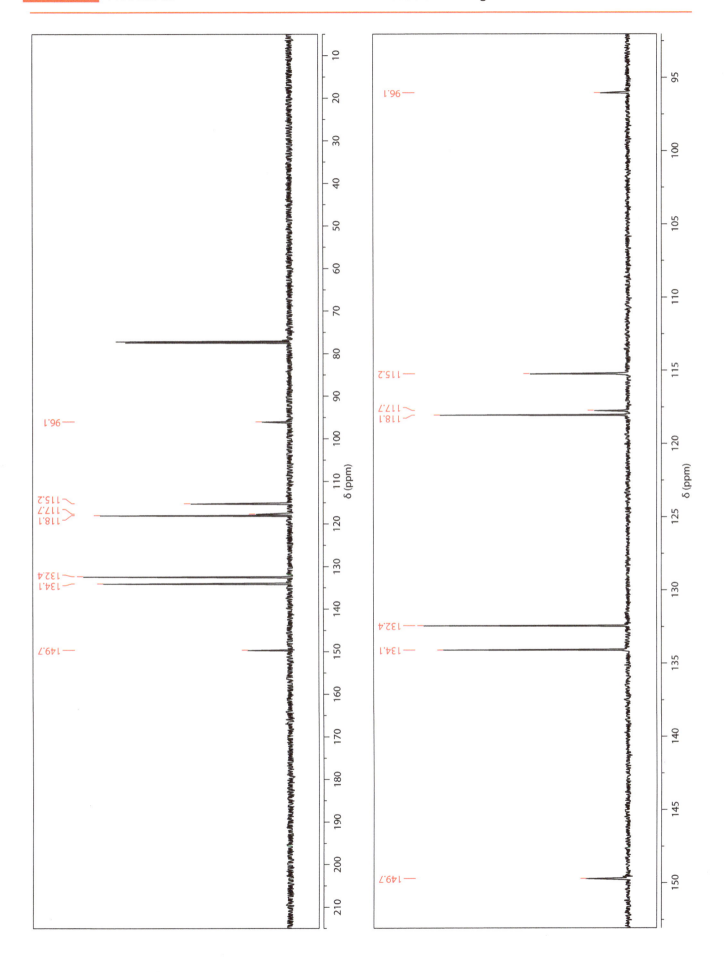

SECTION 3 Problem 66

Determine the structure of the unknown compound below with molecular formula $C_8H_8O_3$.

Spectra acquired in $(CD_3)_2SO$ at 500 MHz (^{1}H) or 125 MHz (^{13}C).

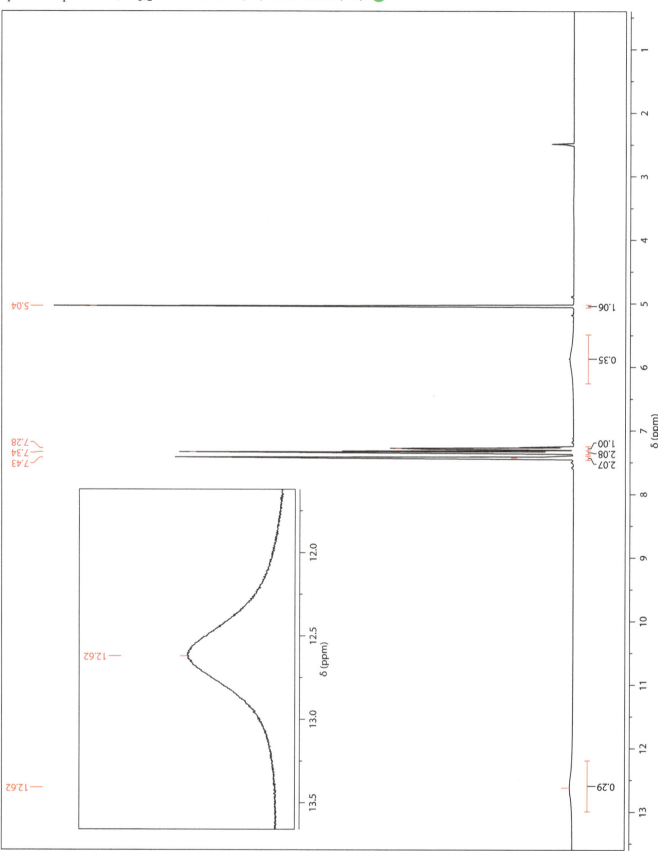

SECTION 3 Problem 66

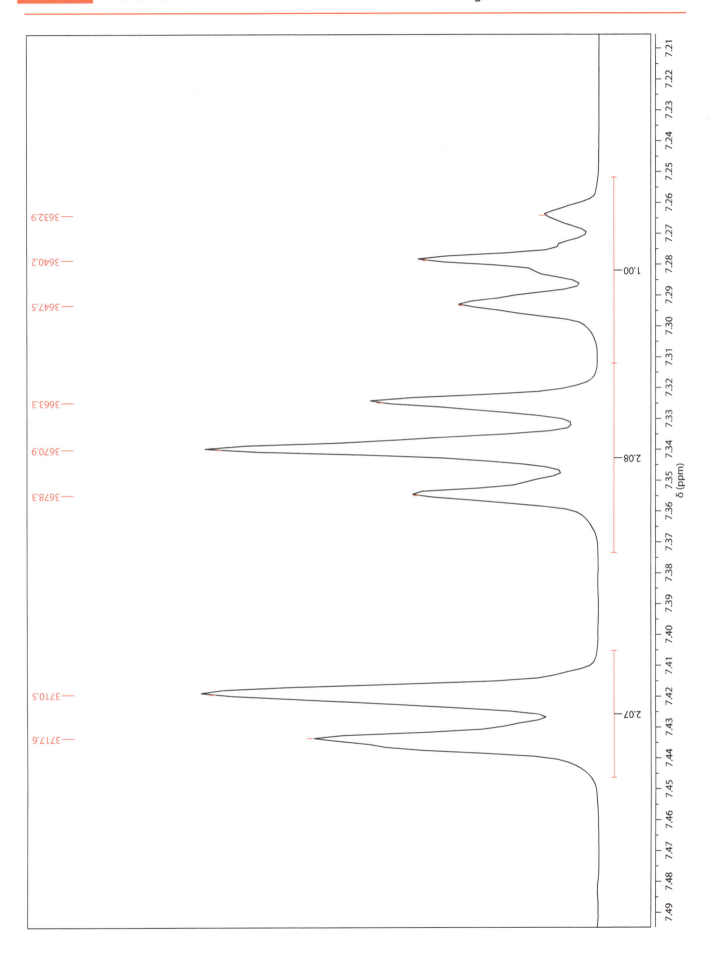

Problem 66

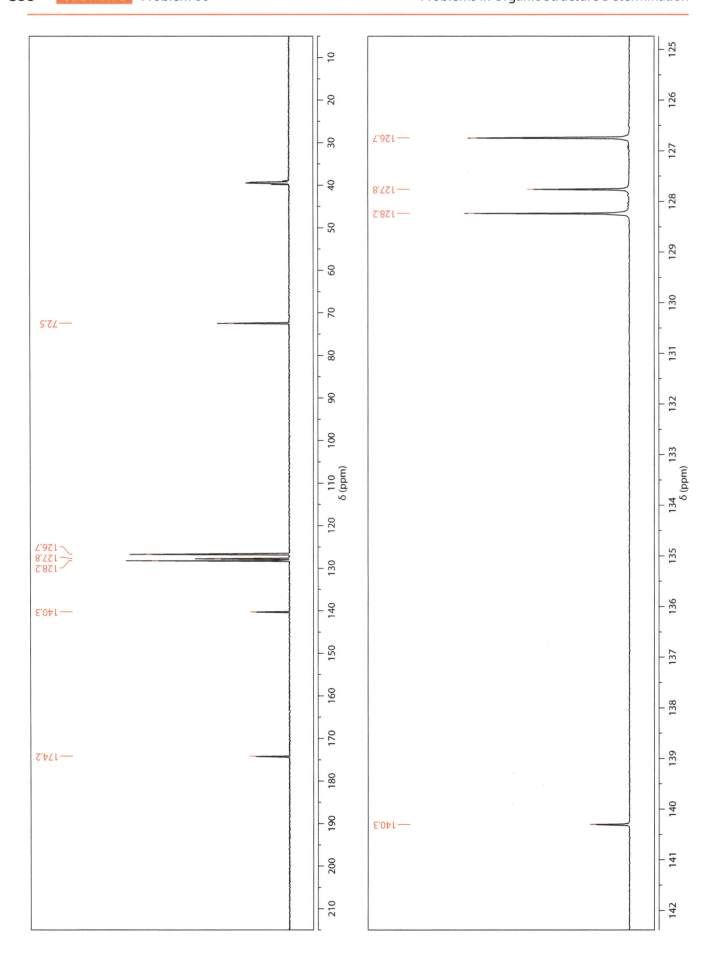

SECTION 3 Problem 67

Determine the structure of the unknown compound below with molecular formula $C_6H_{12}O_2$.

Spectra acquired in $CDCl_3$ at 500 MHz (^{1}H) or 125 MHz (^{13}C).

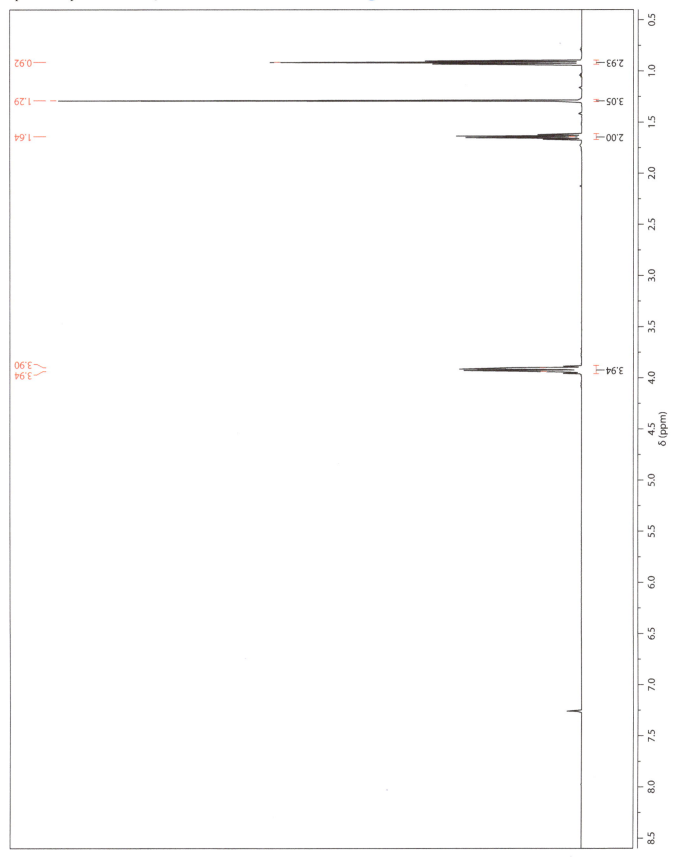

Problem 67

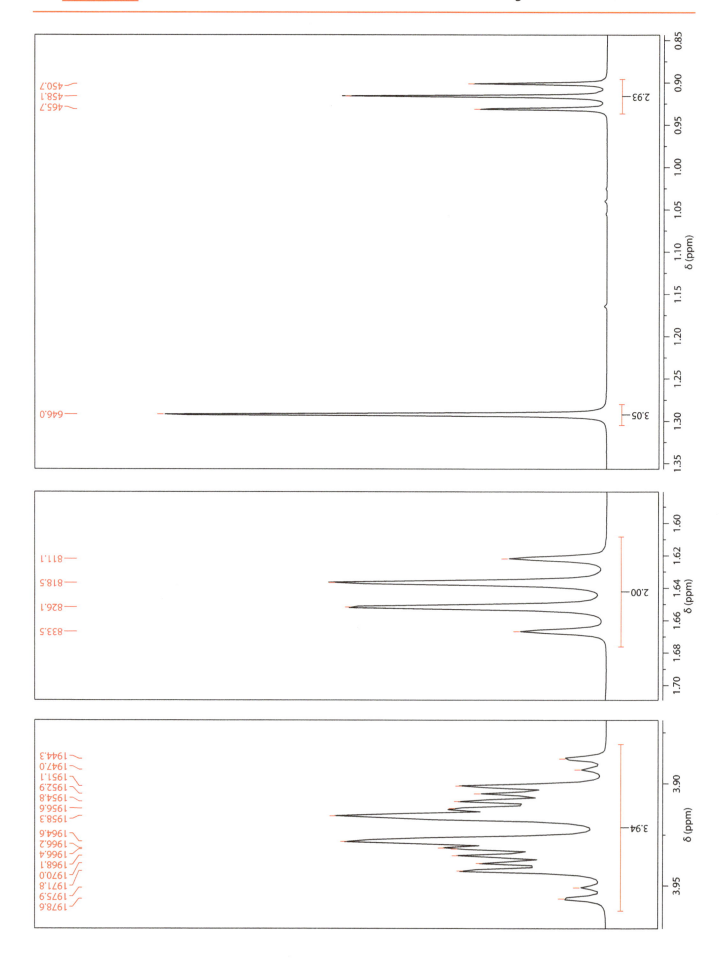

SECTION 3 Problem 67

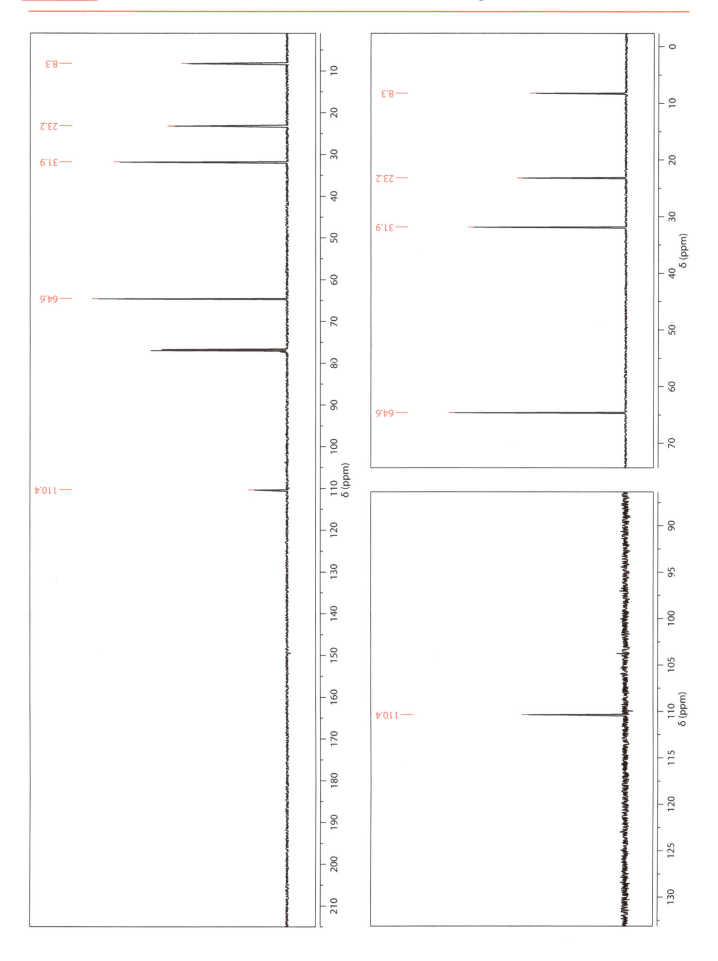

Problem 67

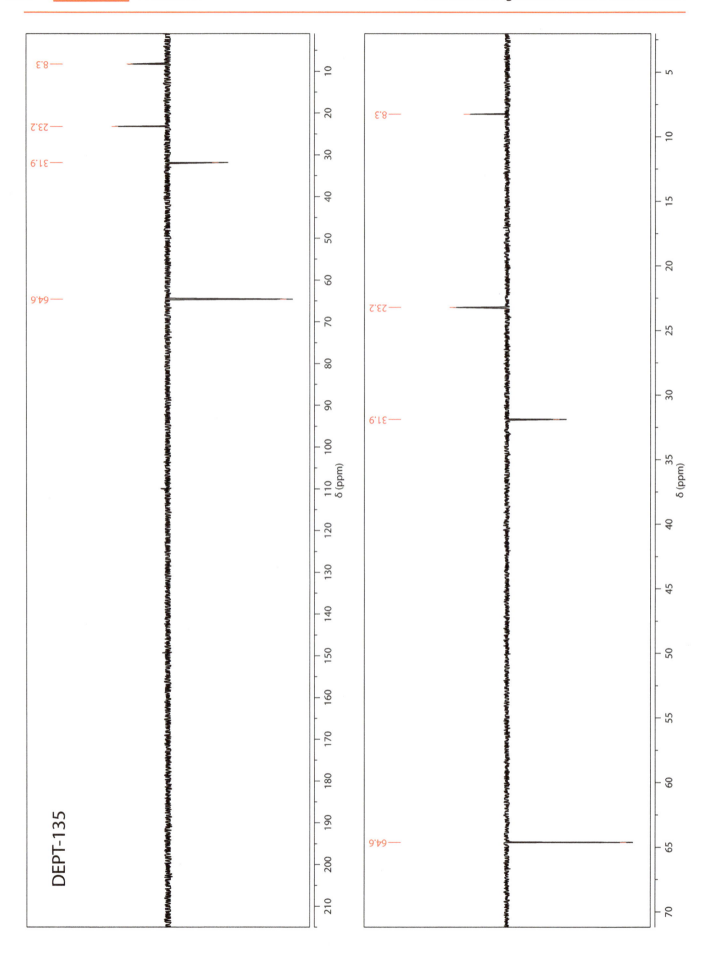

SECTION 3 Problem 68

Determine the structure of the unknown compound below with molecular formula C₉H₁₃NO.

Spectra acquired in (CD₃)₂SO at 500 MHz (¹H) or 125 MHz (¹³C).

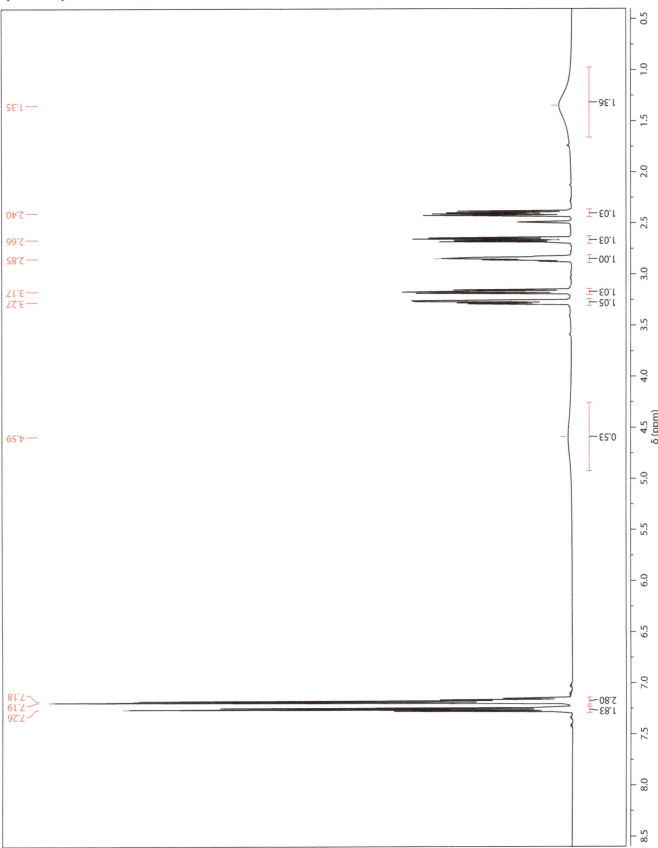

Problem 68

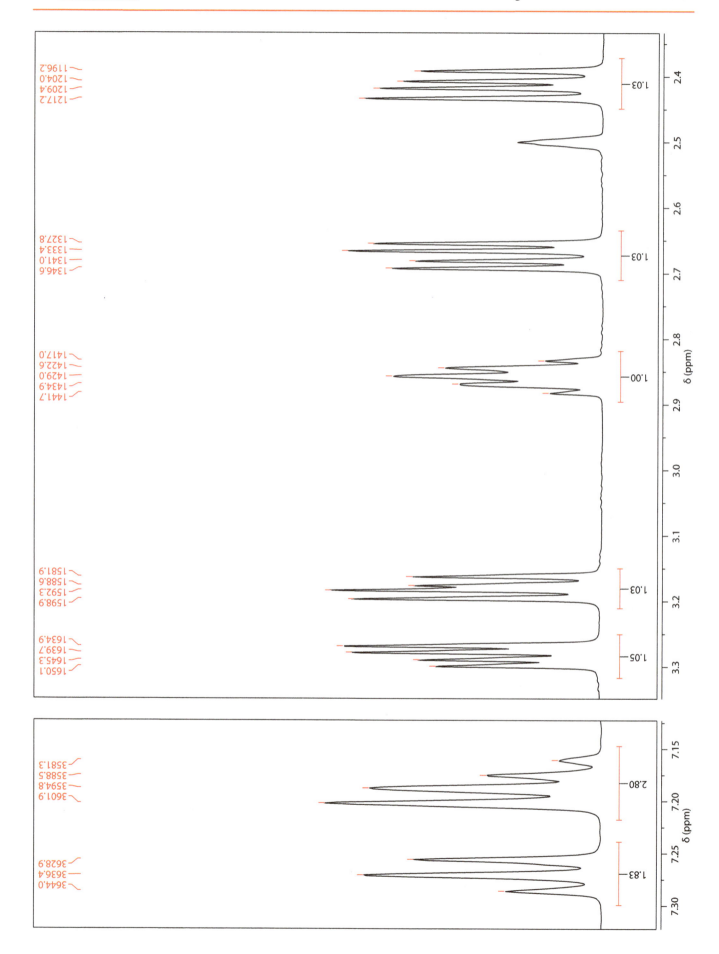

SECTION 3 Problem 68

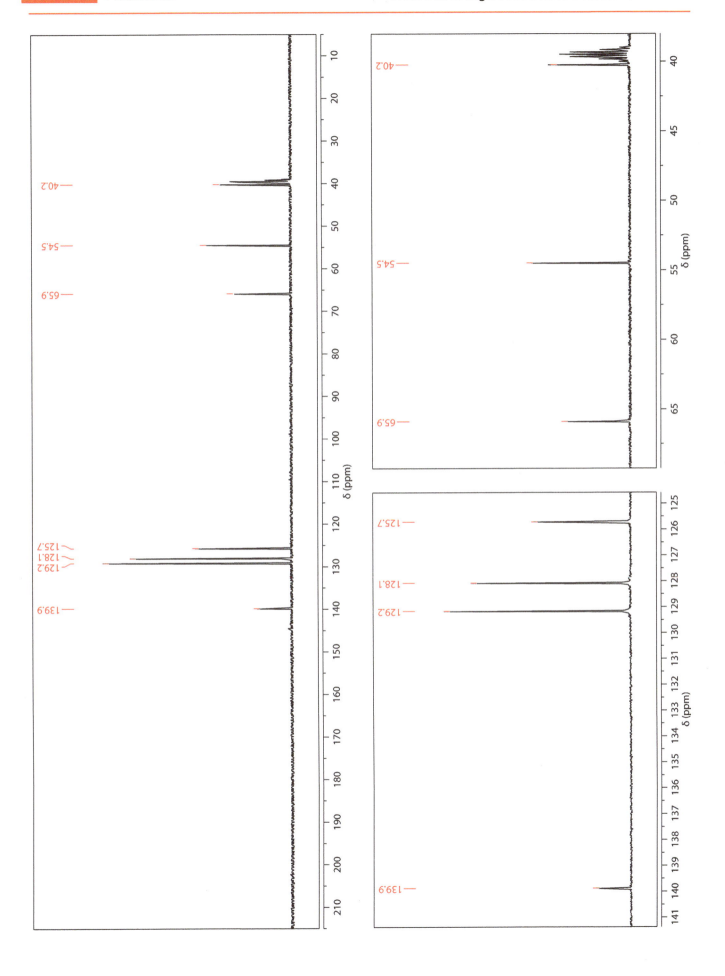

Problem 68

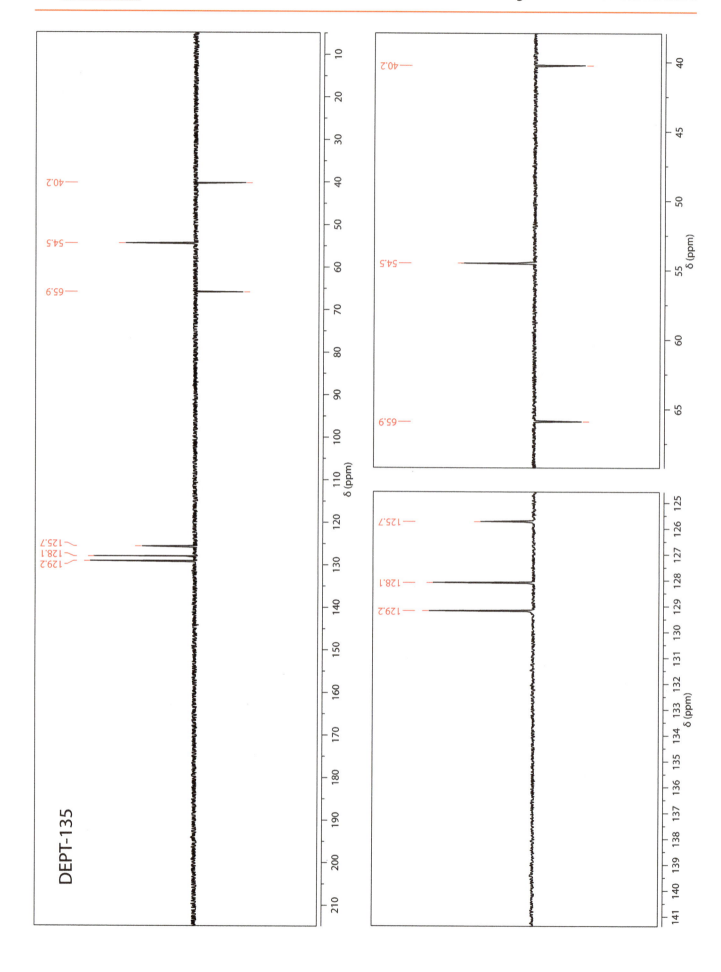

SECTION 3 Problem 69

Determine the structure of the unknown compound below with molecular formula $C_7H_{10}O_2$.

Spectra acquired in $CDCl_3$ at 500 MHz (^1H) or 125 MHz (^{13}C).

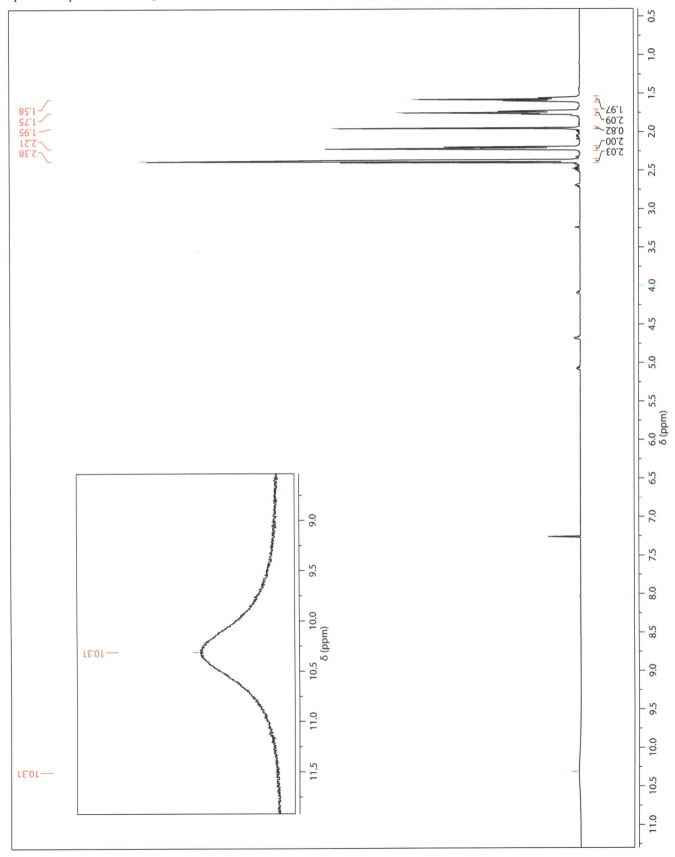

Problem 69

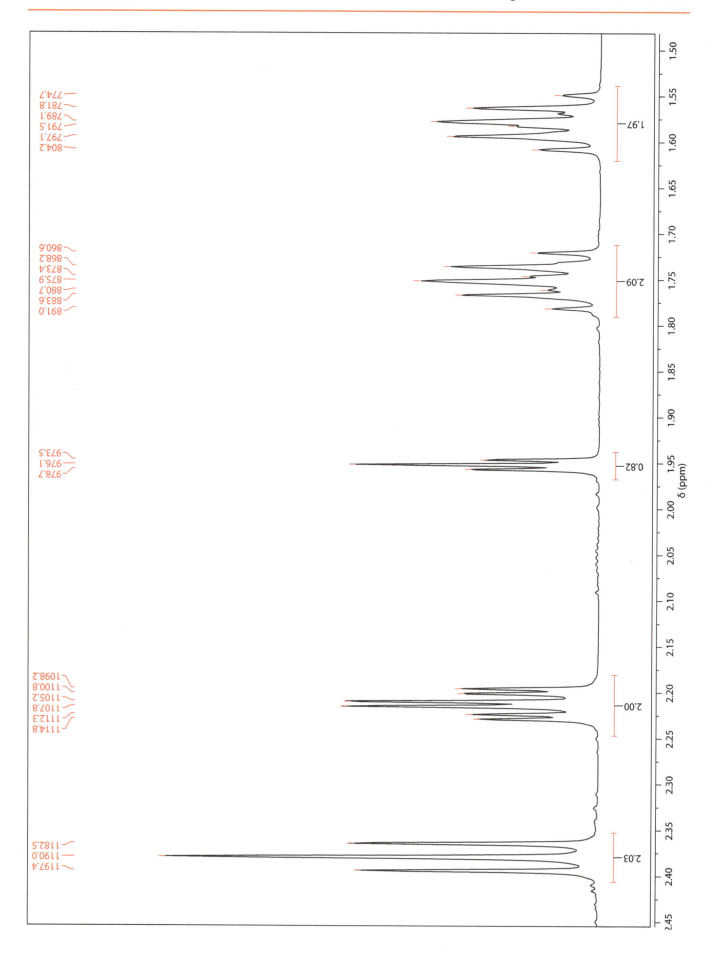

SECTION 3 Problem 69

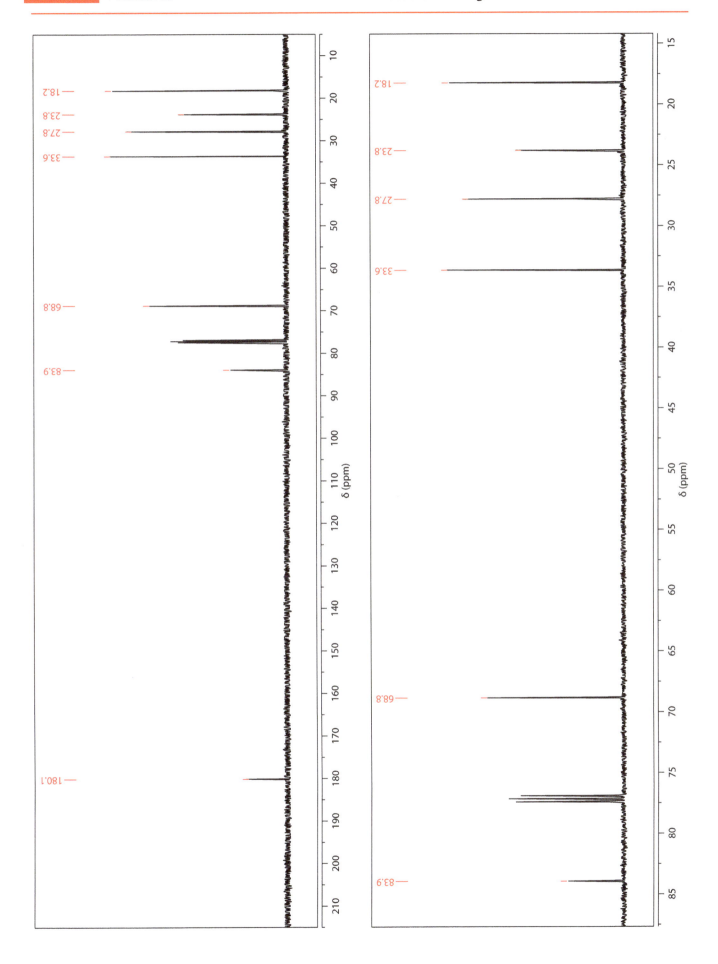

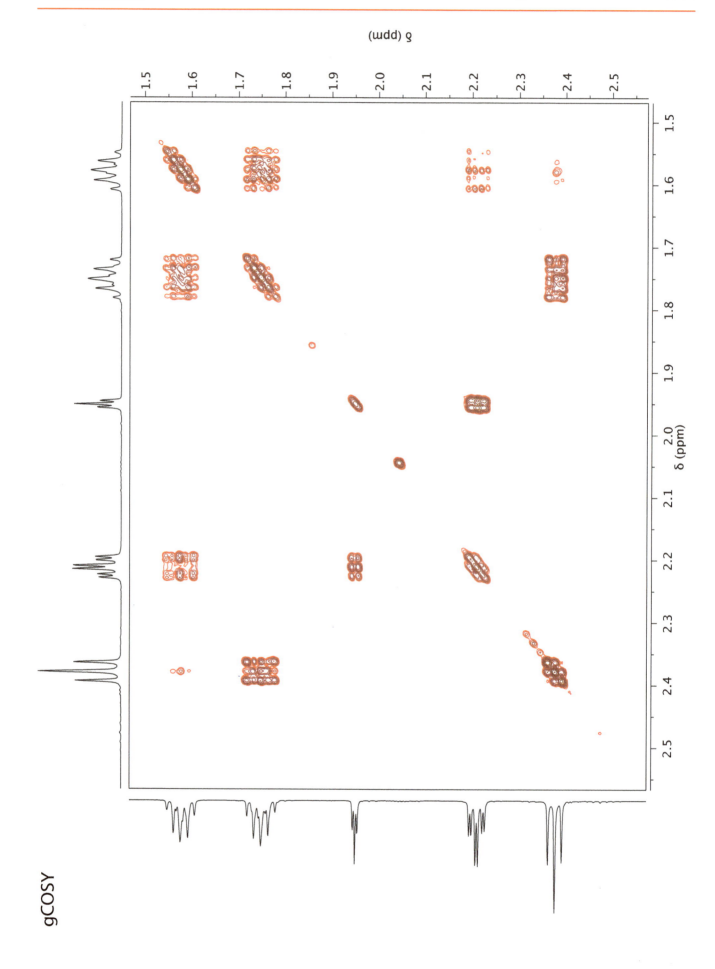

SECTION 3 Problem 69

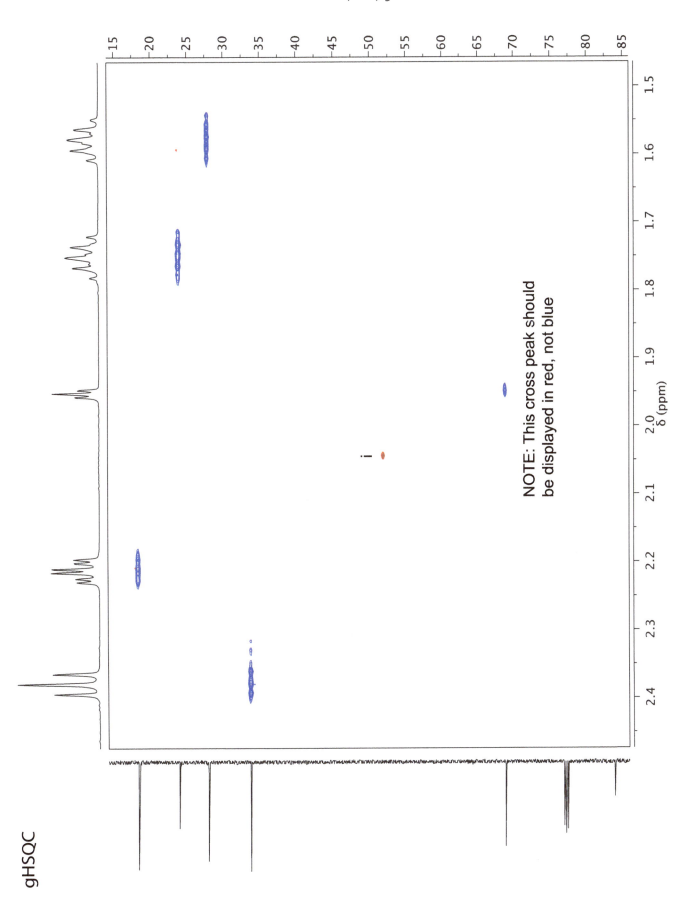

gHSQC

Problem 70

Determine the structure of the unknown compound below with molecular formula $C_5H_8O_2$.

Spectra acquired in $CDCl_3$ at 500 MHz (^1H) or 125 MHz (^{13}C).

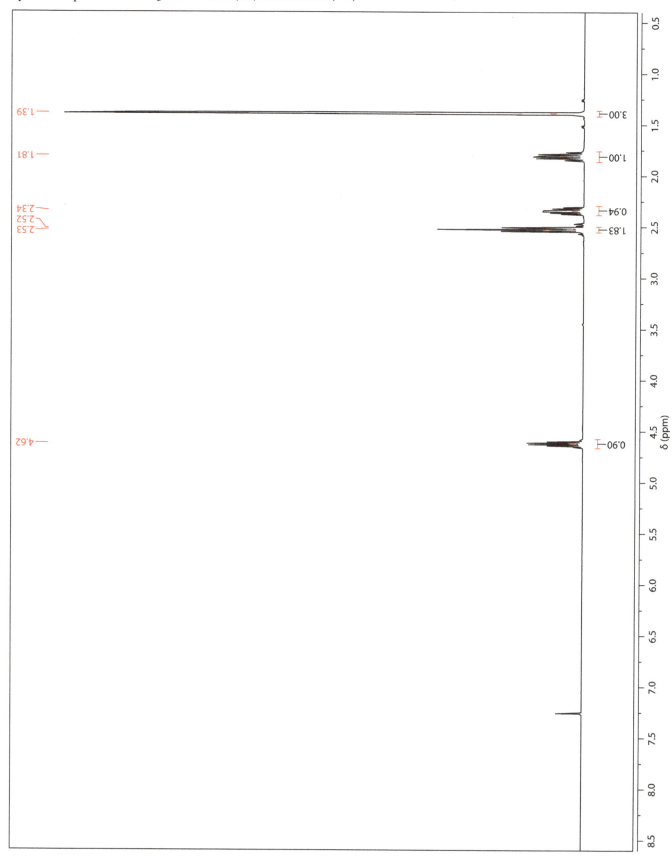

SECTION 3 Problem 70

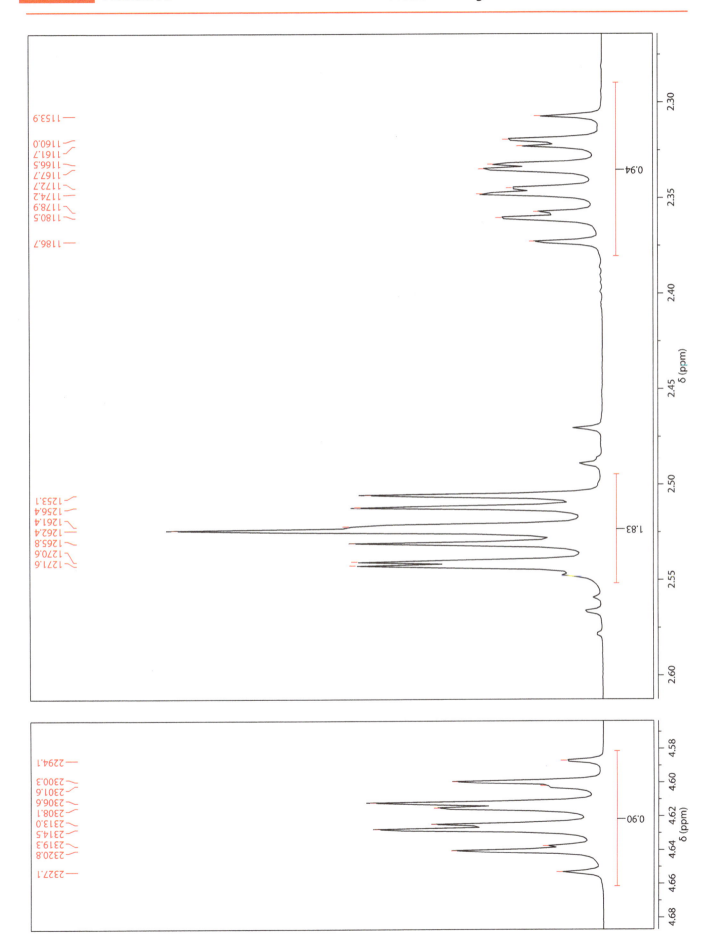

Problem 70

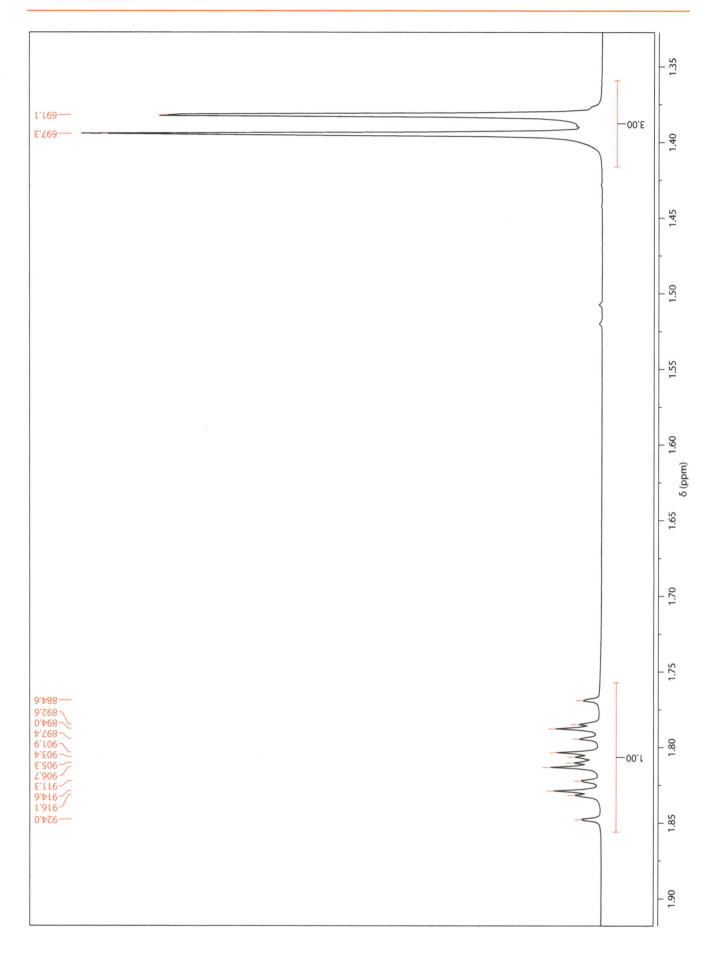

SECTION 3 Problem 70

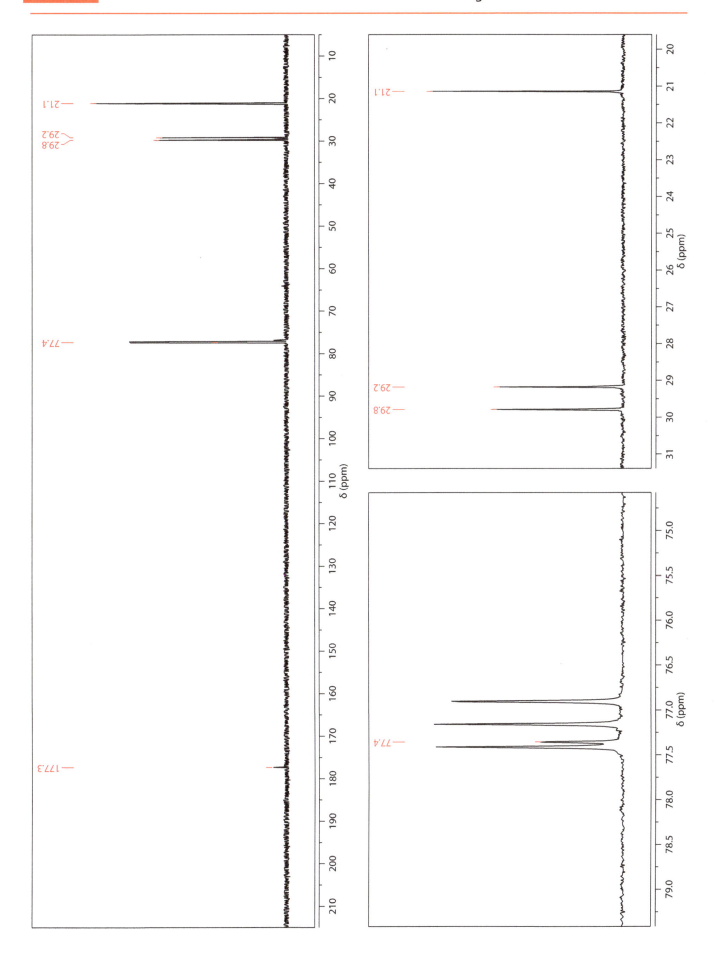

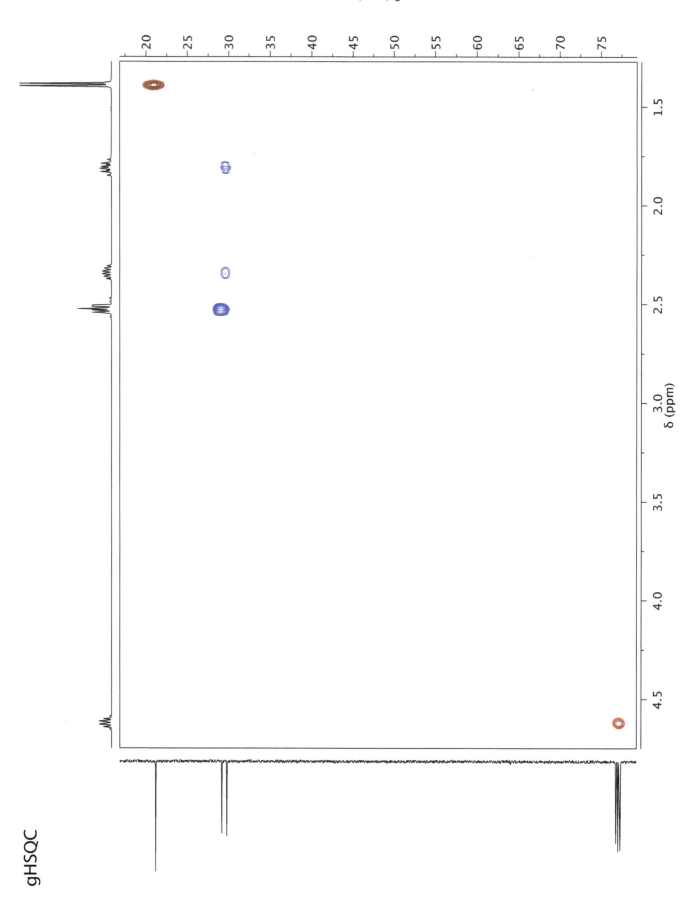

SECTION 3 Problem 71

Determine the structure of the unknown compound below with molecular formula C_8H_7NO.

Spectra acquired in $CDCl_3$ at 500 MHz (^1H) or 125 MHz (^{13}C).

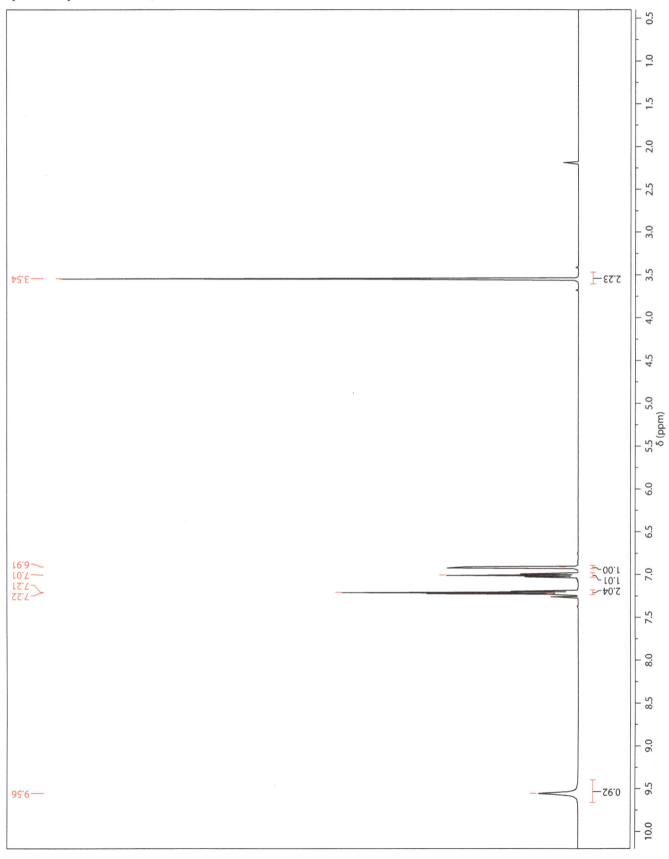

Problem 71

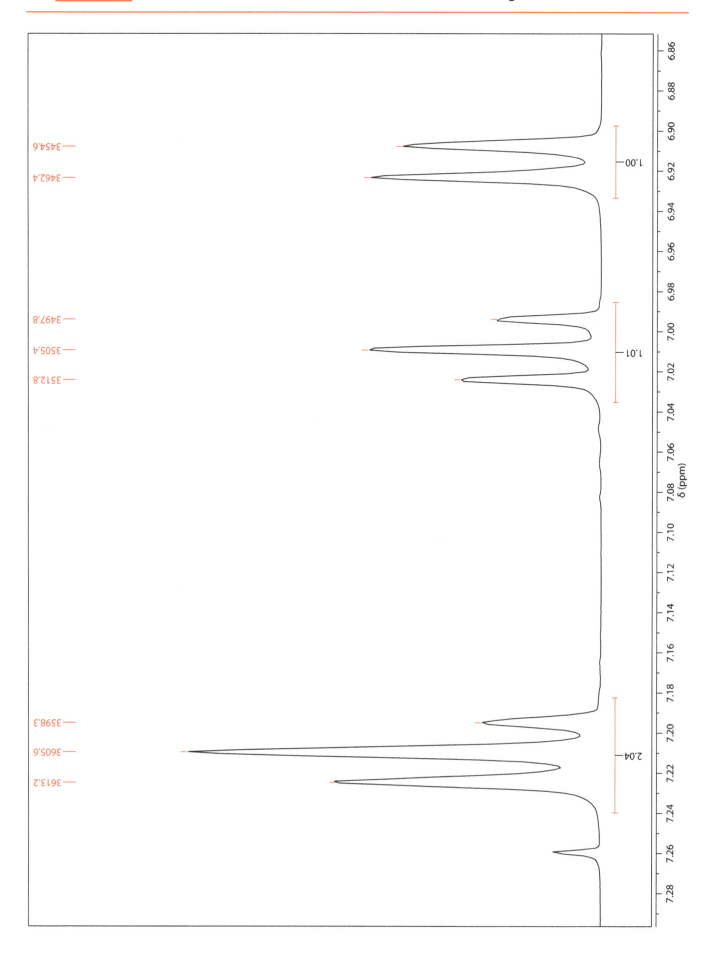

SECTION 3 Problem 71

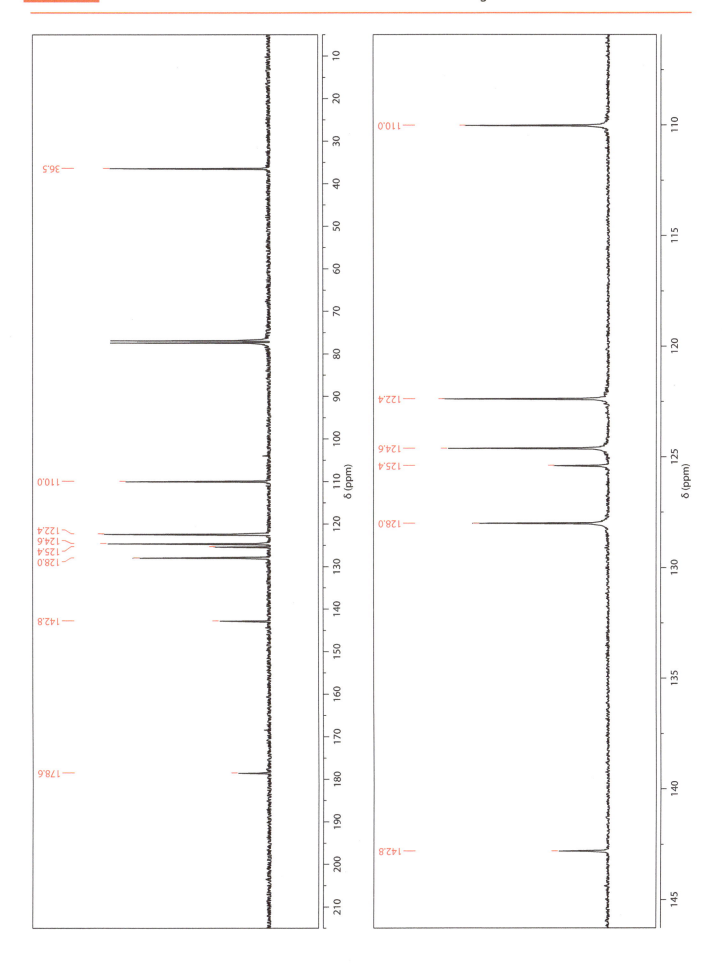

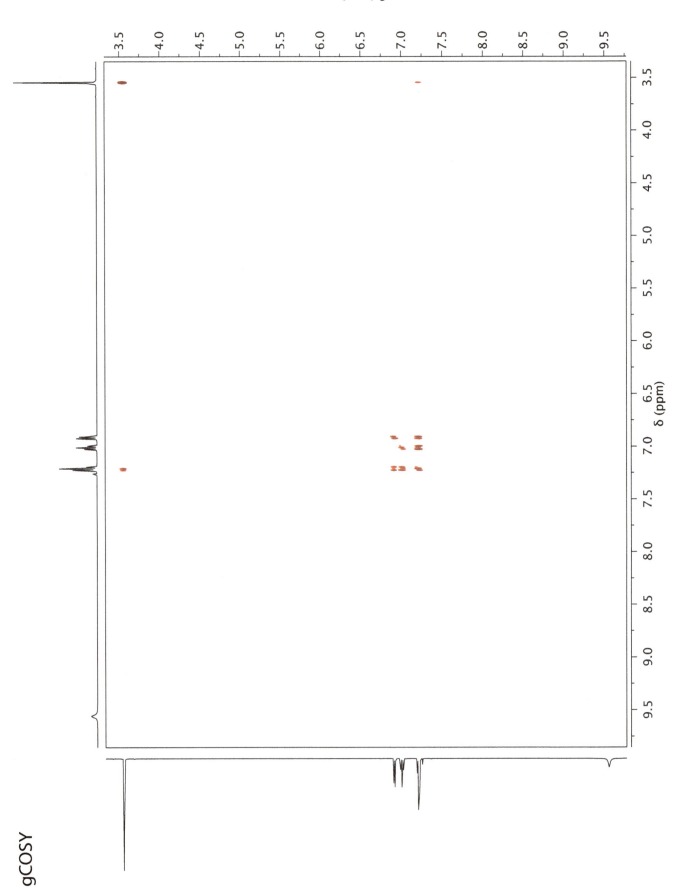

SECTION 3 Problem 71

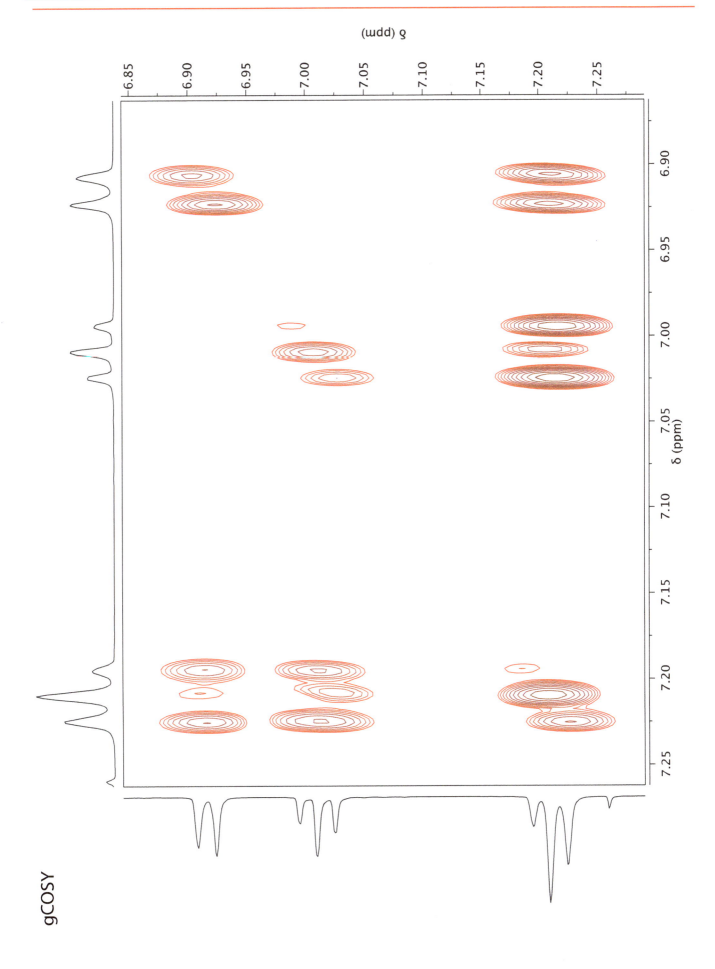

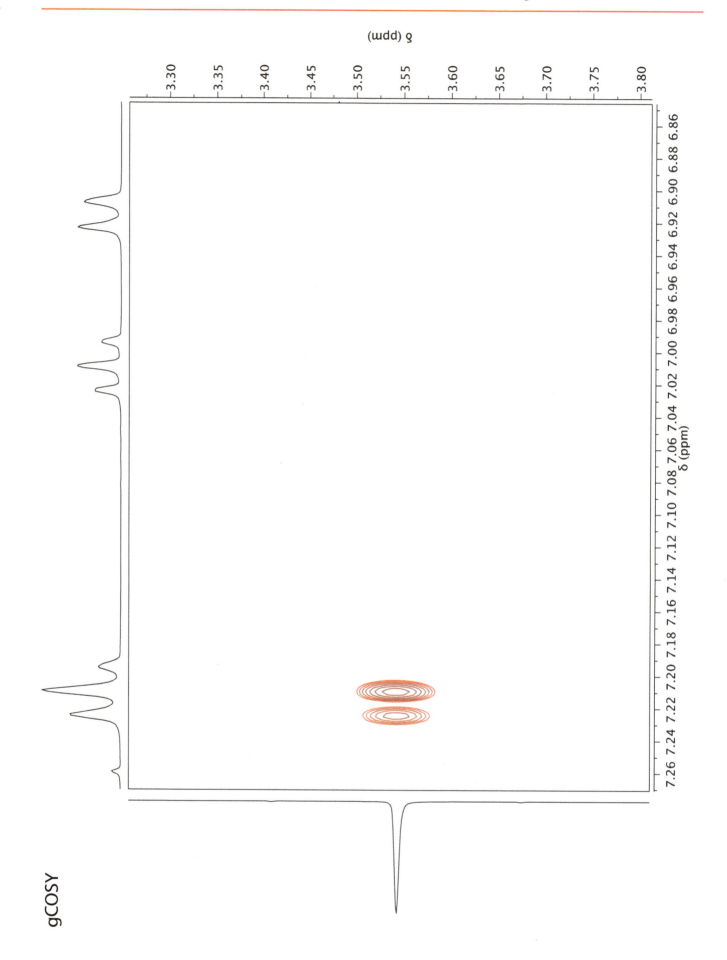

gCOSY

SECTION 3 Problem 71

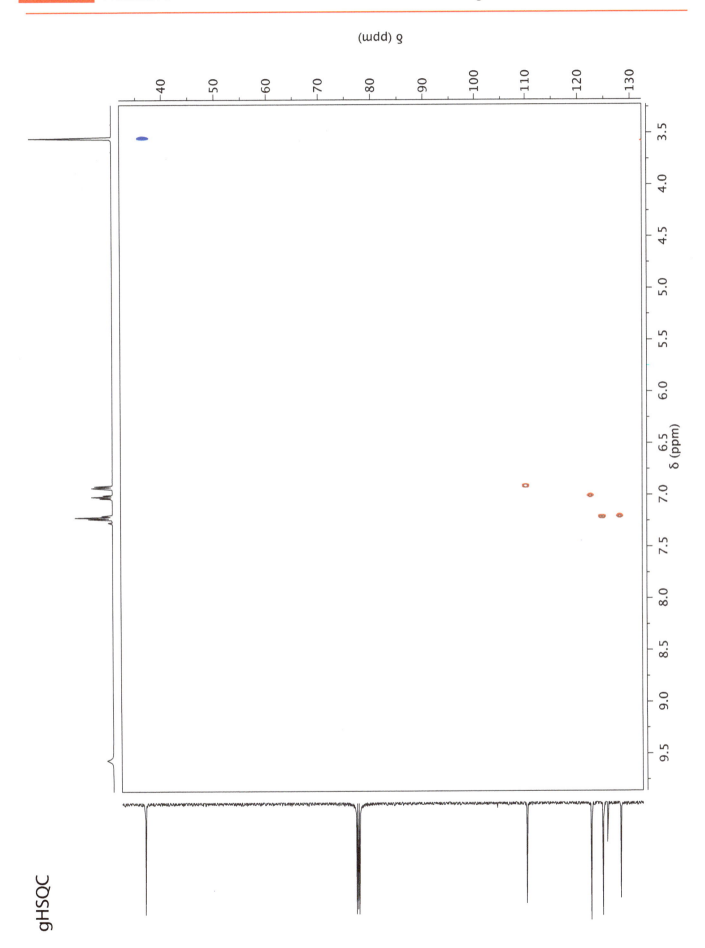

gHSQC

Problem 71

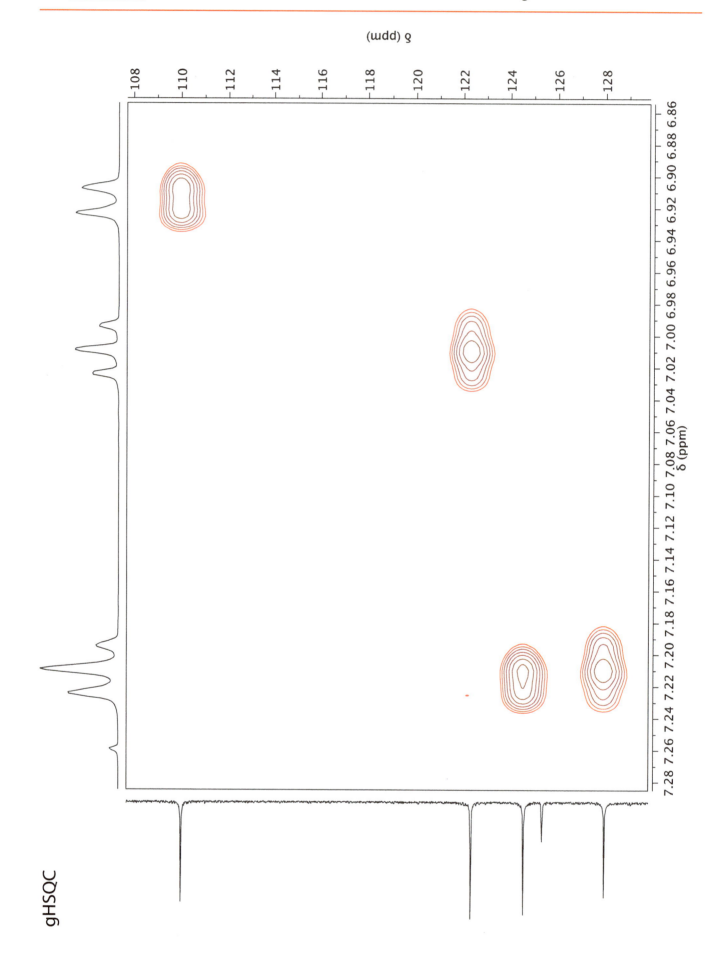

gHSQC

SECTION 3 Problem 72

Determine the structure of the unknown compound below with molecular formula $C_6H_{14}N_2O \cdot HCl$.

Spectra acquired in $(CD_3)_2SO$ at 500 MHz (^1H) or 125 MHz (^{13}C).

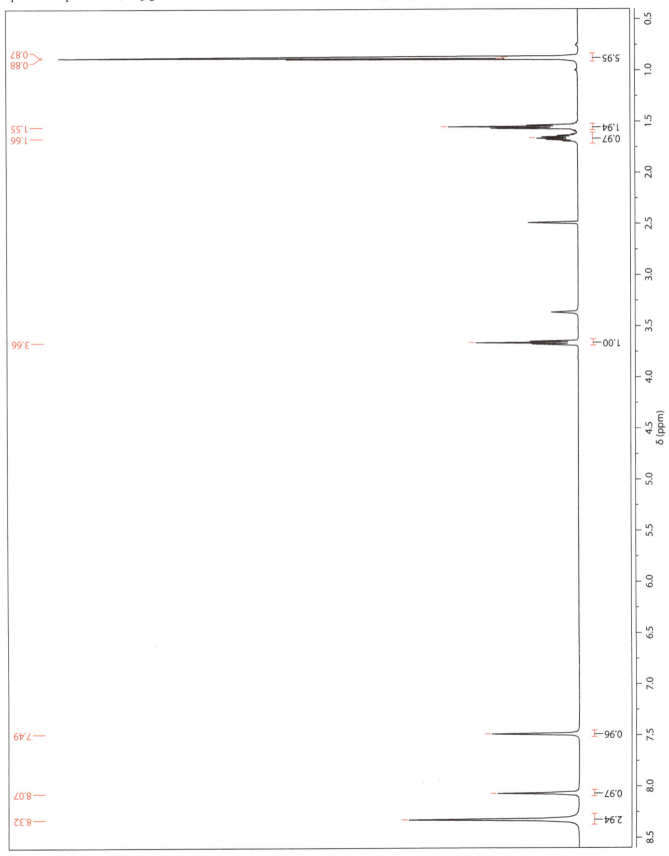

Problem 72

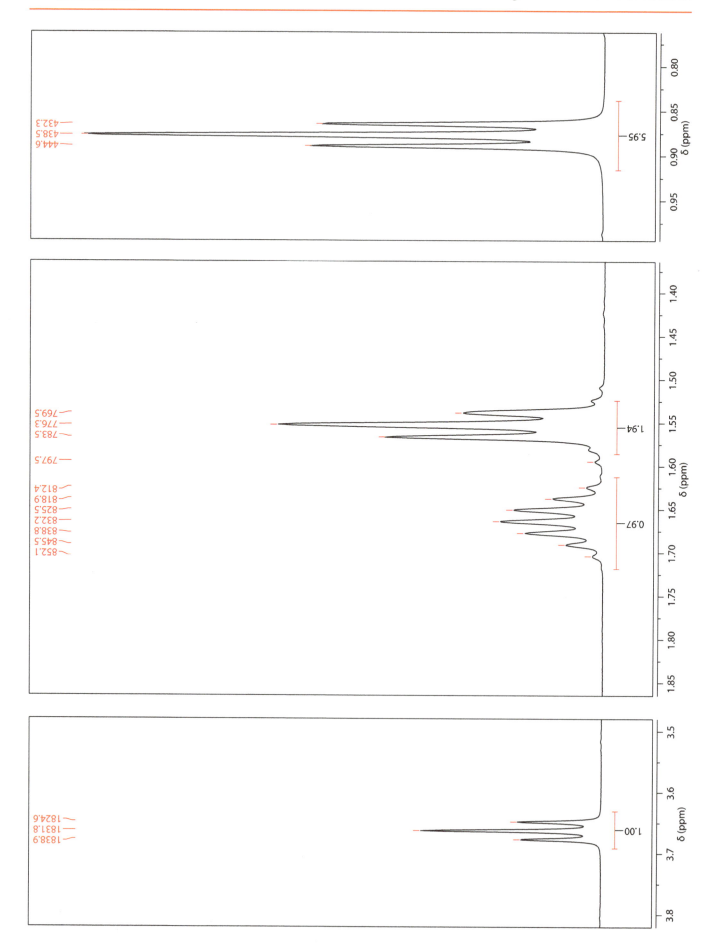

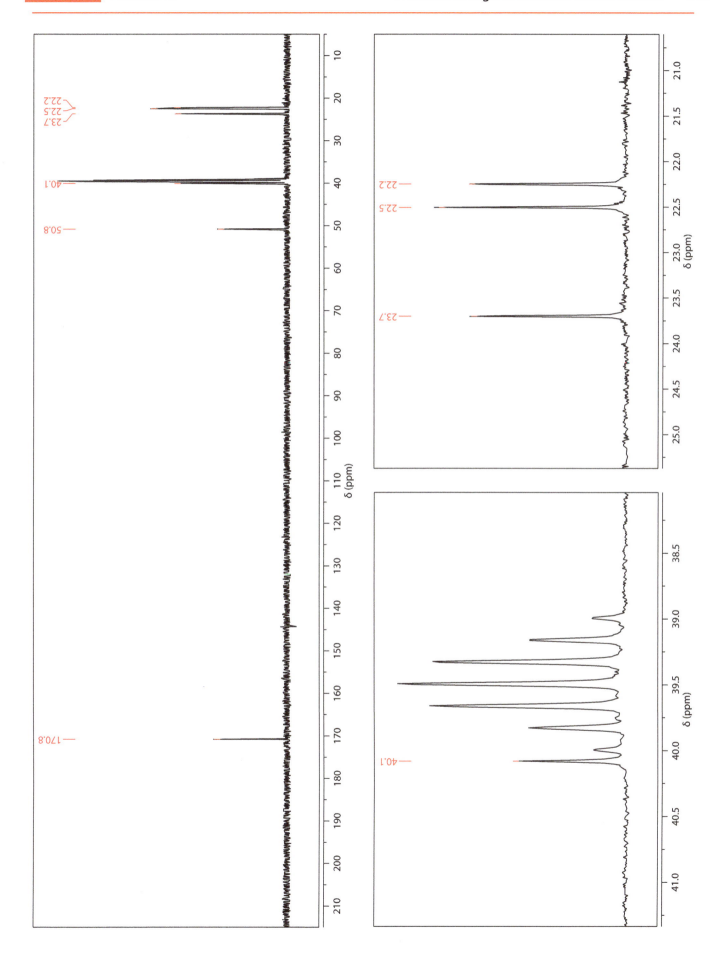

Problem 72

gCOSY

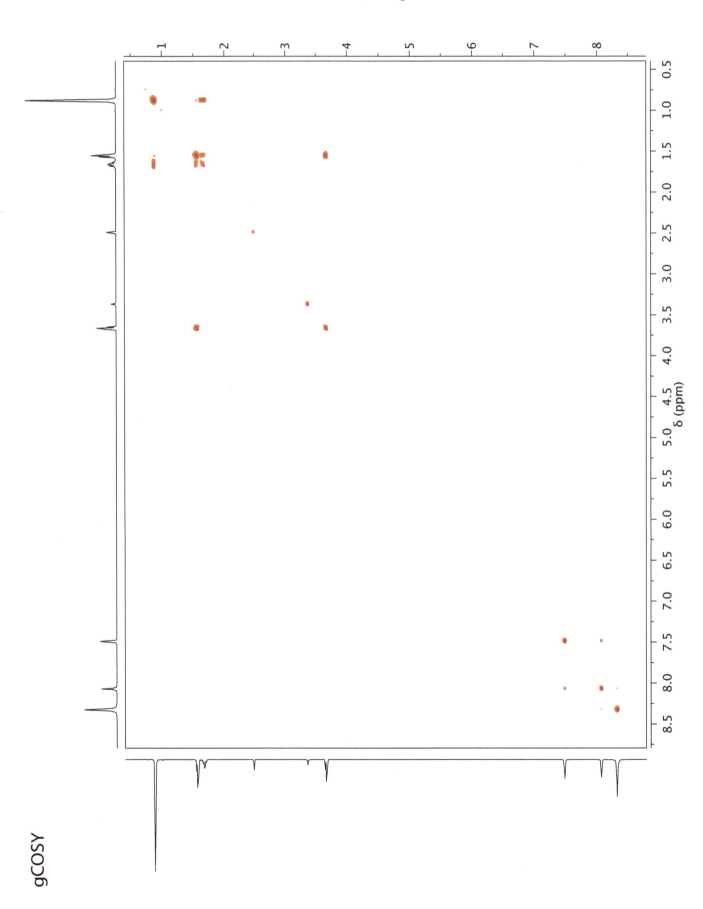

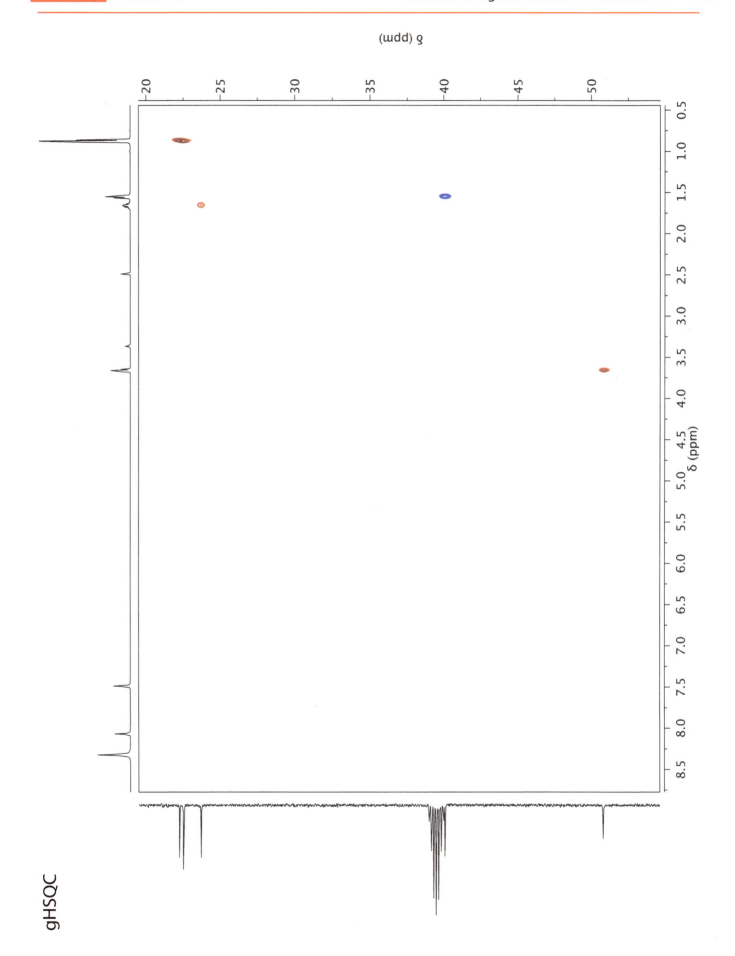

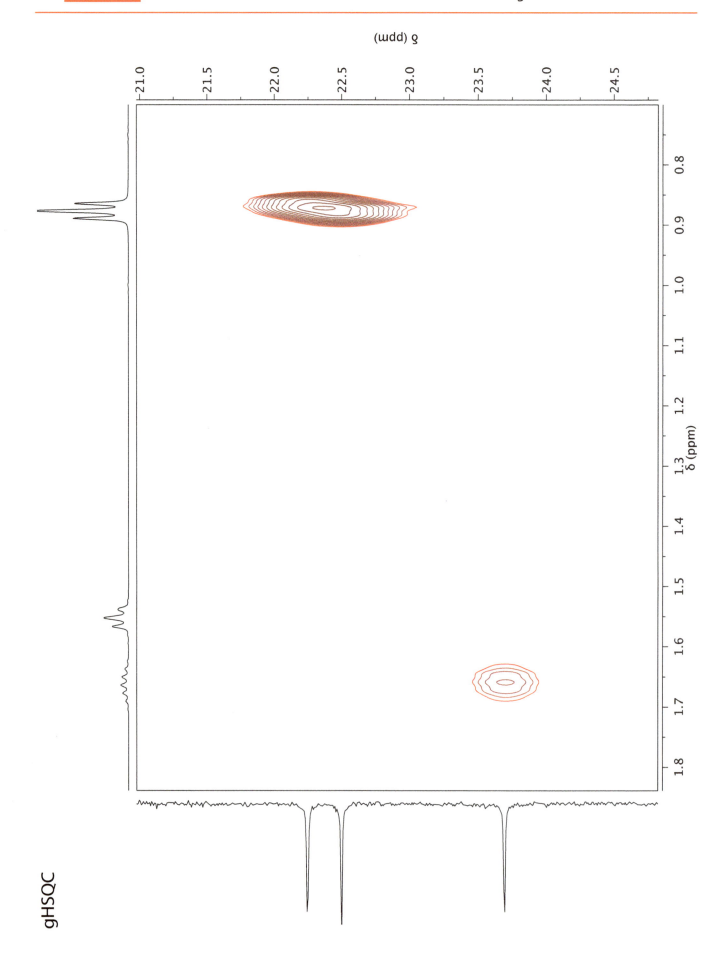

SECTION 3 Problem 72

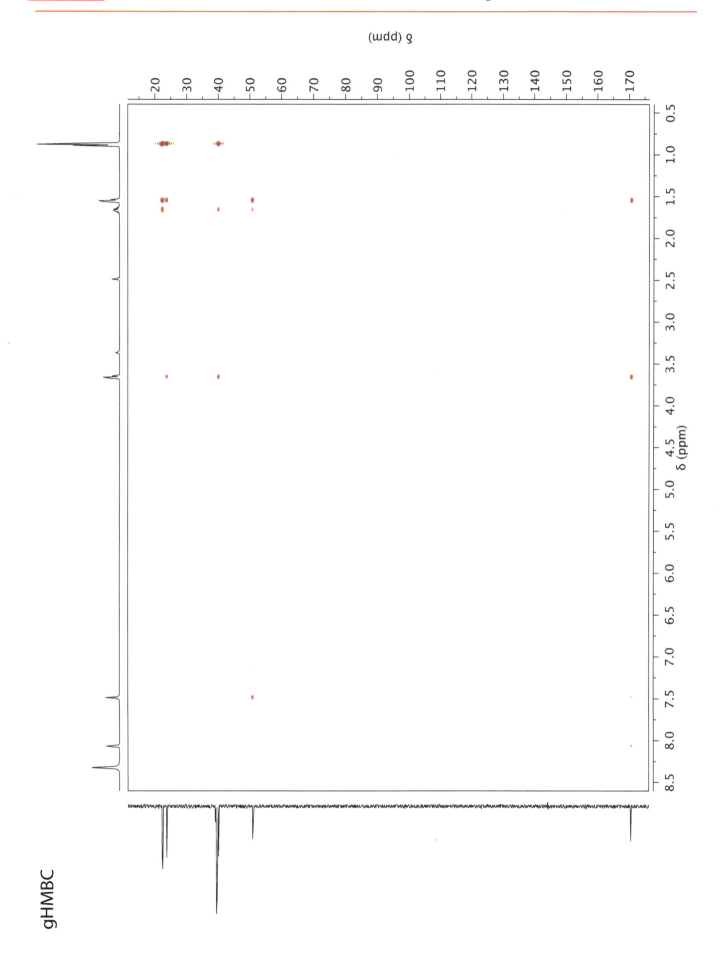

gHMBC

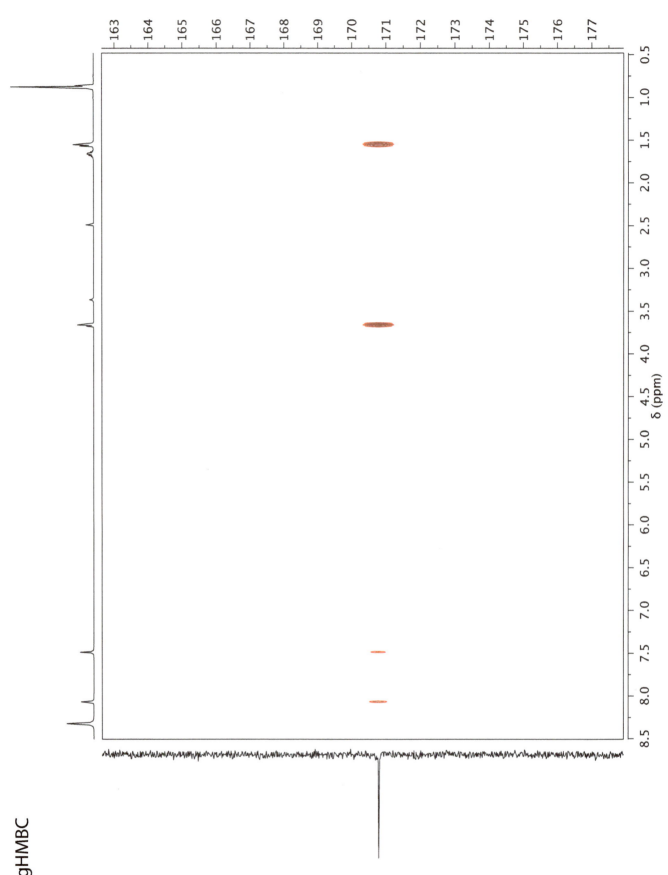

SECTION 3 Problem 72

Problems in Organic Structure Determination

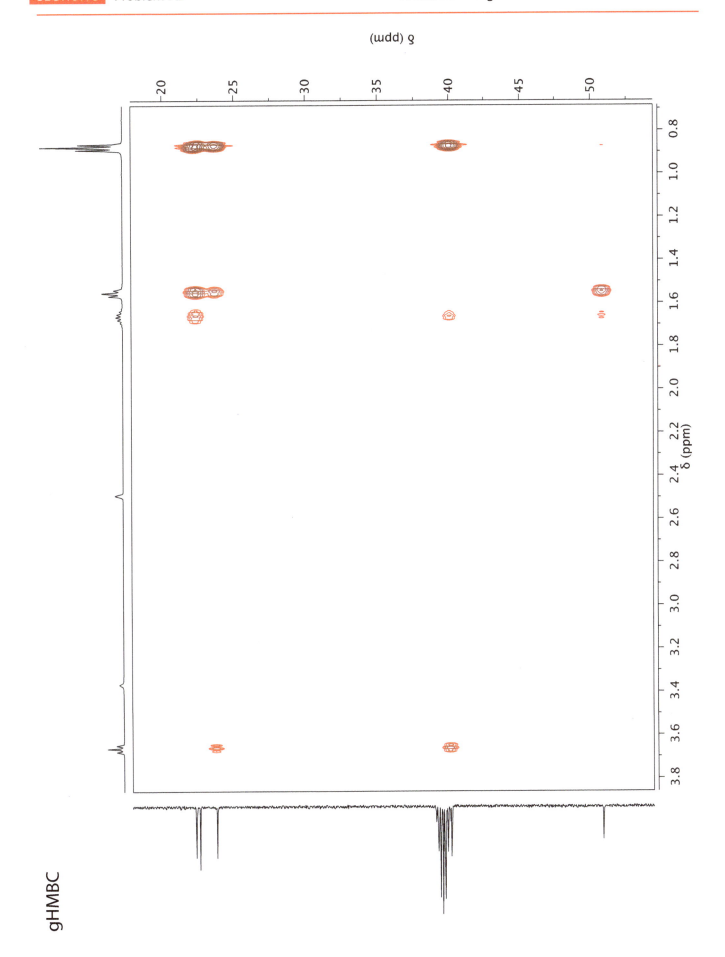

gHMBC

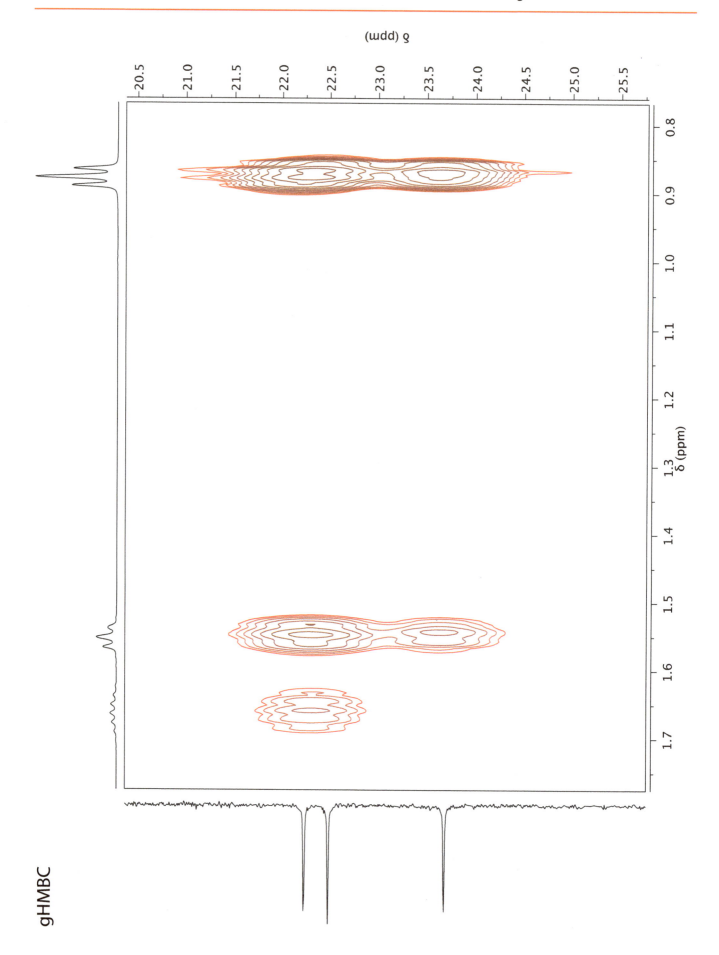

SECTION 3 Problem 73

Determine the structure of the unknown compound below with molecular formula $C_{10}H_9NO_3$.

Spectra acquired in $CDCl_3$ at 500 MHz (^1H) or 125 MHz (^{13}C).

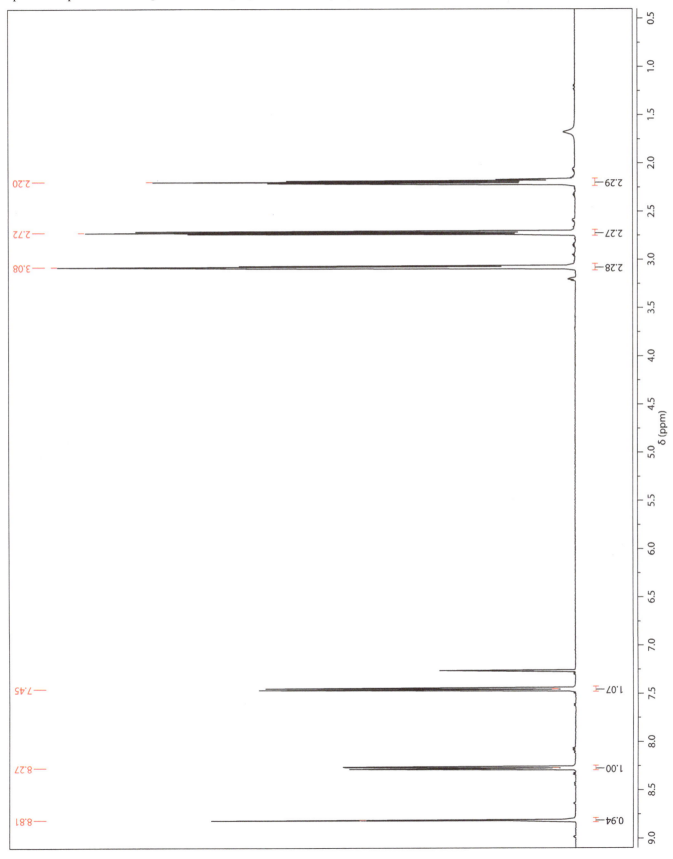

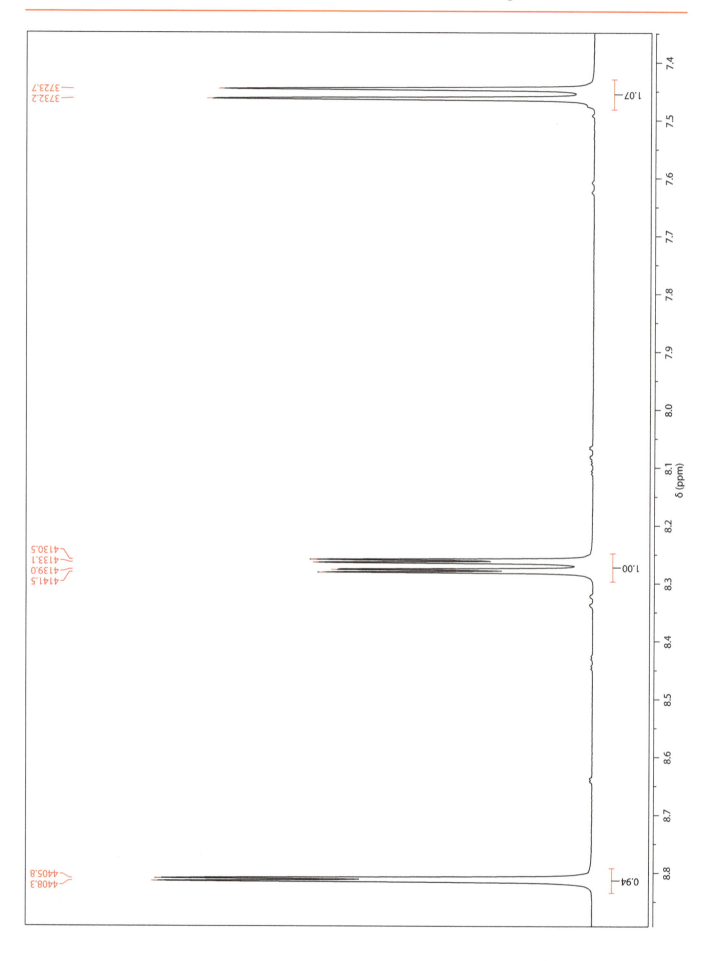

SECTION 3 Problem 73

SECTION 3 Problem 73

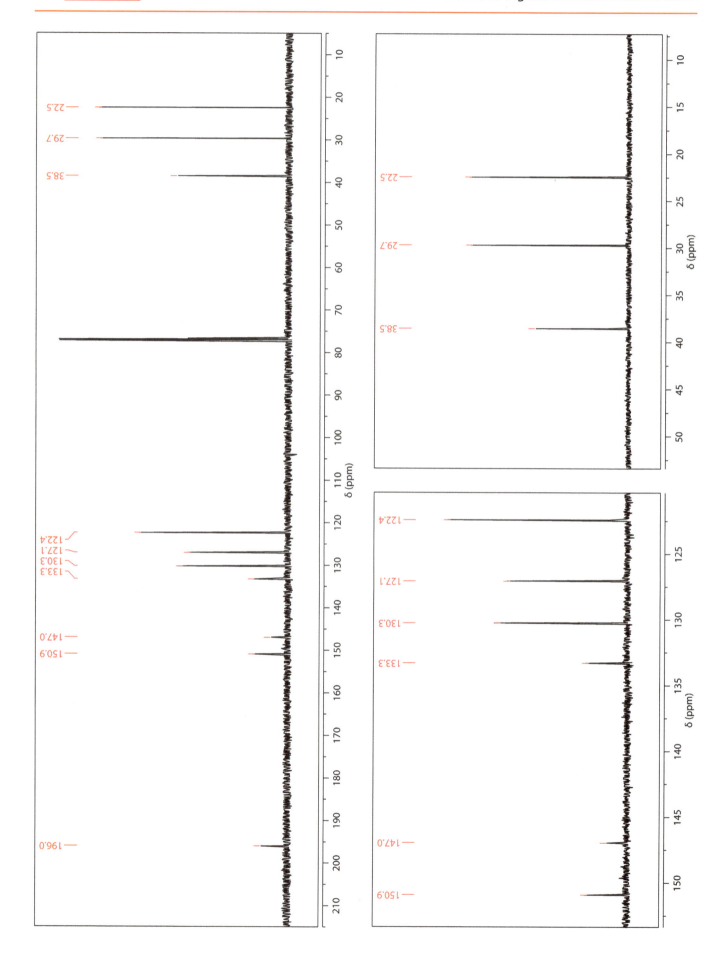

SECTION 3 Problem 73

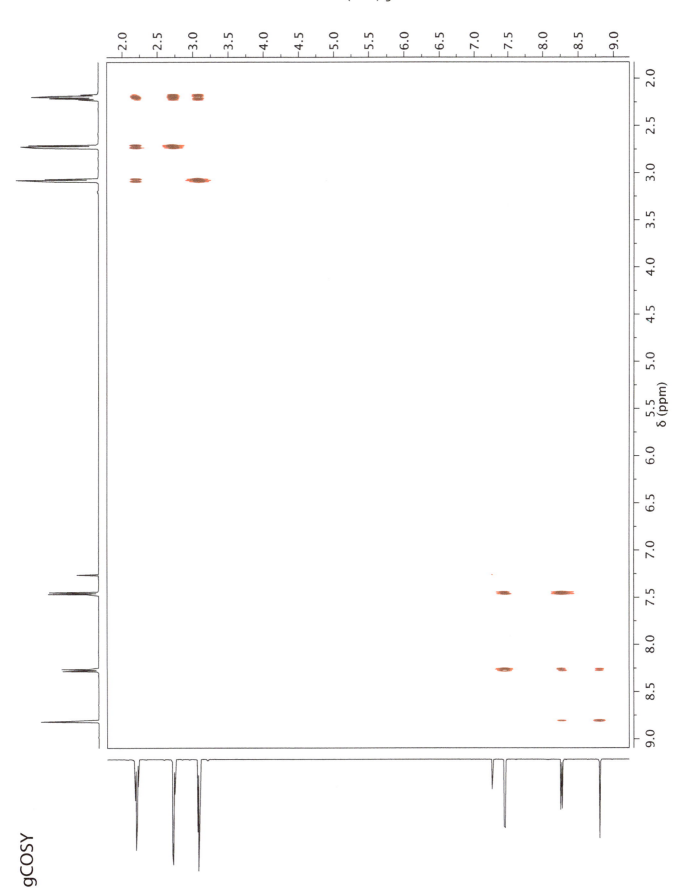

gCOSY

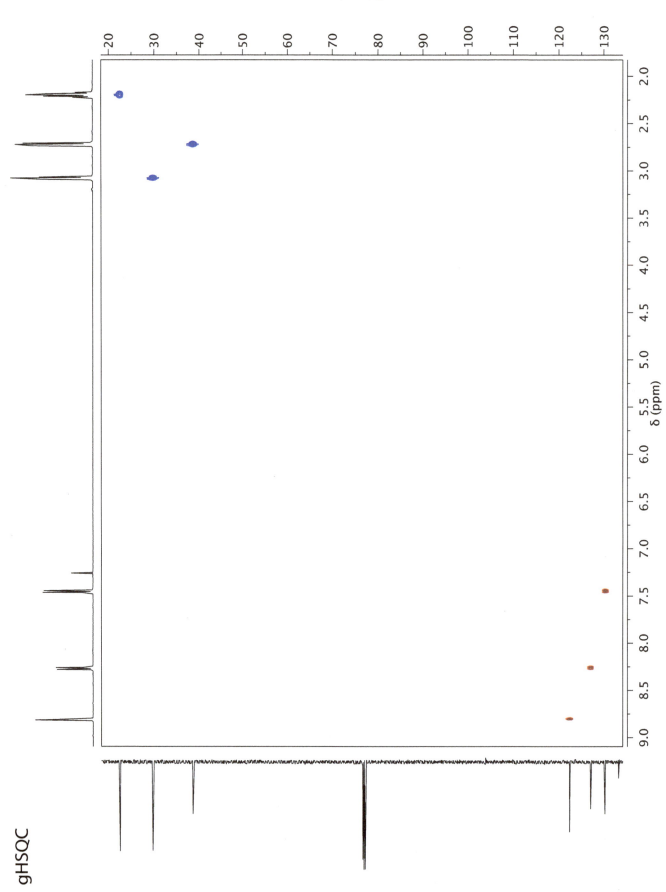

Problem 73

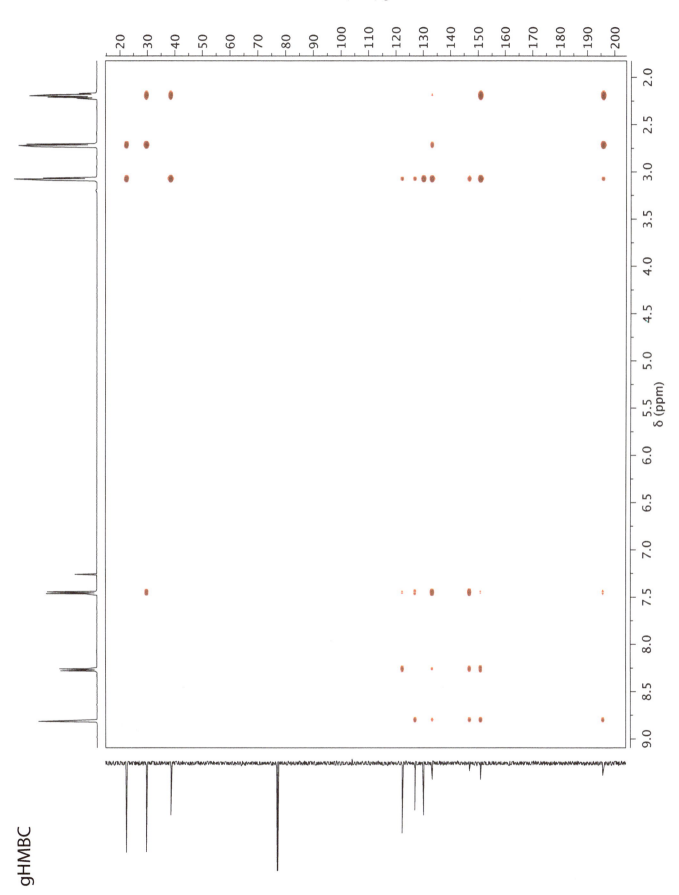

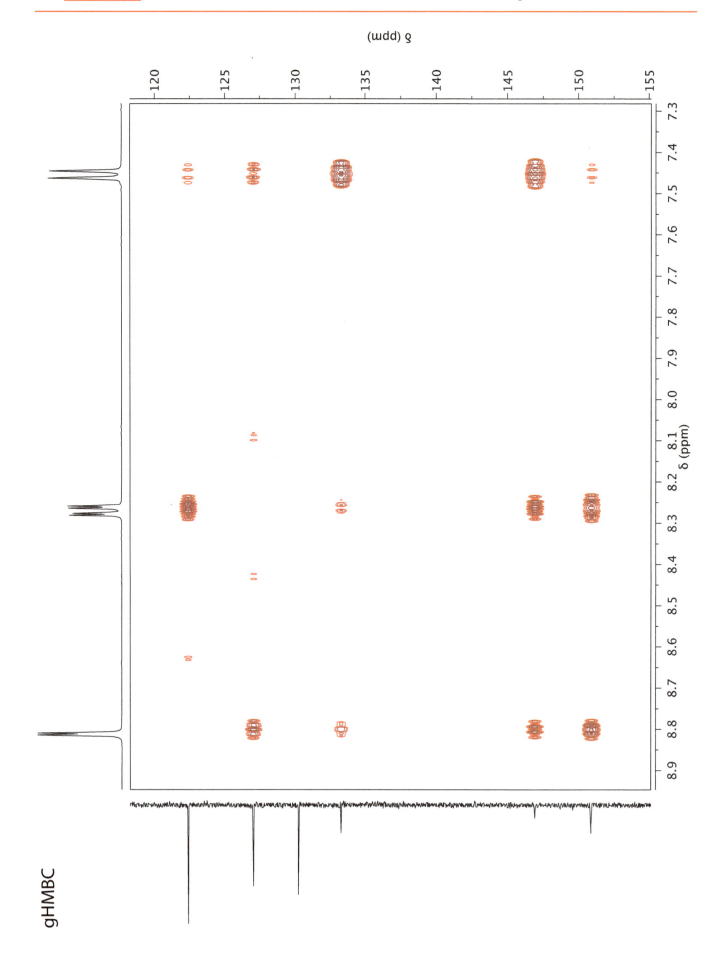

SECTION 3 Problem 73

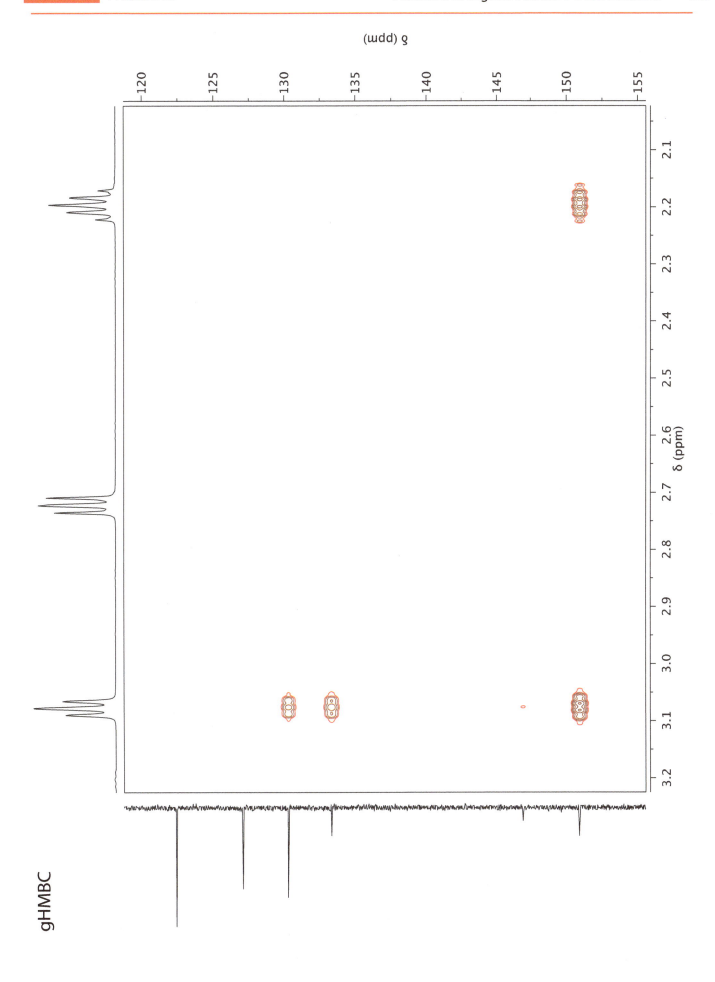

Problem 74

Determine the structure of the unknown compound below with molecular formula $C_9H_{18}O_6$.

Spectra acquired in $(CD_3)_2SO$ at 500 MHz (^1H) or 125 MHz (^{13}C).

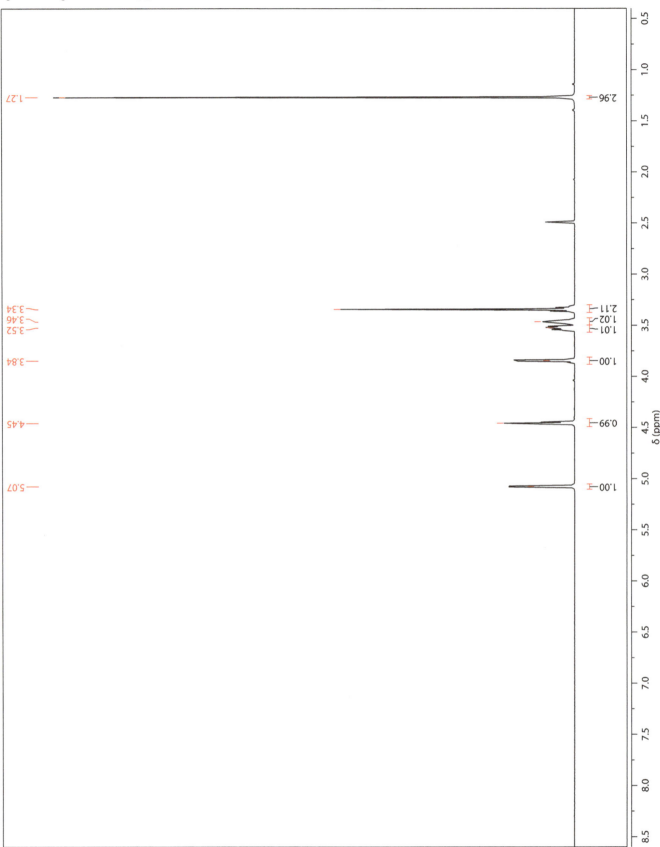

SECTION 3 Problem 74

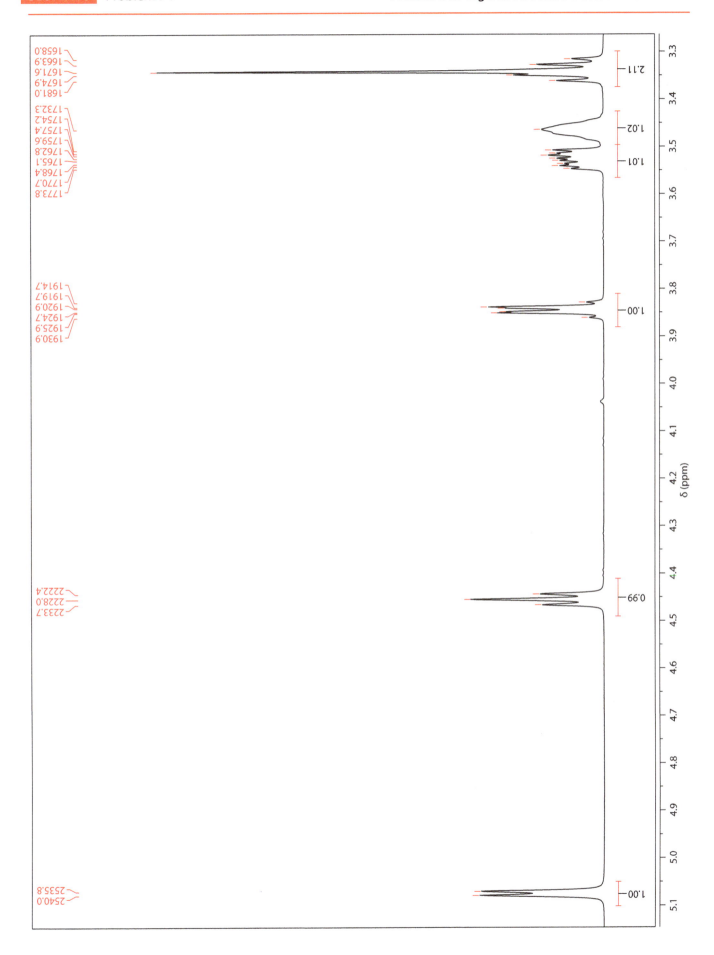

Problem 74

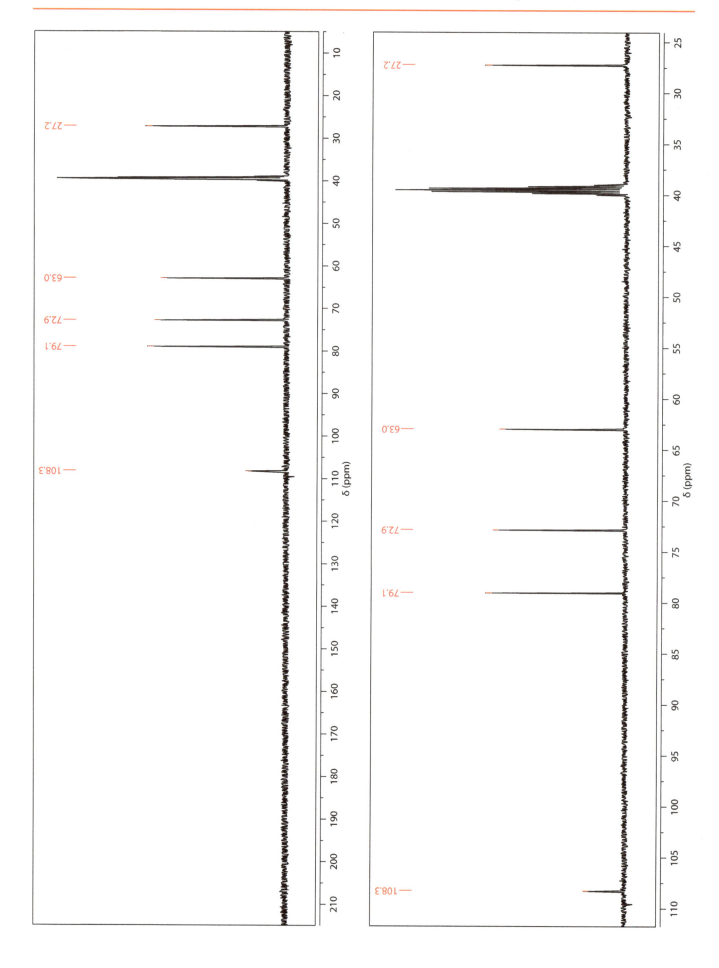

SECTION 3 Problem 74

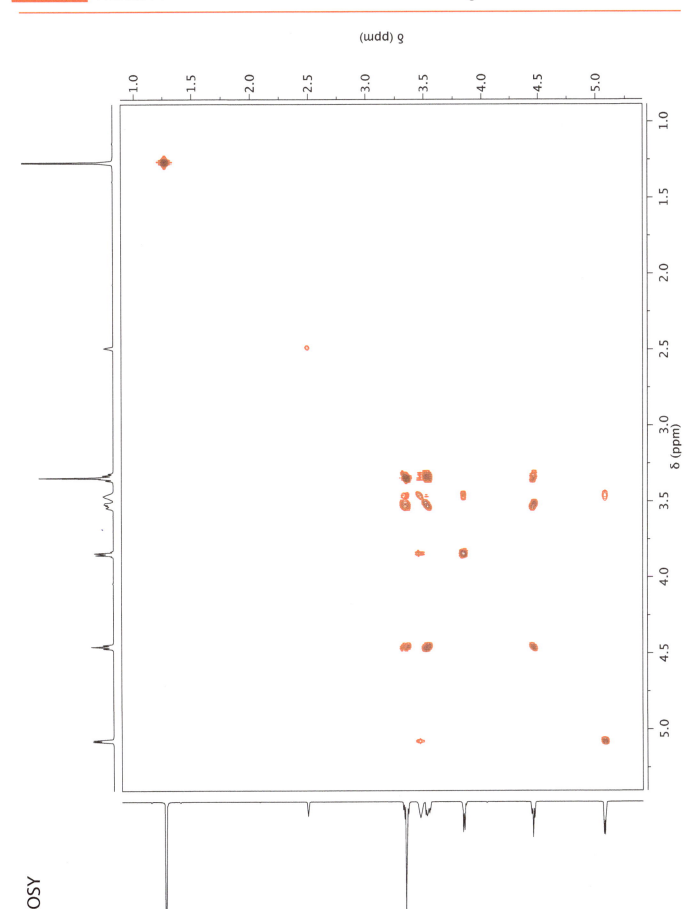

gCOSY

SECTION 3 Problem 74

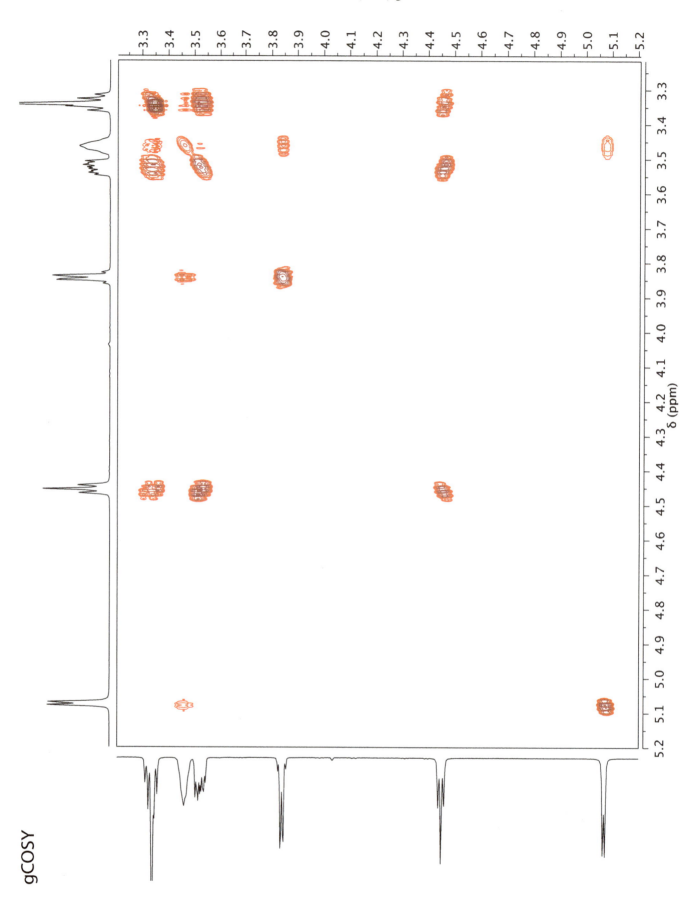

gCOSY

SECTION 3 Problem 74

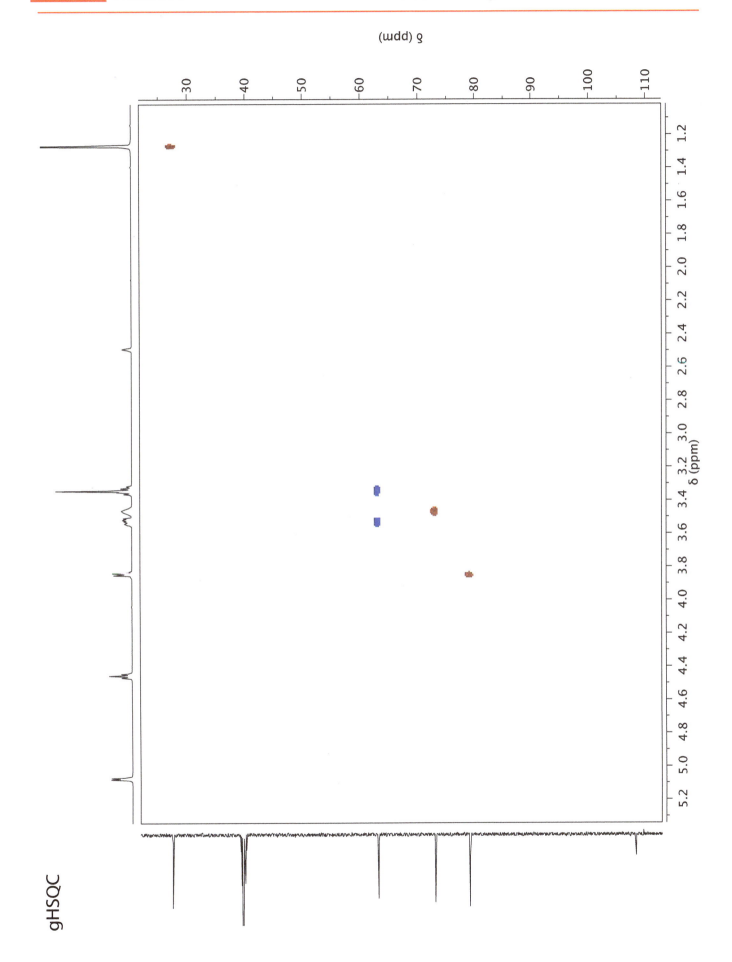

gHSQC

Problem 74

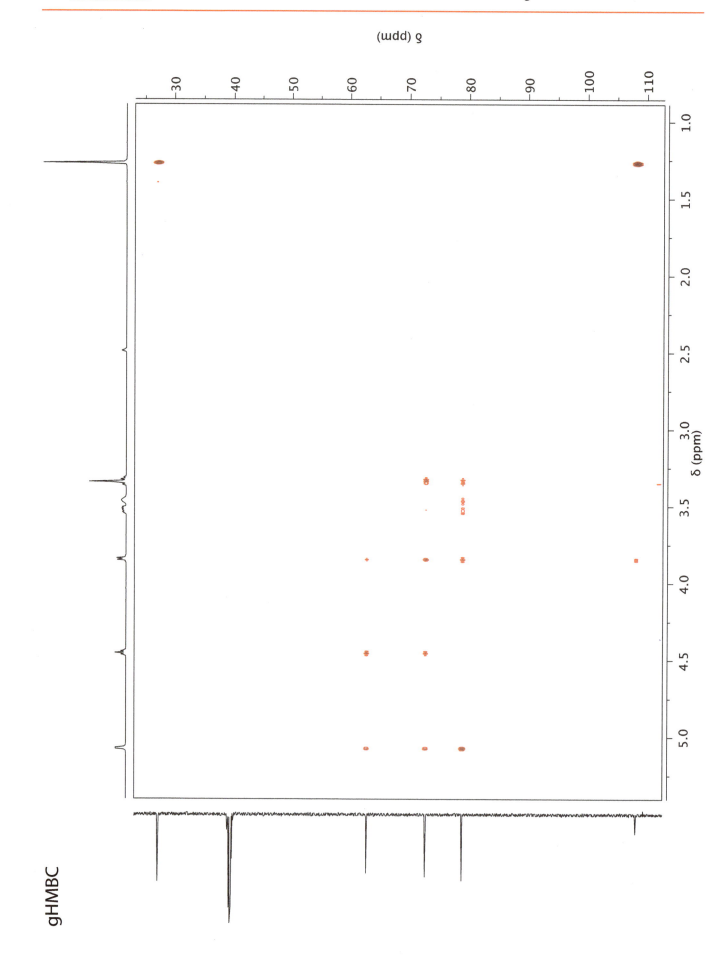

SECTION 3 Problem 74

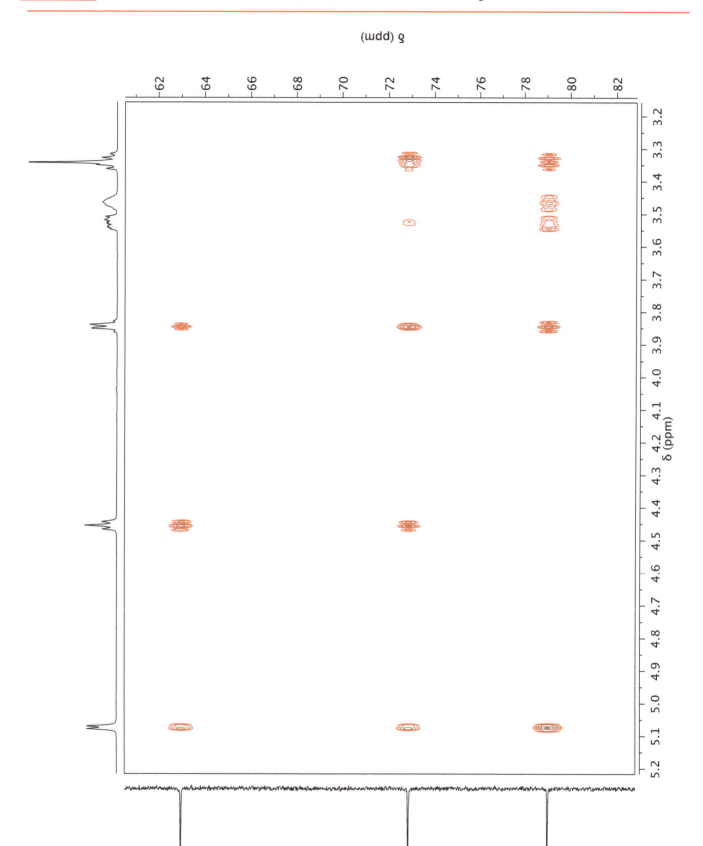

gHMBC

Problem 75

Determine the structure of the unknown compound below with molecular formula $C_{10}H_{11}NO_2$.

Spectra acquired in $CDCl_3$ at 500 MHz (^1H) or 125 MHz (^{13}C).

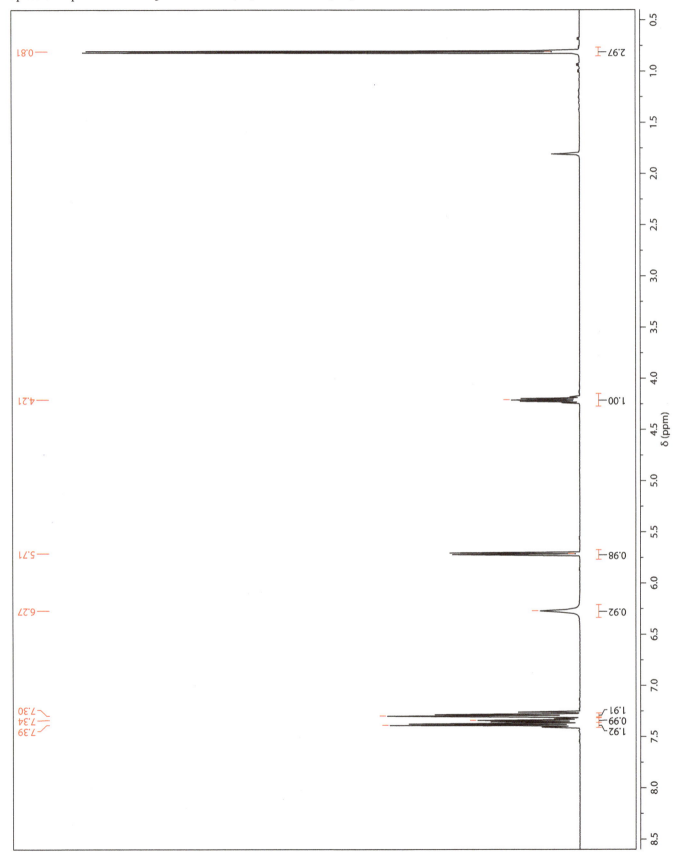

SECTION 3 Problem 75

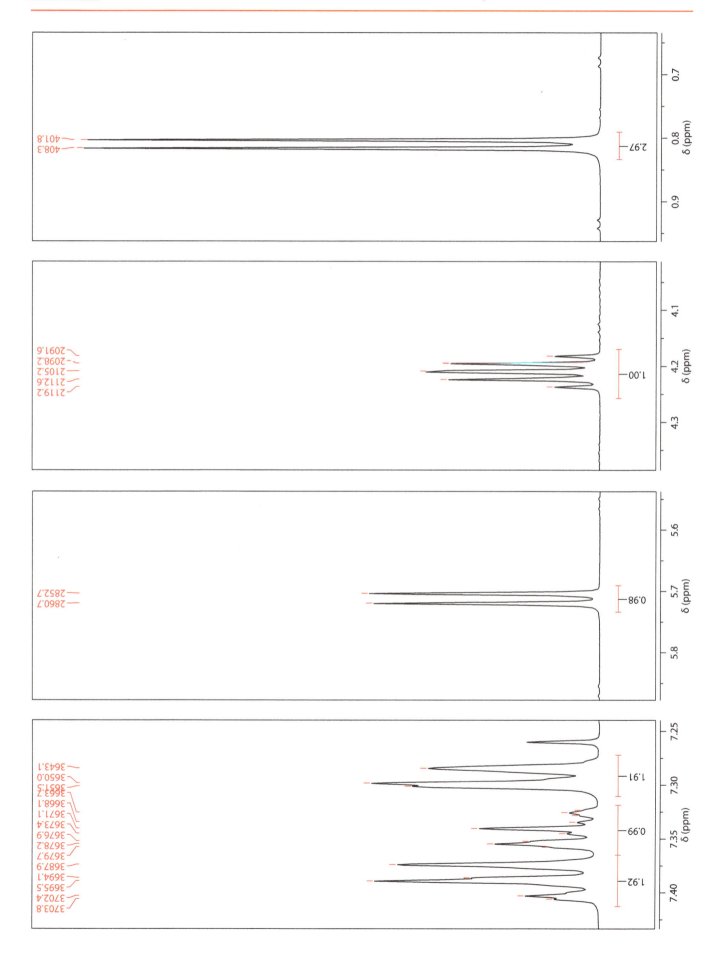

Problem 75

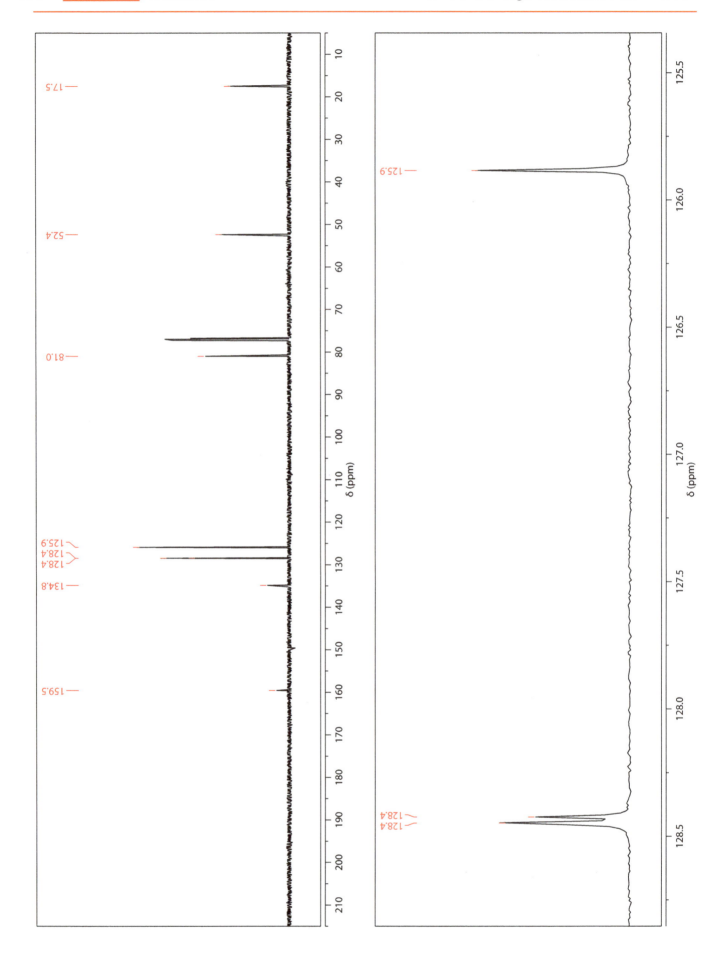

SECTION 3 Problem 75

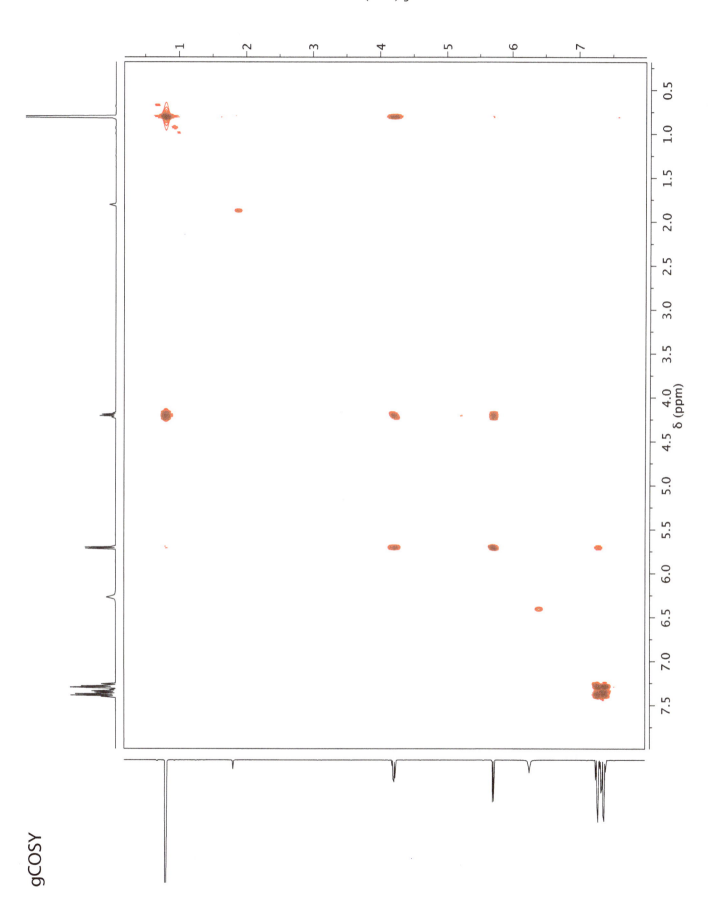

gCOSY

SECTION 3 Problem 75

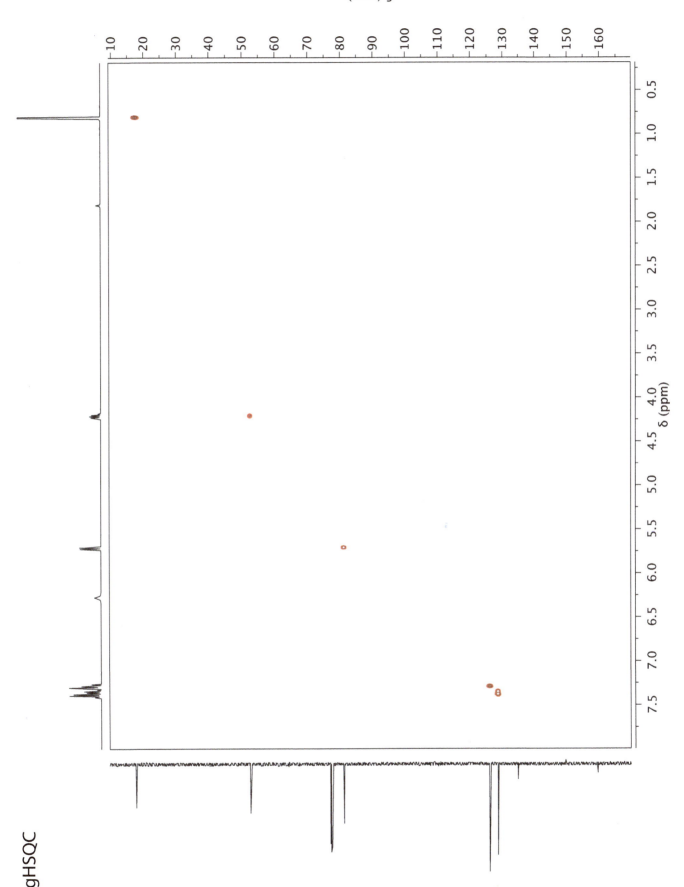

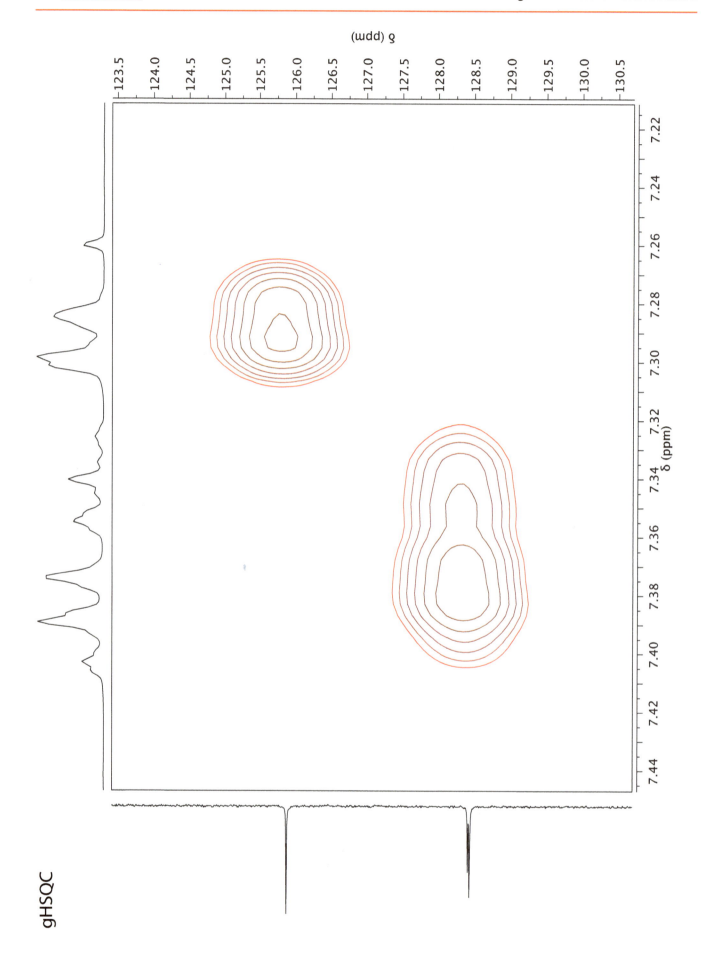

gHSQC

SECTION 3 Problem 75

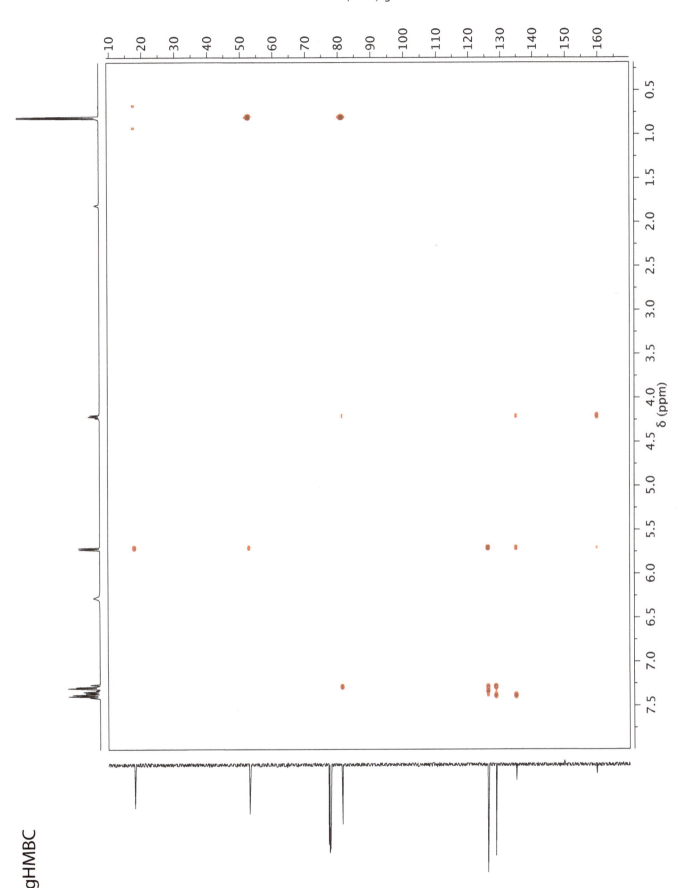

gHMBC

Problem 75

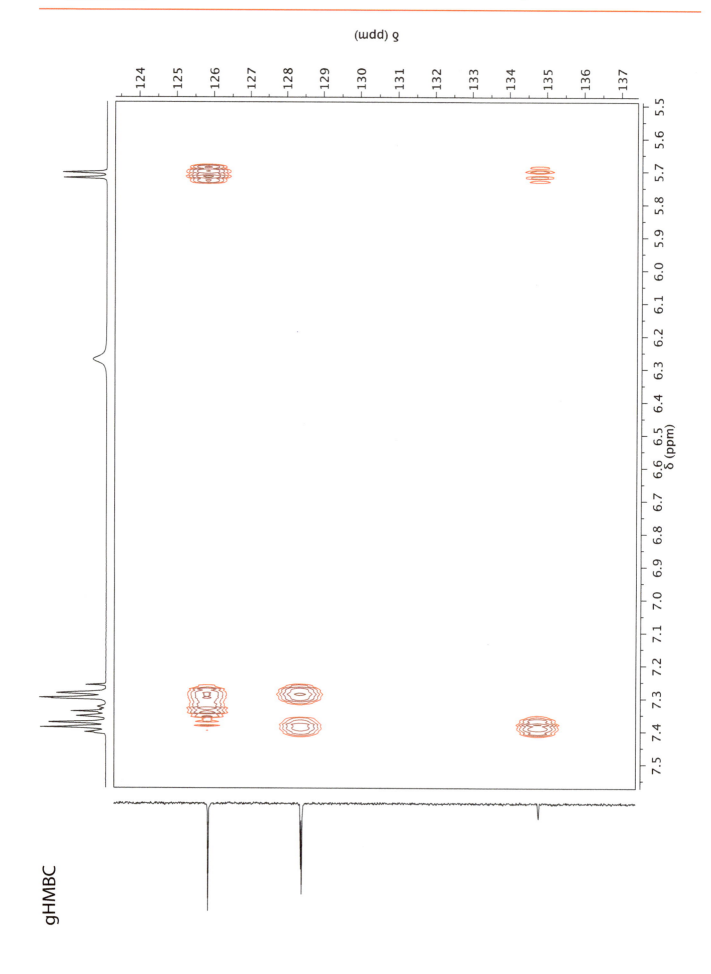

gHMBC

SECTION 3 Problem 76

Determine the structure of the unknown compound below with molecular formula $C_{16}H_{21}NO_6$.

Spectra acquired in $CDCl_3$ at 500 MHz (^1H) or 125 MHz (^{13}C).

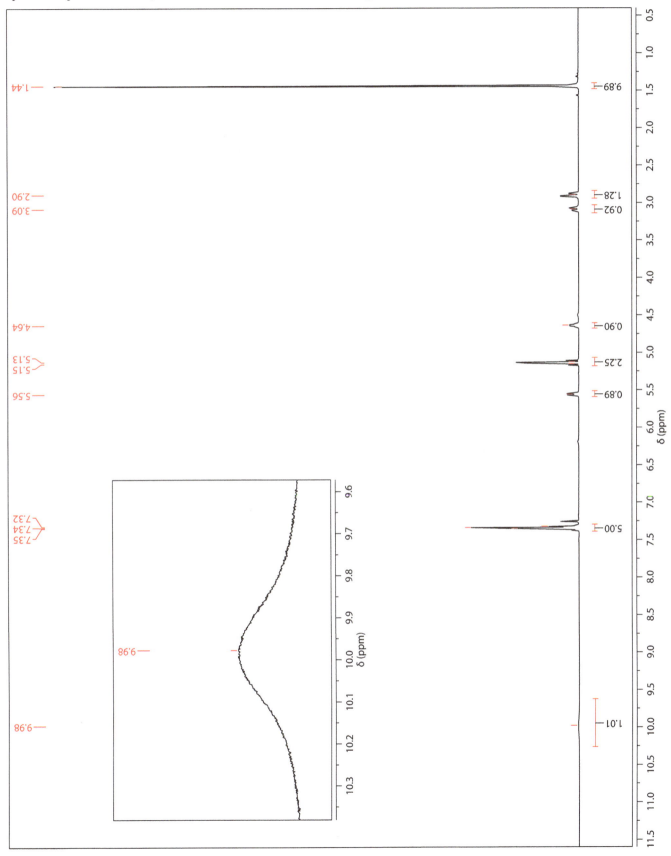

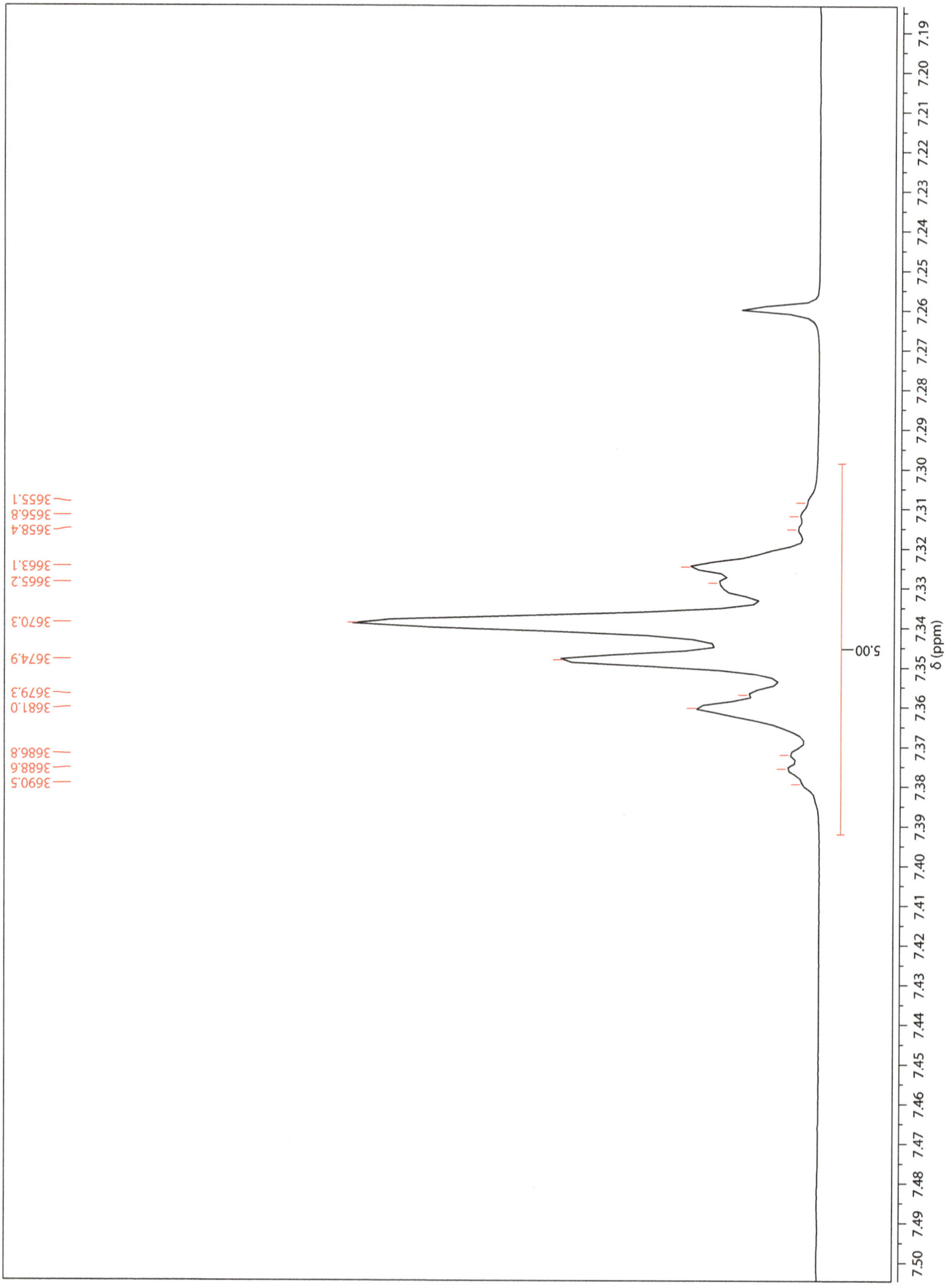

SECTION 3 Problem 76

Frequencies (Hz):
3655.1
3656.8
3658.4
3663.1
3665.2
3670.3
3674.9
3679.3
3681.0
3686.8
3688.6
3690.5

Integration: 5.00

δ (ppm) range: 7.19 – 7.50

Problem 76

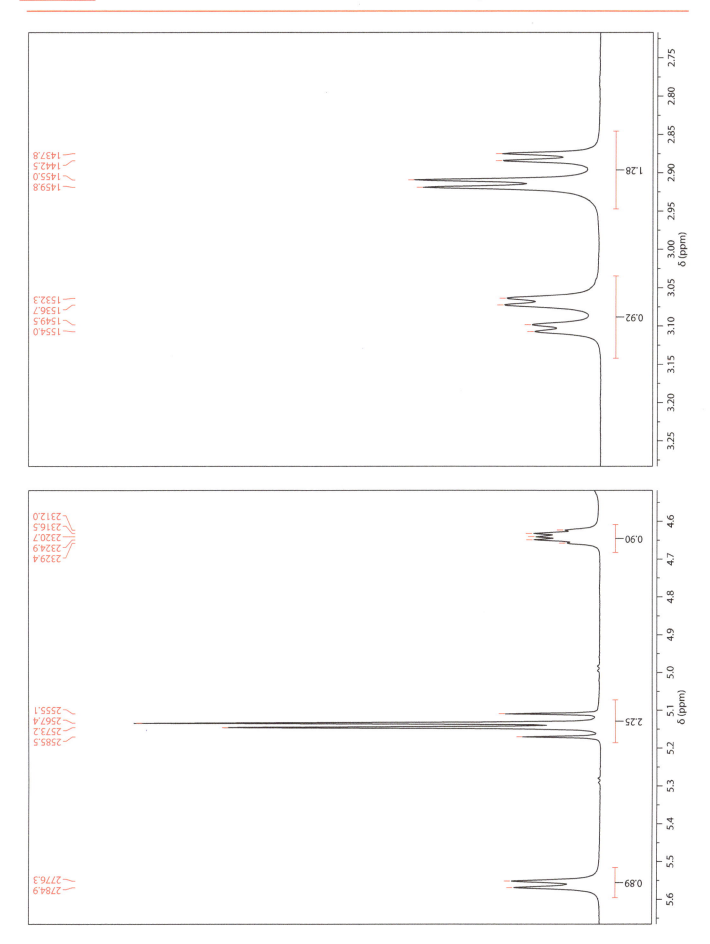

Problem 76

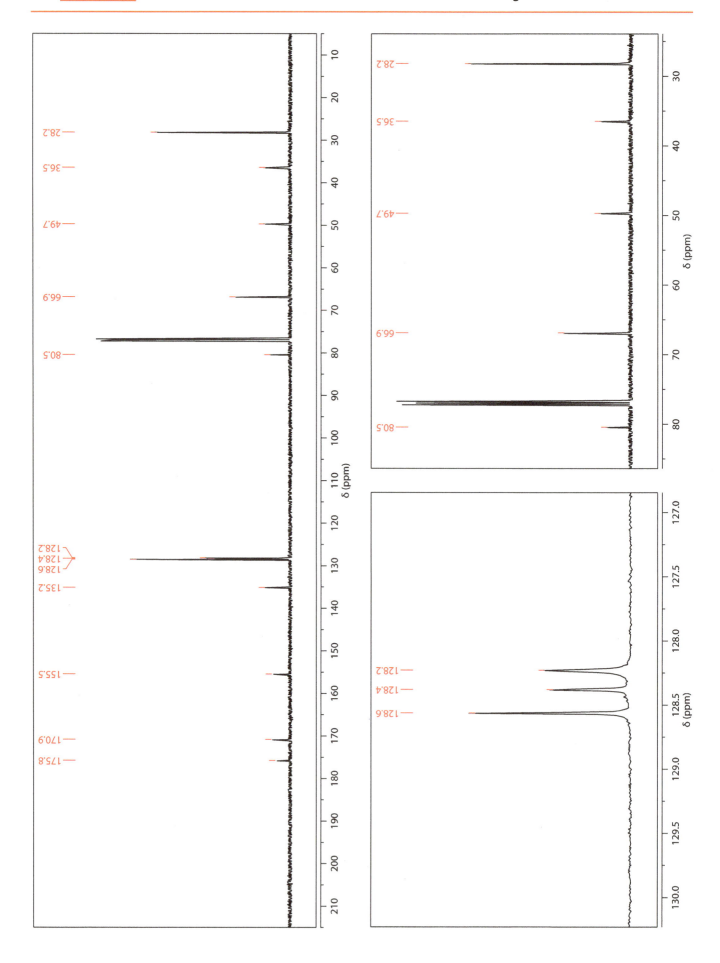

SECTION 3 Problem 76

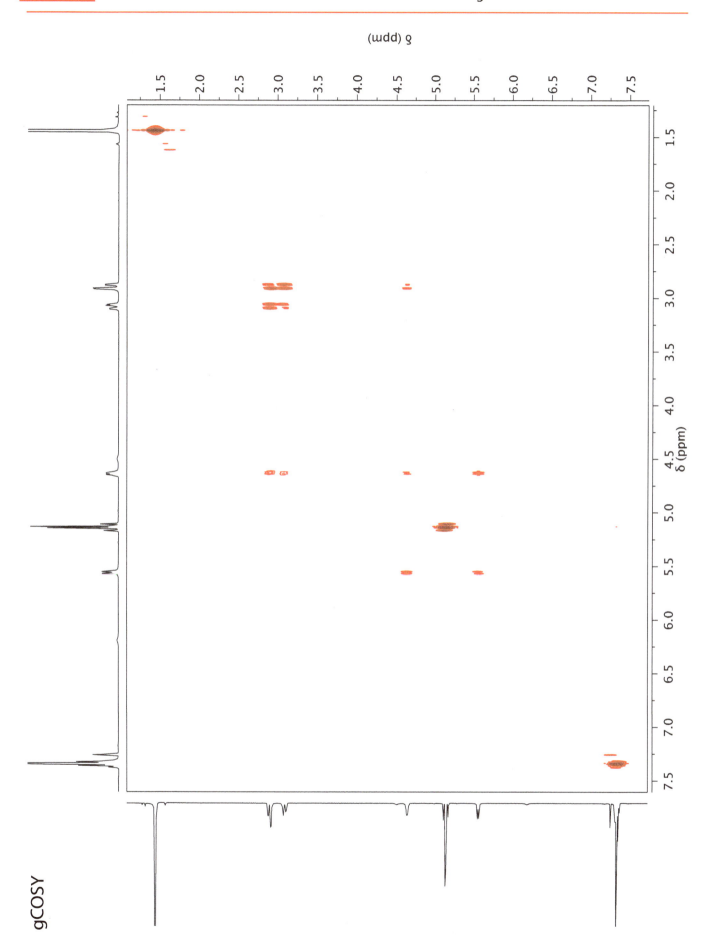

gCOSY

Problem 76

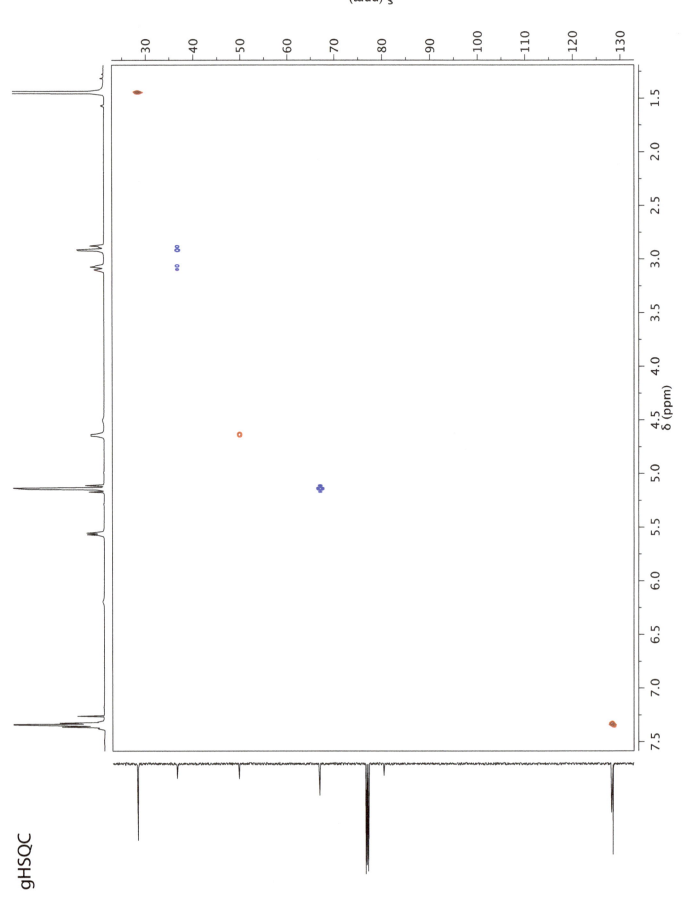

Problem 76

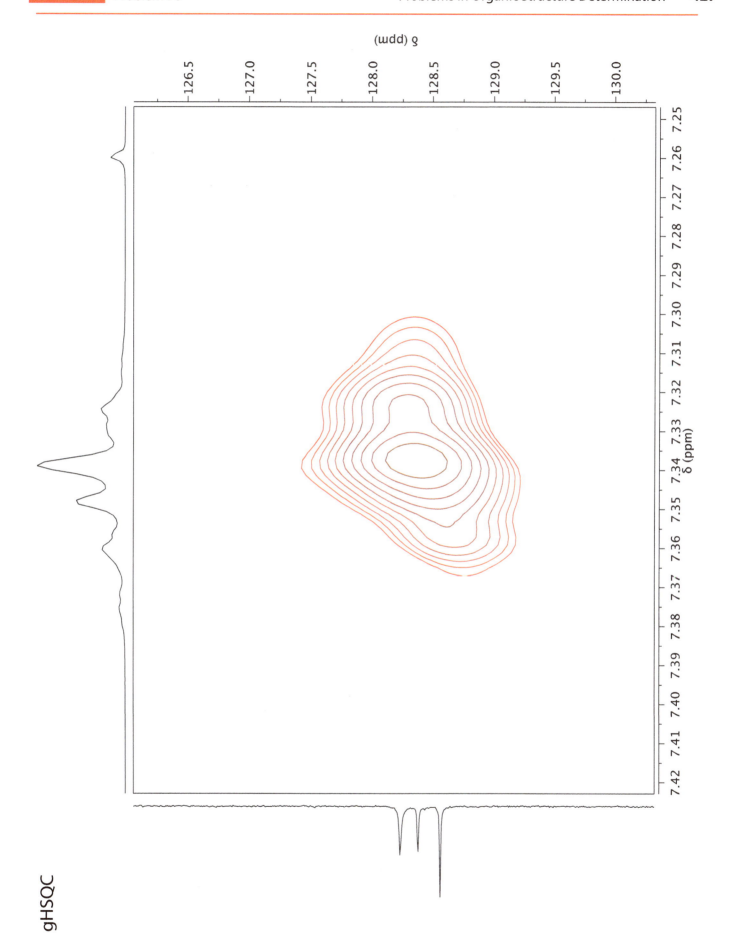

gHSQC

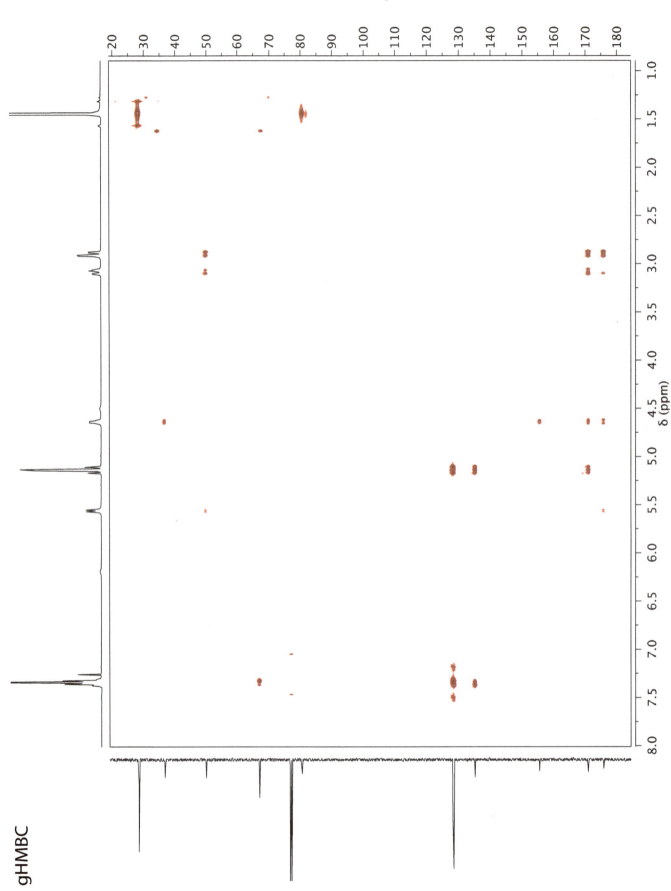

SECTION 3 Problem 77

Determine the structure of the unknown compound below with molecular formula $C_8H_{12}O_4$.

Spectra acquired in $CDCl_3$ at 500 MHz (^1H) or 125 MHz (^{13}C).

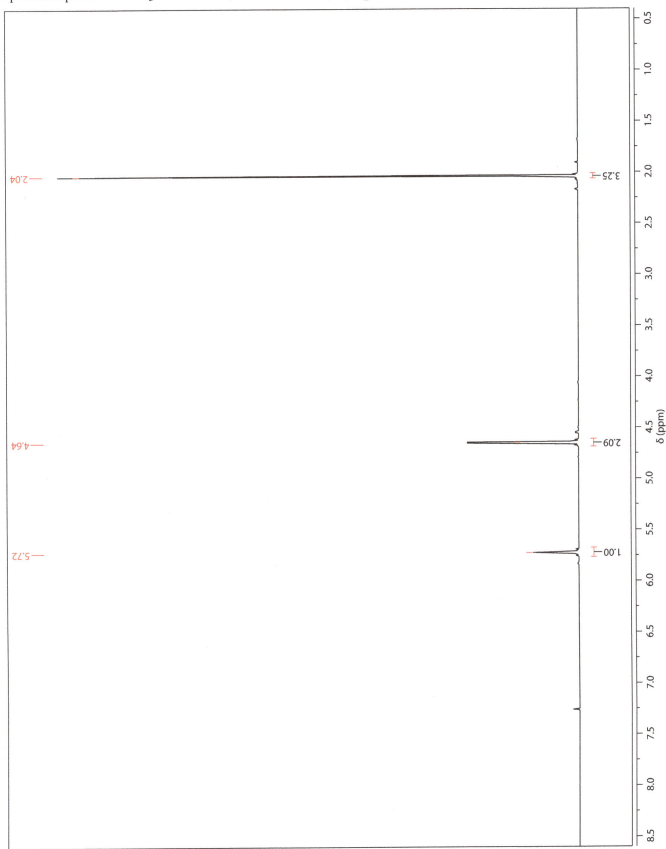

Problem 77

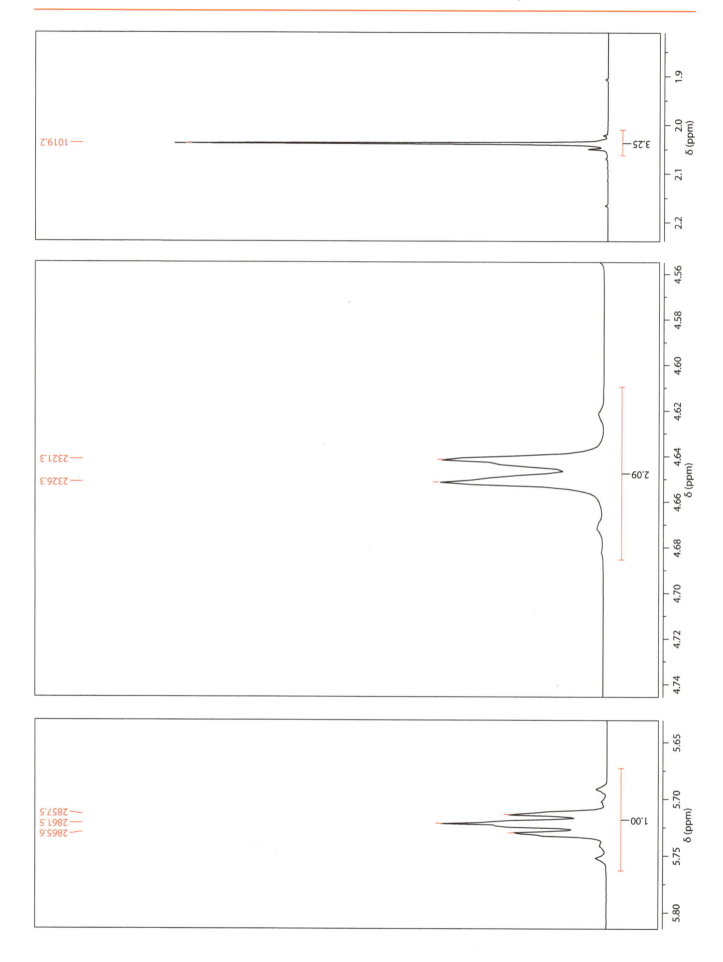

SECTION 3 Problem 77

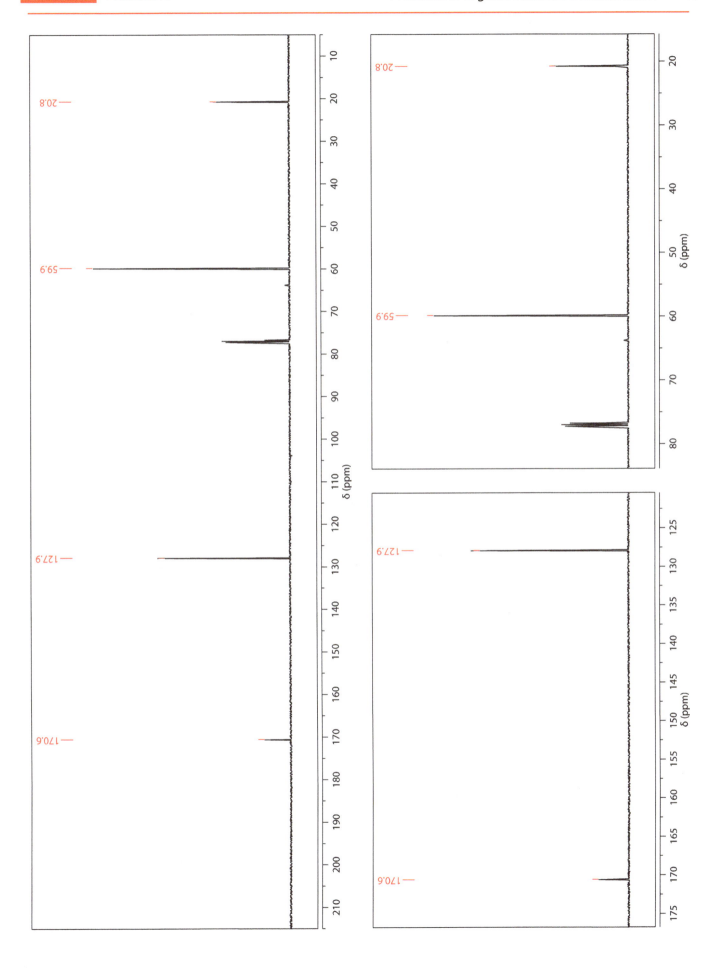

Problem 78

Determine the structure of the unknown compound below with molecular formula $C_{10}H_9NO_2$.

Spectra acquired in $(CD_3)_2SO$ at 500 MHz (^1H) or 125 MHz (^{13}C).

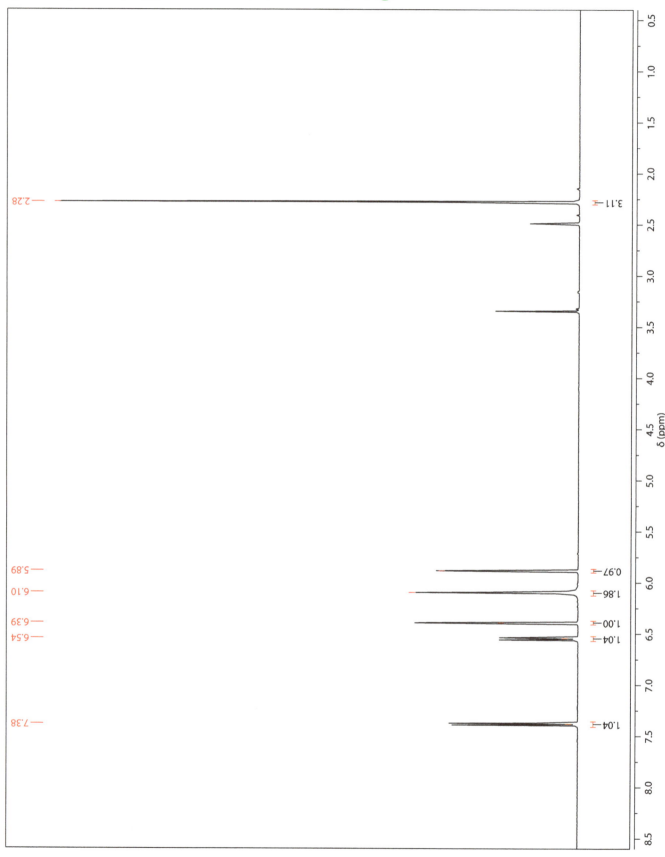

SECTION 3 Problem 78

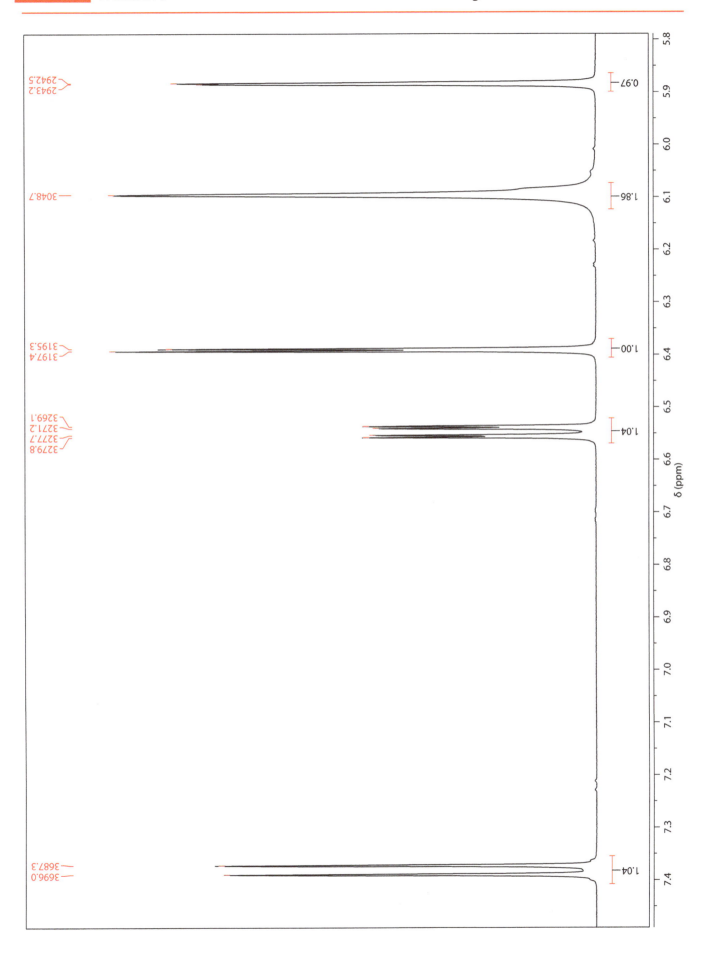

Problem 78

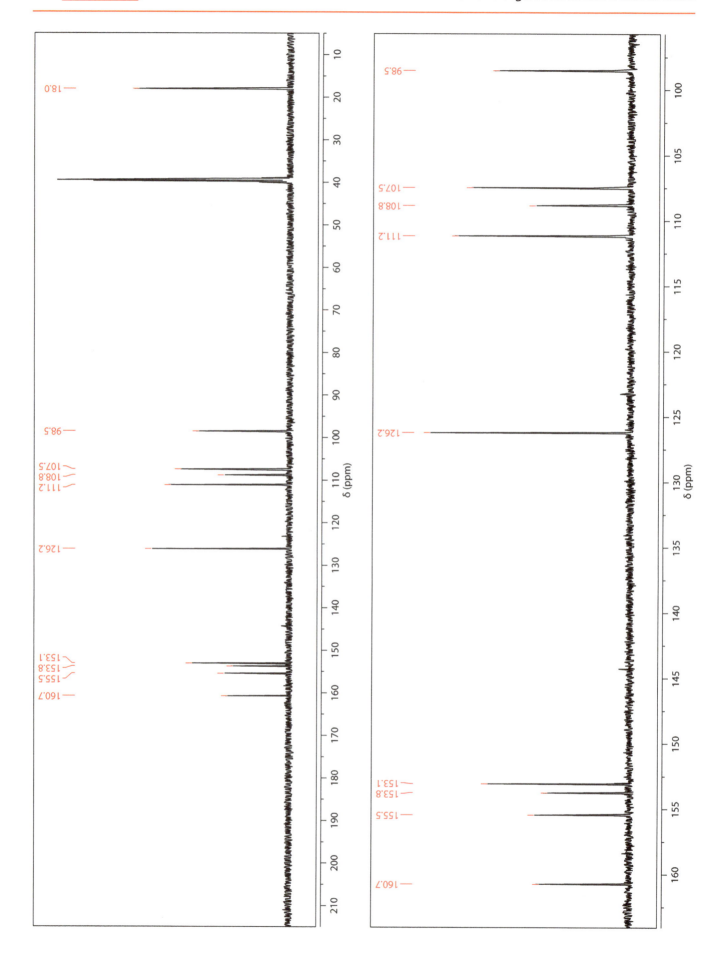

SECTION 3 Problem 78

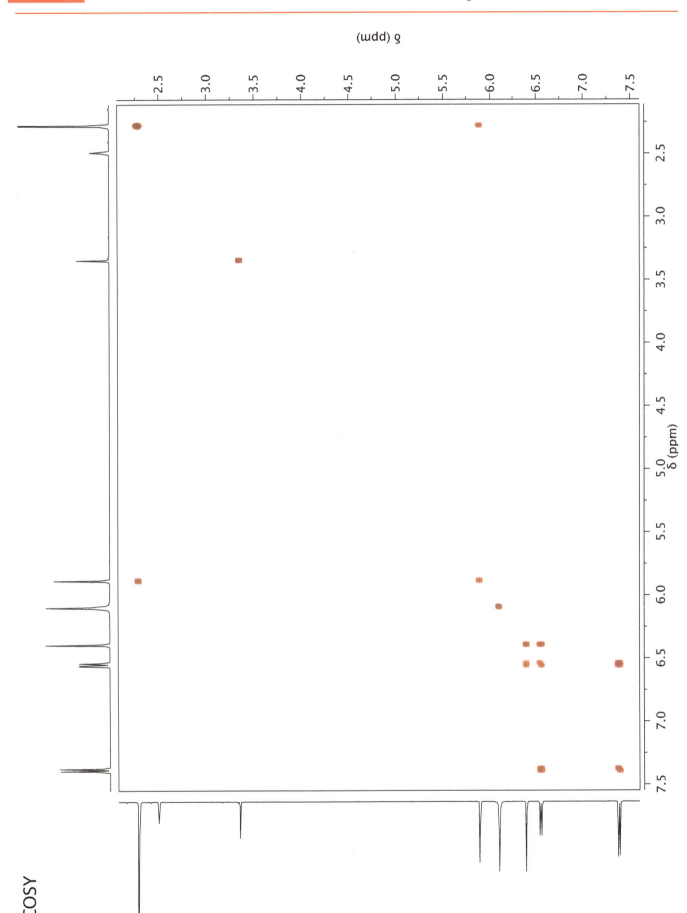

Problem 78

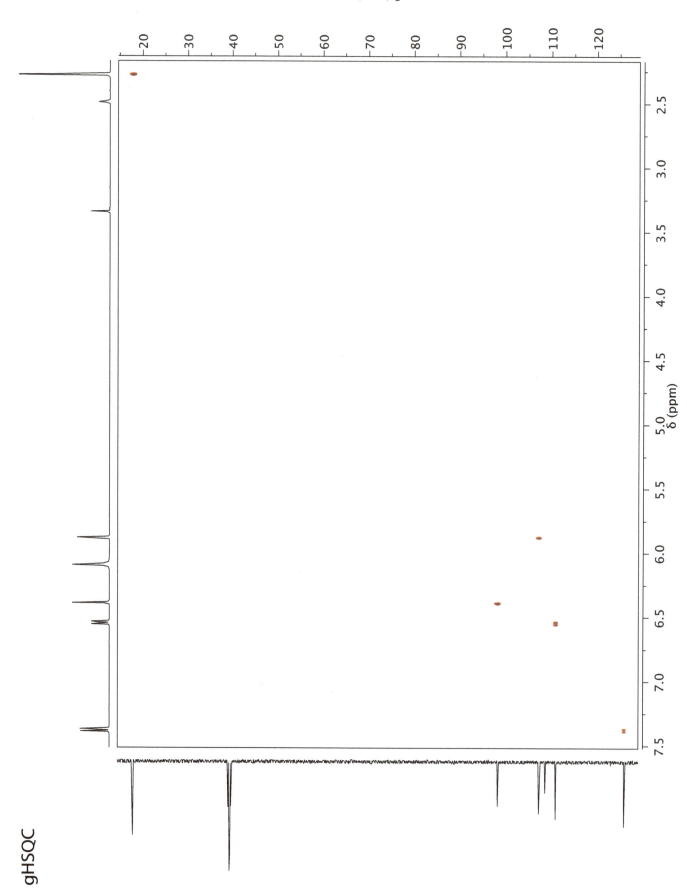

gHSQC

SECTION 3 Problem 78

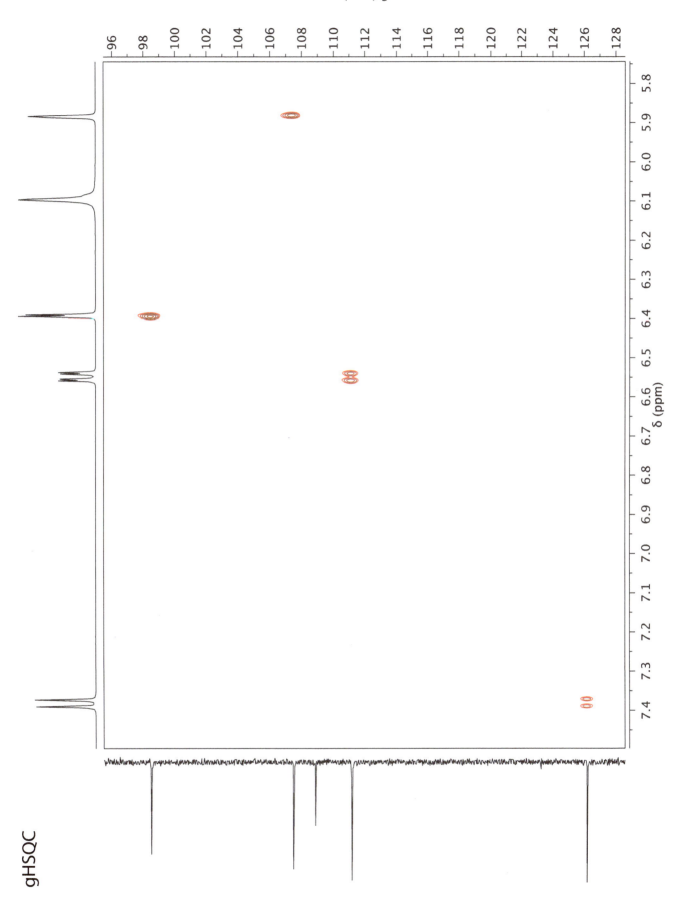

gHSQC

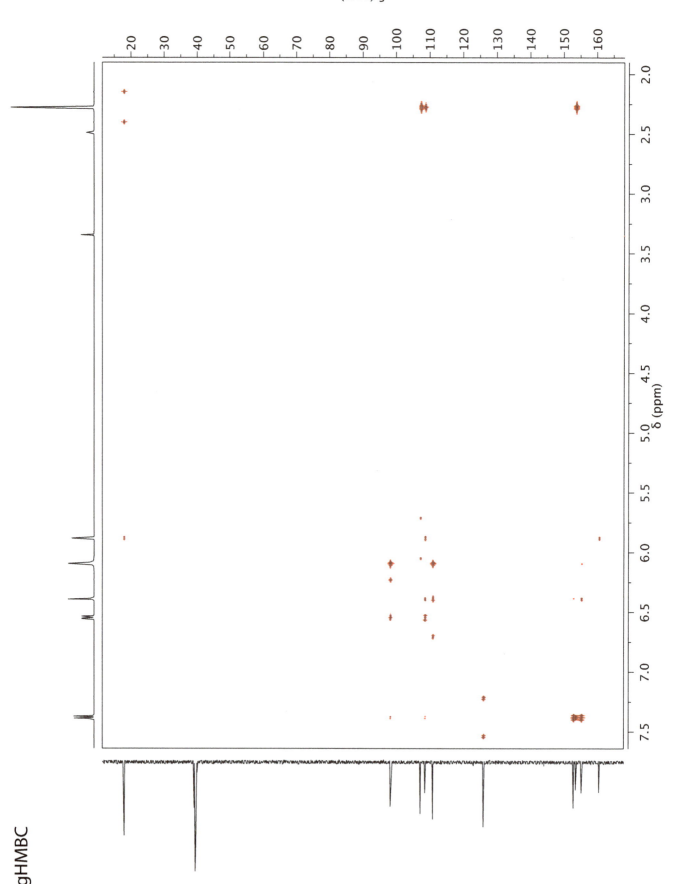

SECTION 3 Problem 78

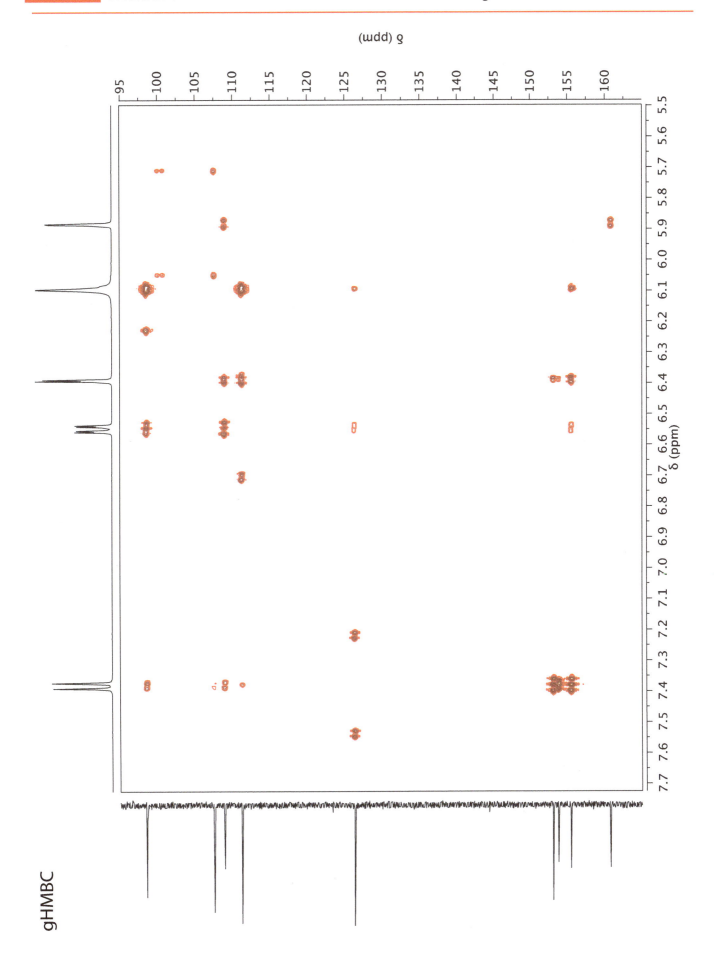

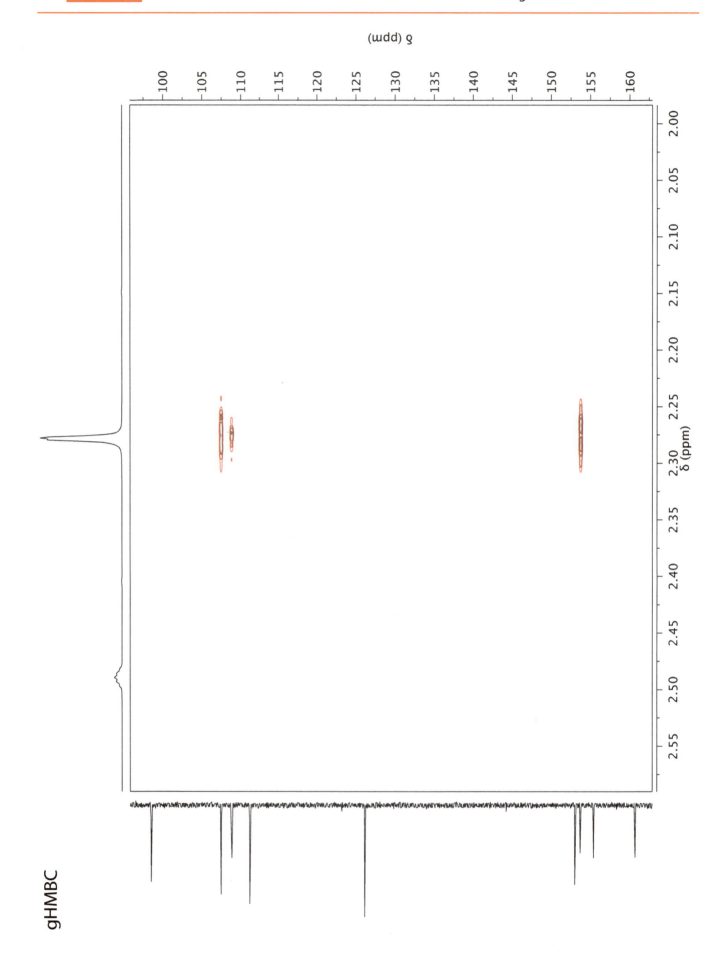

SECTION 3 Problem 78

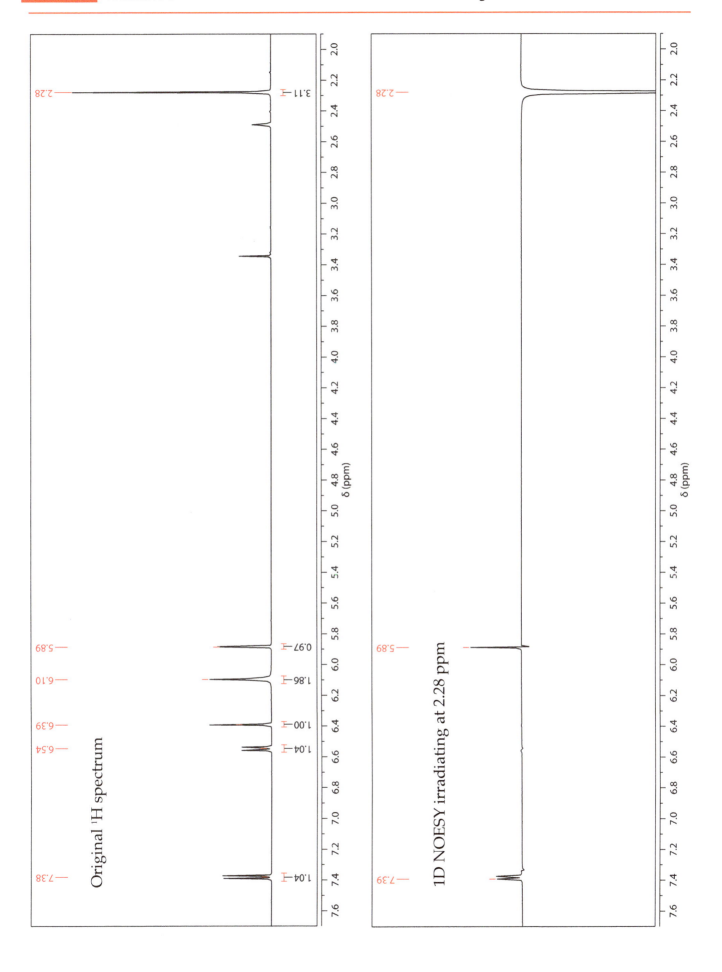

SECTION 3 Problem 79

Determine the structure of the unknown compound below with molecular formula $C_{11}H_{12}ClNO_4$.

Spectra acquired in $(CD_3)_2SO$ at 500 MHz (^1H) or 125 MHz (^{13}C).

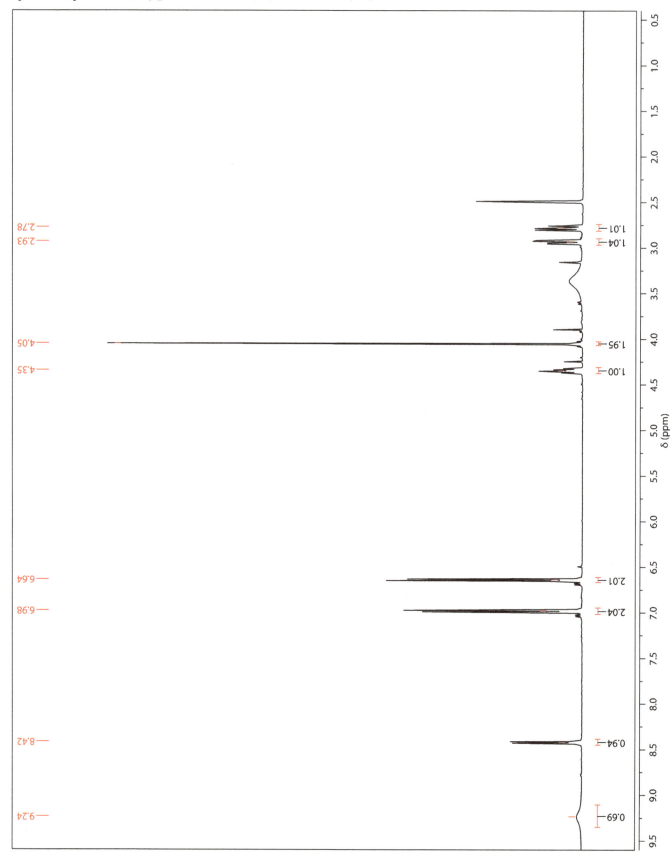

Problem 79

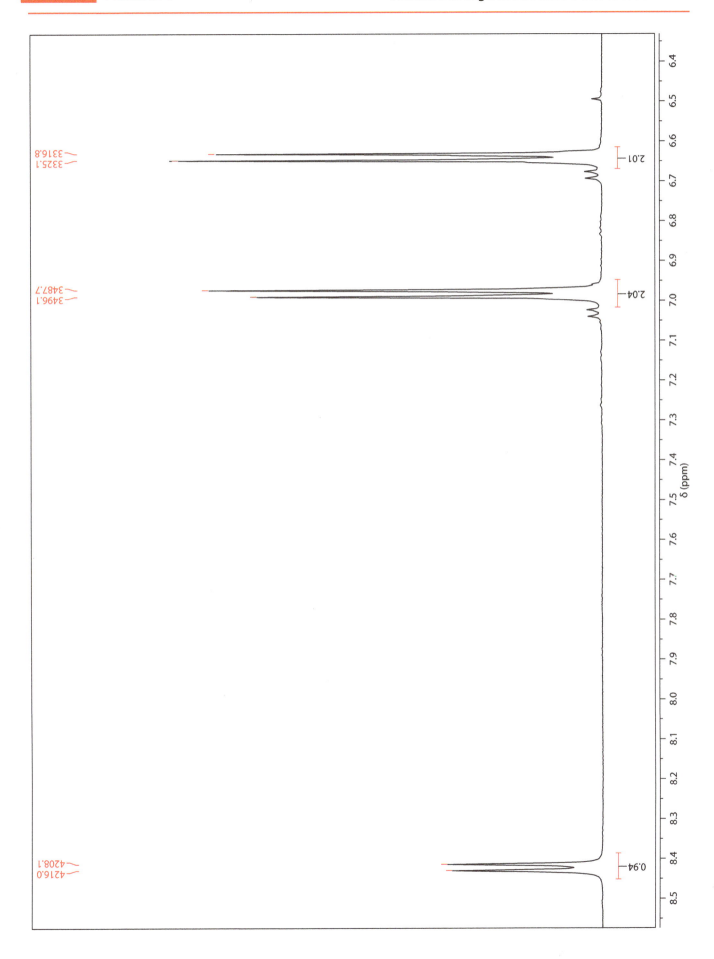

Problem 79

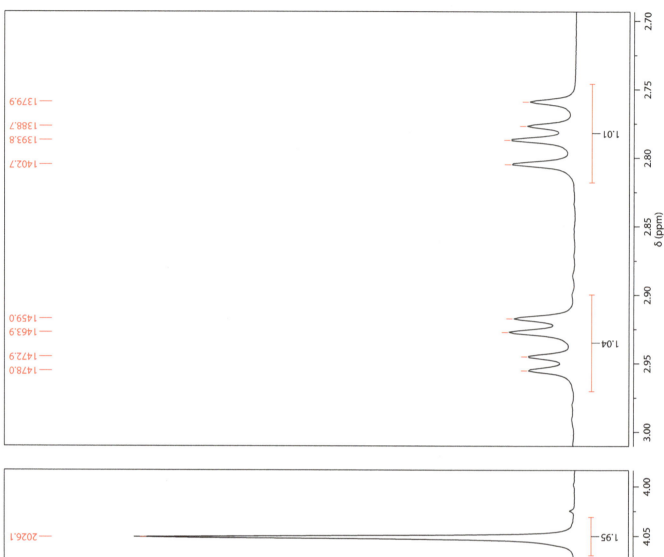
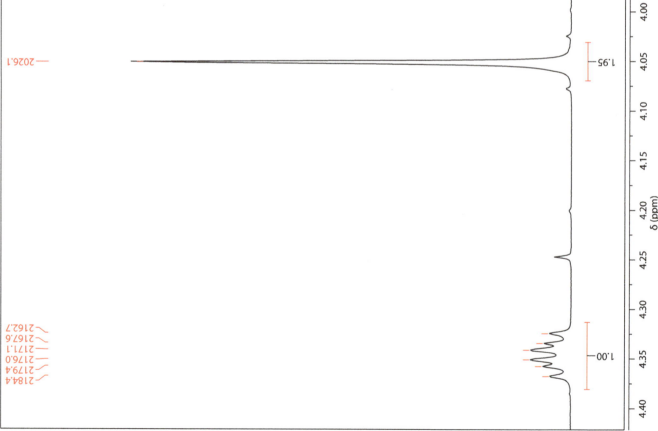

SECTION 3   Problem 79

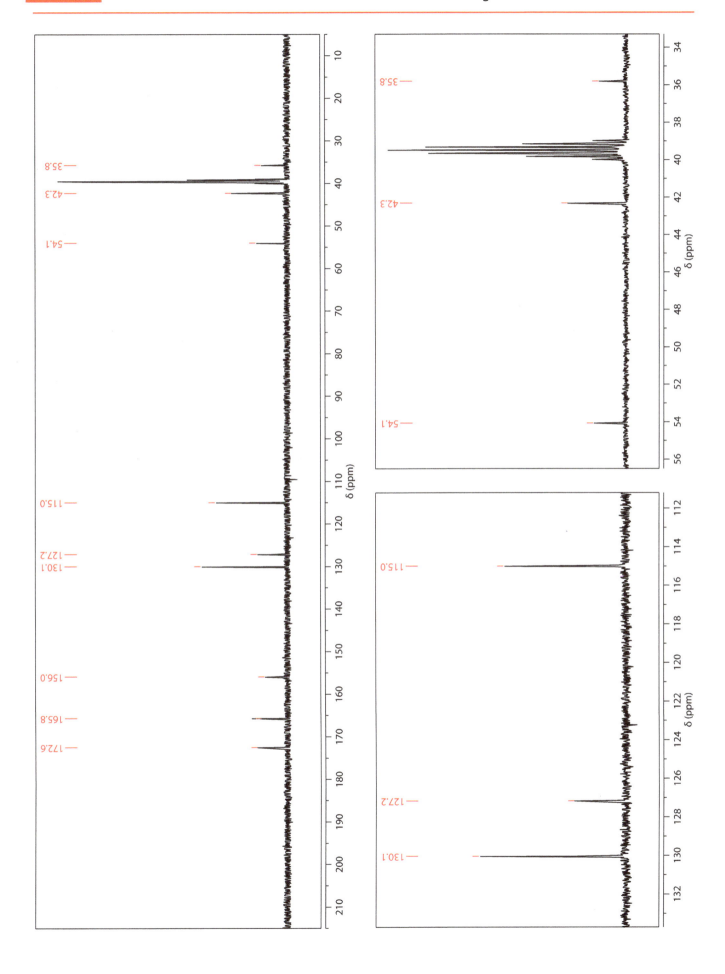

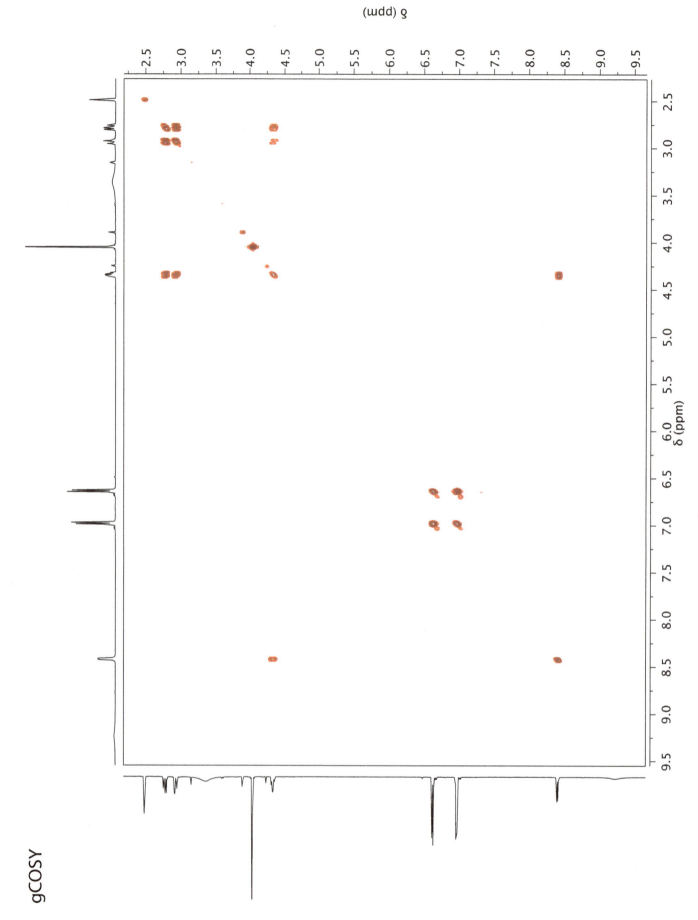

SECTION 3 Problem 79

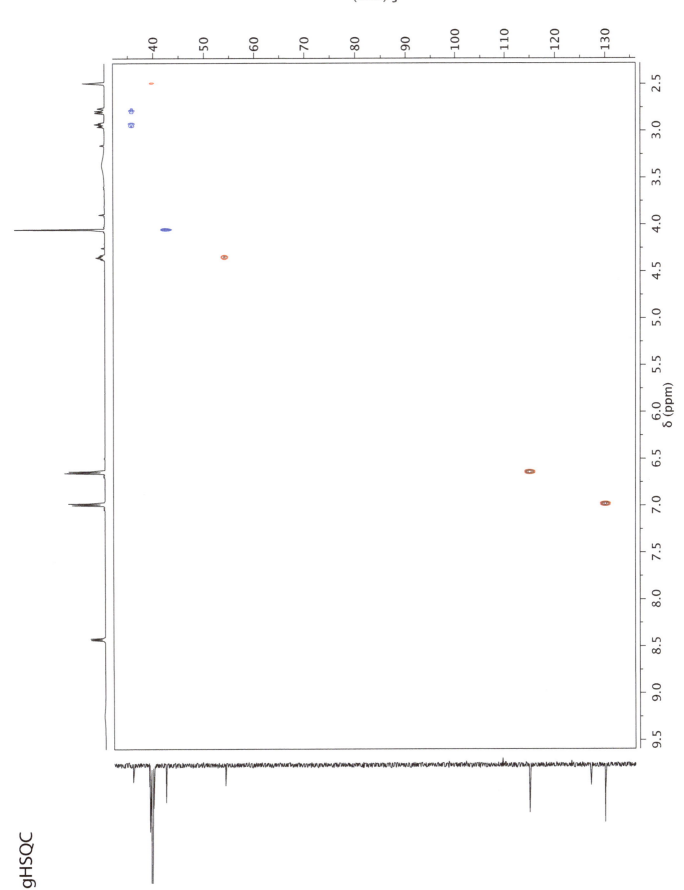

gHSQC

Problem 79

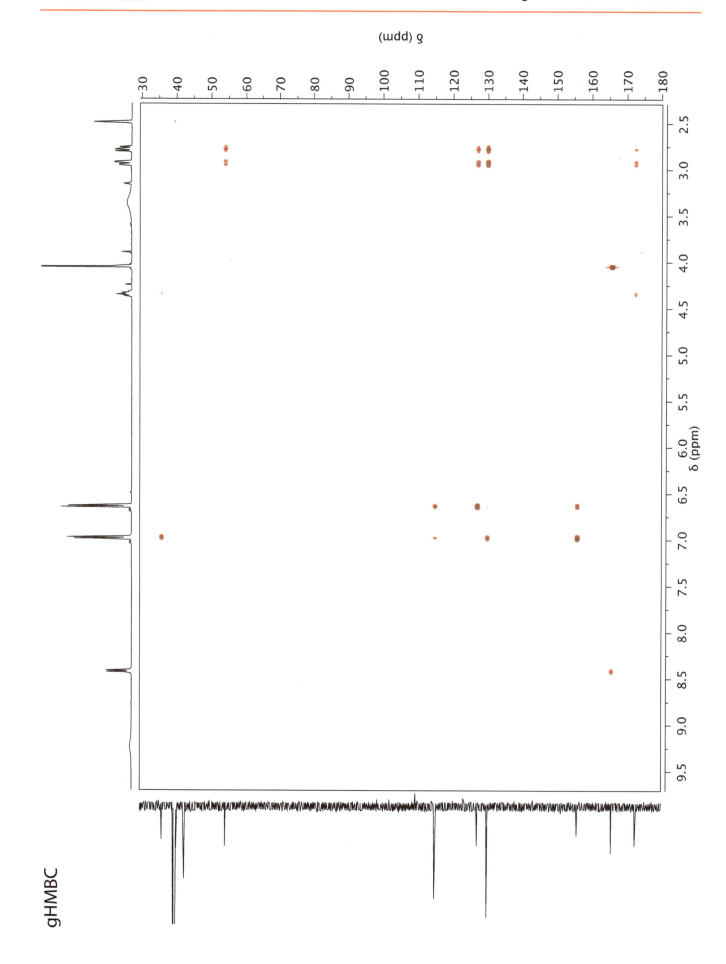

gHMBC

SECTION 3 Problem 80

Determine the structure of the unknown compound below with molecular formula C₉H₉NO₄.

Spectra acquired in CD₃OD at 500 MHz (¹H) or 125 MHz (¹³C).

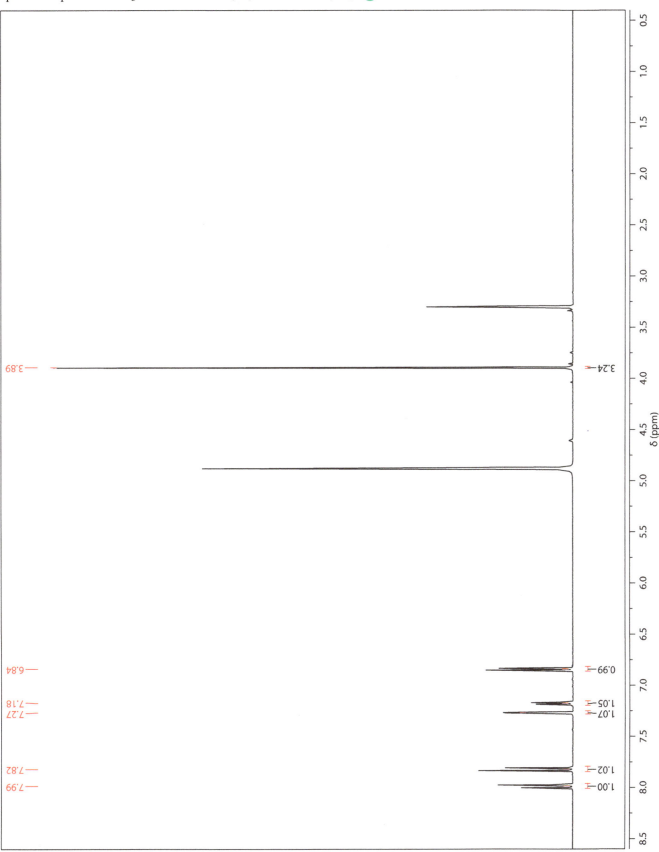

SECTION 3 Problem 80

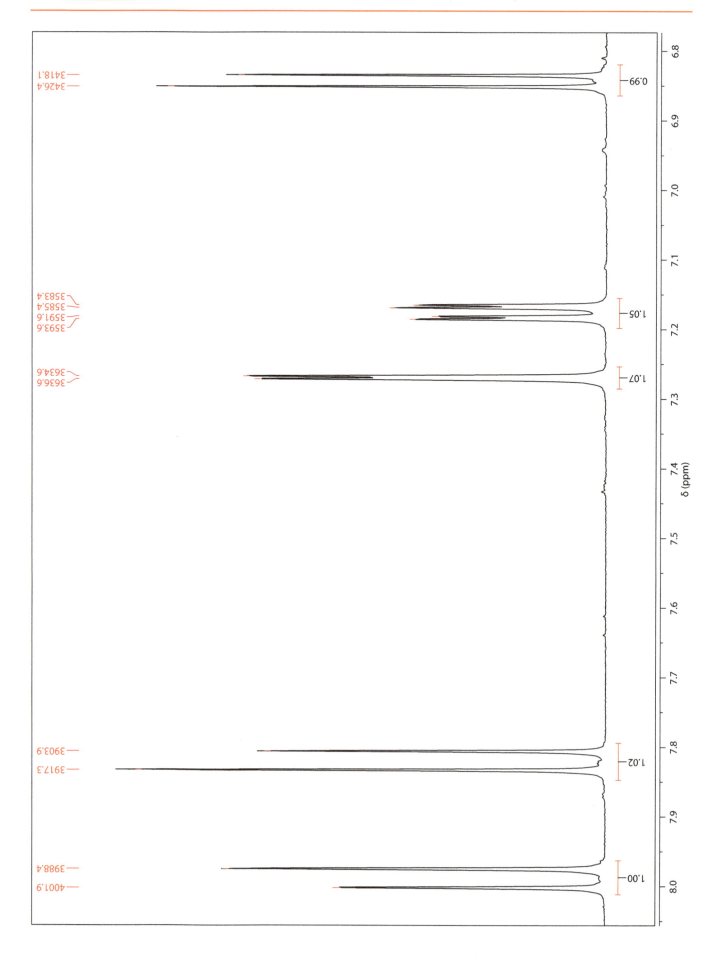

SECTION 3 Problem 80

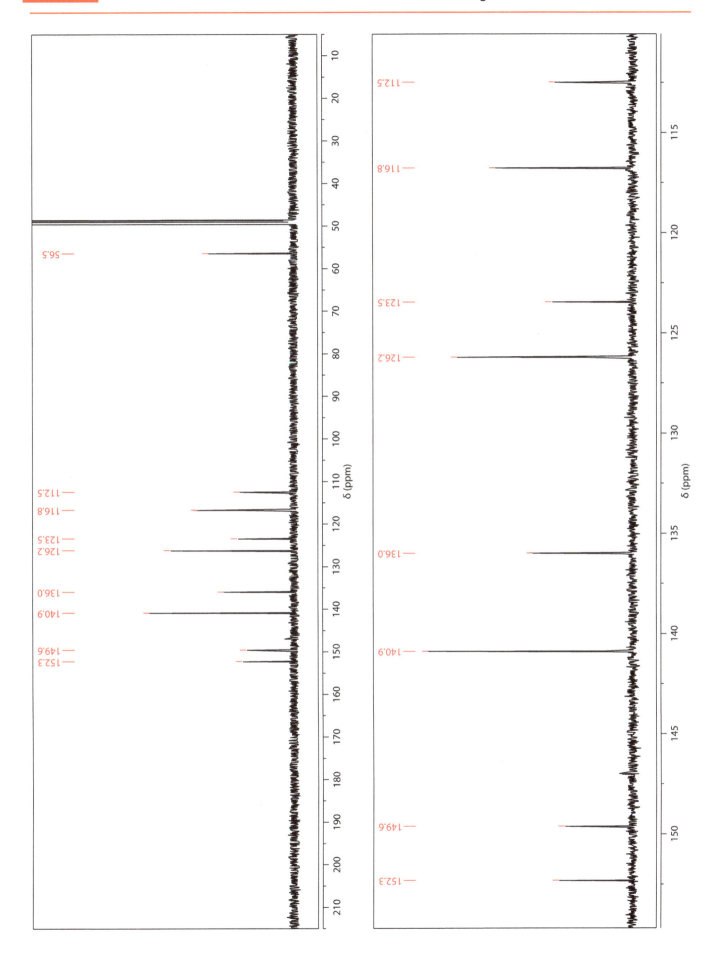

Problem 80

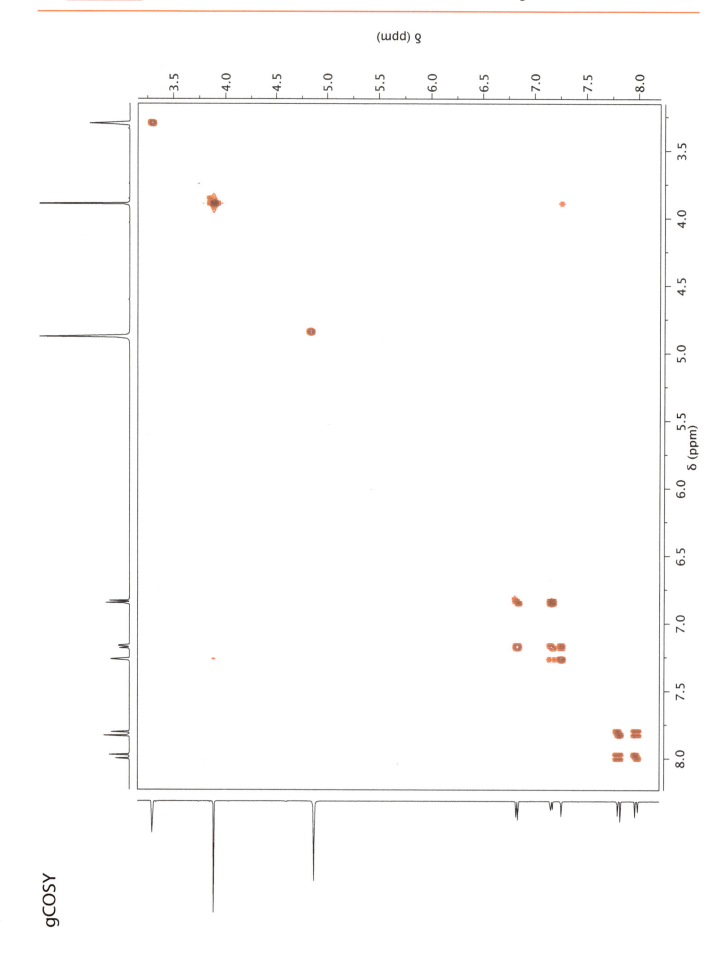

gCOSY

SECTION 3 Problem 80

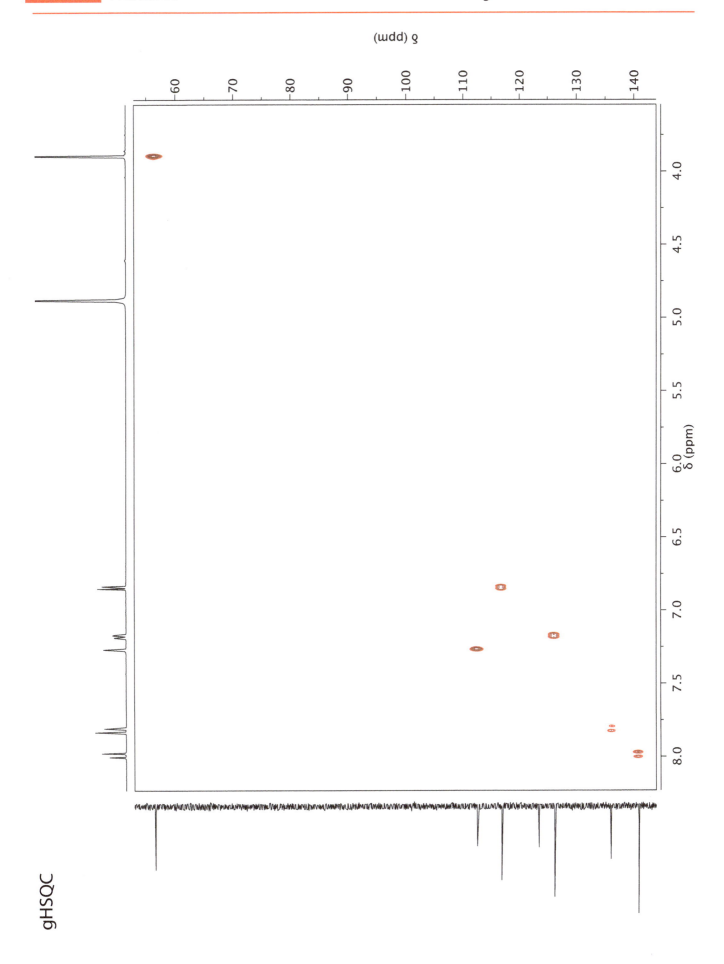

gHSQC

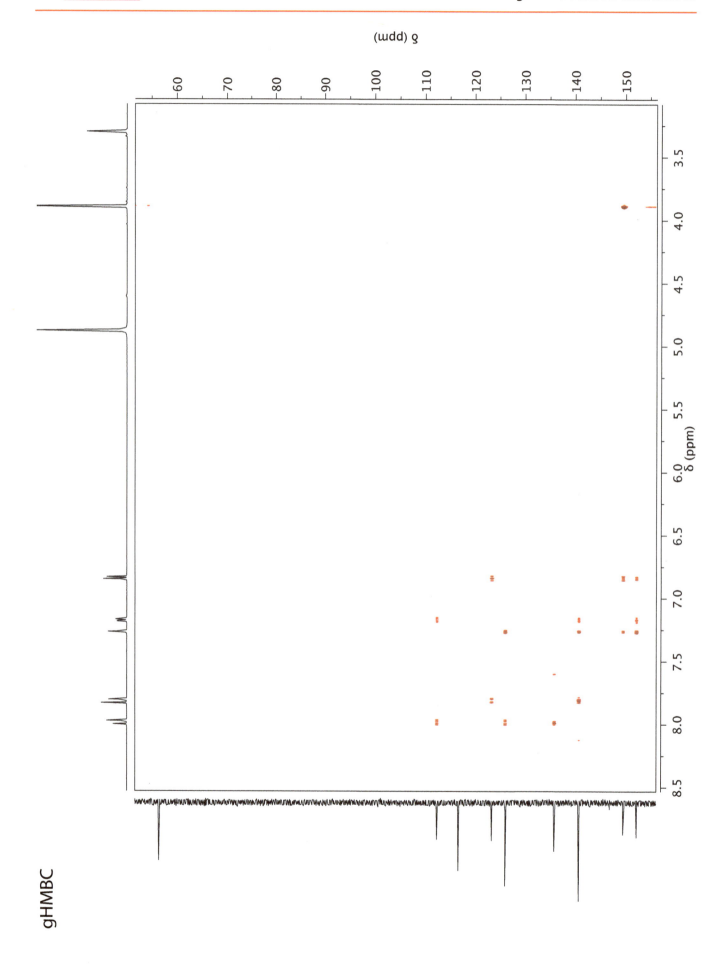

SECTION 3 Problem 80

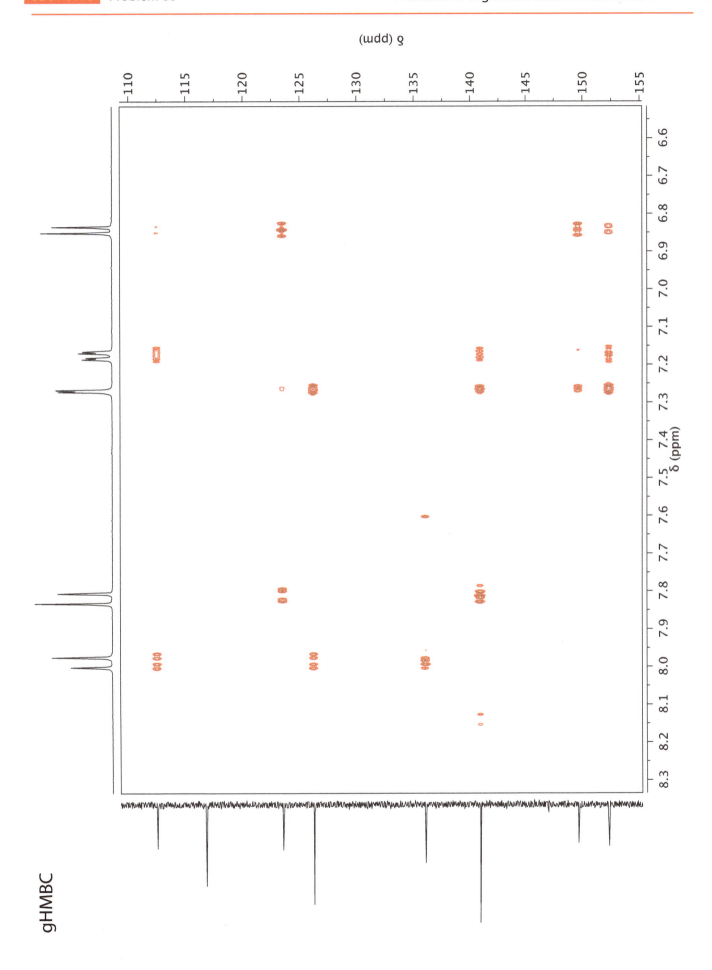

gHMBC

Section 4

Section 4 is a continuation of the *de novo* structure elucidation problems introduced in section 3 using the provided molecular formula and NMR spectra. The difficulty of section 4 problems are greatly increased over those in section 3, and will require the careful integration of all experimental data.

LEARNING OBJECTIVES

- Integrate data from all of the provided NMR experiments and molecular formula to assign chemical structures.
- Integrate the use of ^{15}N gHMBC data for functional group determination and substructure connectivity.

EXPERIMENTS INCLUDED

^{1}H, ^{13}C, DEPT-135, gCOSY, gHSQC, gHMBC, ^{15}N-HMBC and variable temperature ^{1}H spectra.

Note: All gHSQC experiments are multiplicity edited with cross peaks deriving from CH_2 groups depicted in blue, and cross peaks deriving from CH and CH_3 groups depicted in red.

TYPES OF MOLECULES

This section contains complex molecules that contain polyaromatics, heterocycles, stereogenic centers and fluorine. There are a number of examples of nitrogen containing molecules that can be assigned using the included ^{1}H–^{15}N HMBC data.

STRATEGIES FOR SUCCESS

The molecular formula for each compound is provided. A helpful strategy for most of these problems will be to determine the degrees of unsaturation using the following equation:

$$DOU = \frac{(2 \times \#C + 2) - \#H - \#X + \#N}{2}$$

where X= Cl, Br, F, I. Furthermore, a number of problems in this section contain hints that may help determine functional groups that do not have NMR signatures.

Because the compounds in this section are structurally more complex it is important to validate proposed substructure connectivities using all available 2D data. For example, if a gCOSY correltaion is observed, are the corresponding gHMBC correlations present, and are the ^{1}H multiplicities consistent with this assignment?

LEGEND

Spectrum annotations:

i = impurity.

 = worked problem in the answer key.

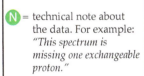 = technical note about the data. For example: *"This spectrum is missing one exchangeable proton."*

H = hint to assist in solving the problem. For example: *"This molecule would have an IR stretch at 2240 cm^{-1}."*

Problem 81

Determine the structure of the unknown compound below with molecular formula $C_9H_{14}O$.

Spectra acquired in $CDCl_3$ at 500 MHz (^{1}H) or 125 MHz (^{13}C).

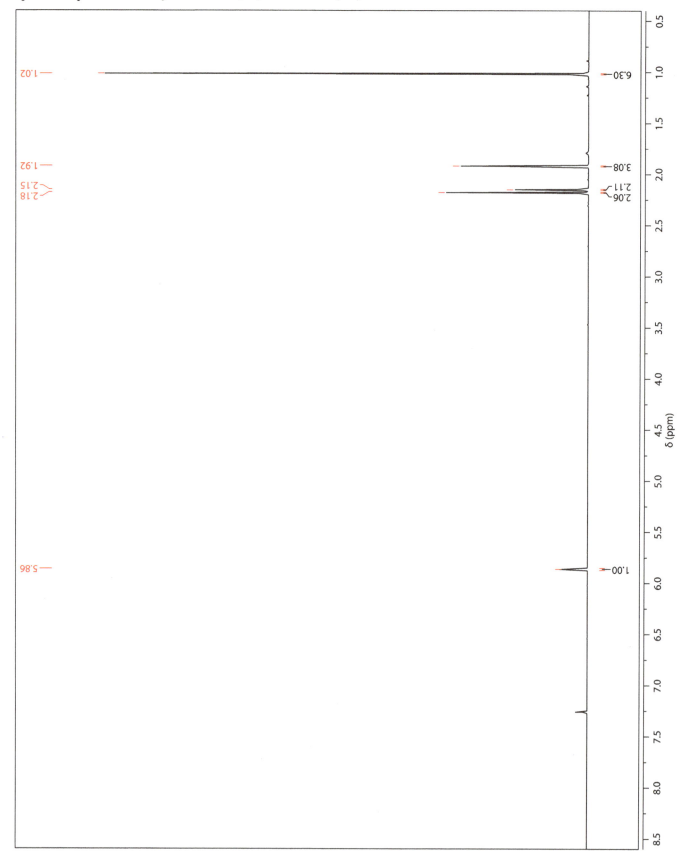

SECTION 4 Problem 81

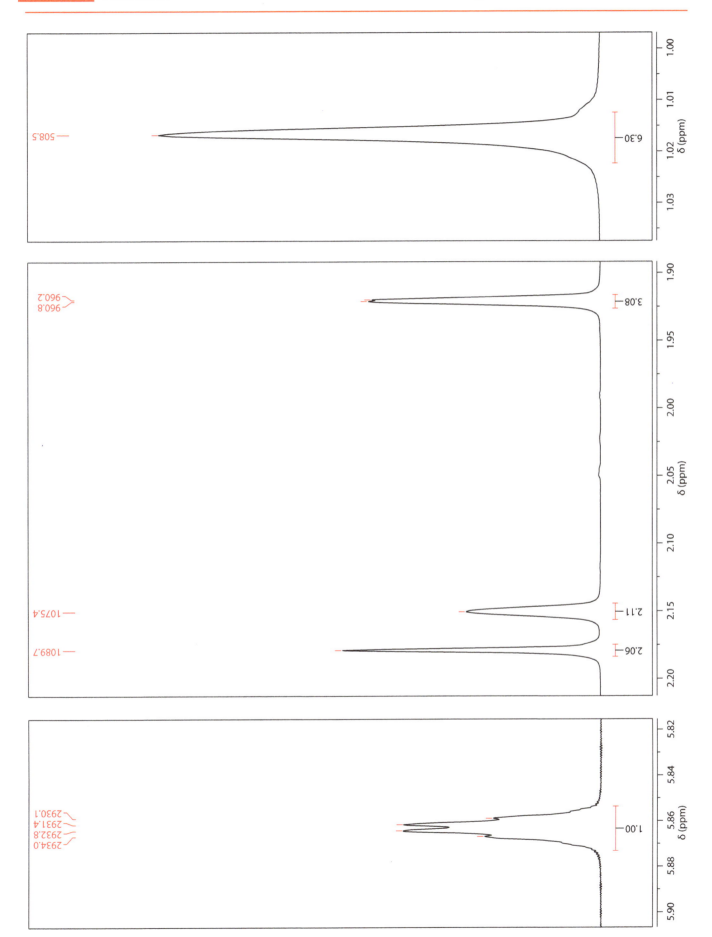

SECTION 4 Problem 81

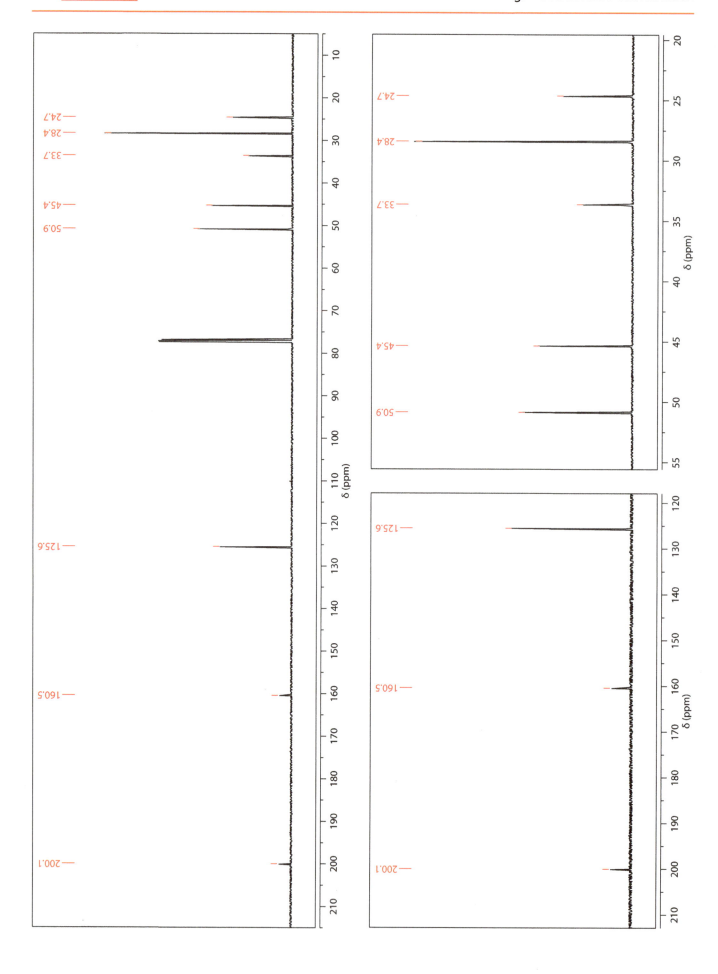

SECTION 4 Problem 81

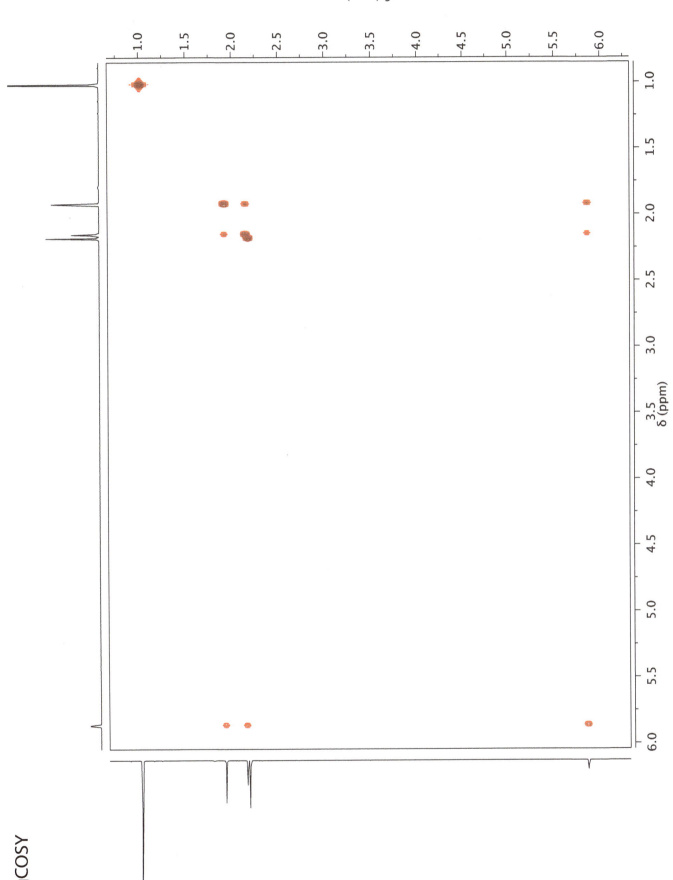

gCOSY

Problem 81

gCOSY

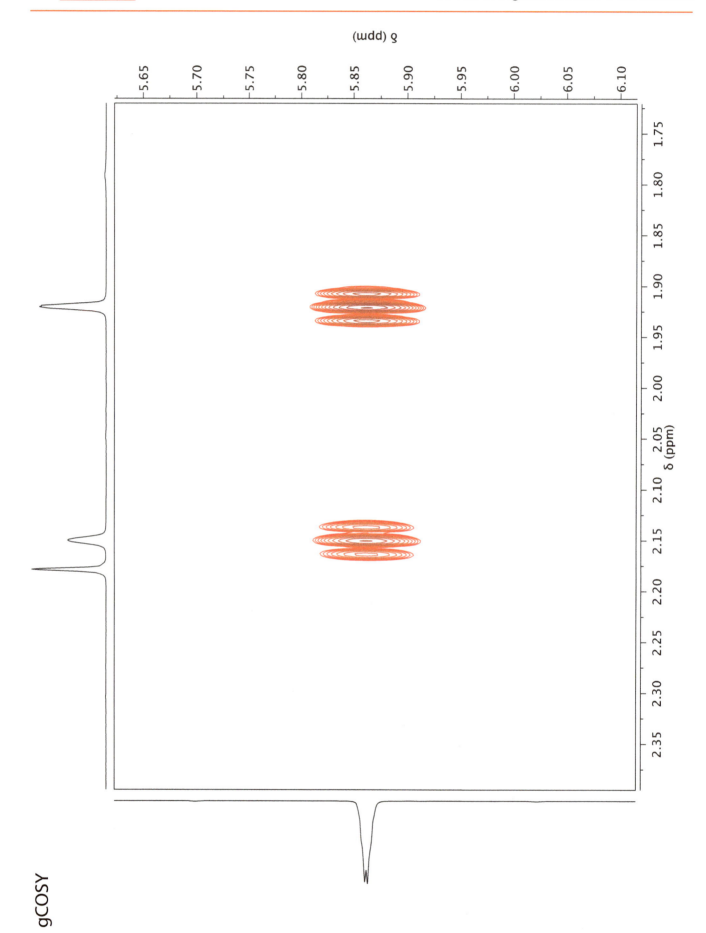

SECTION 4 Problem 81

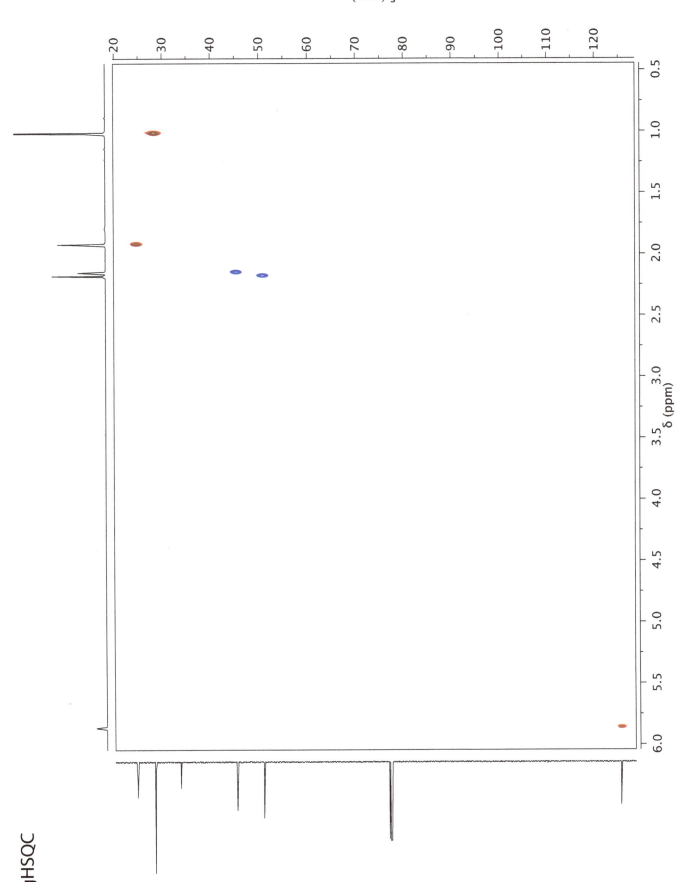

Problem 81

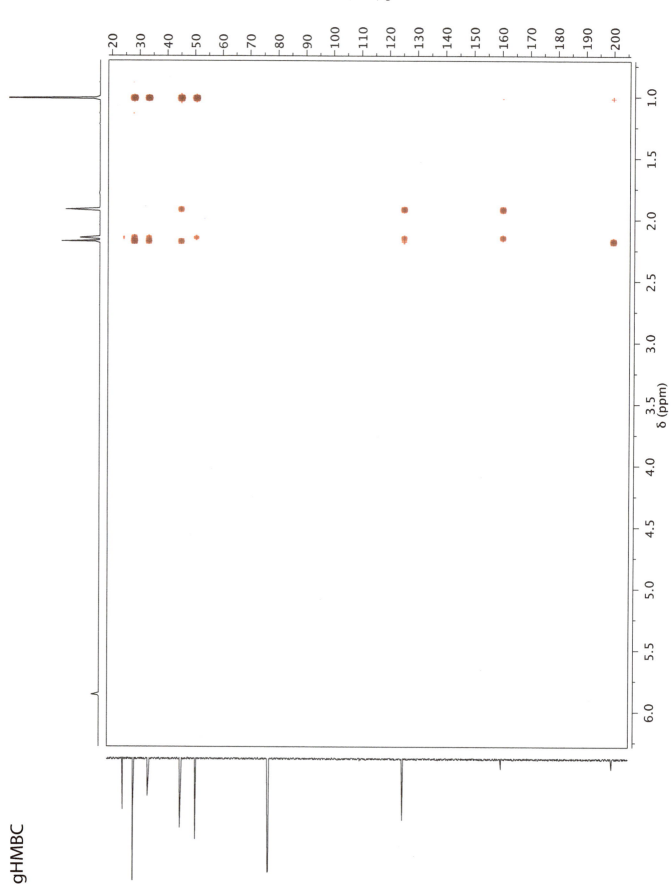

SECTION 4 Problem 81

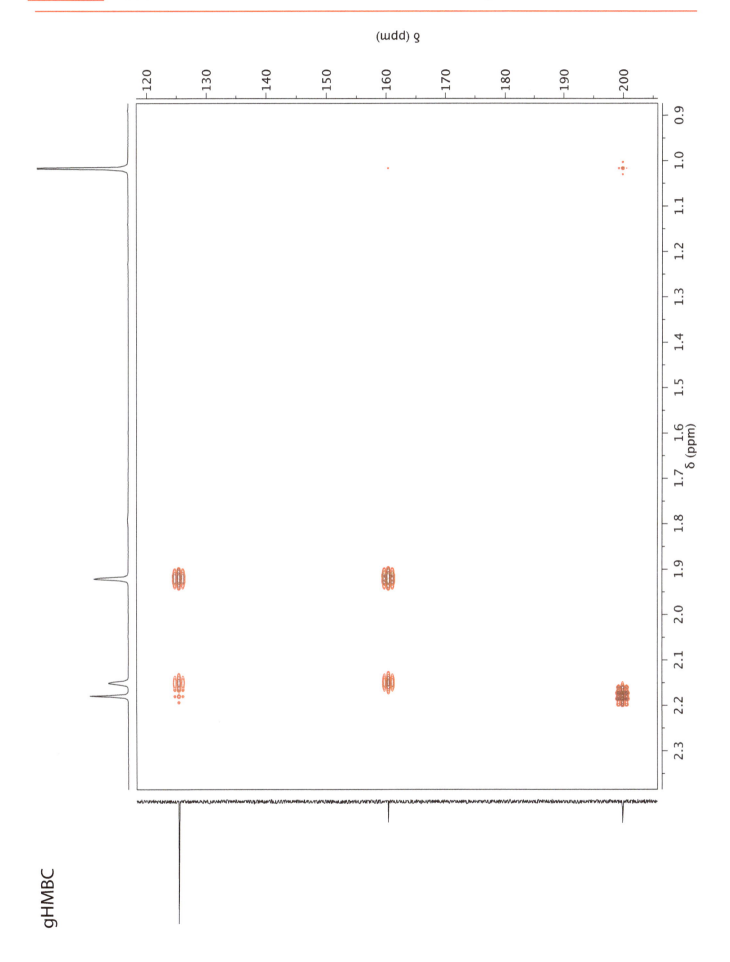

gHMBC

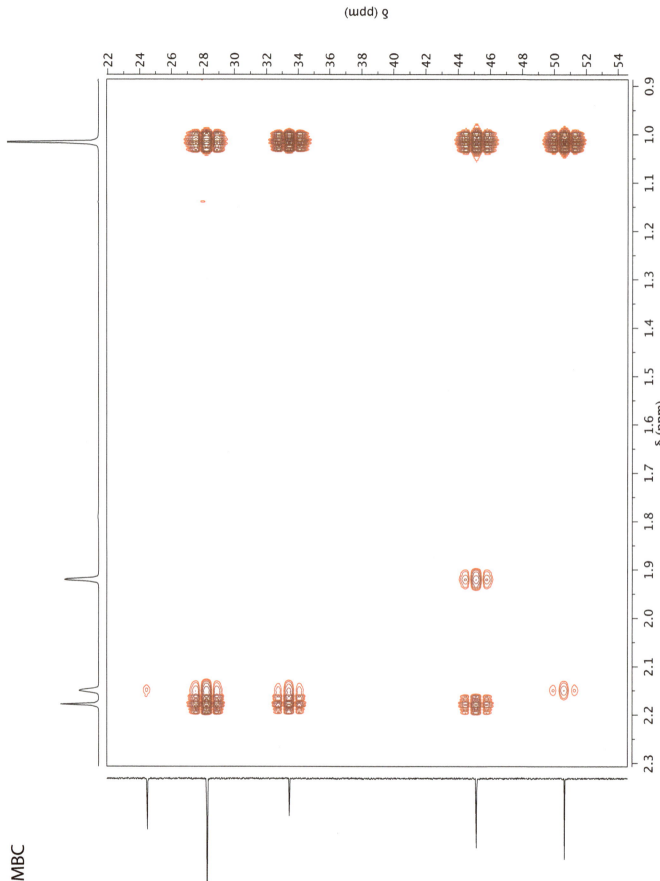

gHMBC

SECTION 4 Problem 82

Determine the structure of the unknown compound below with molecular formula $C_6H_6O_3$.

Spectra acquired in $CDCl_3$ at 500 MHz (^{1}H) or 125 MHz (^{13}C).

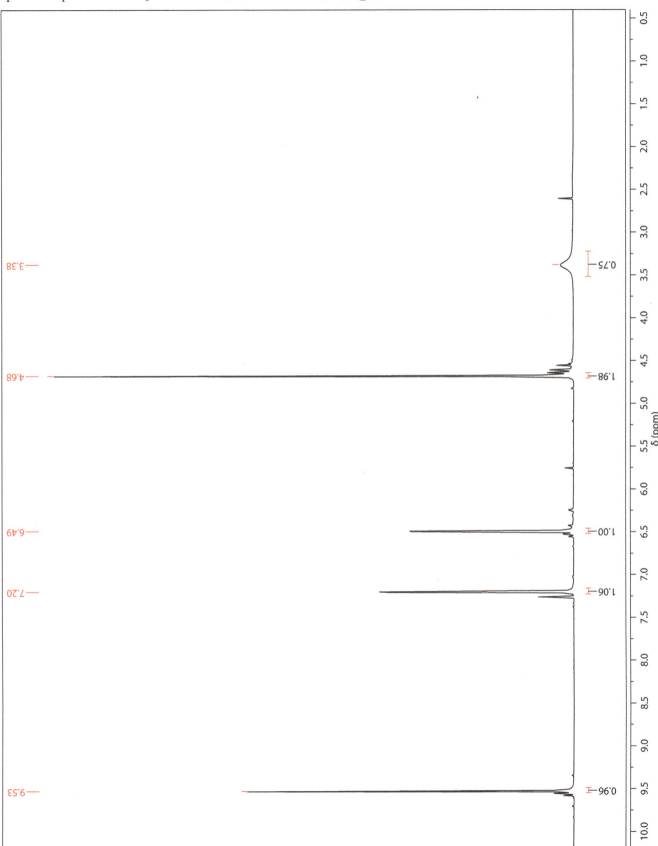

SECTION 4 Problem 82

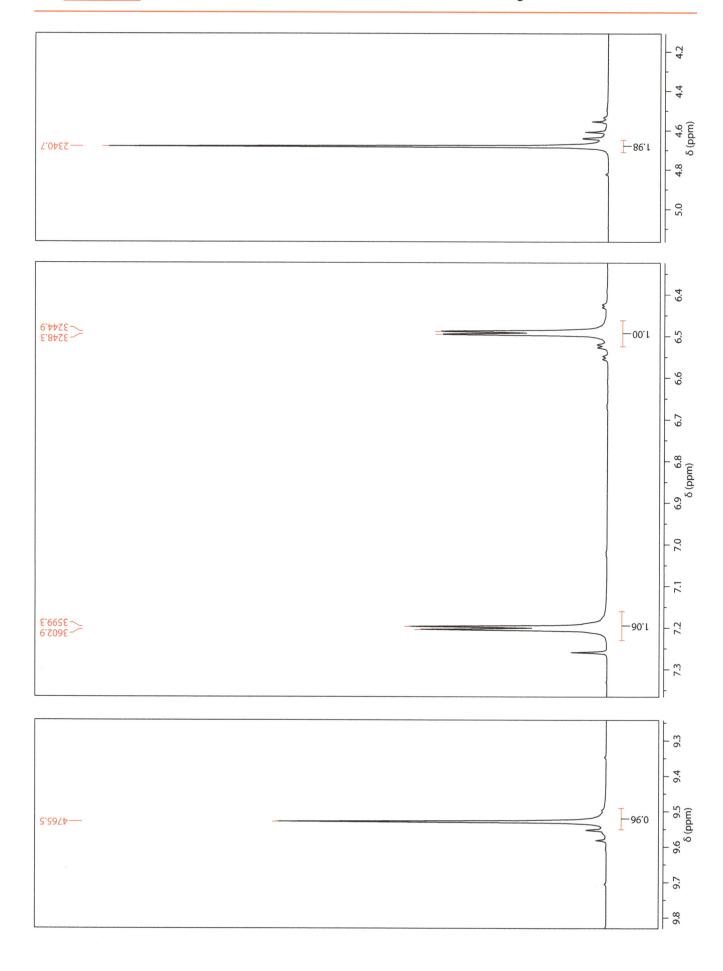

SECTION 4 Problem 82

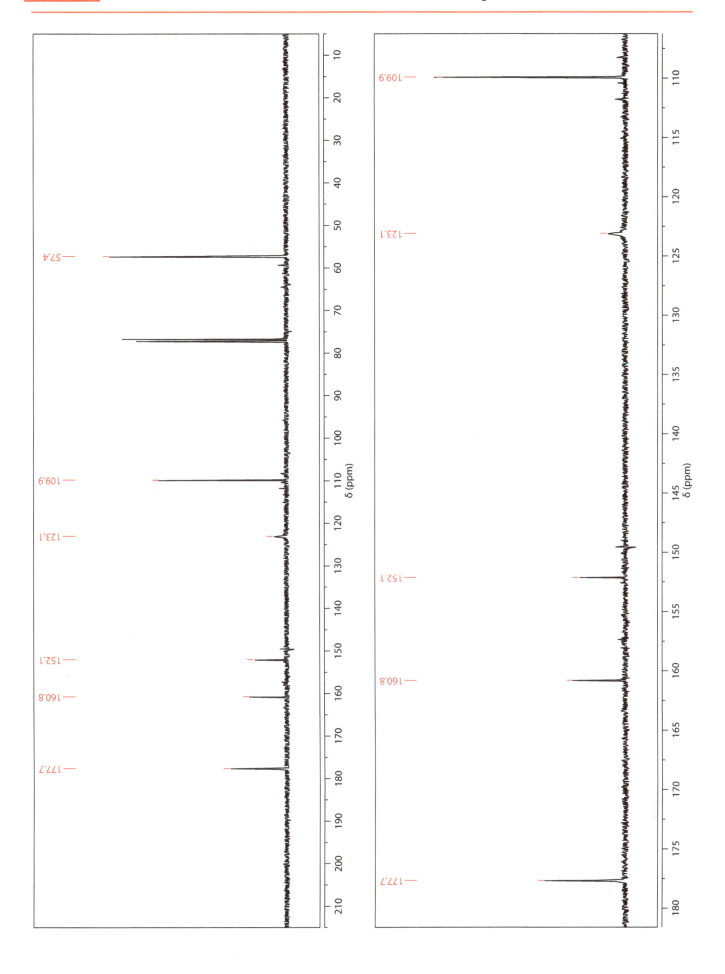

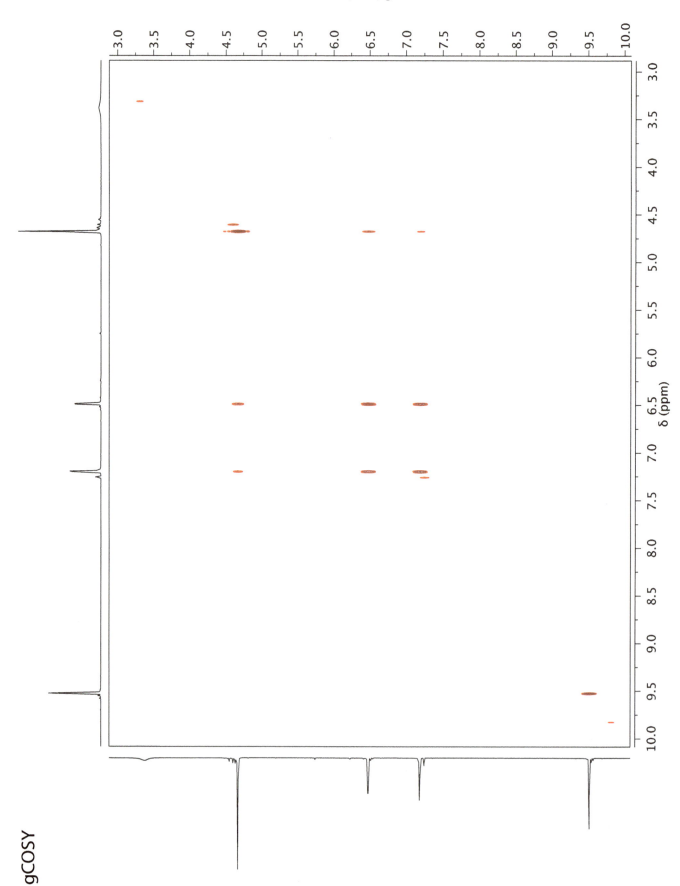

Problem 82

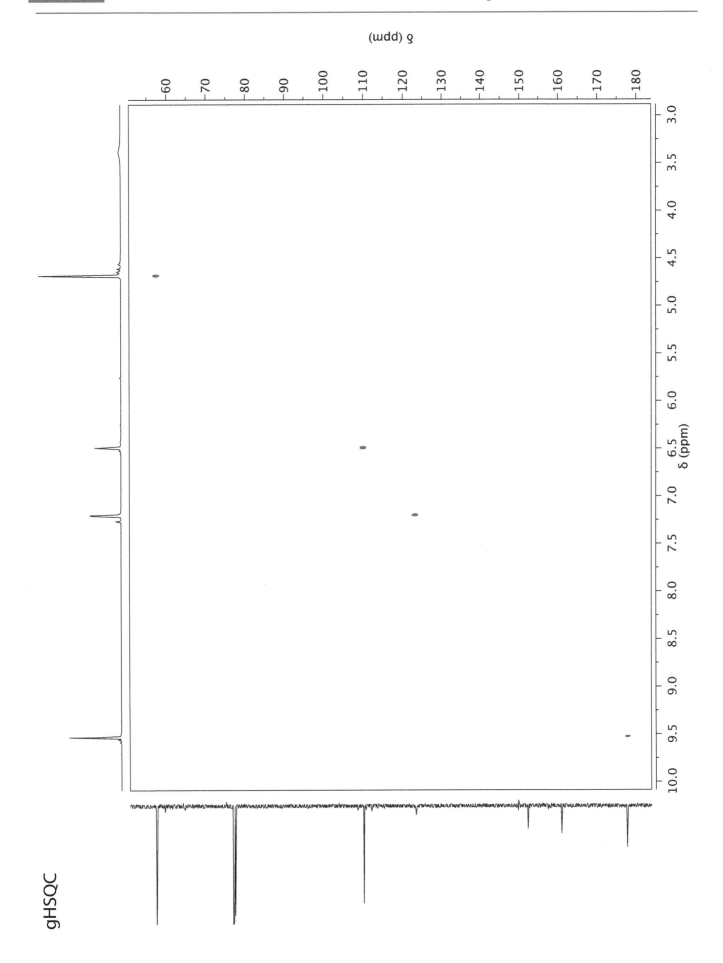

Problem 82

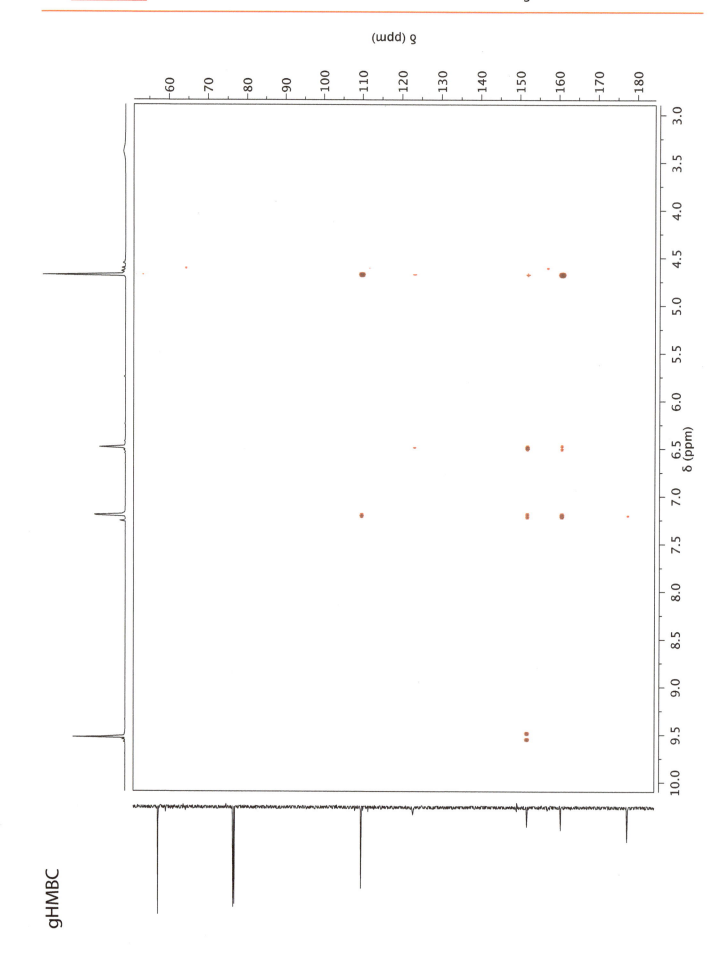

gHMBC

SECTION 4 Problem 83

Determine the structure of the unknown compound below with molecular formula C₁₀H₁₅NO.

Spectra acquired in CDCl₃ at 500 MHz (¹H) or 125 MHz (¹³C).

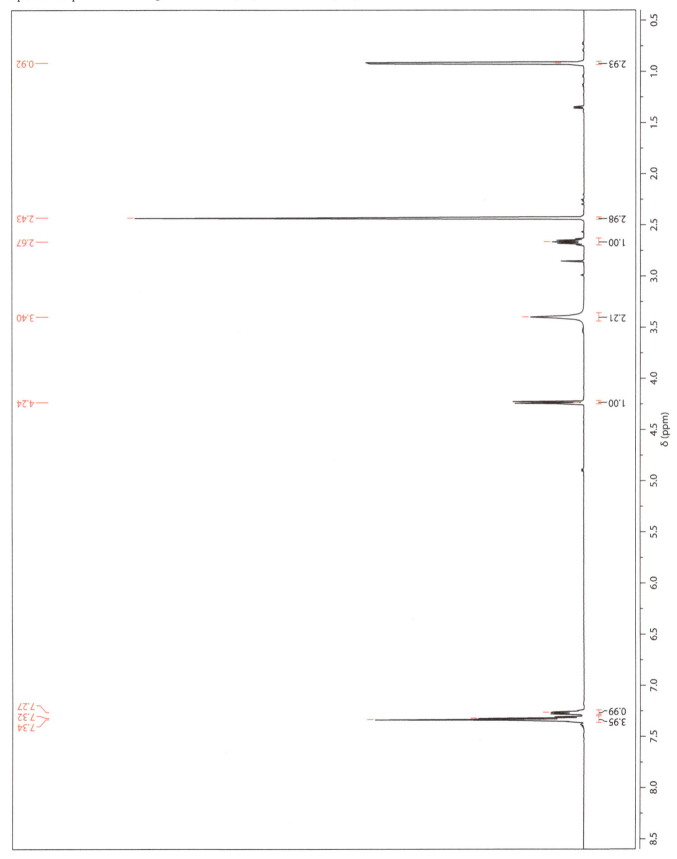

SECTION 4 Problem 83

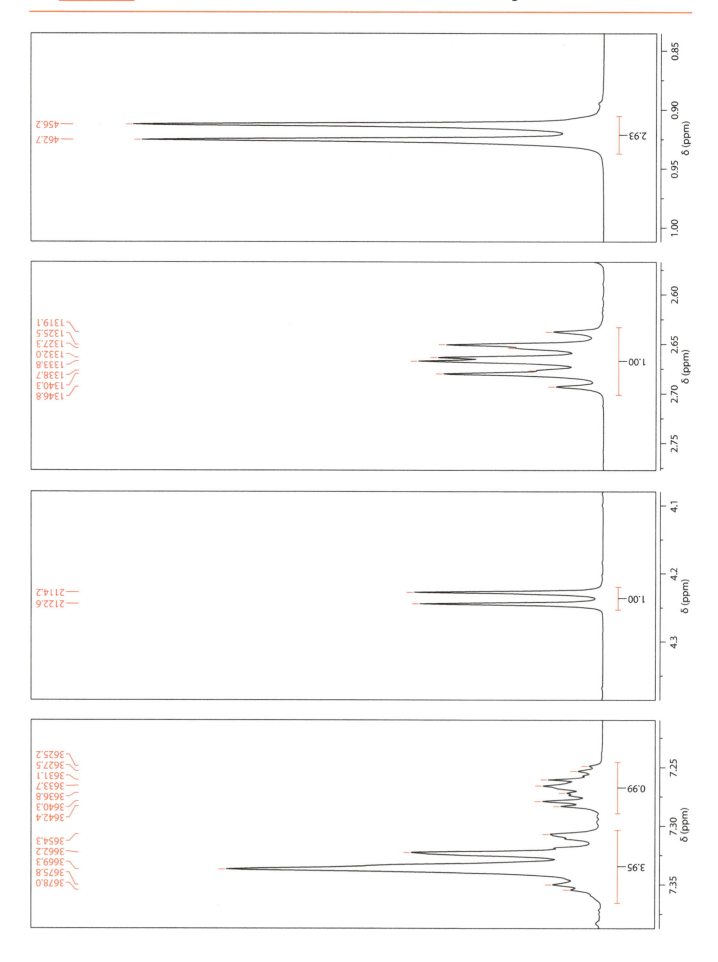

SECTION 4 Problem 83

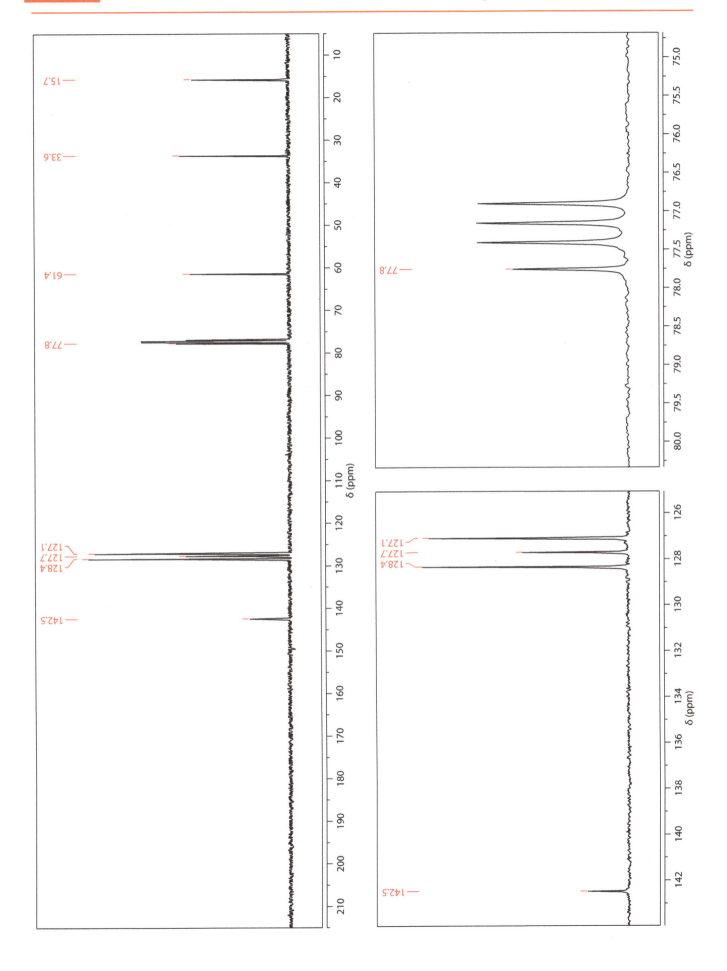

Problem 83

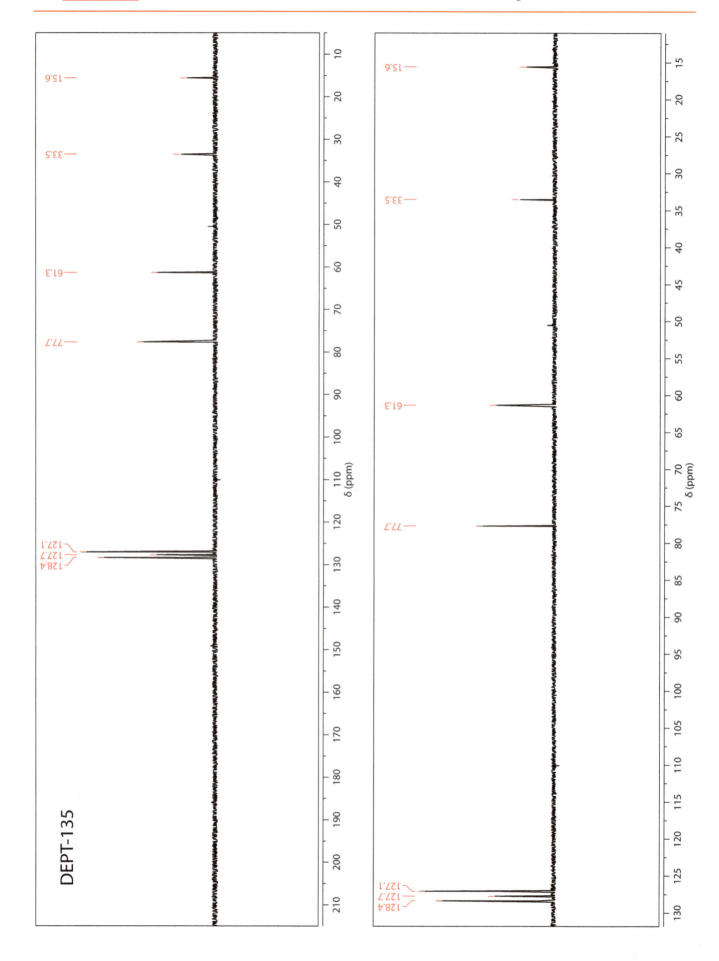

DEPT-135

SECTION 4 Problem 83

gCOSY

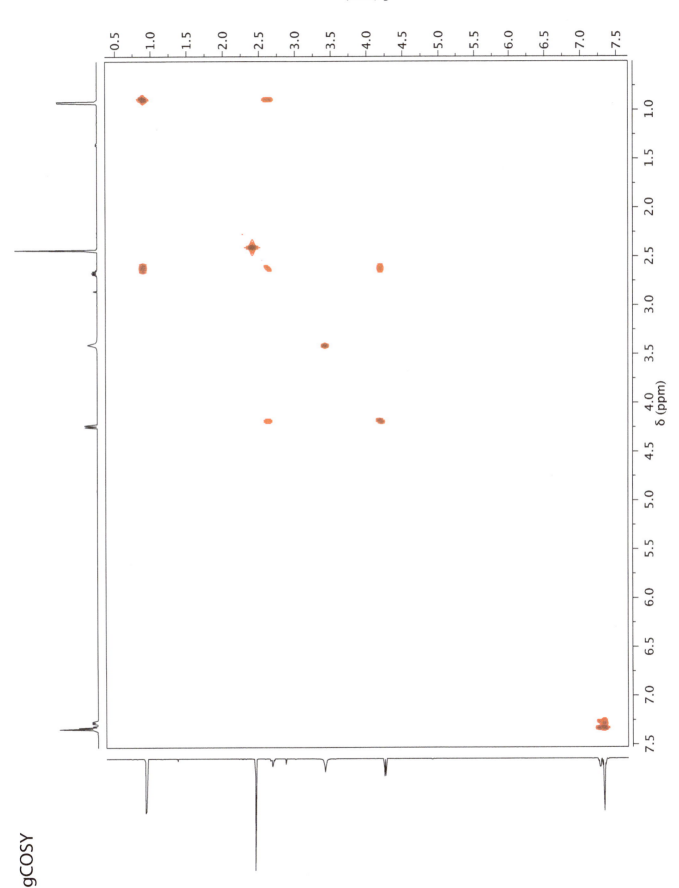

Problem 83

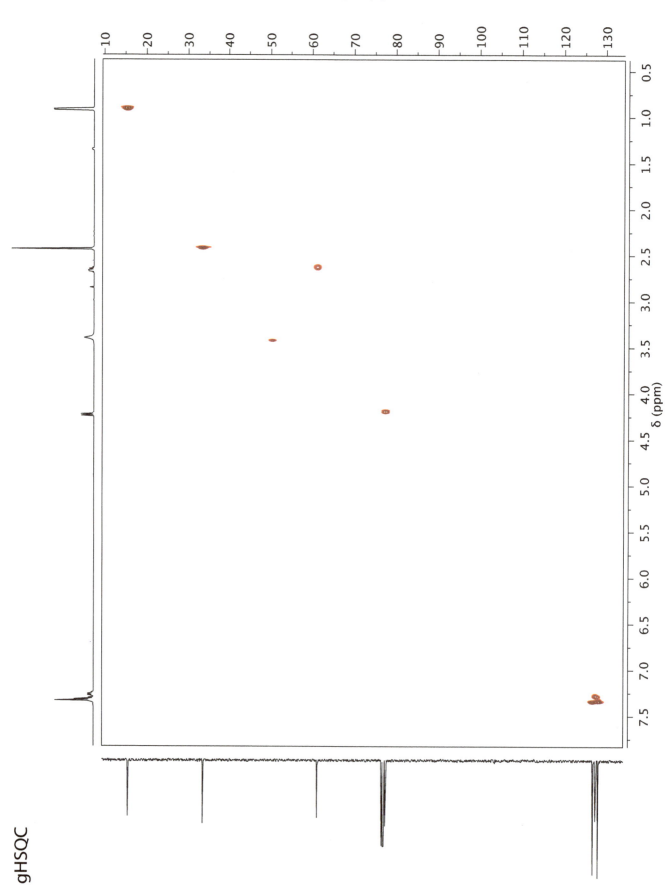

gHSQC

SECTION 4 Problem 83

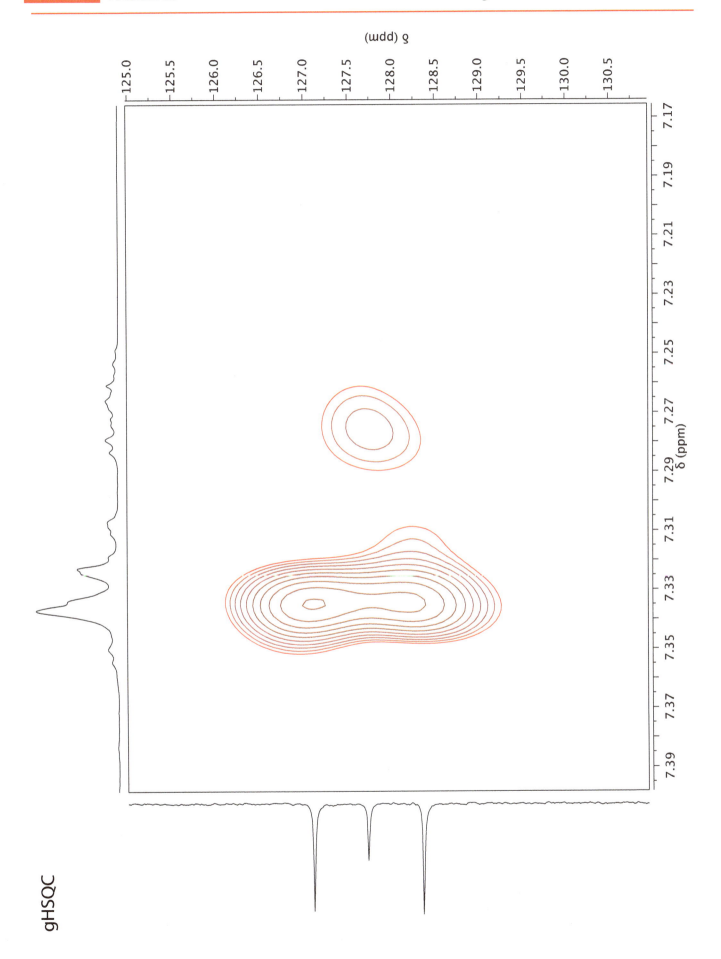

gHSQC

Problem 83

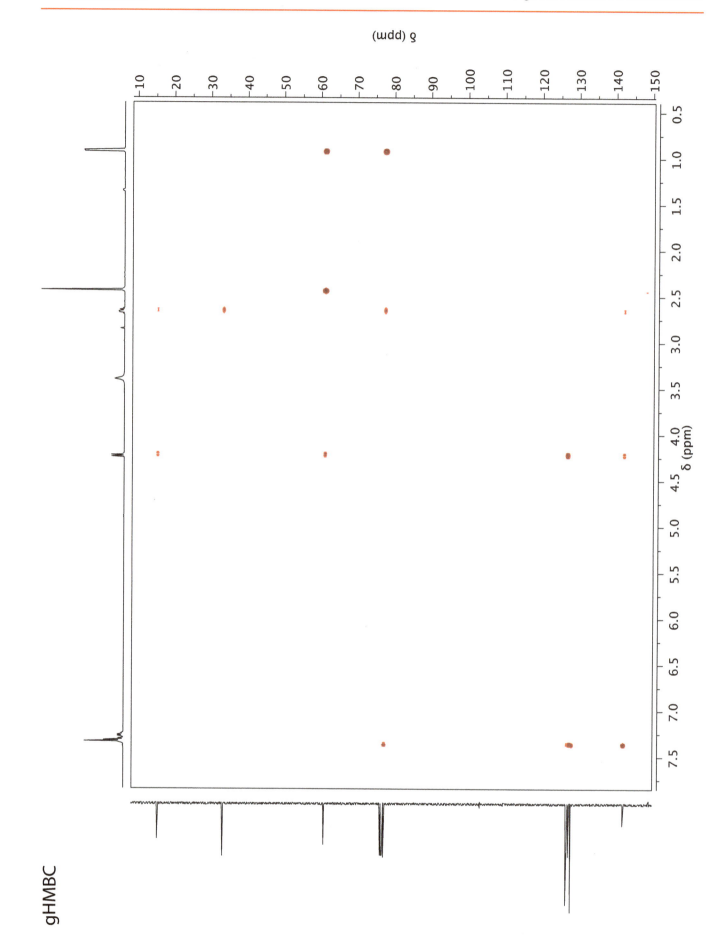

gHMBC

SECTION 4 Problem 83

Problems in Organic Structure Determination

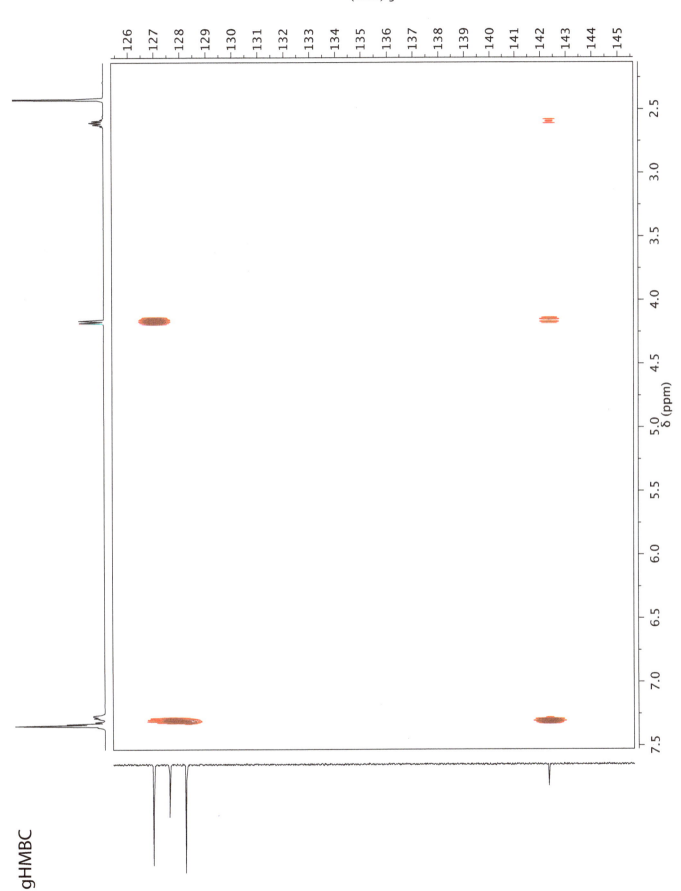

gHMBC

Problem 83

gHMBC

SECTION 4 Problem 84

Problems in Organic Structure Determination

Determine the structure of the unknown compound below with molecular formula $C_{17}H_{19}ClN_2S \cdot HCl$.

Spectra acquired in $(CD_3)_2SO$ at 500 MHz (^1H) or 125 MHz (^{13}C).

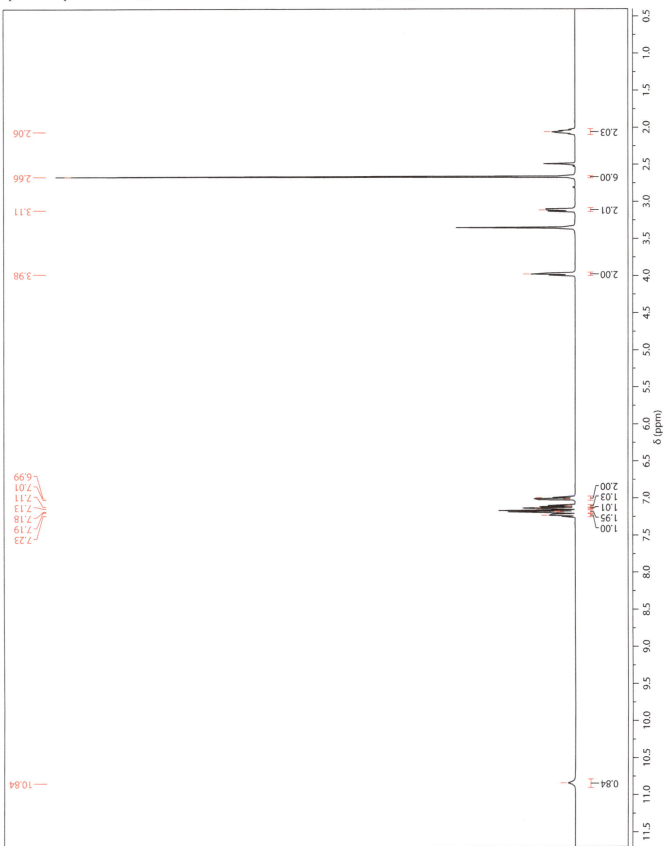

SECTION 4 Problem 84

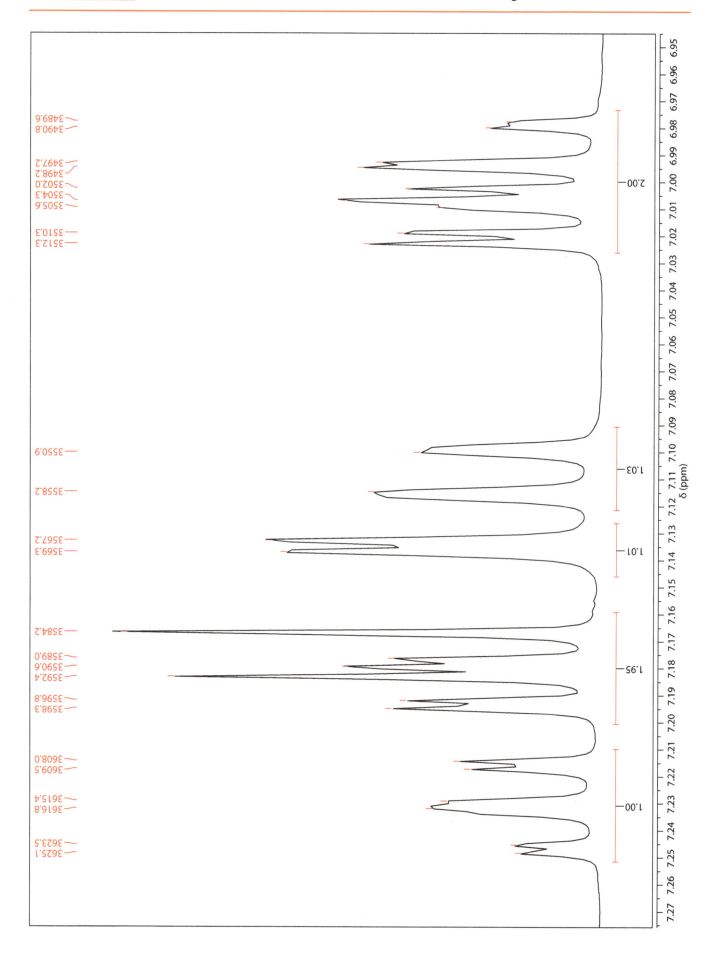

SECTION 4 Problem 84

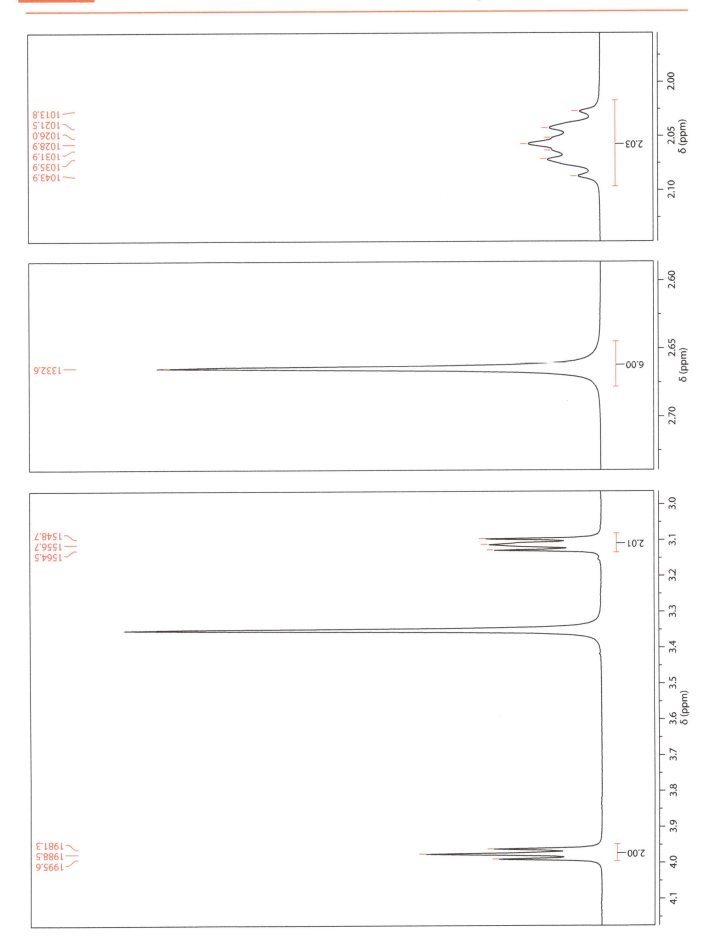

Problem 84

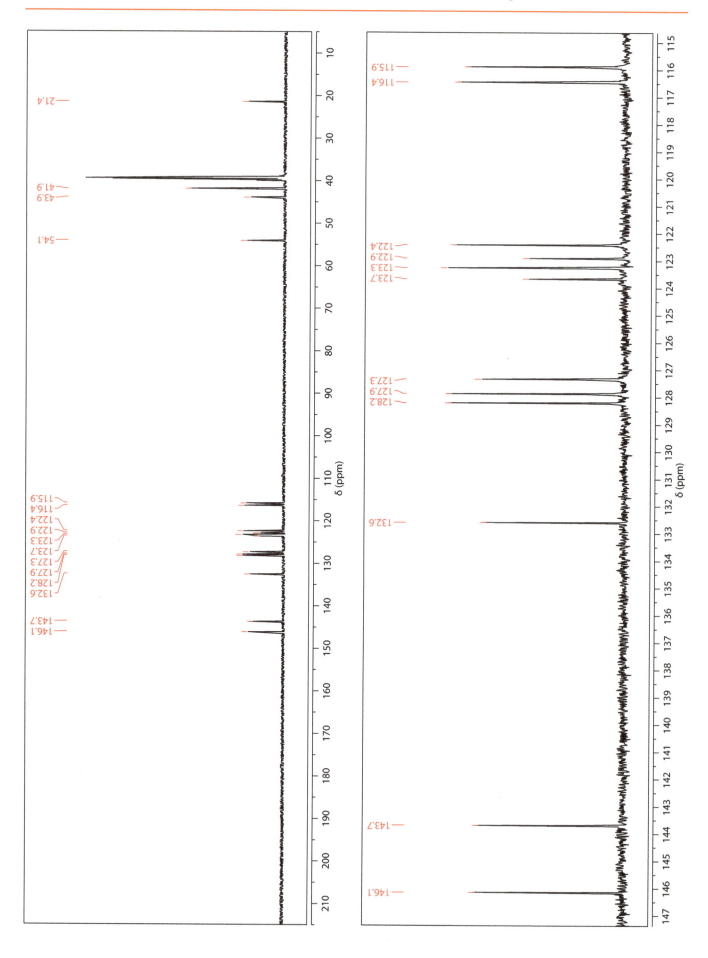

SECTION 4 Problem 84

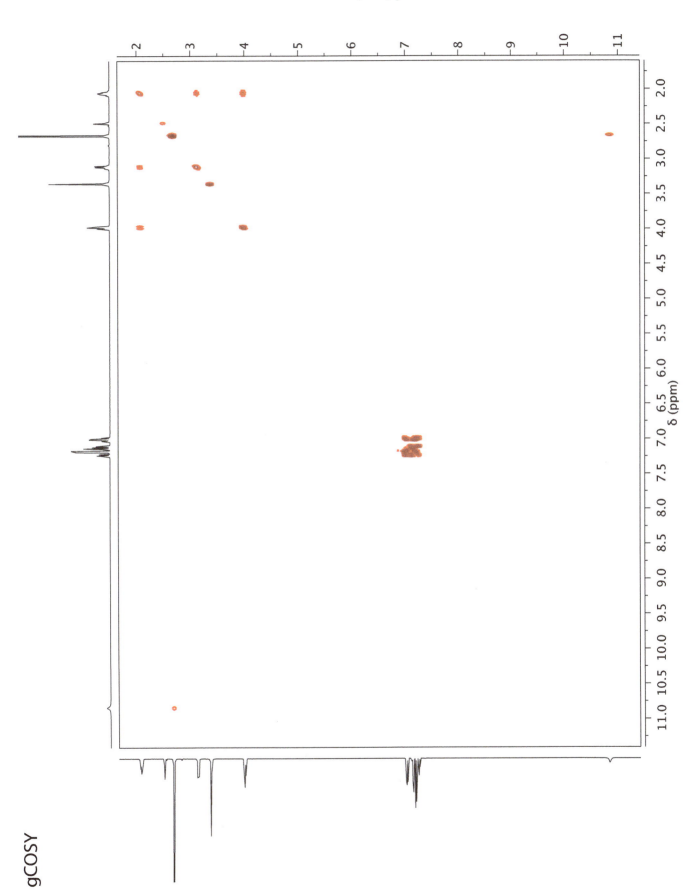

gCOSY

SECTION 4 Problem 84

gCOSY

SECTION 4 Problem 84

gCOSY

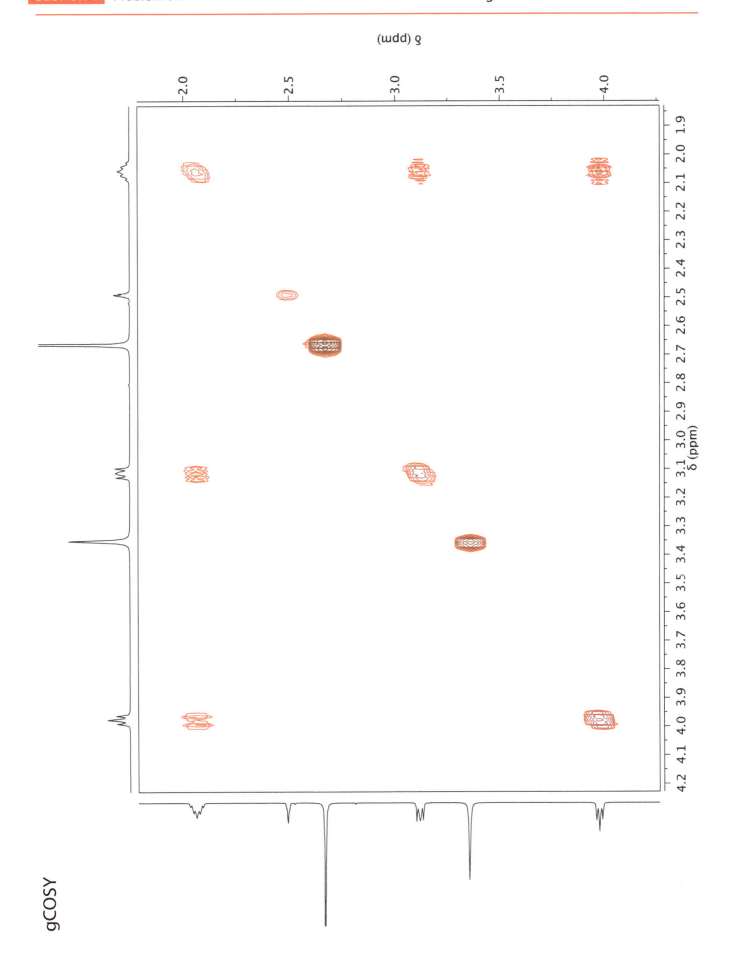

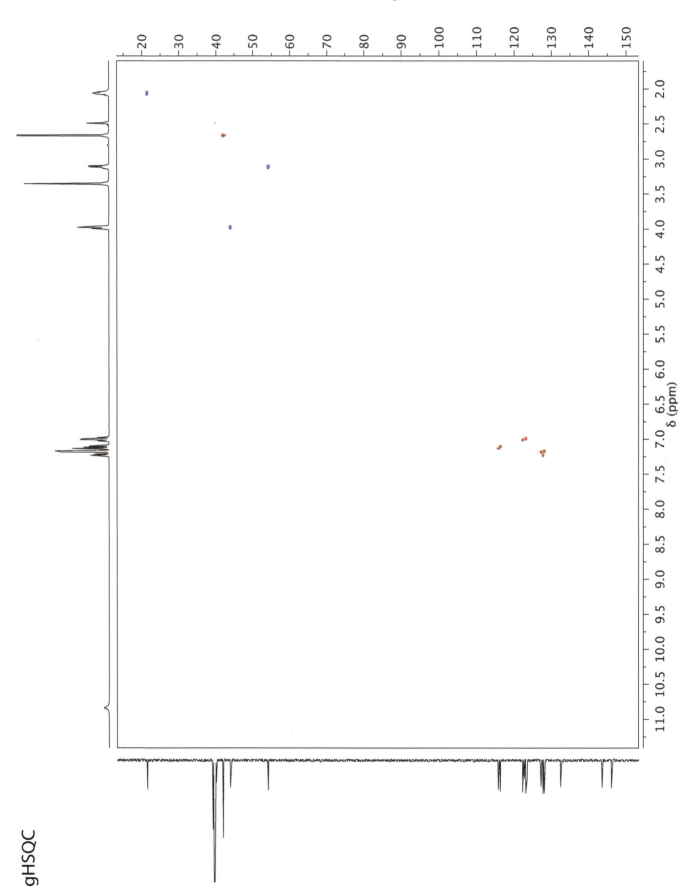

SECTION 4 Problem 84

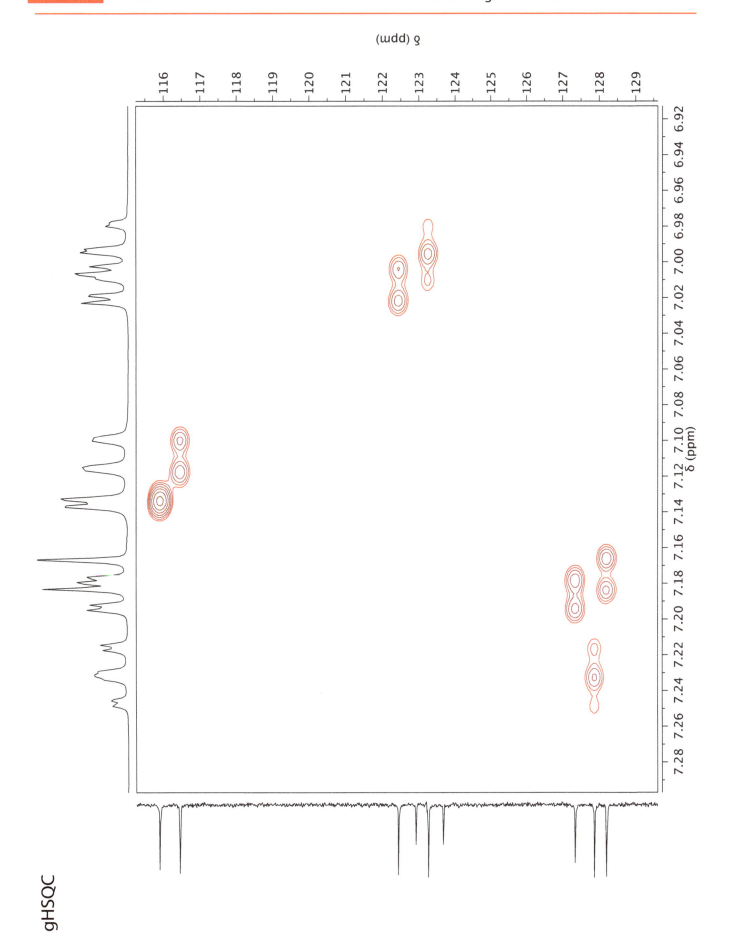

gHSQC

Problem 84

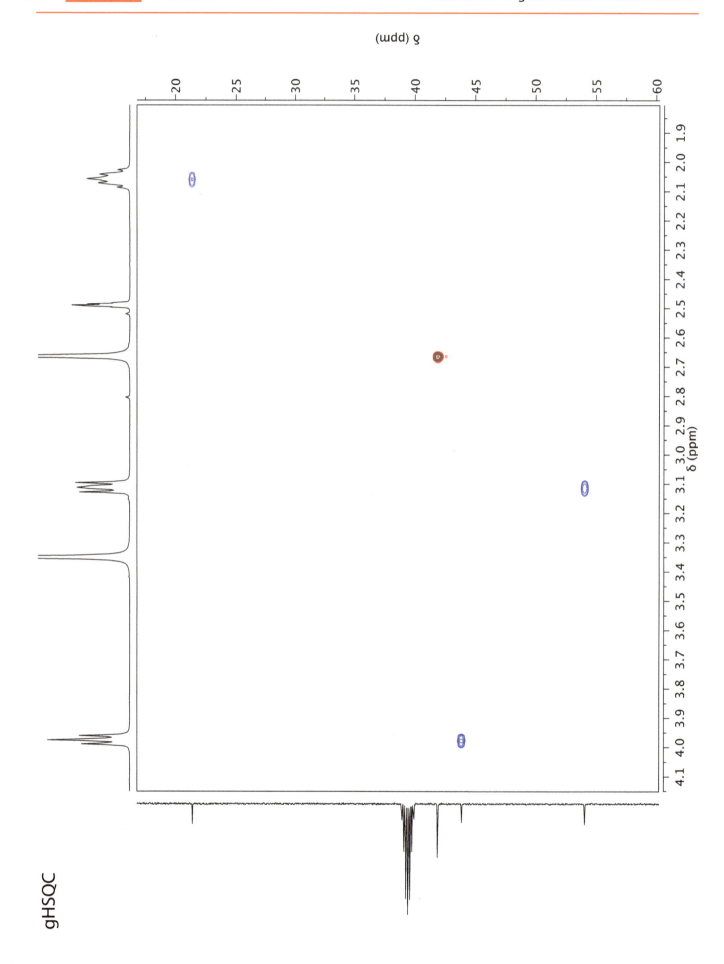

gHSQC

SECTION 4 Problem 84

Problems in Organic Structure Determination

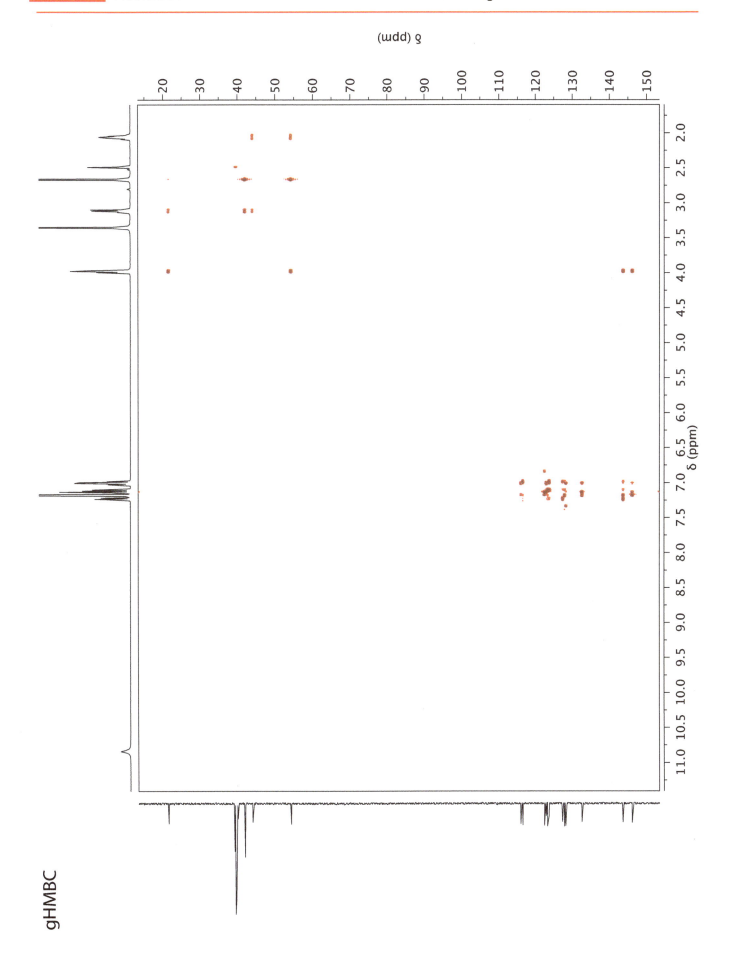

gHMBC

Problem 84

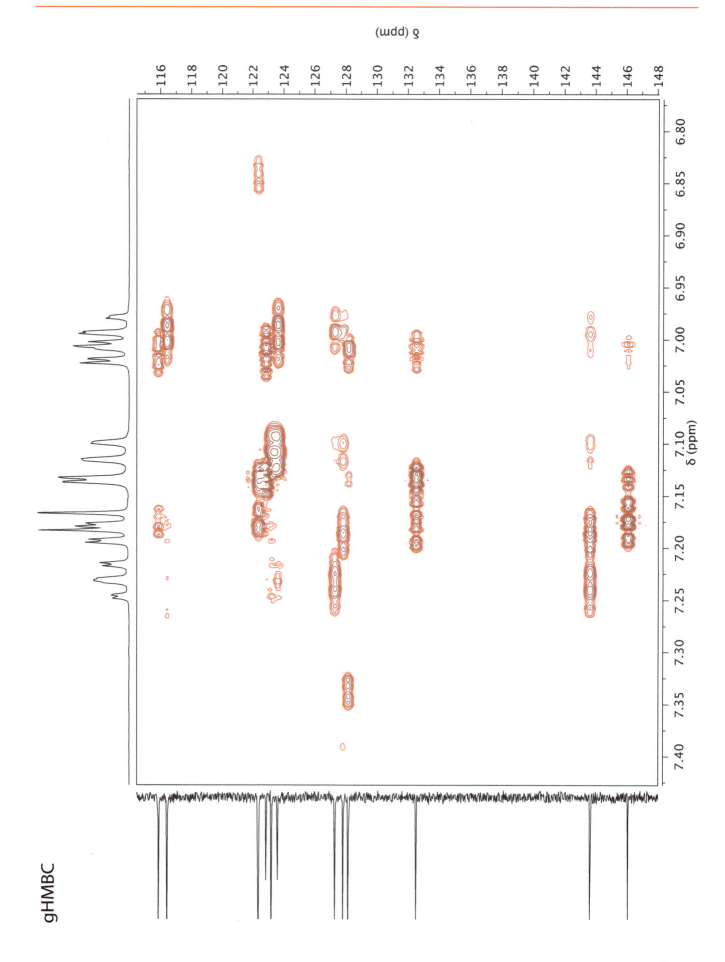

SECTION 4 Problem 84

Problems in Organic Structure Determination

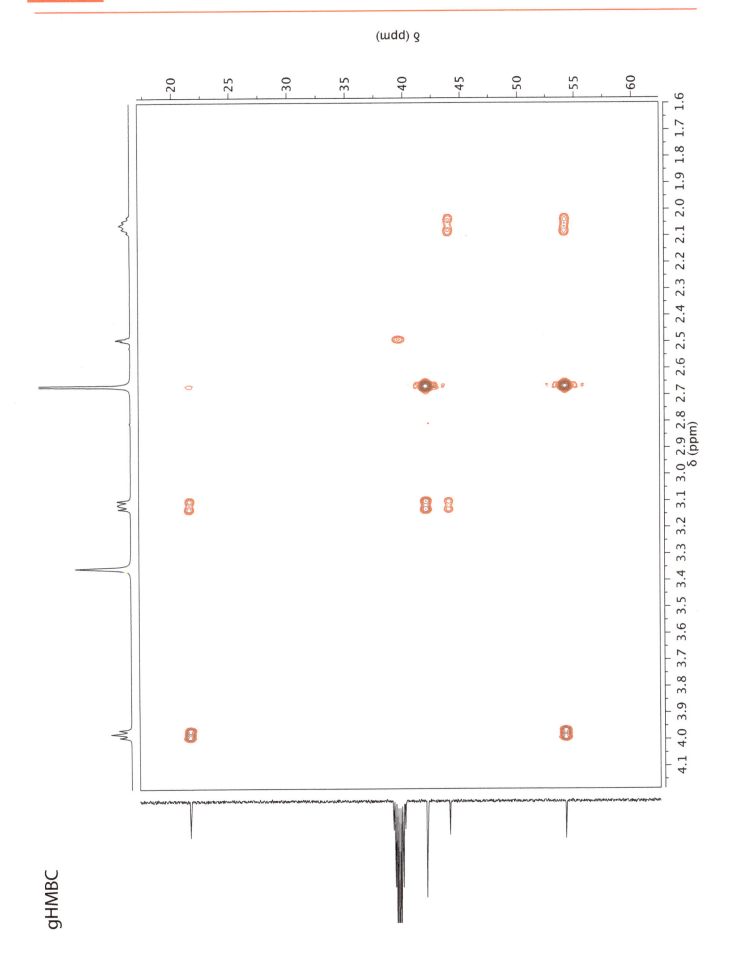

gHMBC

Determine the structure of the unknown compound below with molecular formula $C_{10}H_{13}N$.

Spectra acquired in $CDCl_3$ at 500 MHz (^1H) or 125 MHz (^{13}C).

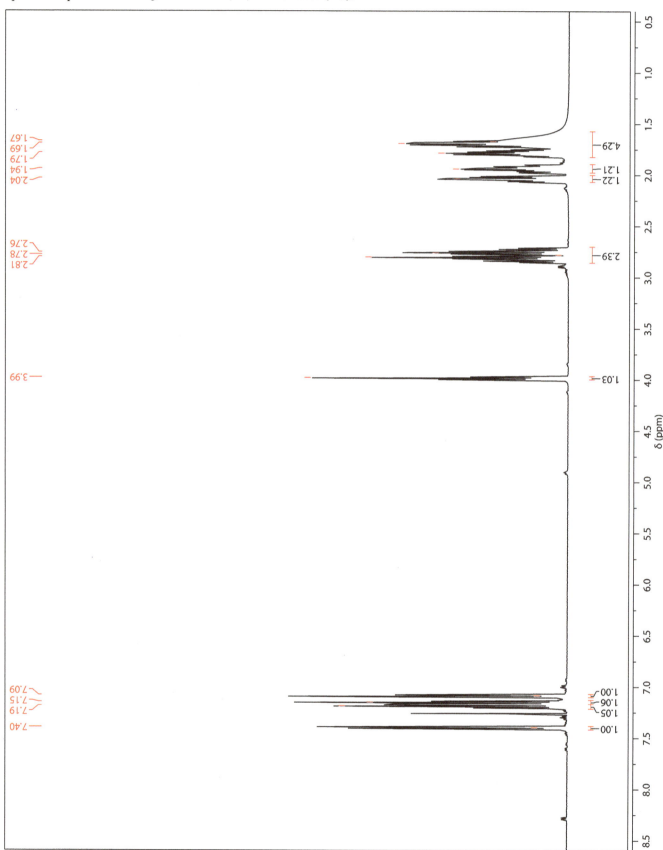

SECTION 4 Problem 85

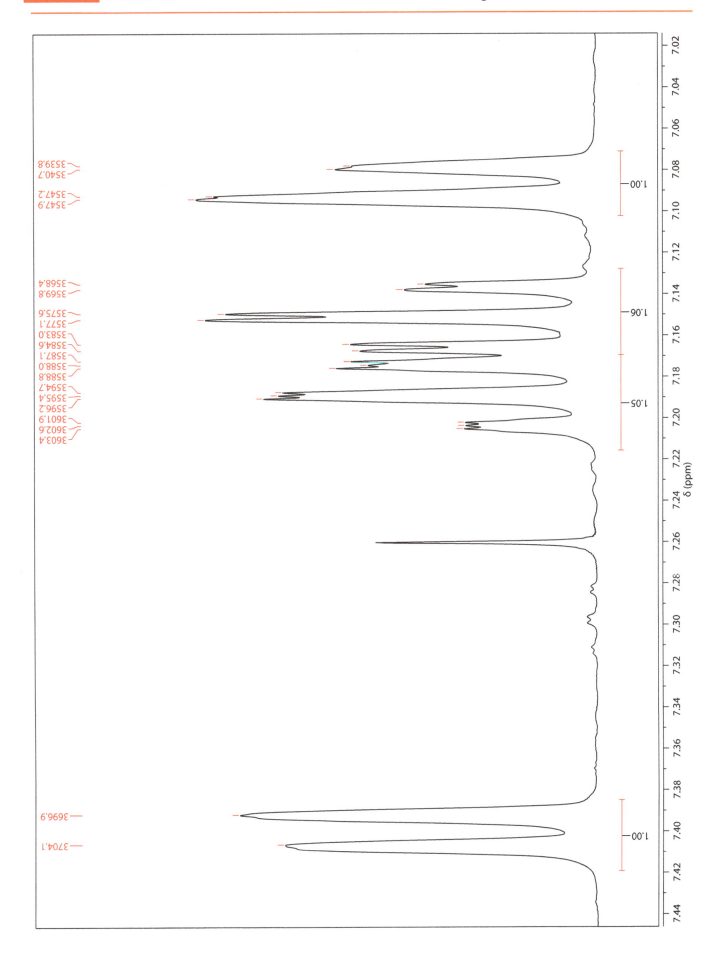

Problem 85

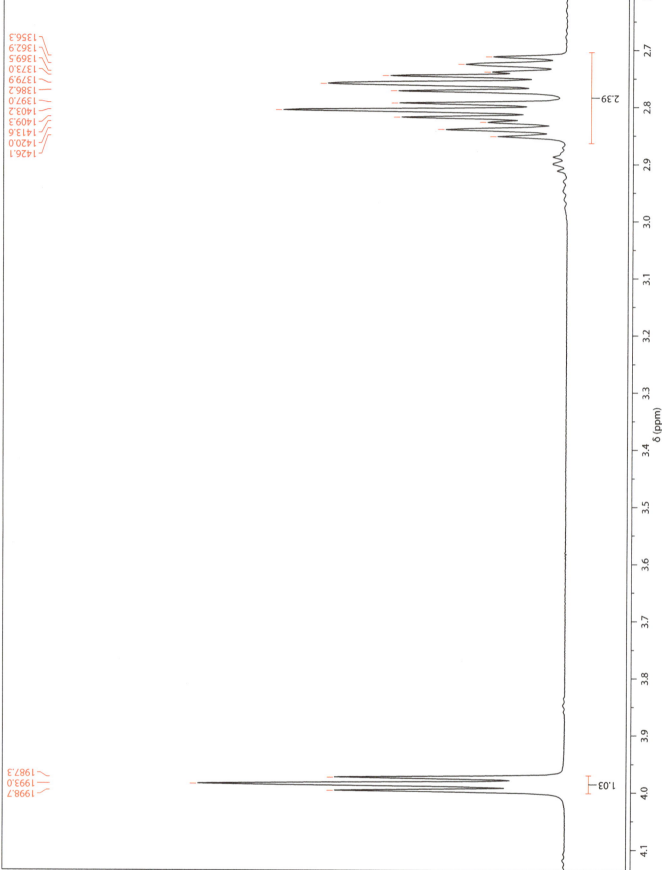

SECTION 4 Problem 85

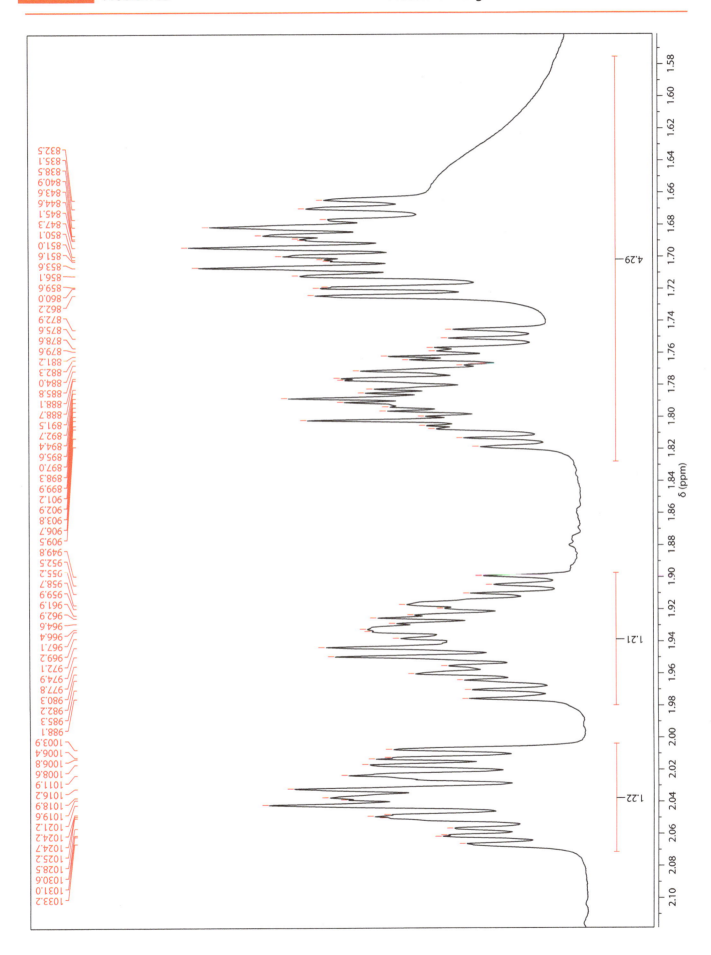

SECTION 4 Problem 85

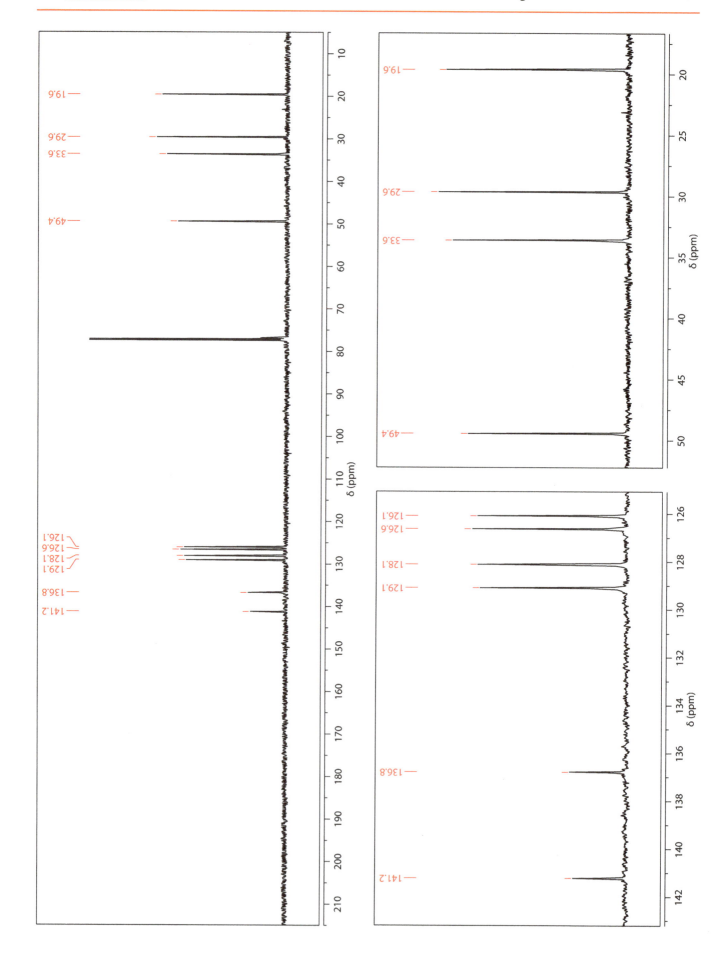

SECTION 4   Problem 85

Problems in Organic Structure Determination

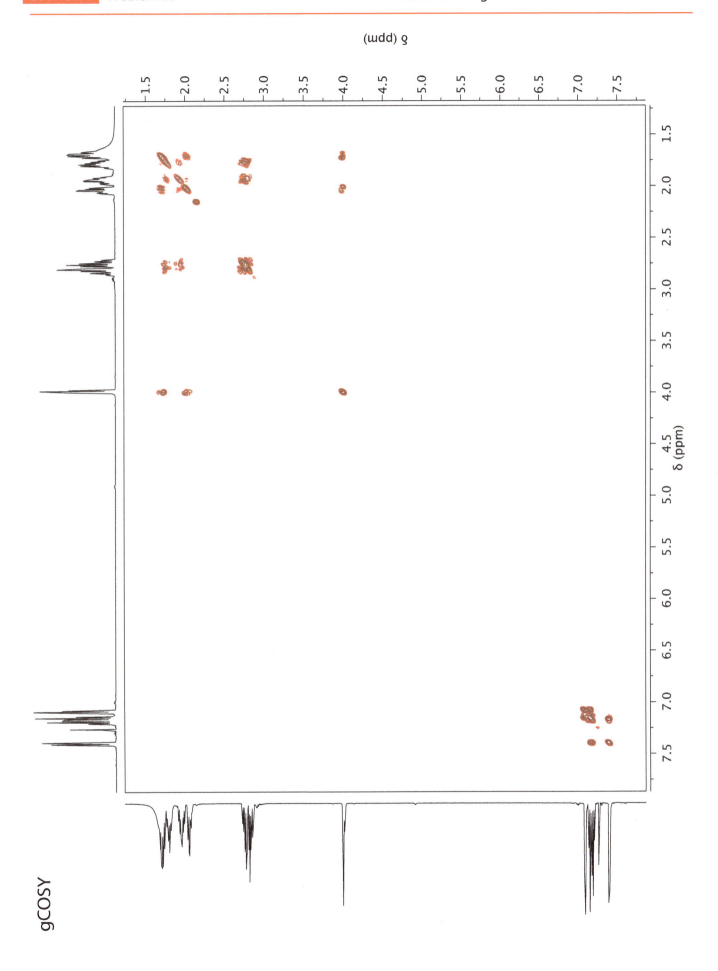

gCOSY

Problem 85

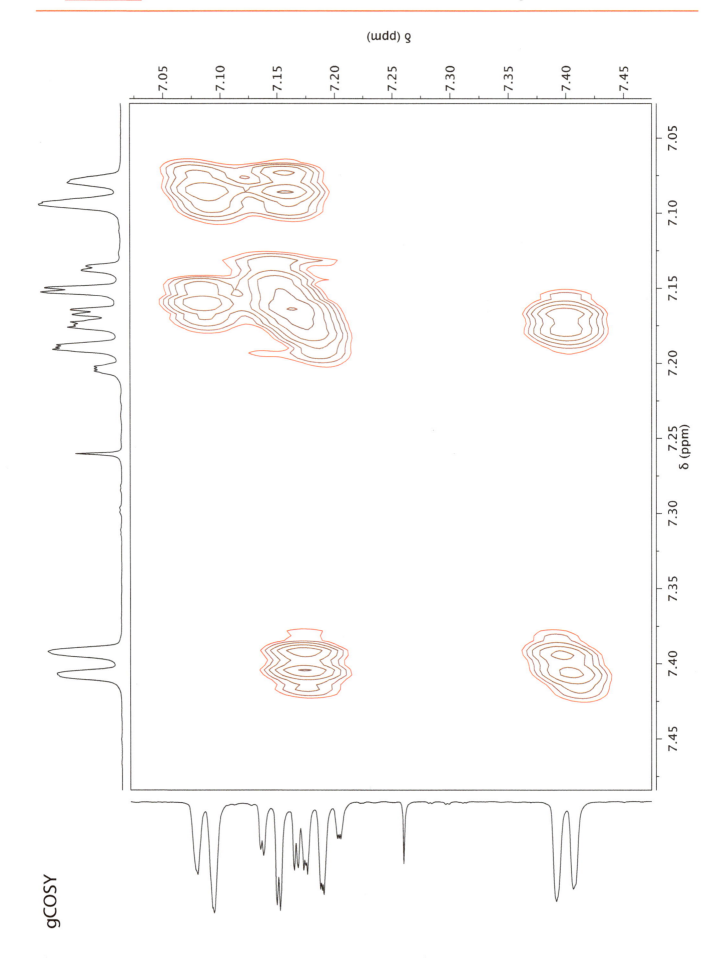

gCOSY

SECTION 4 Problem 85

gCOSY

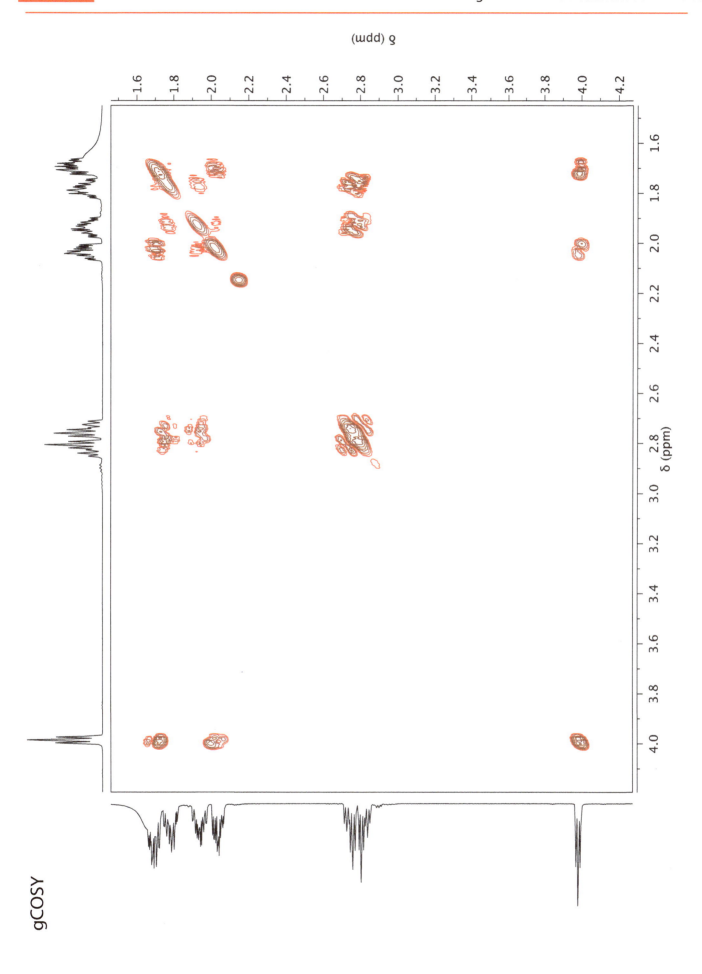

Problem 85

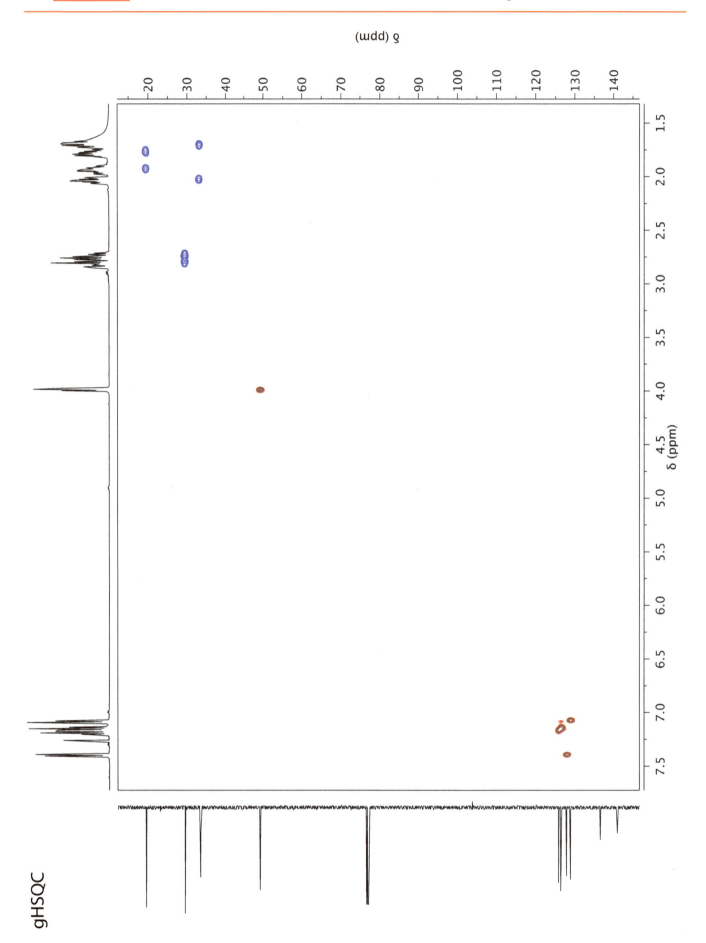

Problem 85

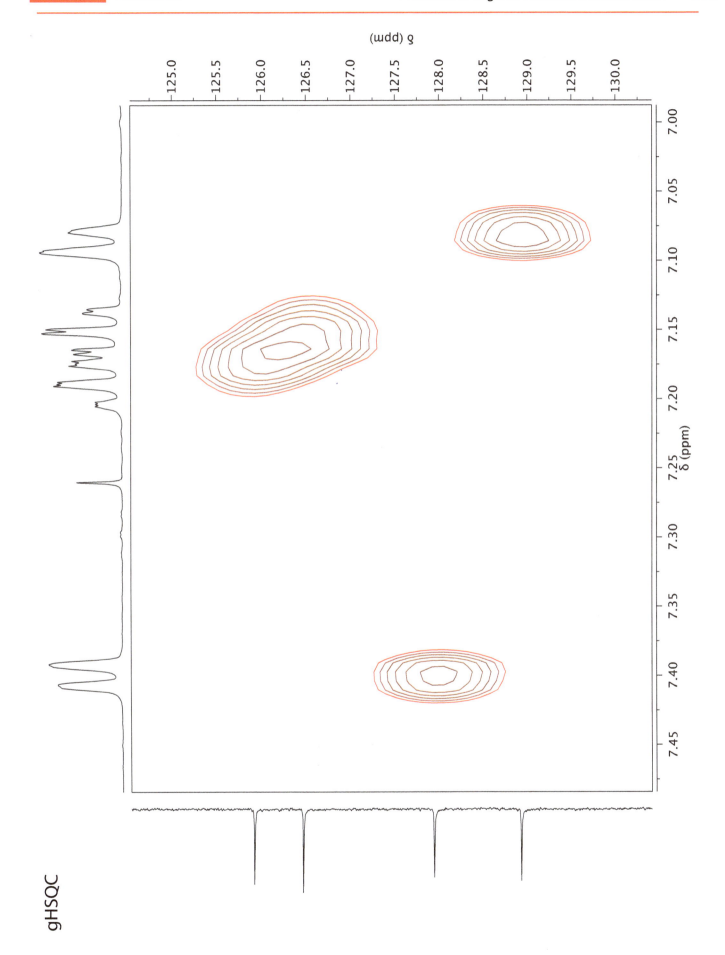

Problem 85

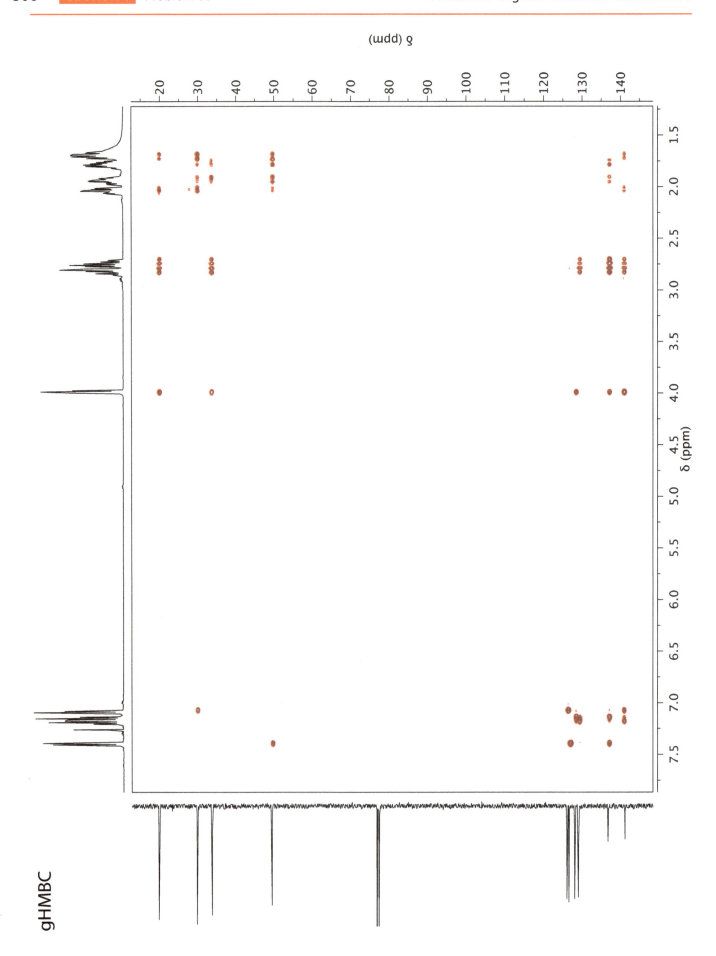

gHMBC

SECTION 4 Problem 85

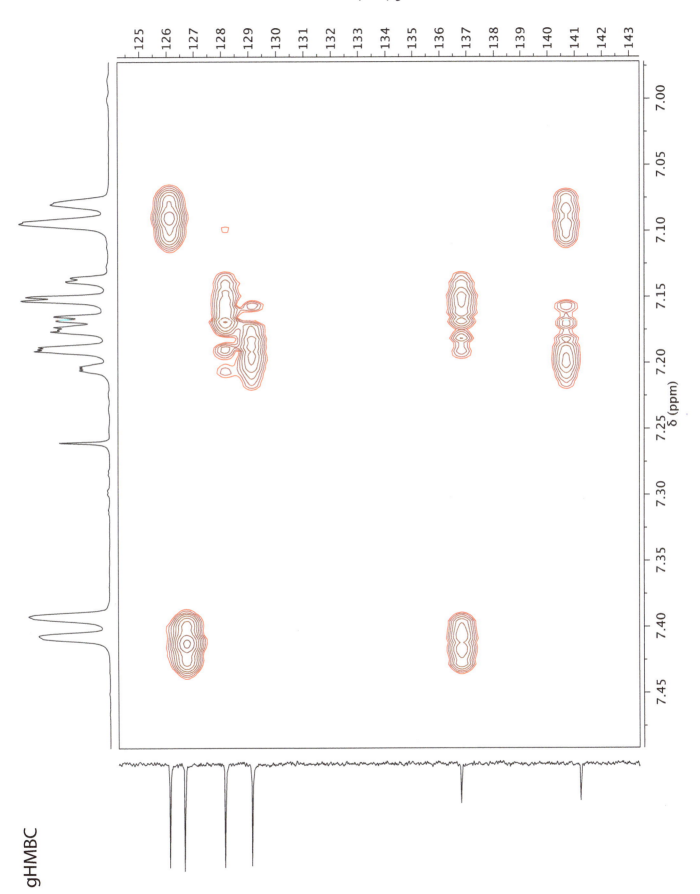

Determine the structure of the unknown compound below with molecular formula $C_{14}H_{22}N_2O_3$.

Spectra acquired in CD_3OD at 500 MHz (^1H) or 125 MHz (^{13}C).

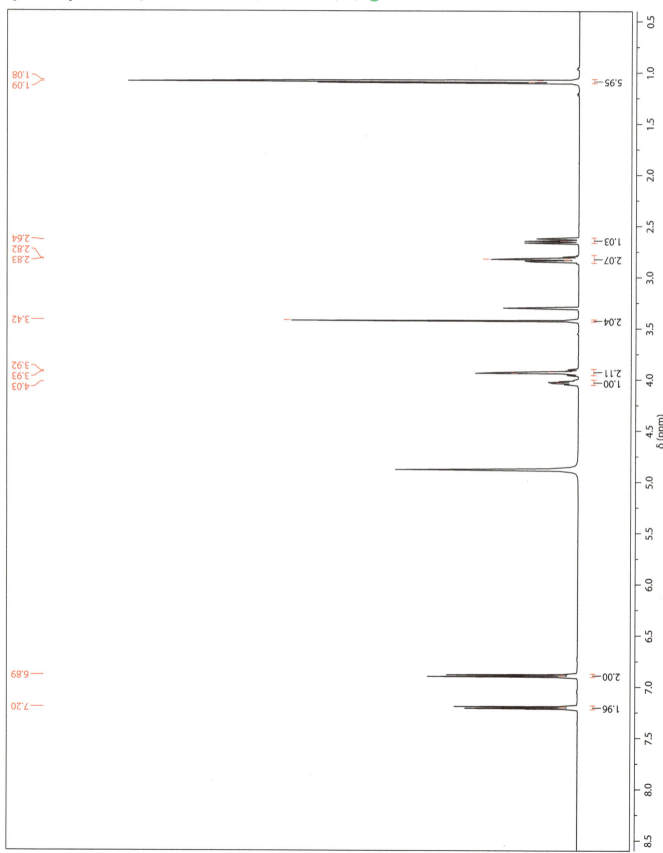

SECTION 4 Problem 86

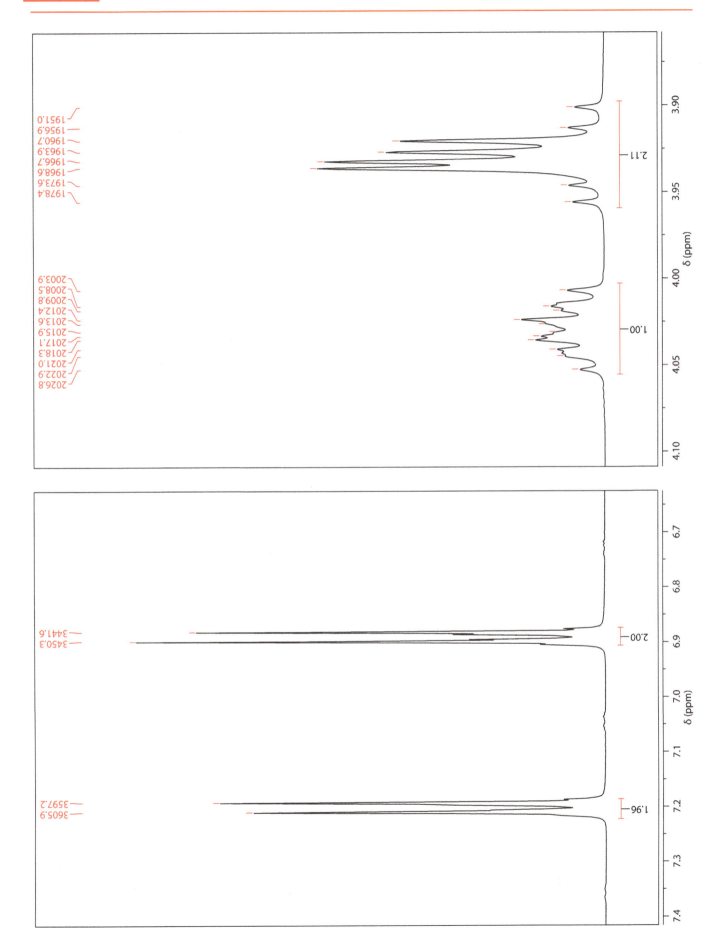

SECTION 4 Problem 86

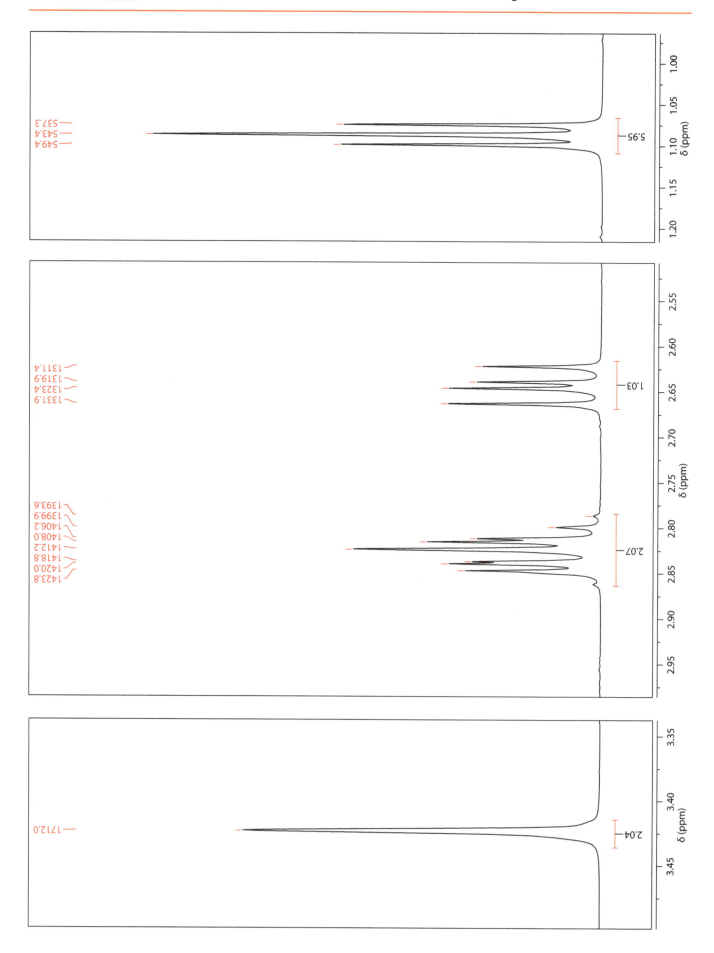

SECTION 4 Problem 86

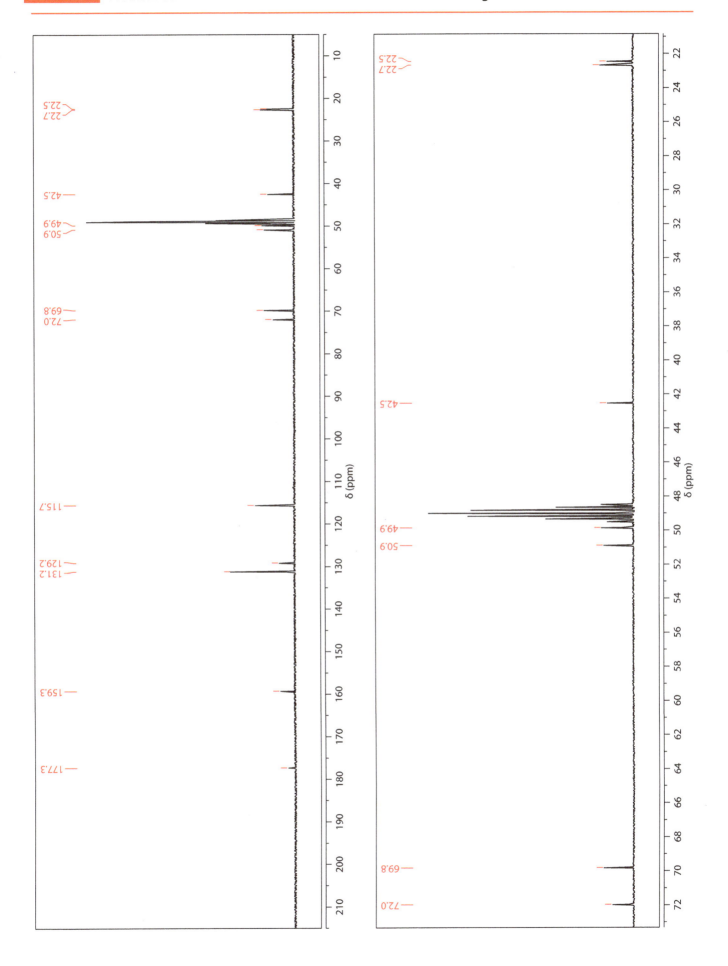

Problem 86

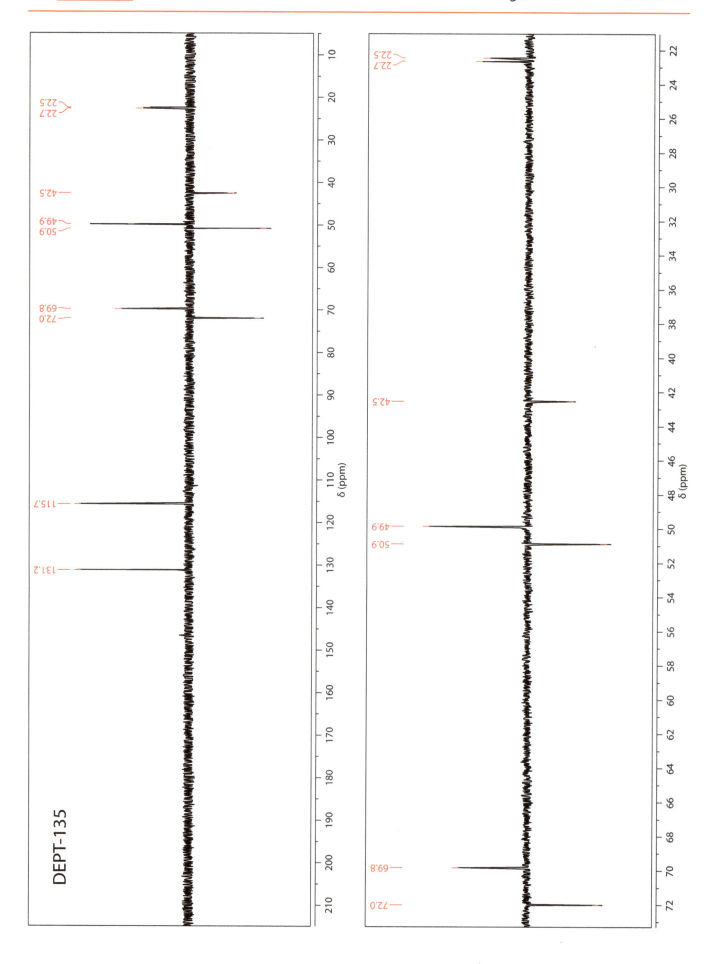

DEPT-135

SECTION 4 Problem 86

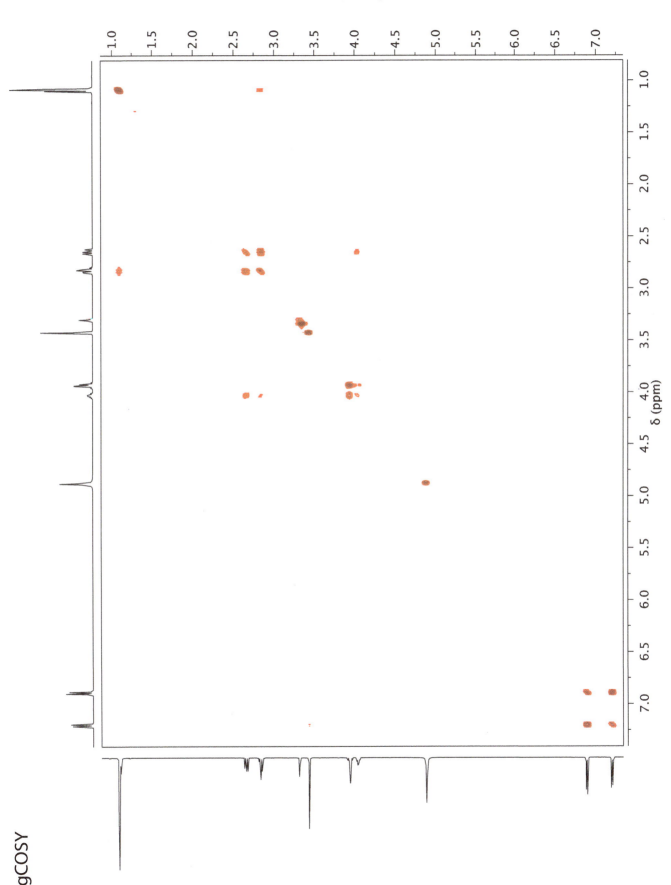

gCOSY

Problem 86

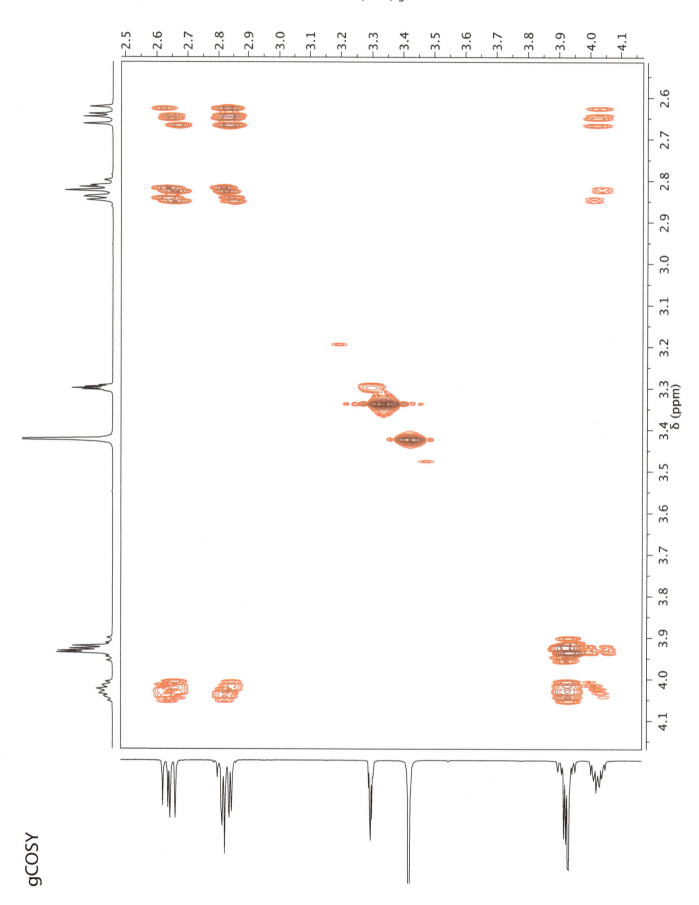

Problem 86

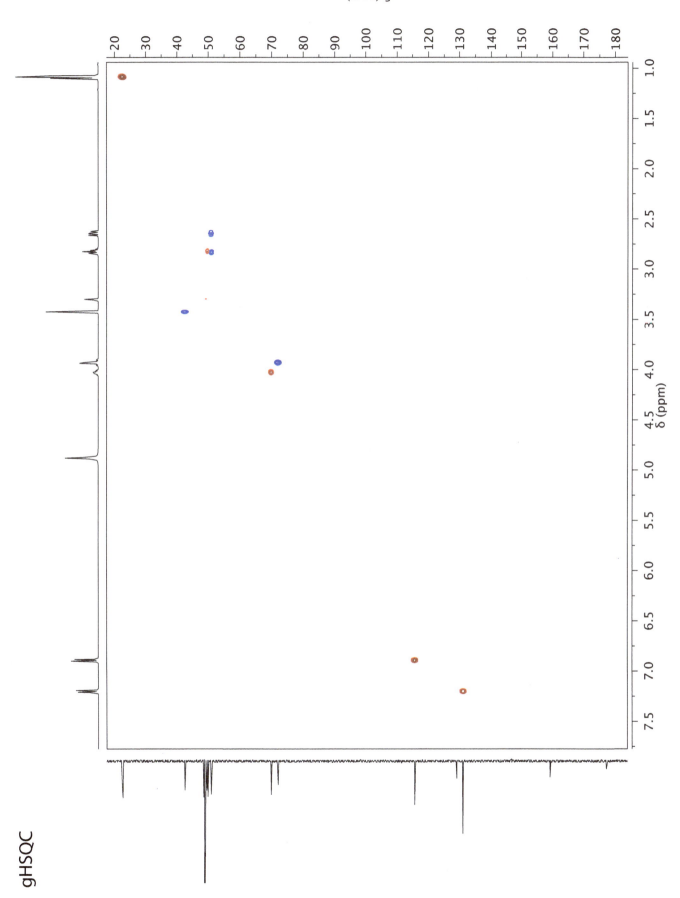

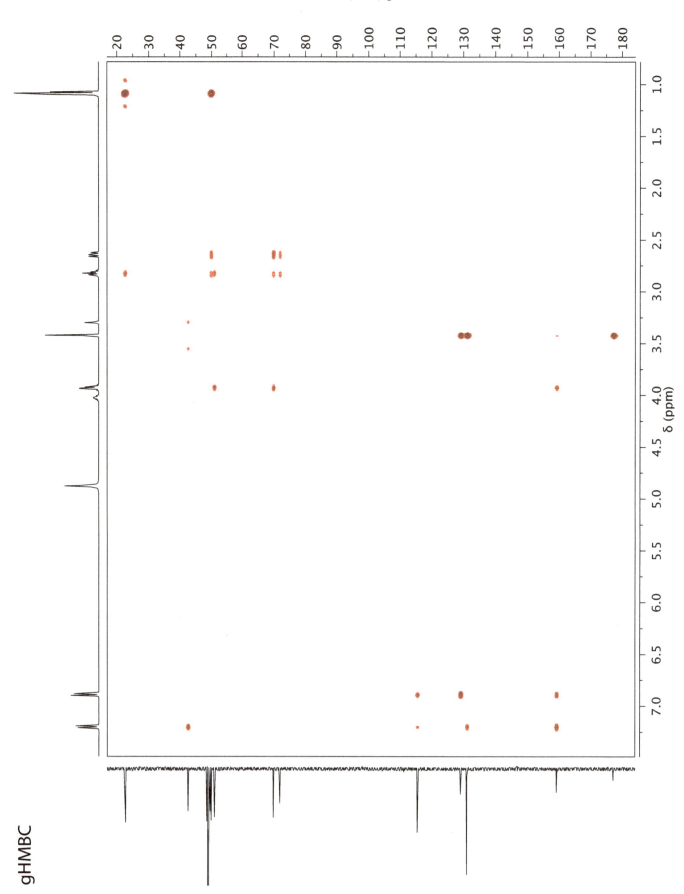

SECTION 4 Problem 86

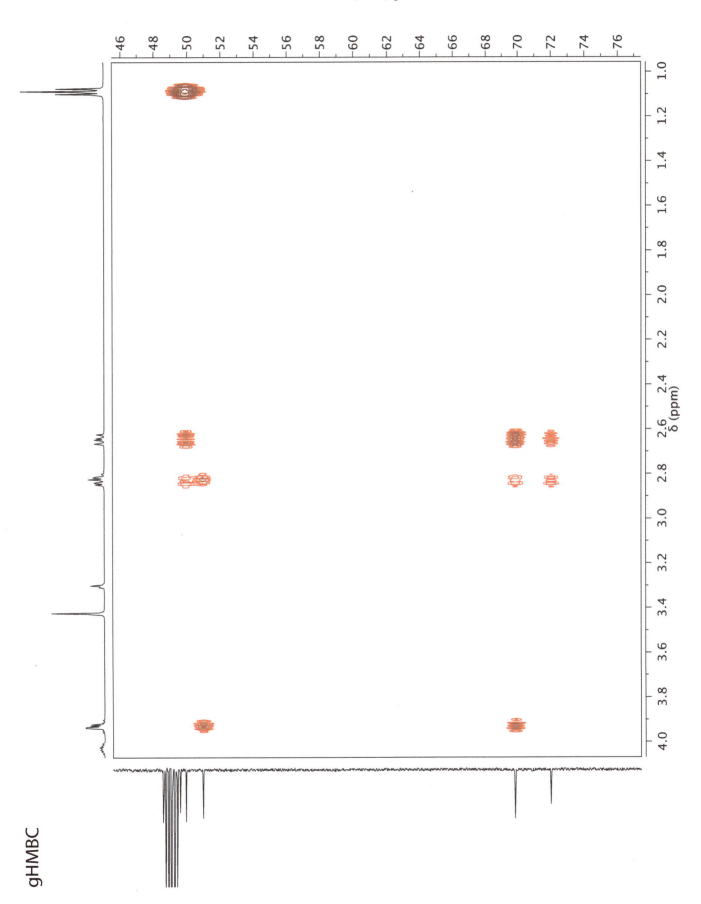

gHMBC

Problem 87

Determine the structure of the unknown compound below with molecular formula $C_{12}H_{10}N_2$.

Spectra acquired in $(CD_3)_2SO$ at 500 MHz (^1H) or 125 MHz (^{13}C).

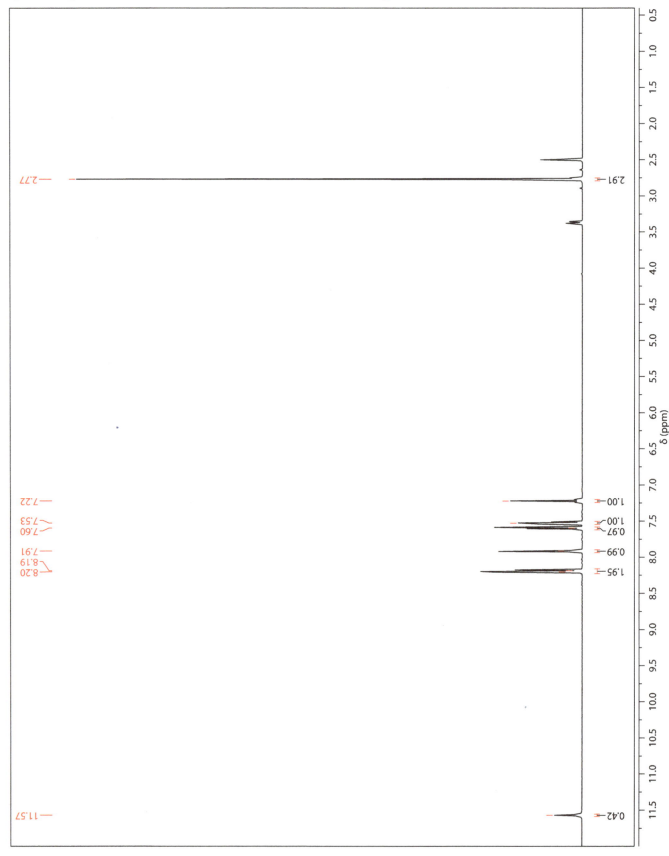

SECTION 4 Problem 87

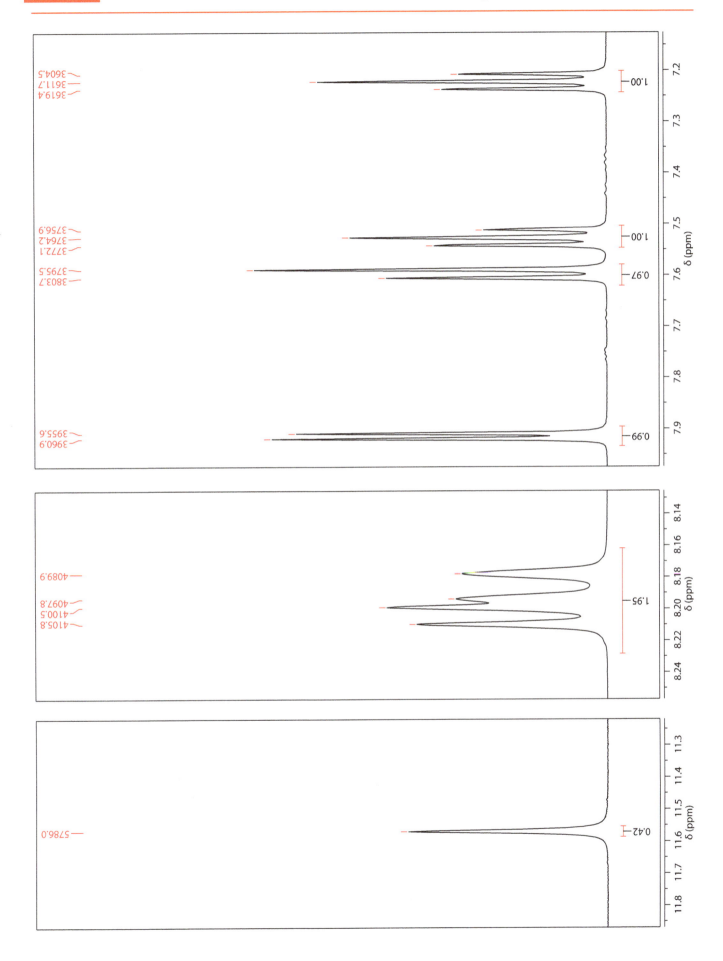

Problem 87

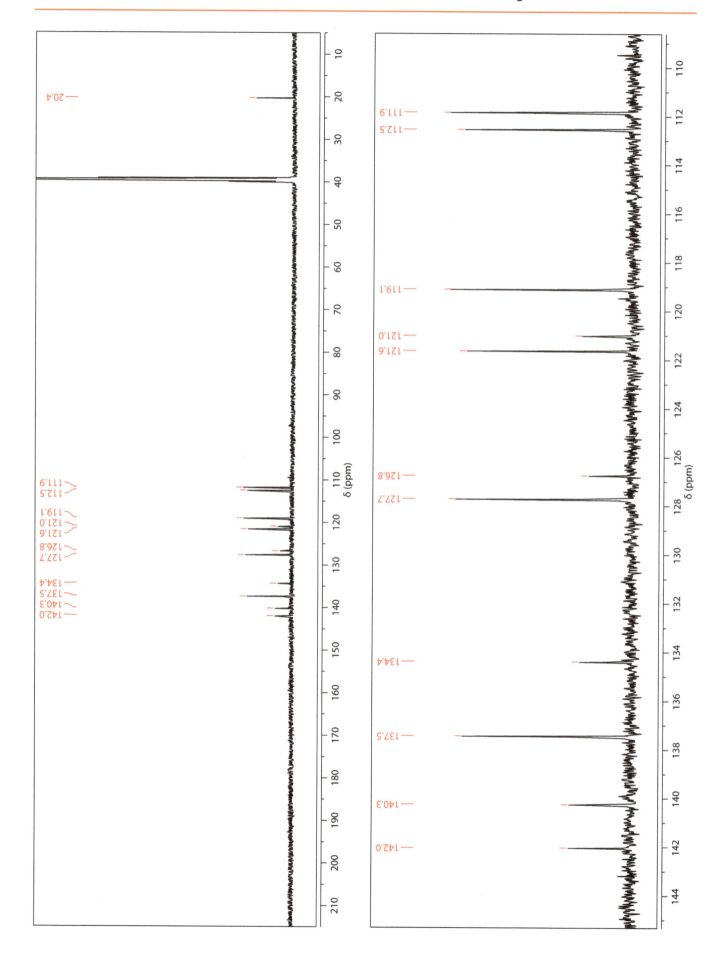

SECTION 4 Problem 87

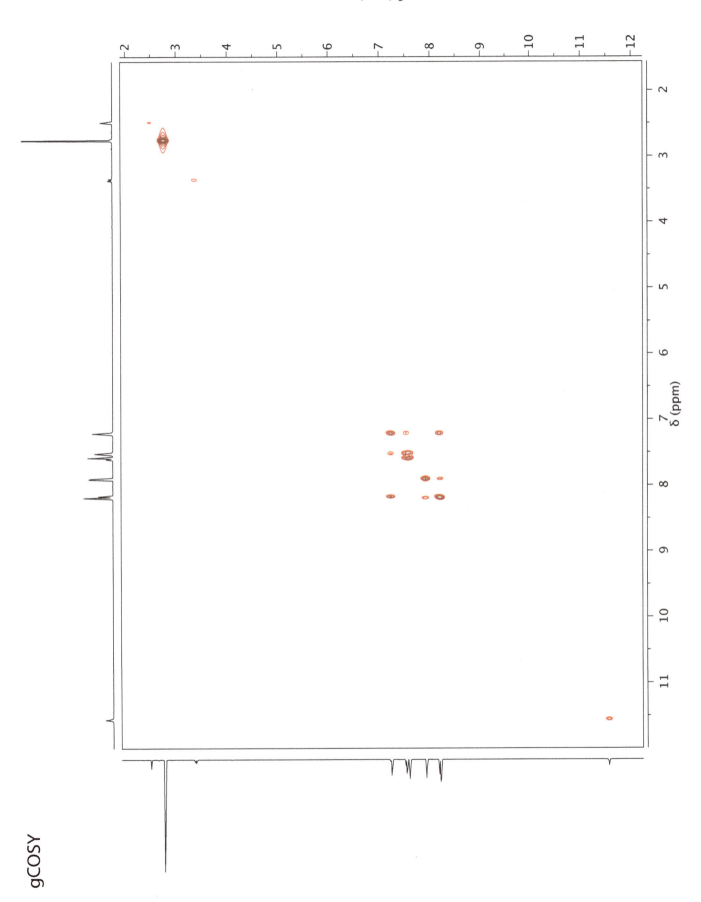

gCOSY

SECTION 4 Problem 87

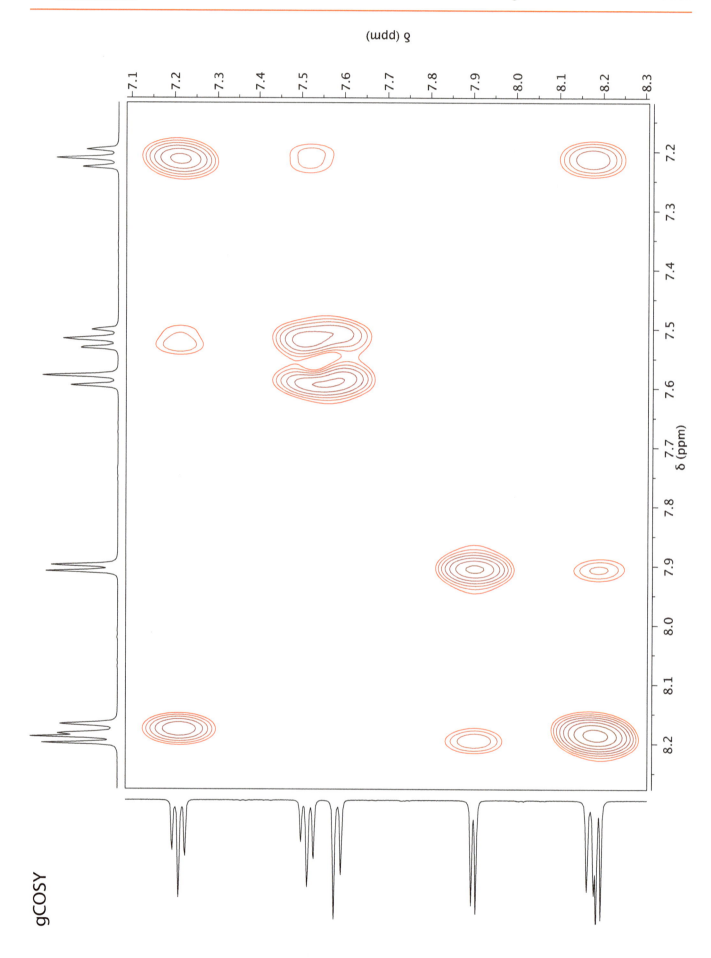

gCOSY

SECTION 4 Problem 87

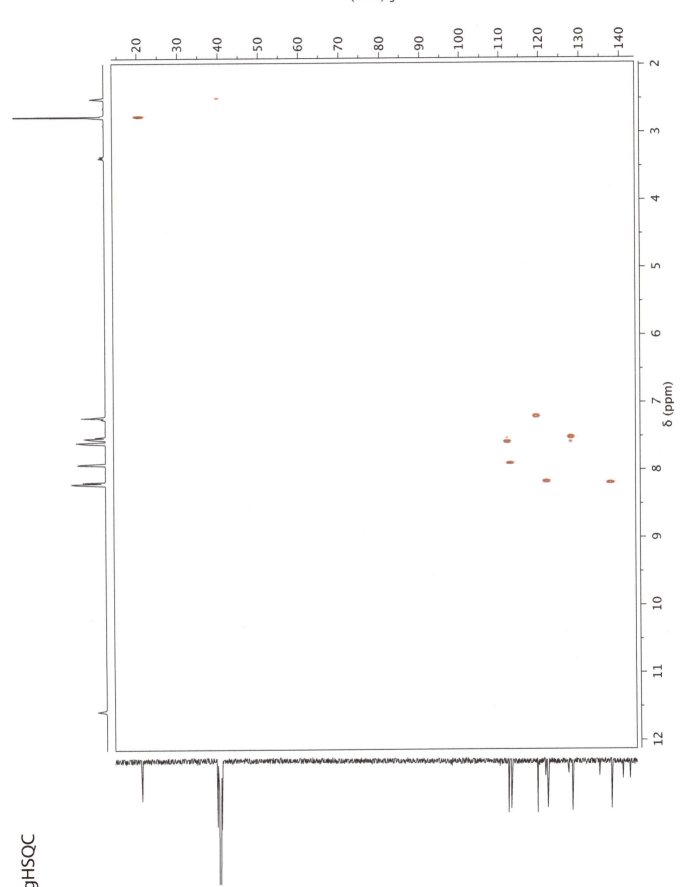

gHSQC

Problem 87

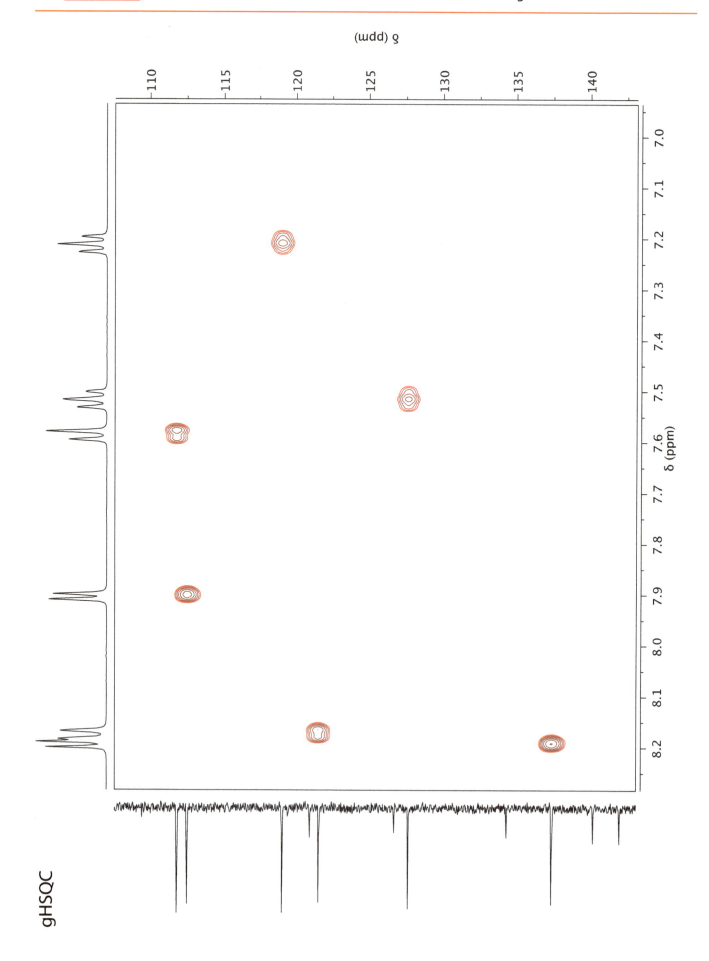

SECTION 4 Problem 87 — Problems in Organic Structure Determination

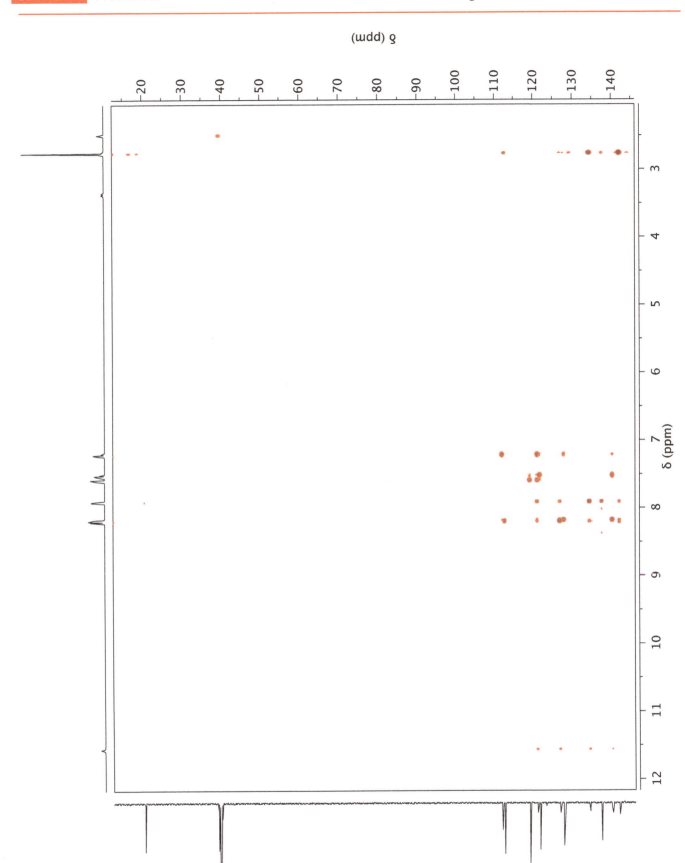

gHMBC

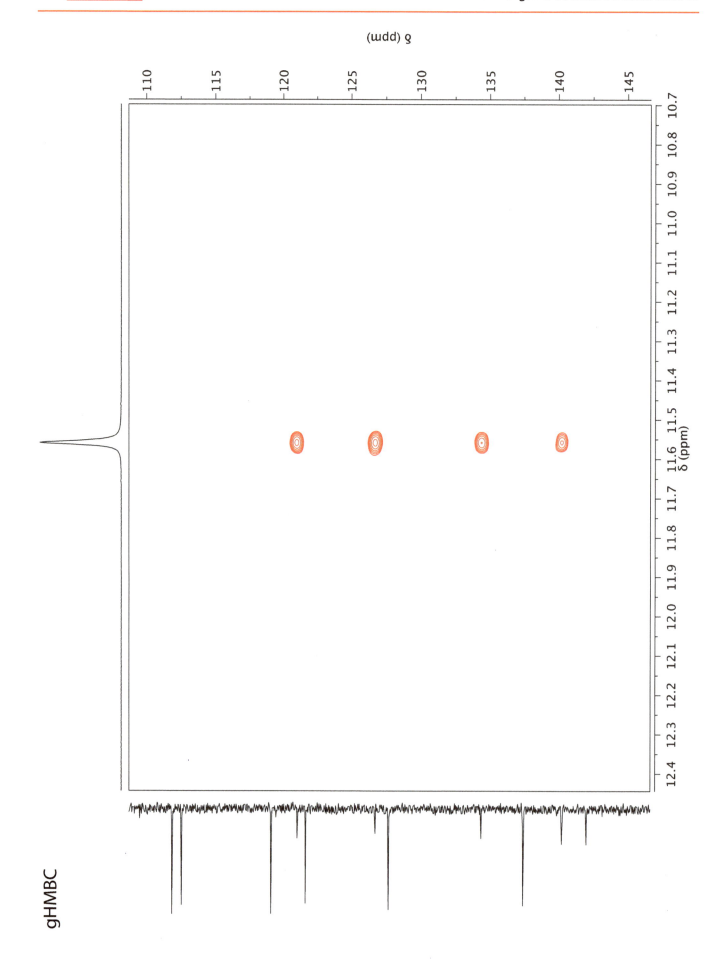

SECTION 4 Problem 87 — Problems in Organic Structure Determination

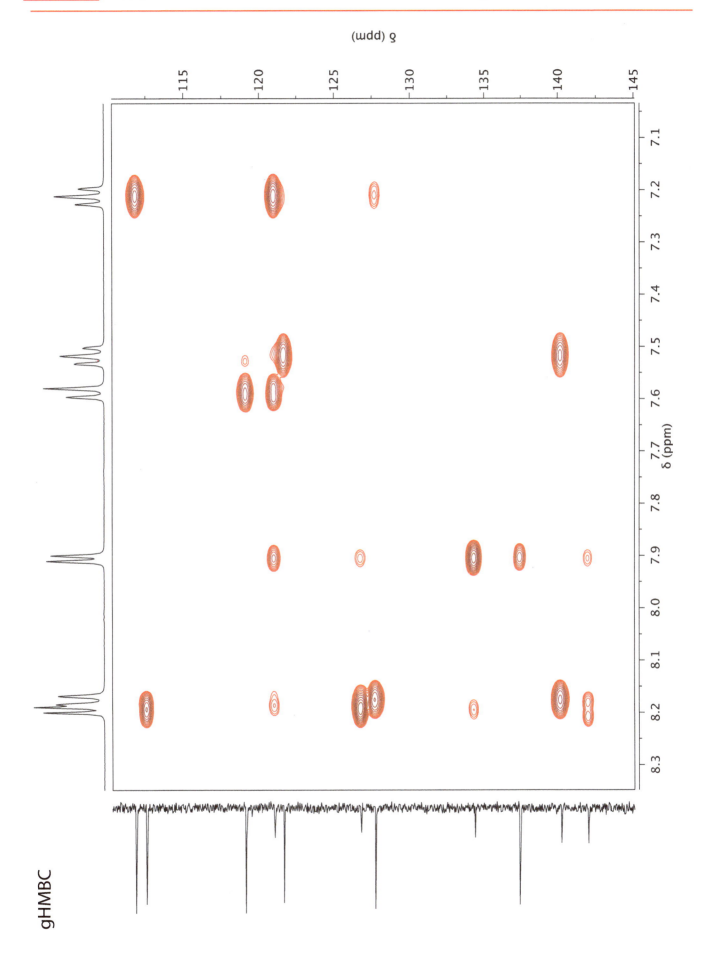

Problem 87

gHMBC

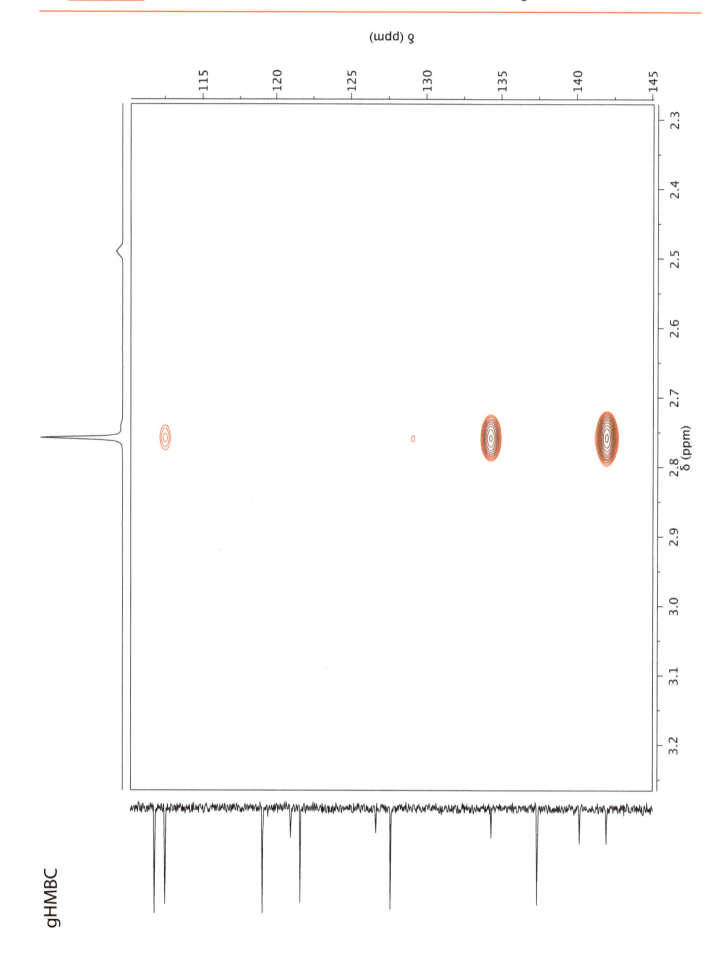

SECTION 4 Problem 87

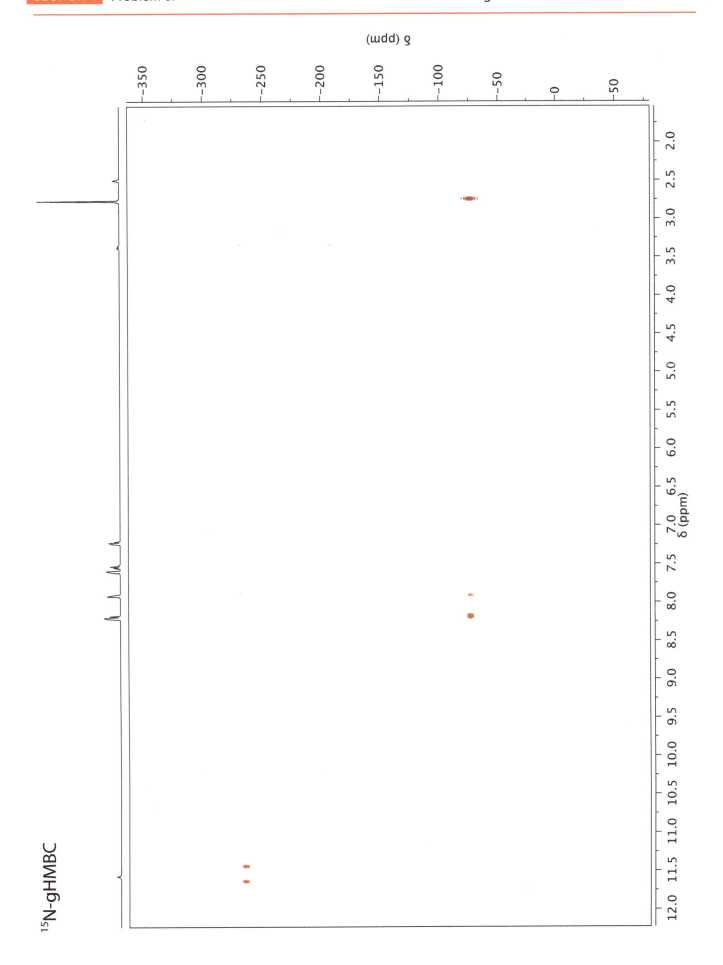

SECTION 4 Problem 88

Determine the structure of the unknown compound below with molecular formula $C_{18}H_{22}N_2 \cdot HCl$.

Spectra acquired in $(CD_3)_2SO$ at 500 MHz (^1H) or 125 MHz (^{13}C).

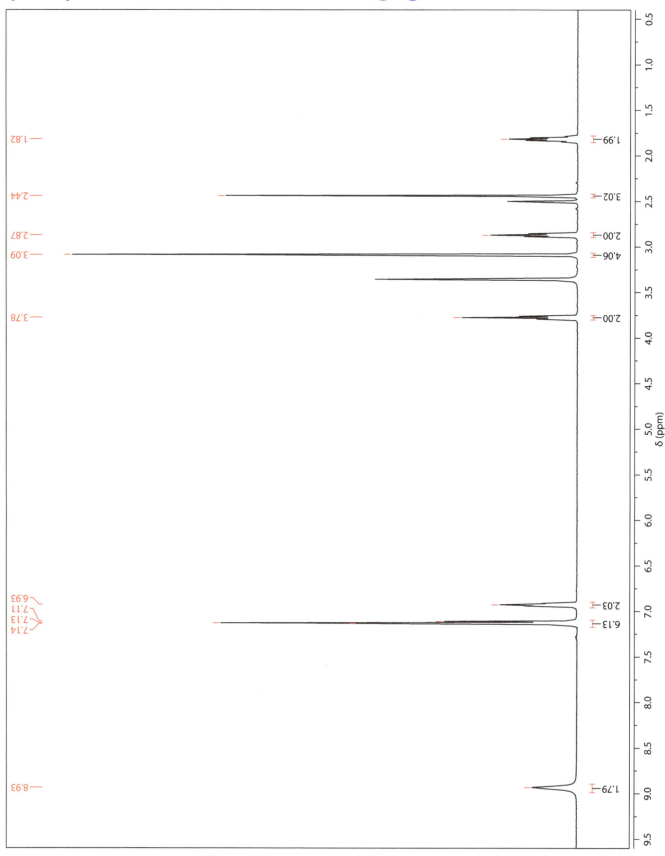

SECTION 4 Problem 88

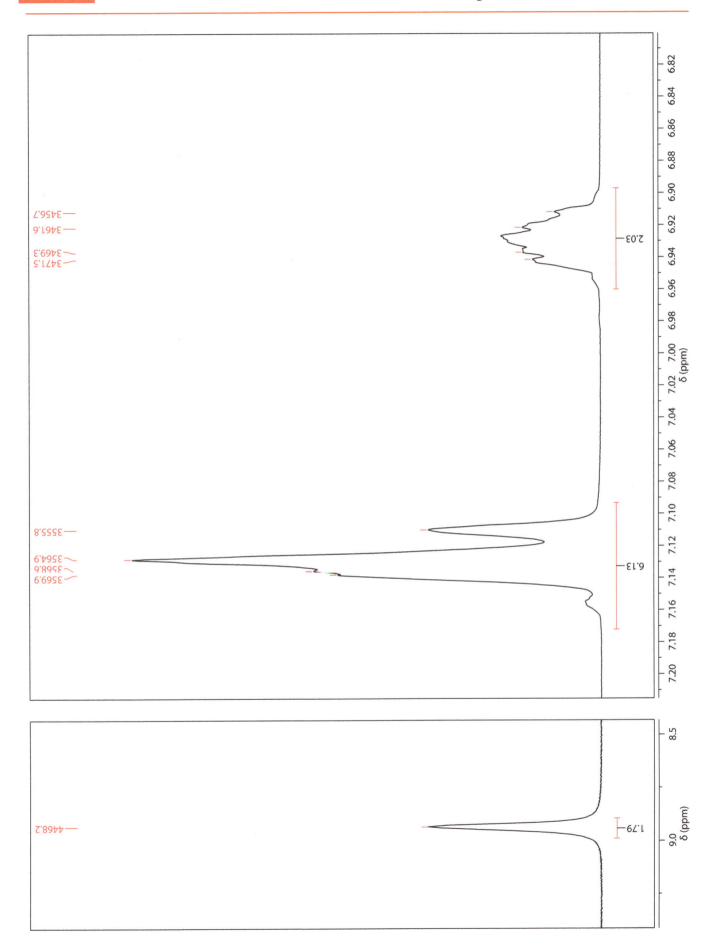

Problem 88

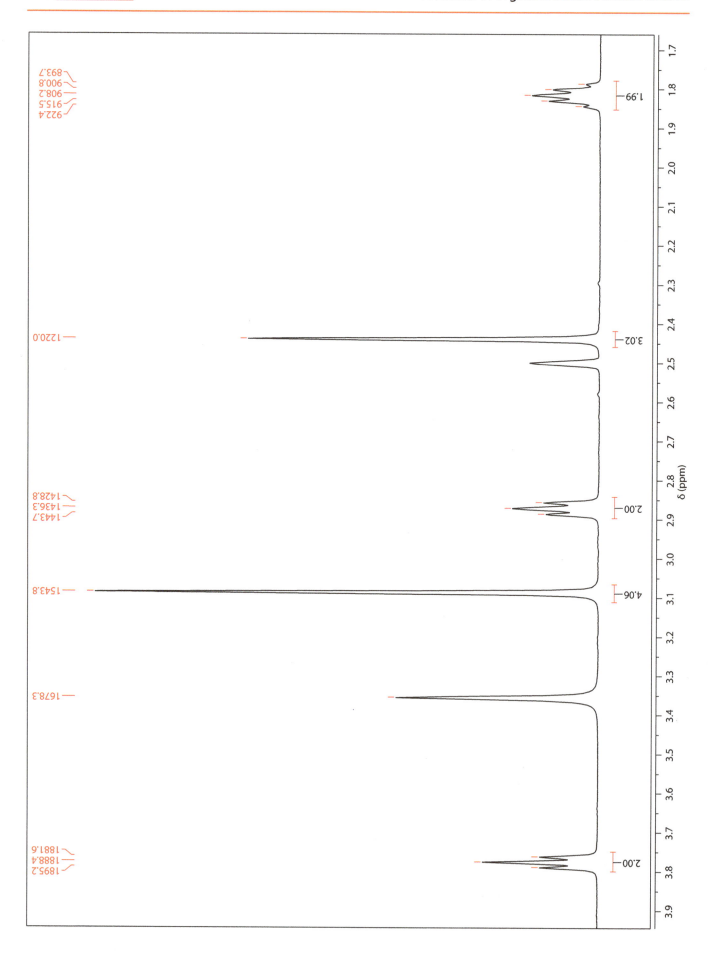

SECTION 4 Problem 88

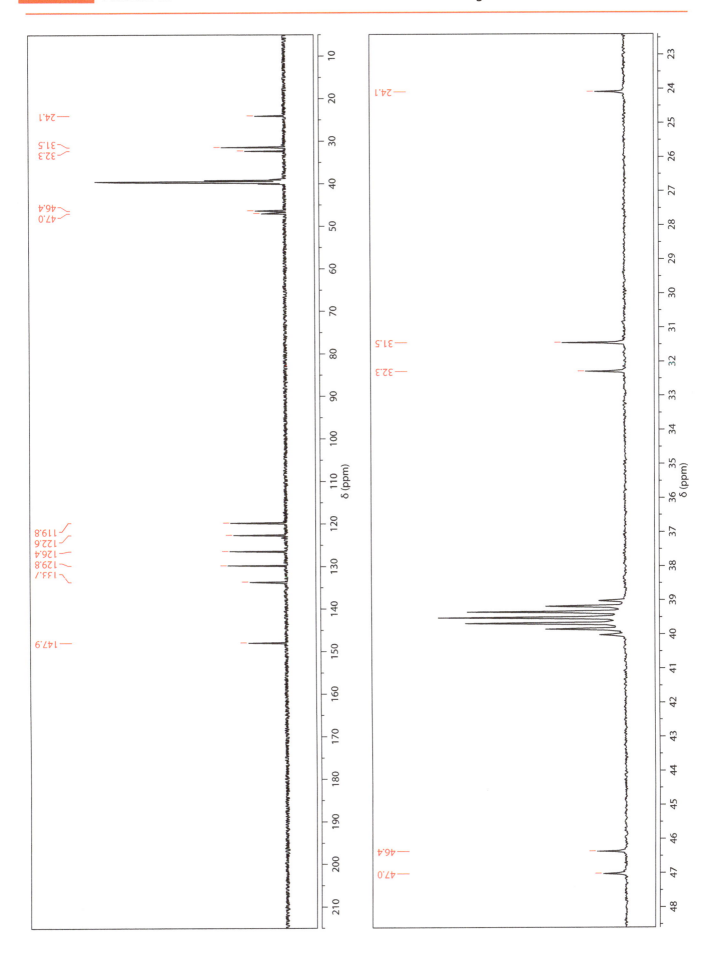

Problem 88

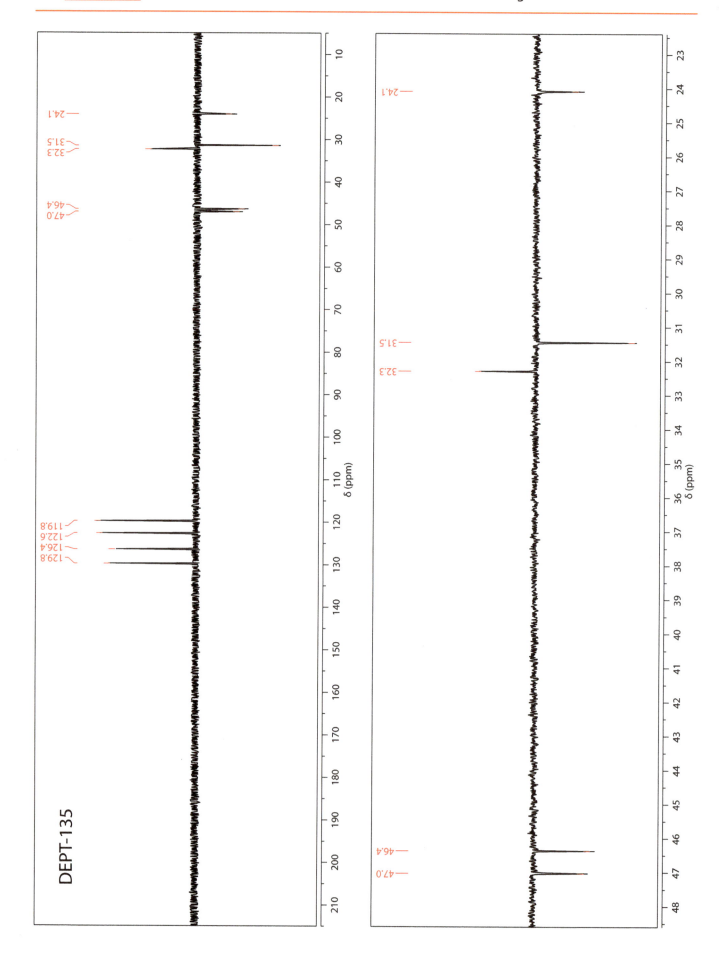

SECTION 4 Problem 88

Problems in Organic Structure Determination

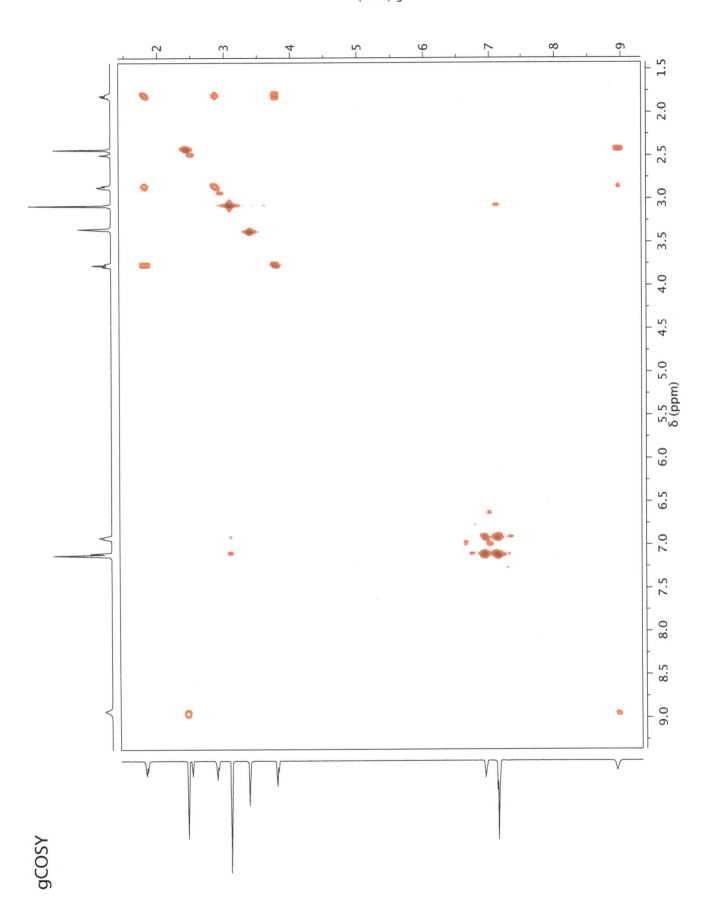

gCOSY

Problem 88

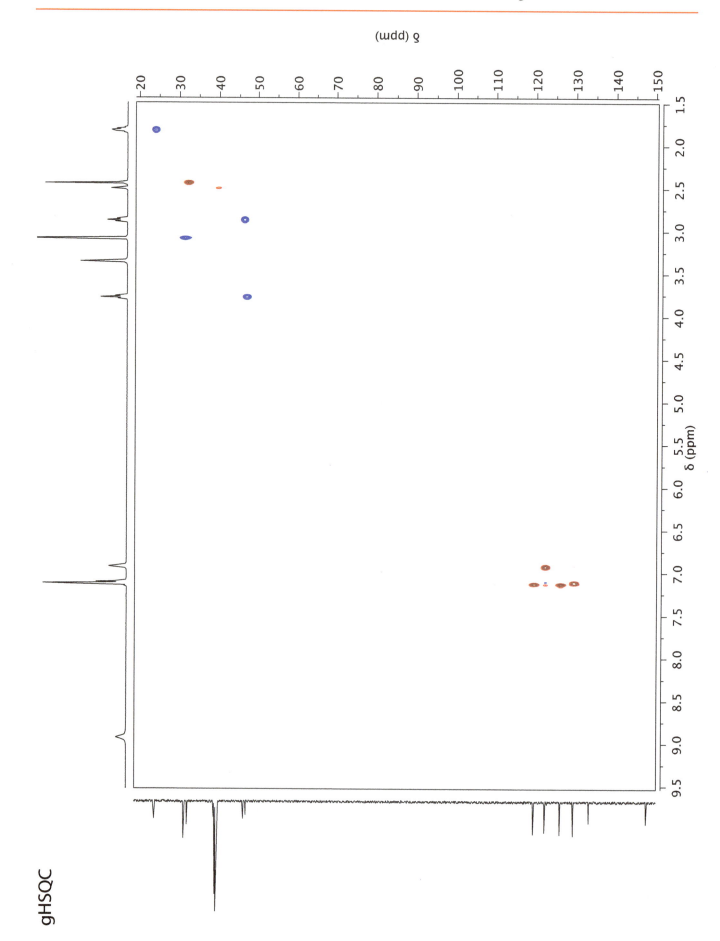

Problem 88

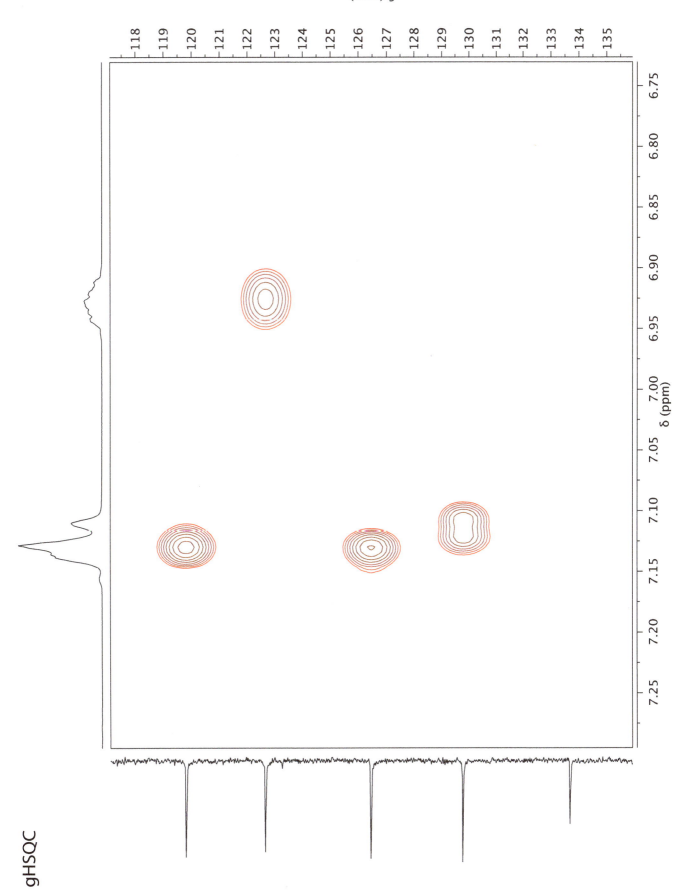

SECTION 4 Problem 88

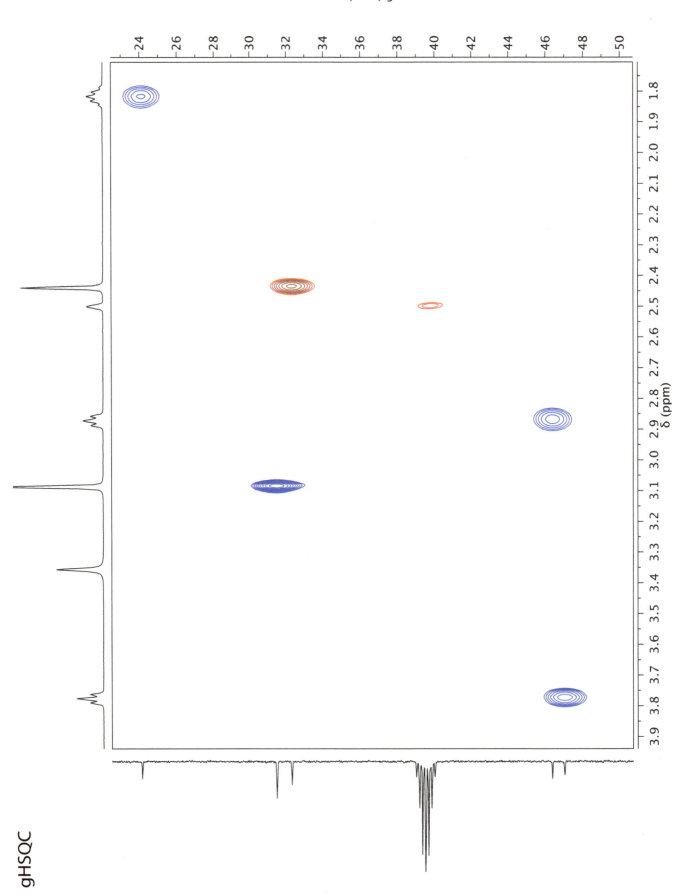

gHSQC

SECTION 4 Problem 88

Problems in Organic Structure Determination

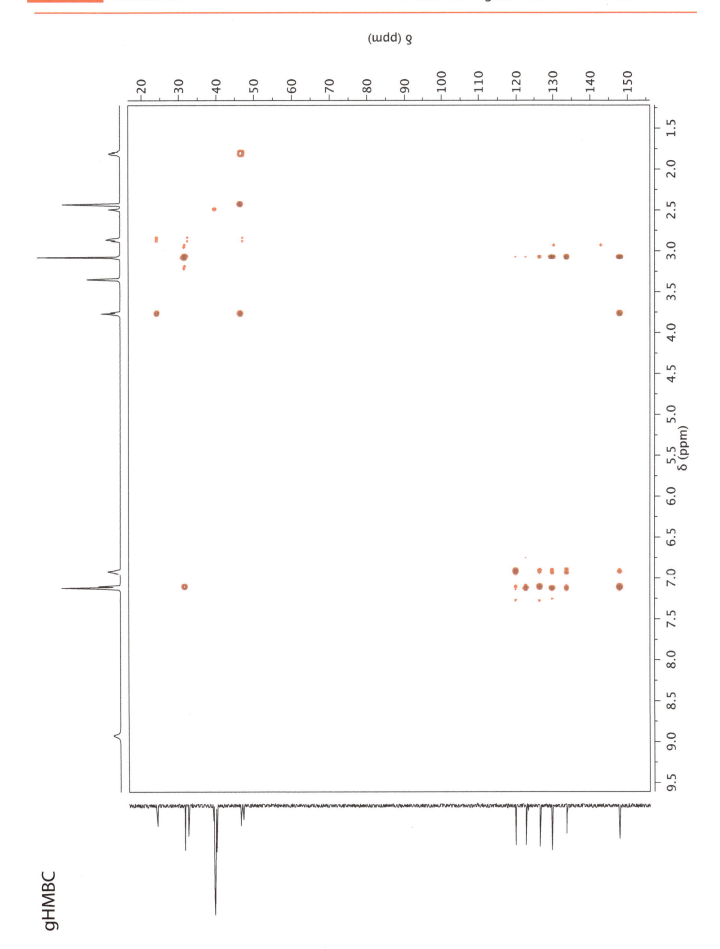

gHMBC

SECTION 4 Problem 88

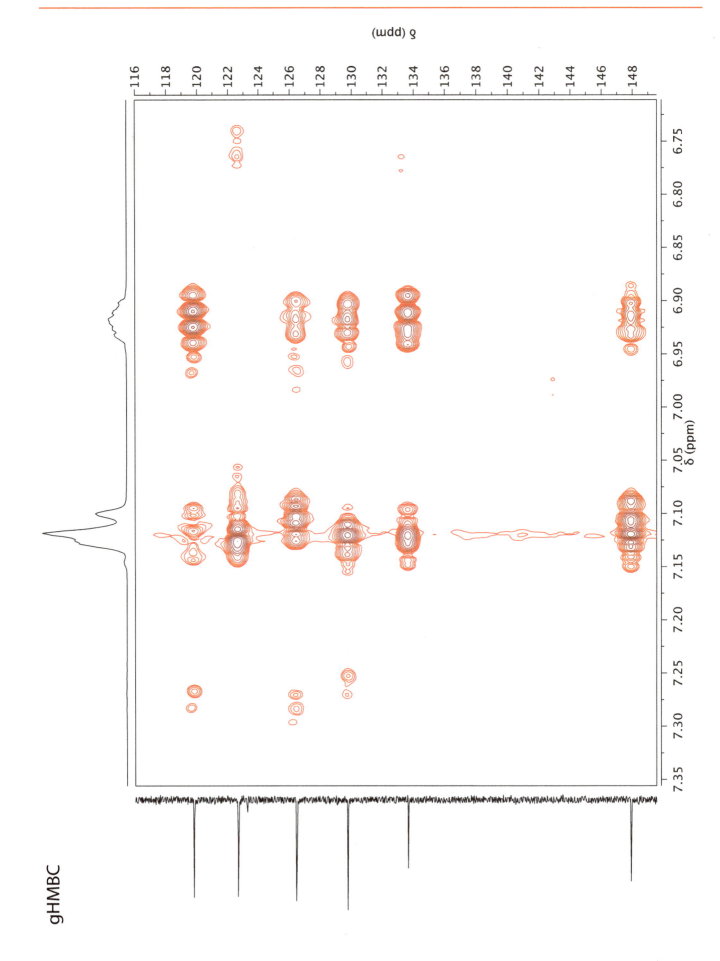

gHMBC

Problem 88

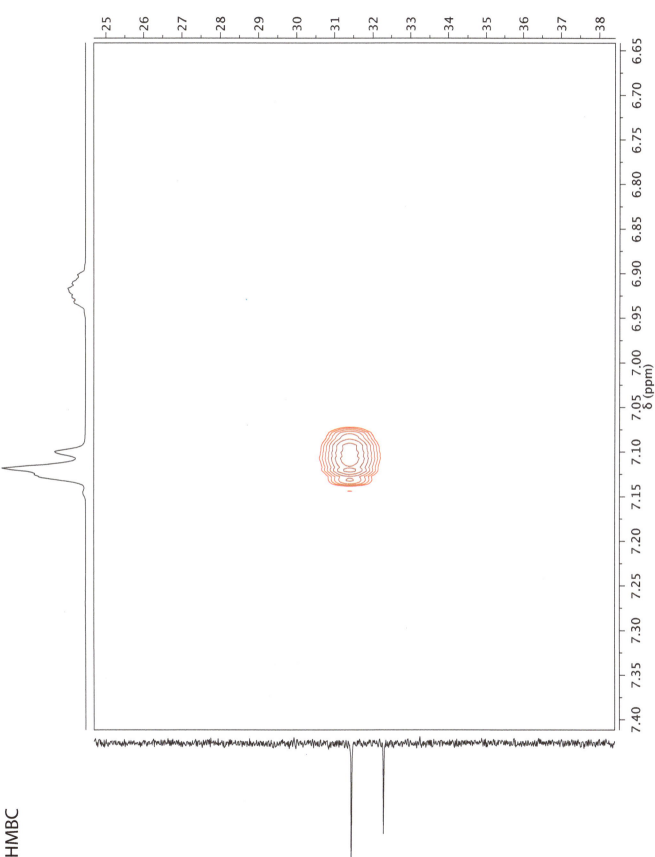

gHMBC

SECTION 4   Problem 88

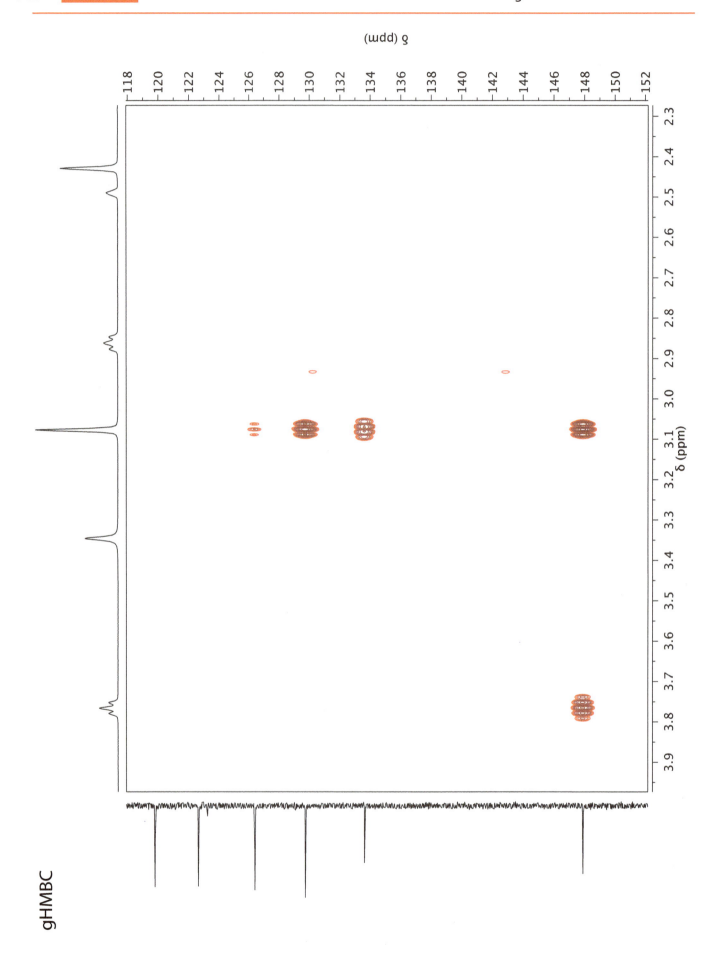

gHMBC

SECTION 4 Problem 88

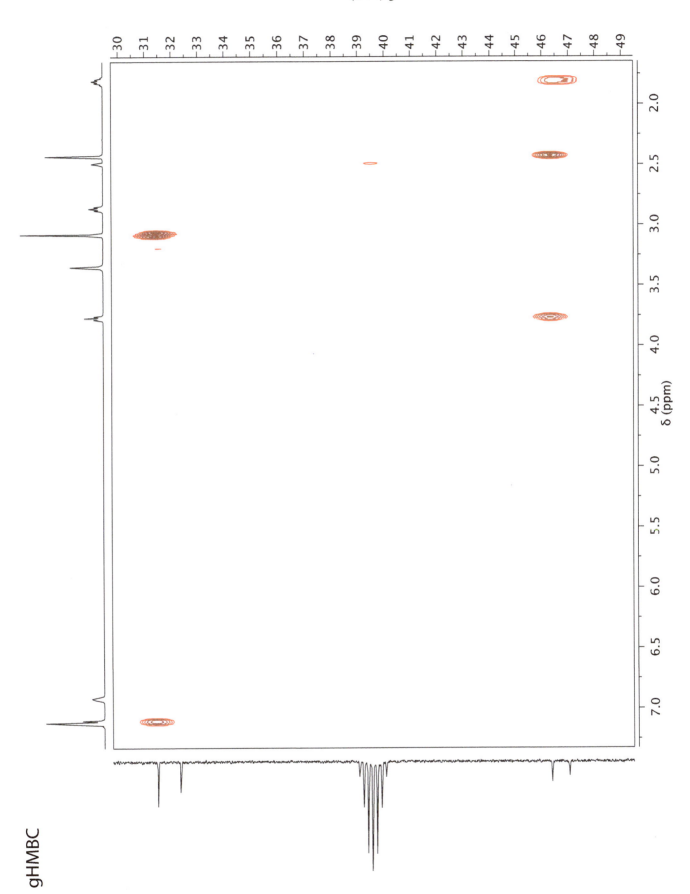

gHMBC

Problem 88

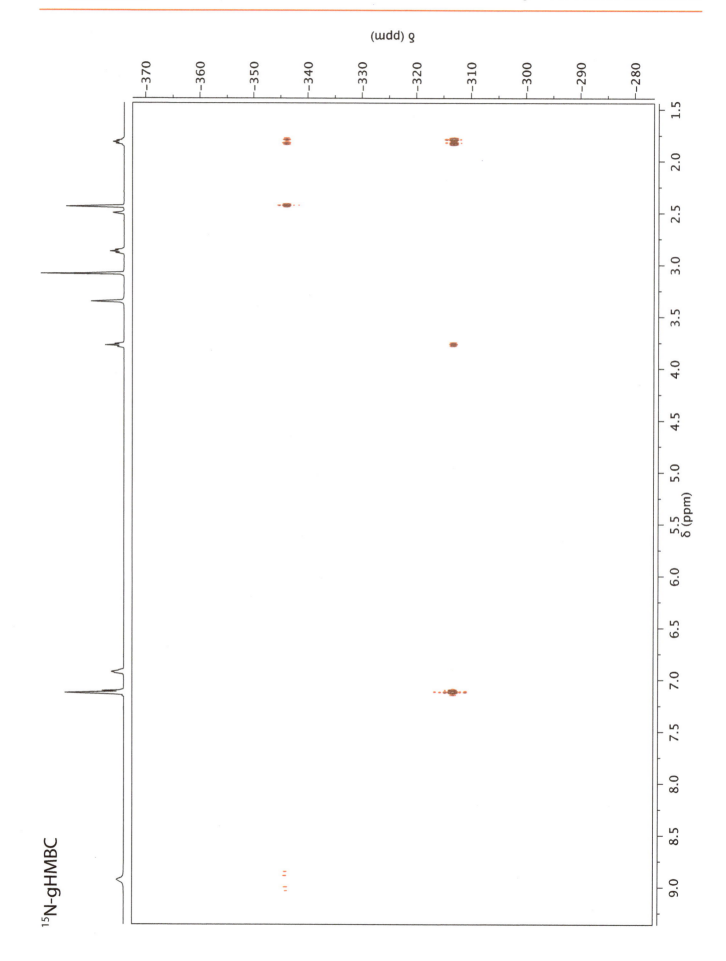

¹⁵N-gHMBC

Problem 88

Problems in Organic Structure Determination

¹⁵N-gHMBC

SECTION 4 Problem 89

Determine the structure of the unknown compound below with molecular formula $C_{17}H_{26}ClNO_2$.

Spectra acquired in $CDCl_3$ at 500 MHz (^{1}H) or 125 MHz (^{13}C).

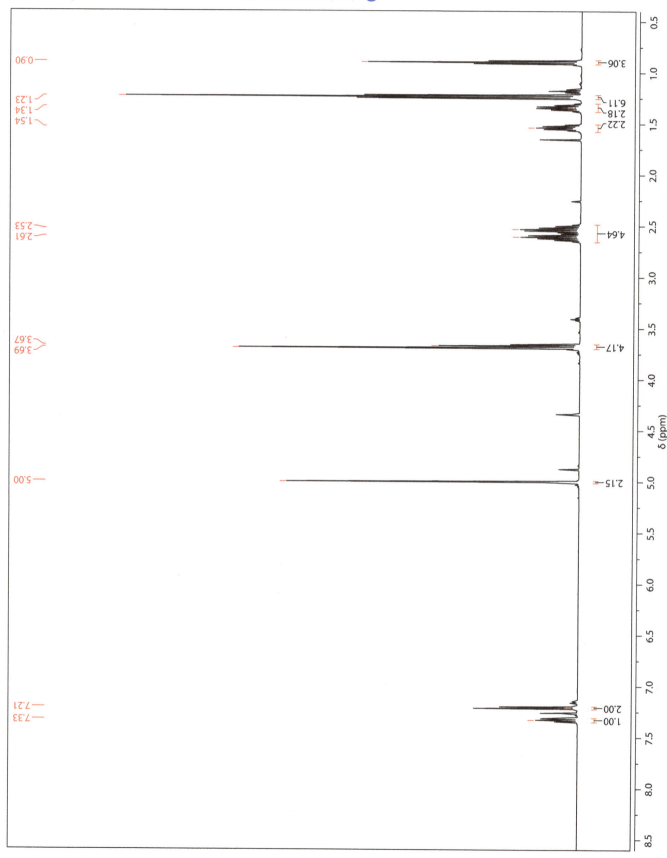

SECTION 4 Problem 89

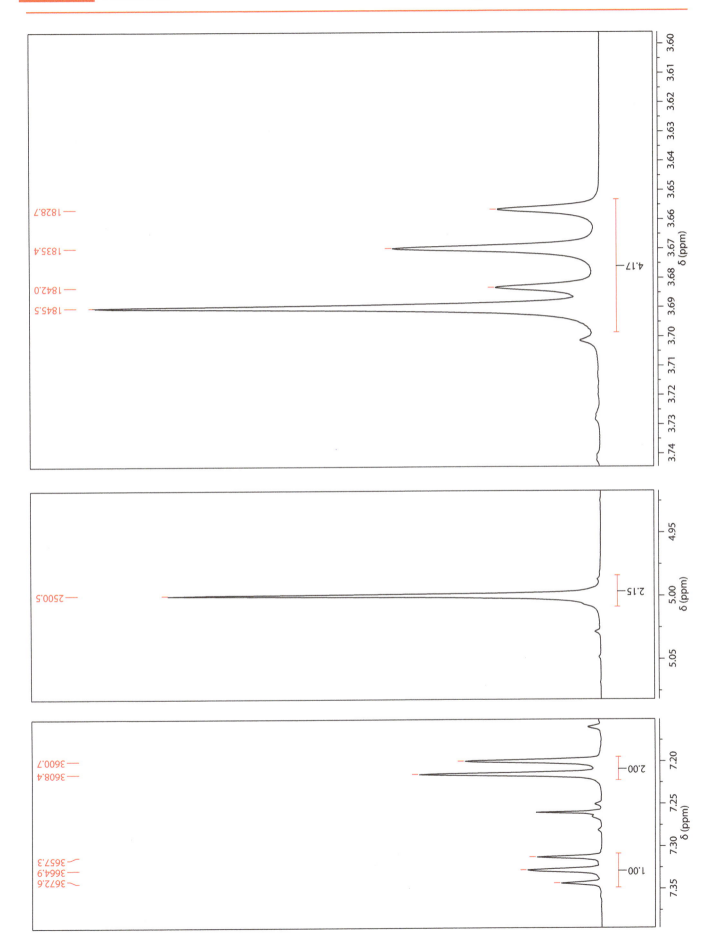

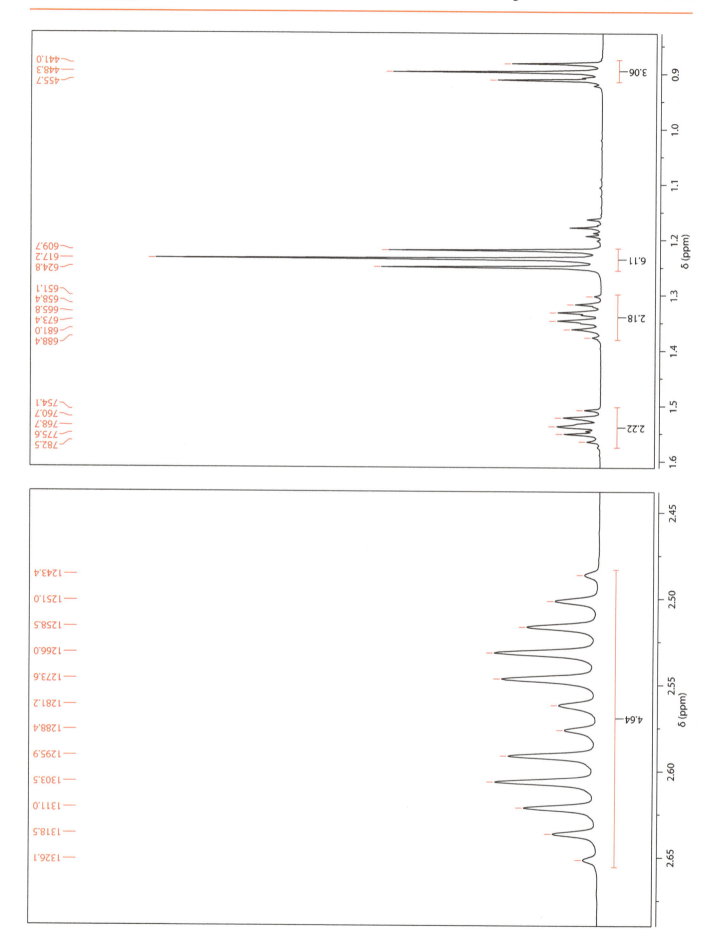

SECTION 4 Problem 89

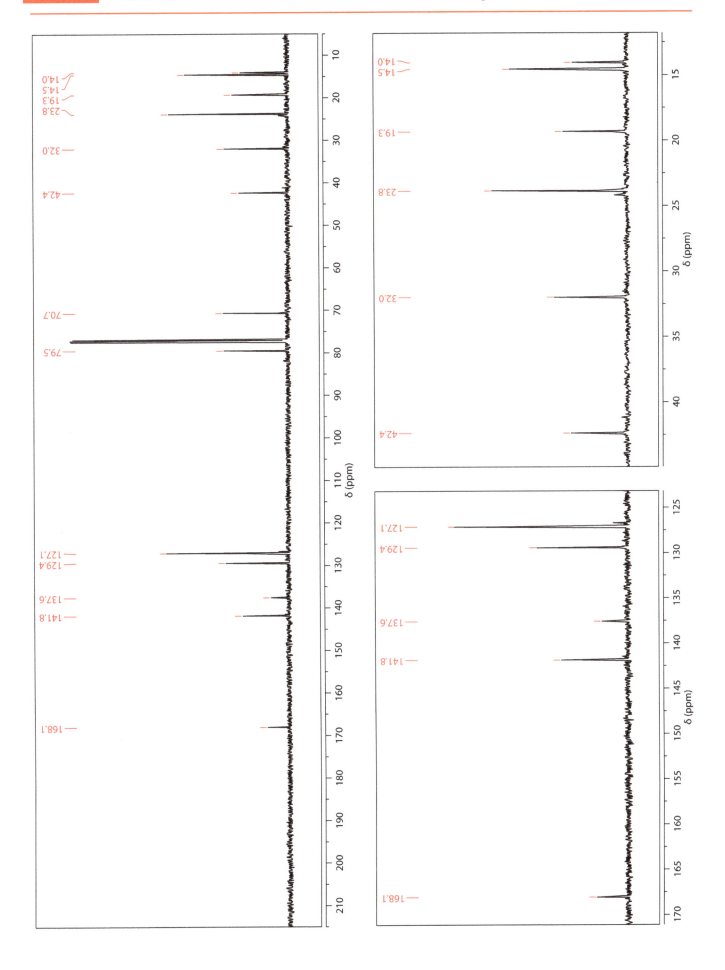

Problem 89

gCOSY

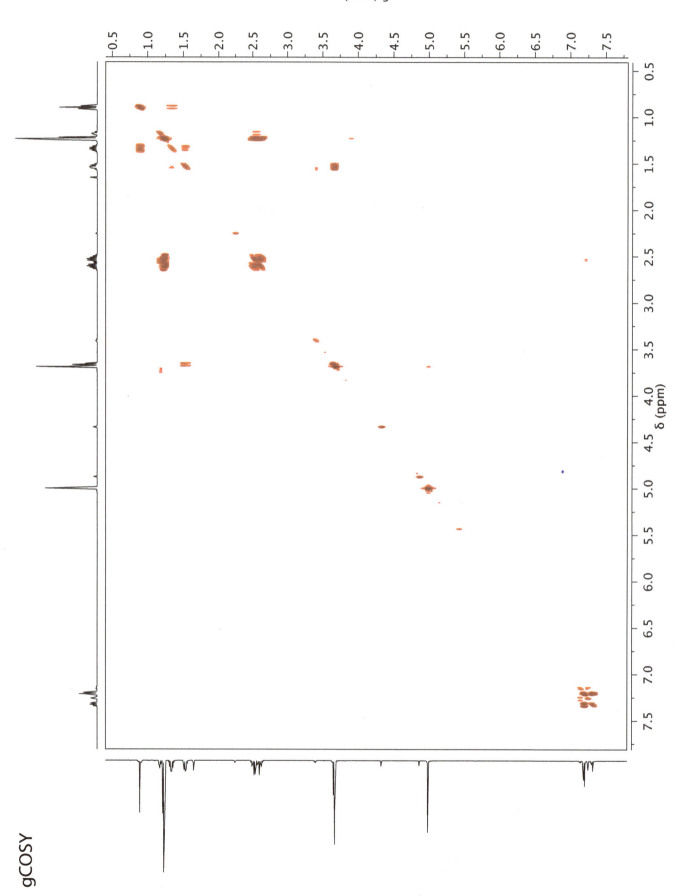

SECTION 4 Problem 89

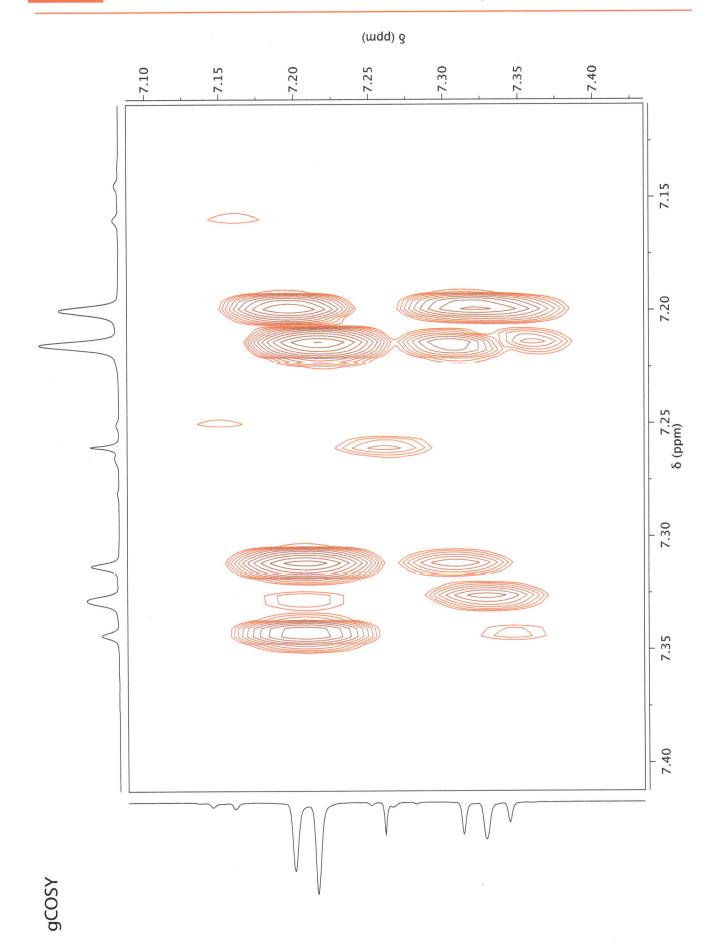

gCOSY

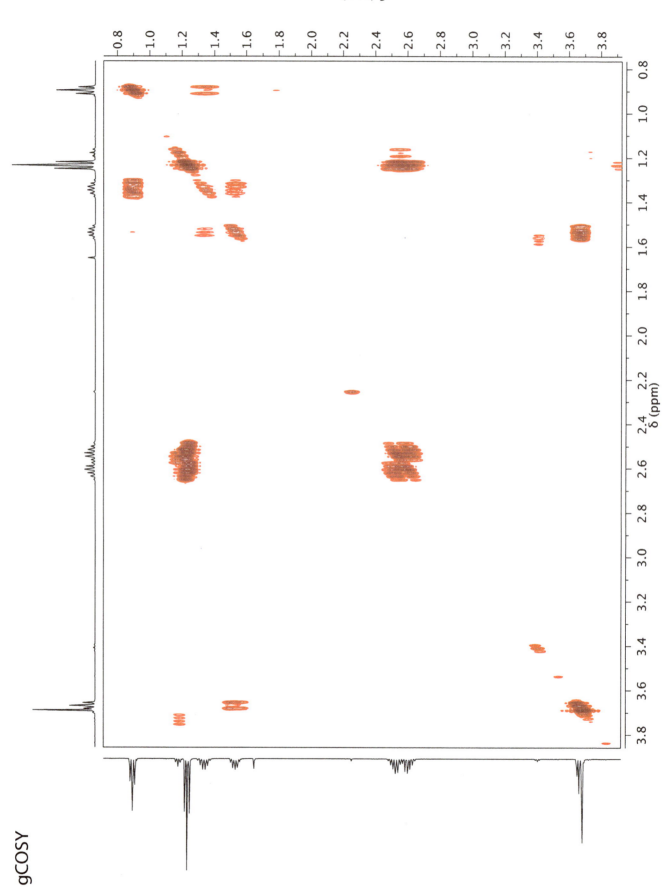

gCOSY

SECTION 4 Problem 89

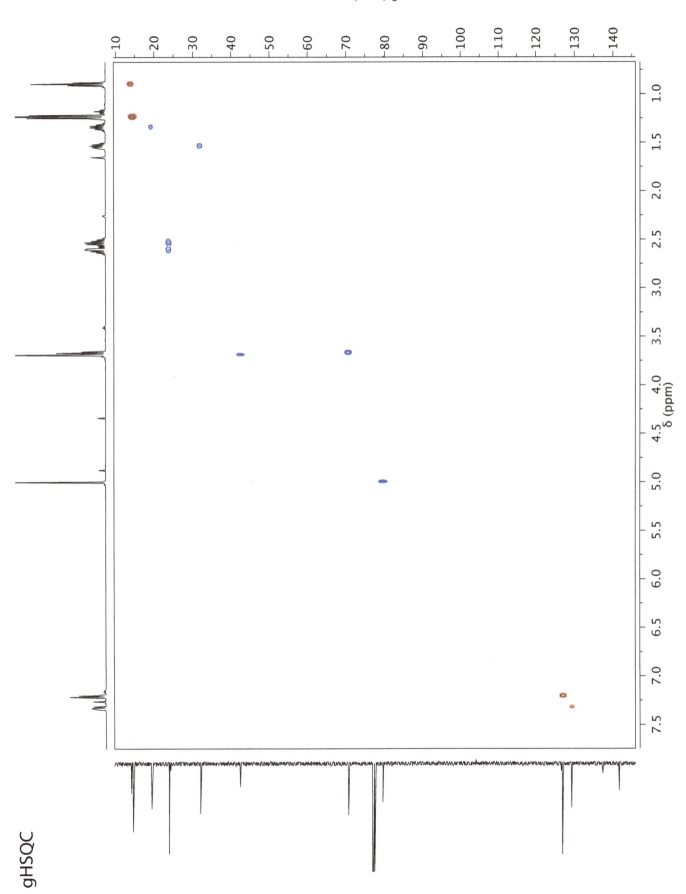

gHSQC

Problem 89

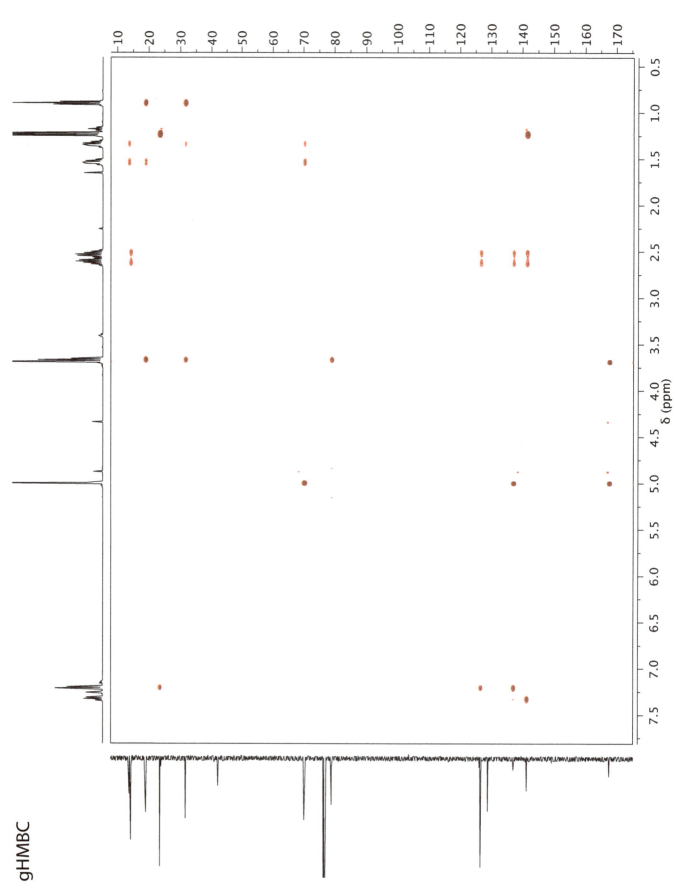

gHMBC

SECTION 4 Problem 89

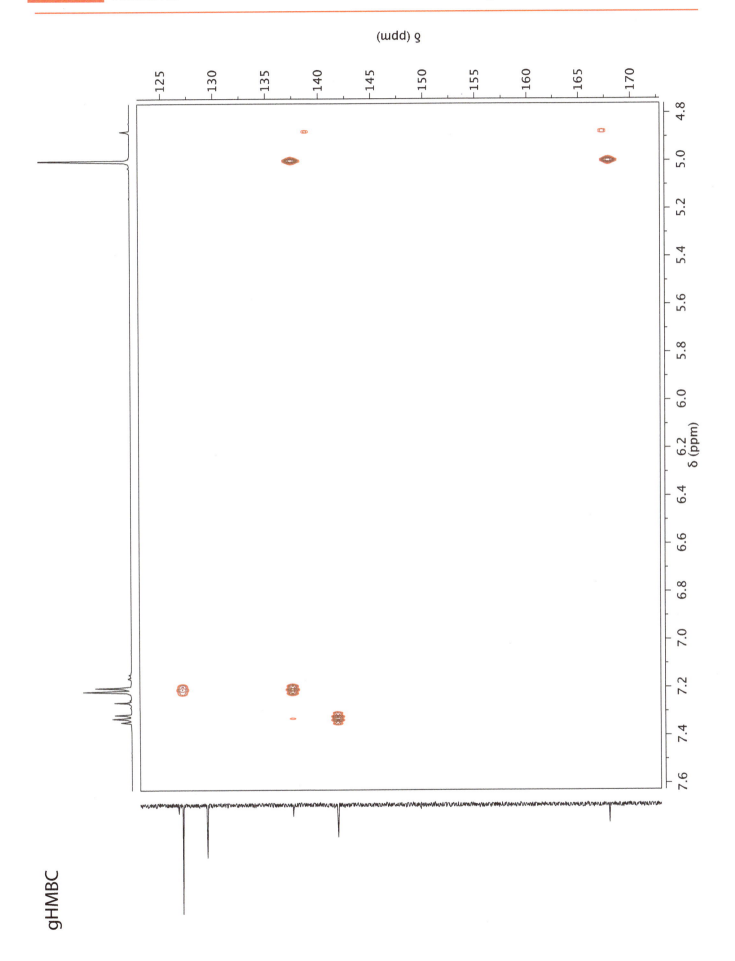

gHMBC

Problem 89

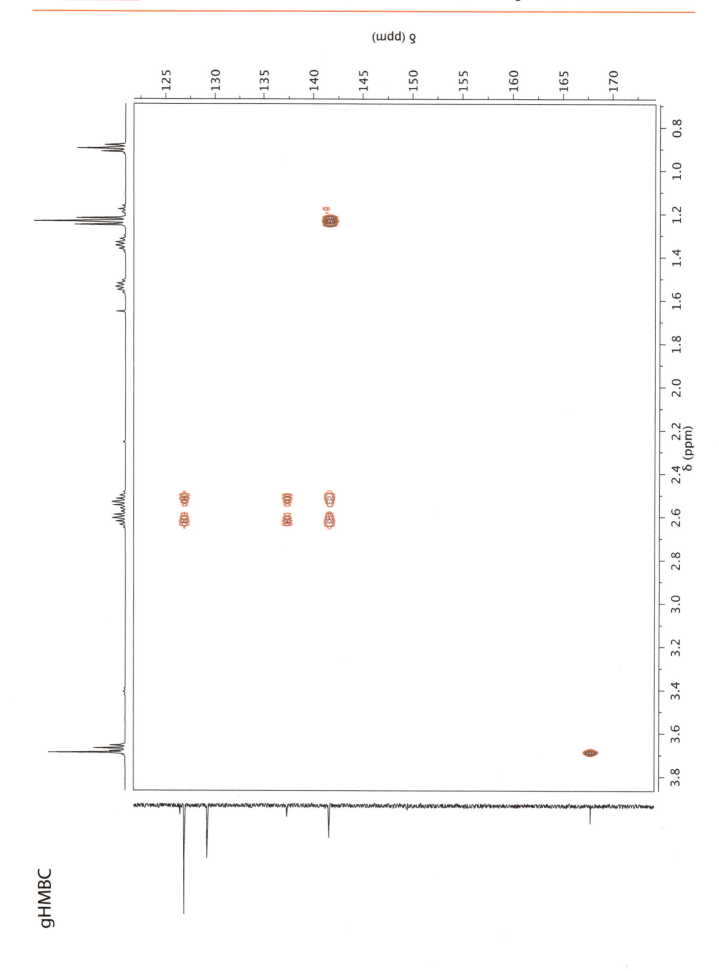

gHMBC

SECTION 4 Problem 89

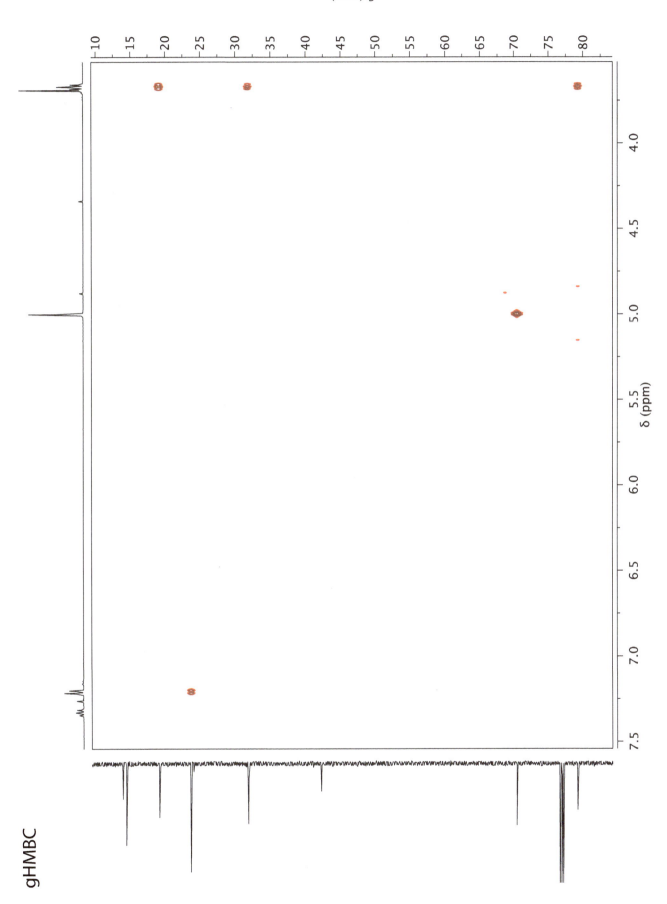

gHMBC

Problem 89

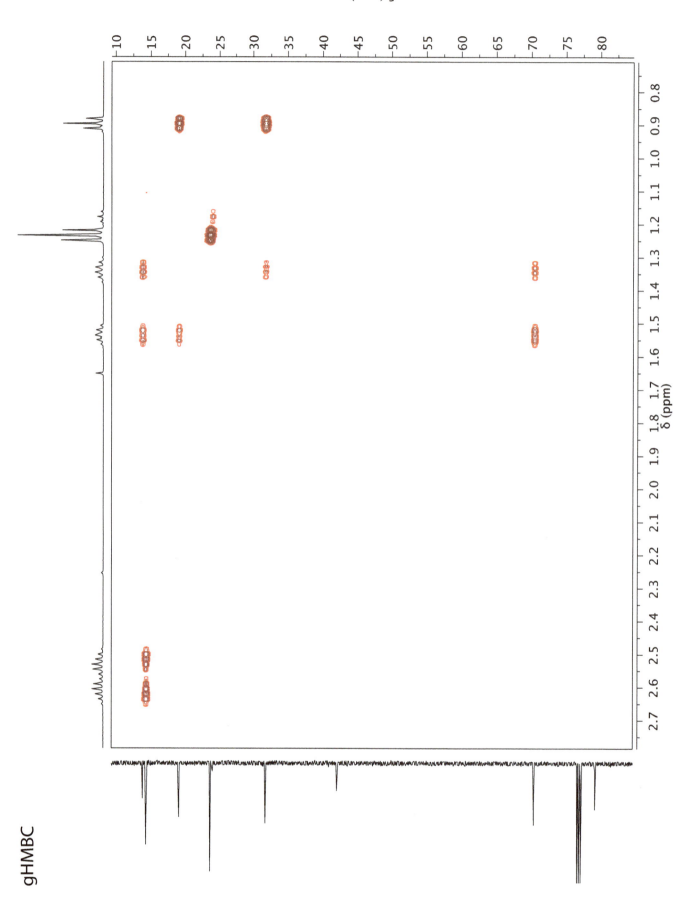

gHMBC

SECTION 4 Problem 90 — Problems in Organic Structure Determination

Determine the structure of the unknown compound below with molecular formula $C_{18}H_{16}O_8$.

Spectra acquired in $(CD_3)_2SO$ at 500 MHz (^1H) or 125 MHz (^{13}C).

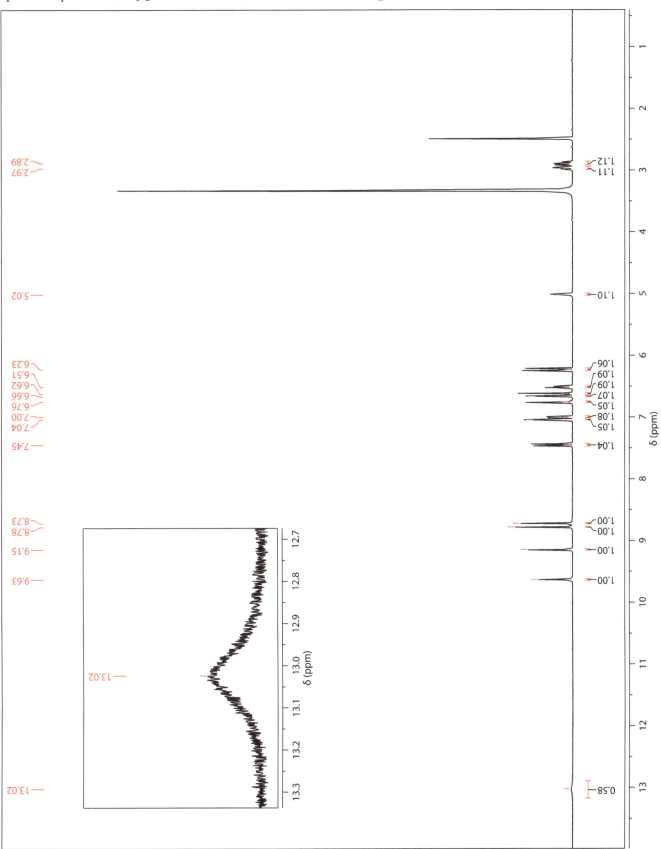

SECTION 4 Problem 90

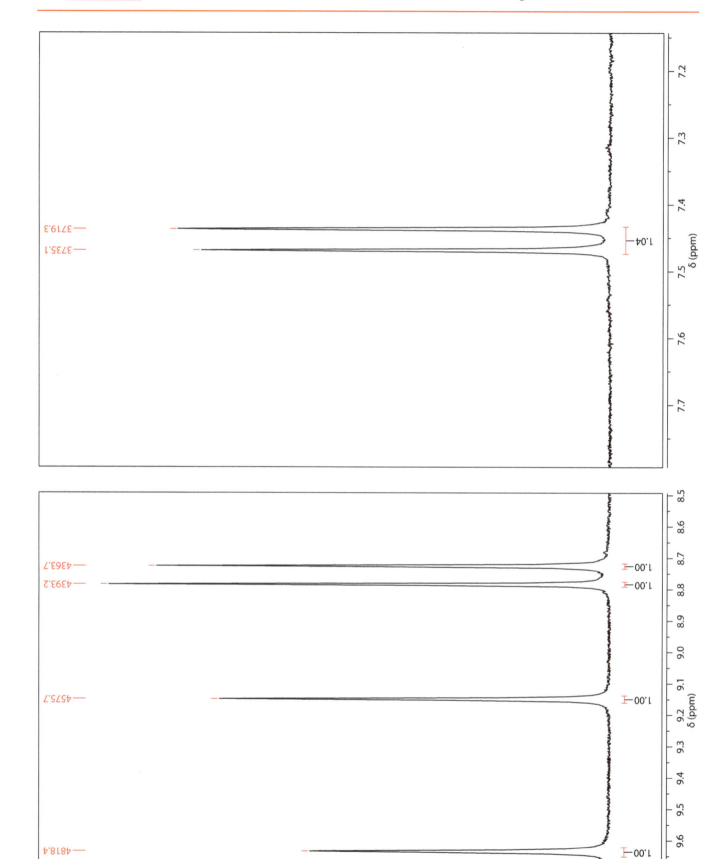

SECTION 4 Problem 90

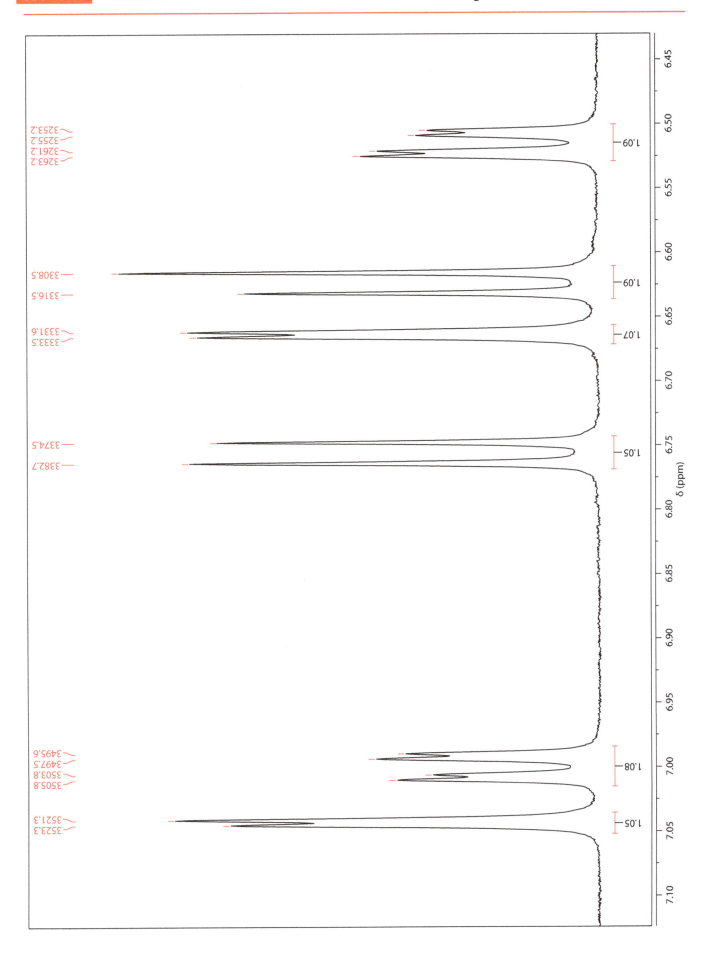

SECTION 4 Problem 90

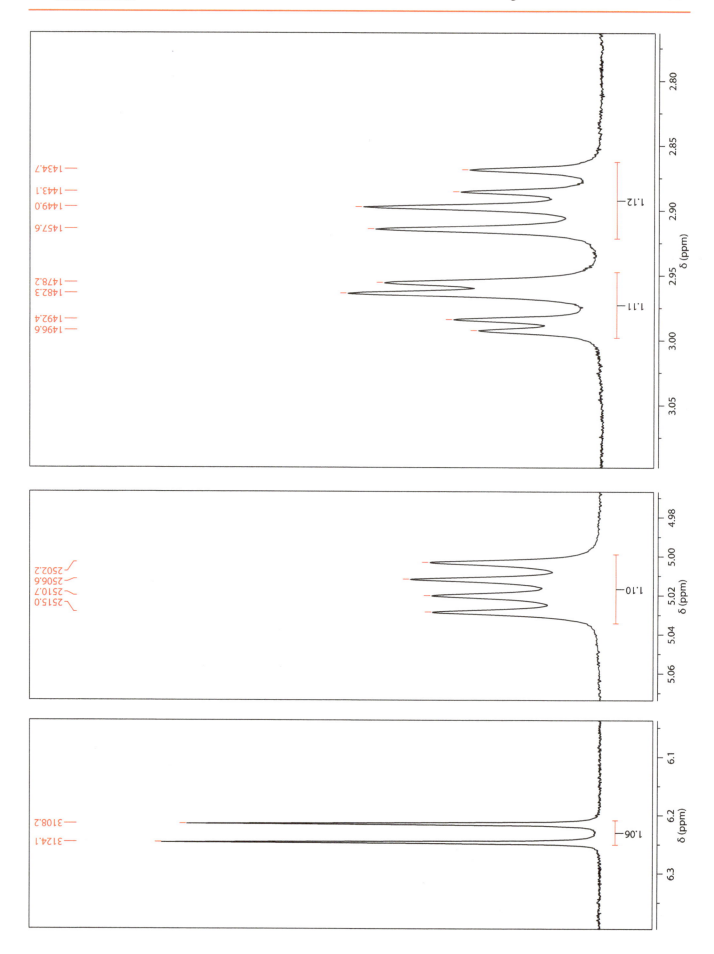

SECTION 4 Problem 90

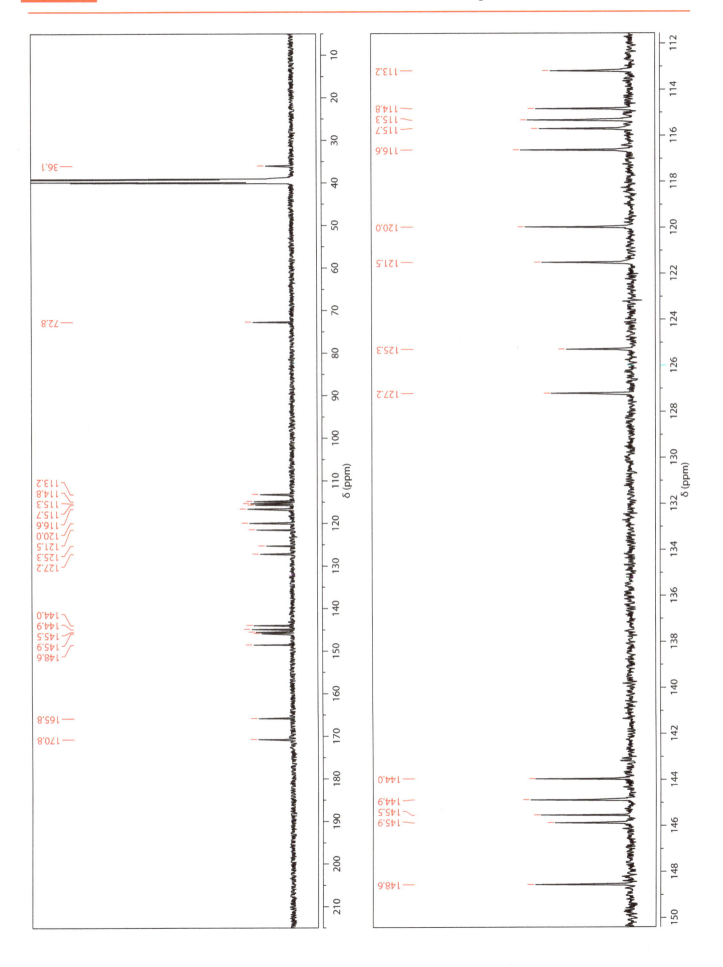

SECTION 4 Problem 90

Problems in Organic Structure Determination

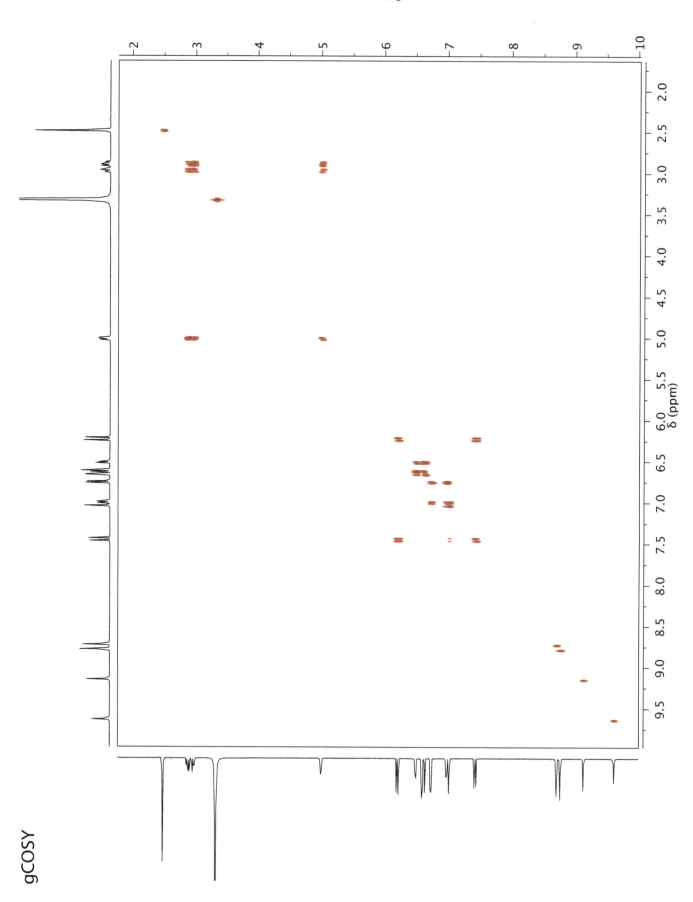

gCOSY

SECTION 4 Problem 90

gCOSY

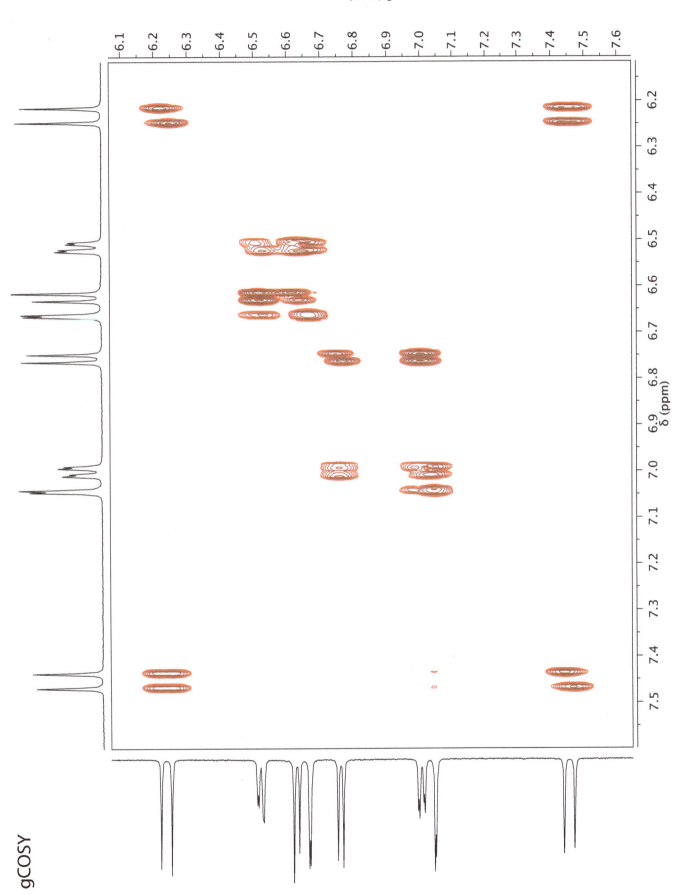

Problem 90

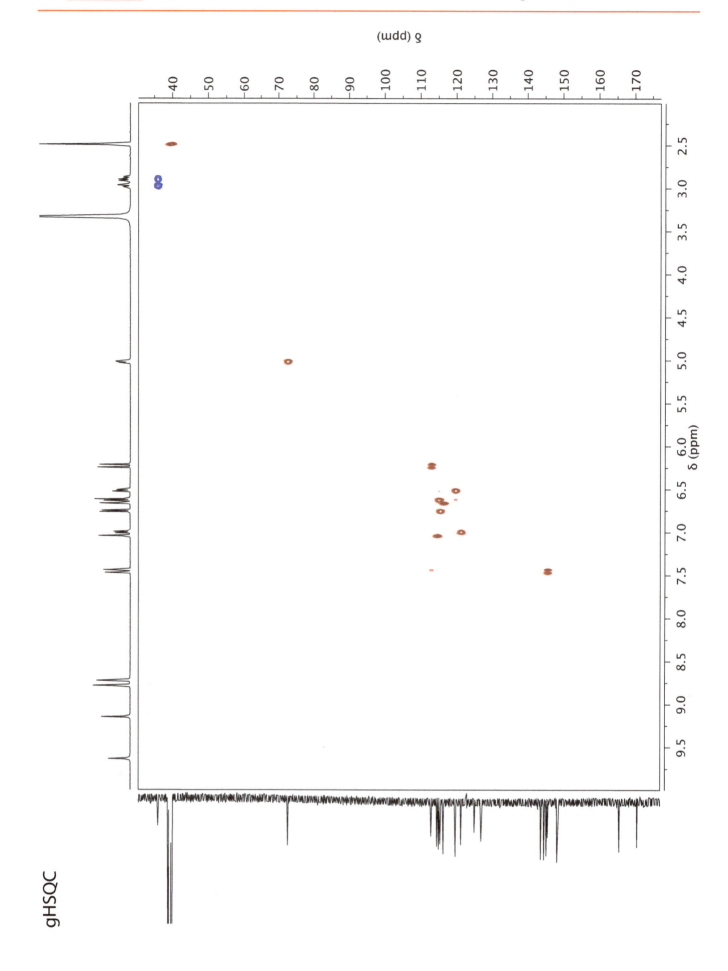

gHSQC

SECTION 4 Problem 90

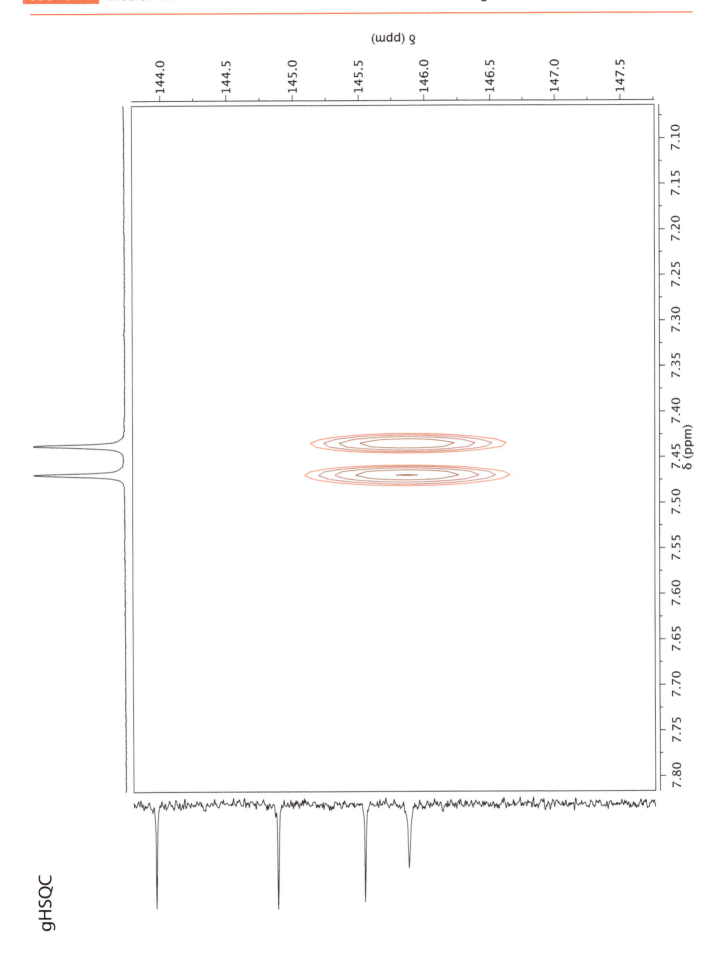

gHSQC

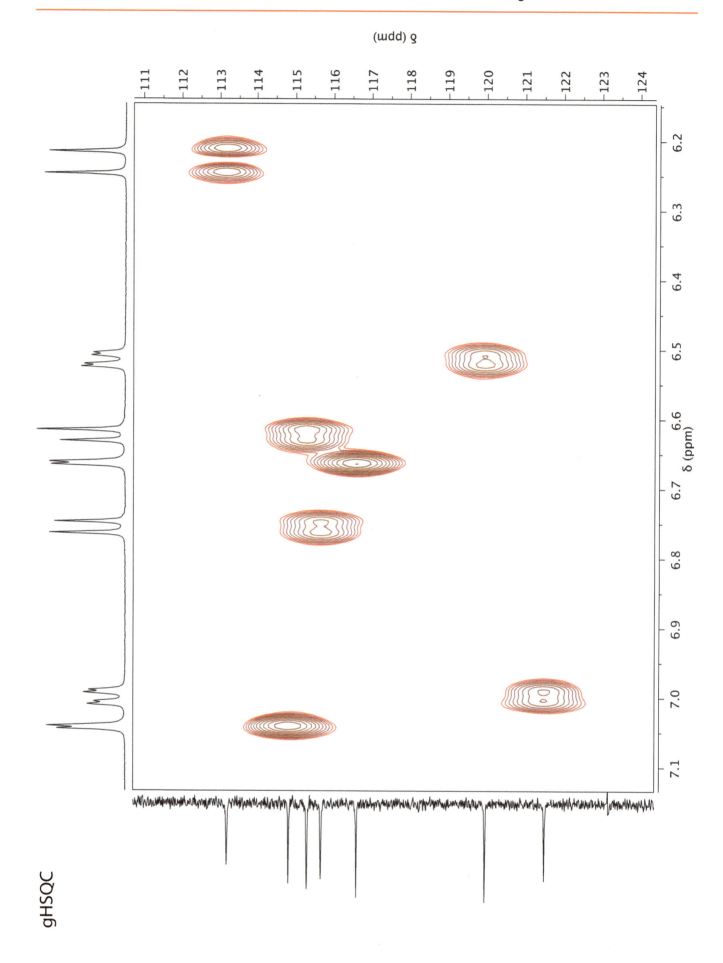

SECTION 4 Problem 90

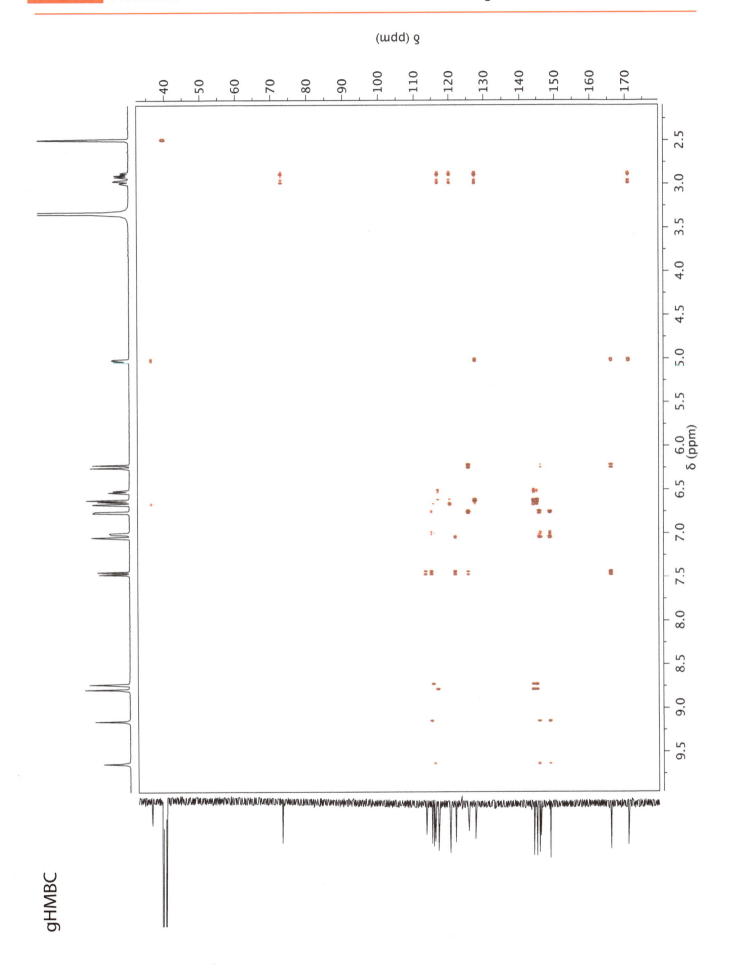

SECTION 4 Problem 90

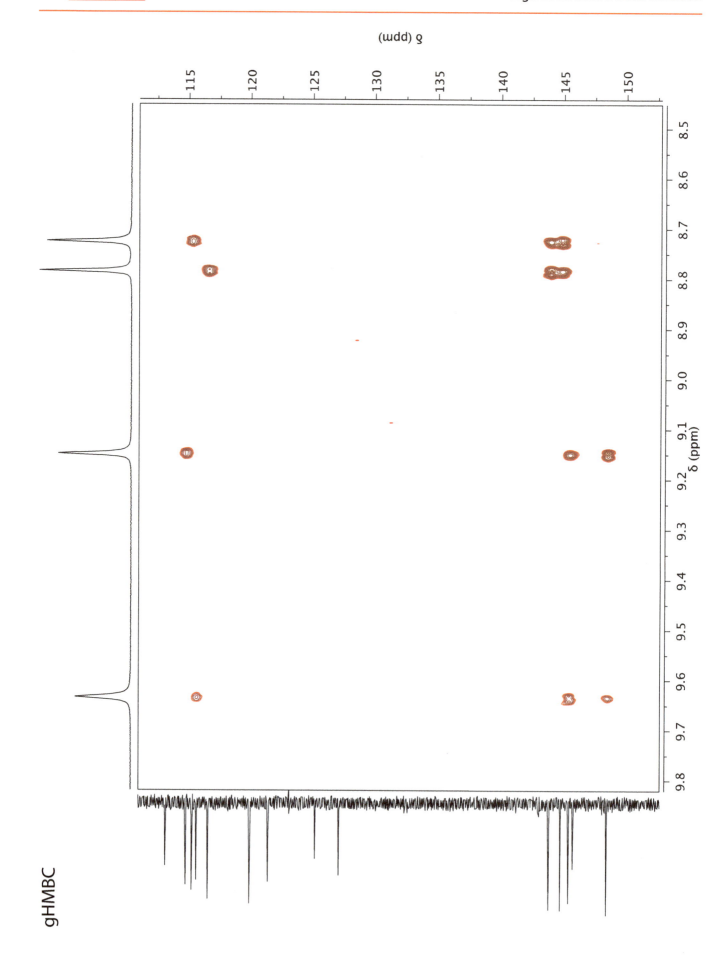

SECTION 4 Problem 90

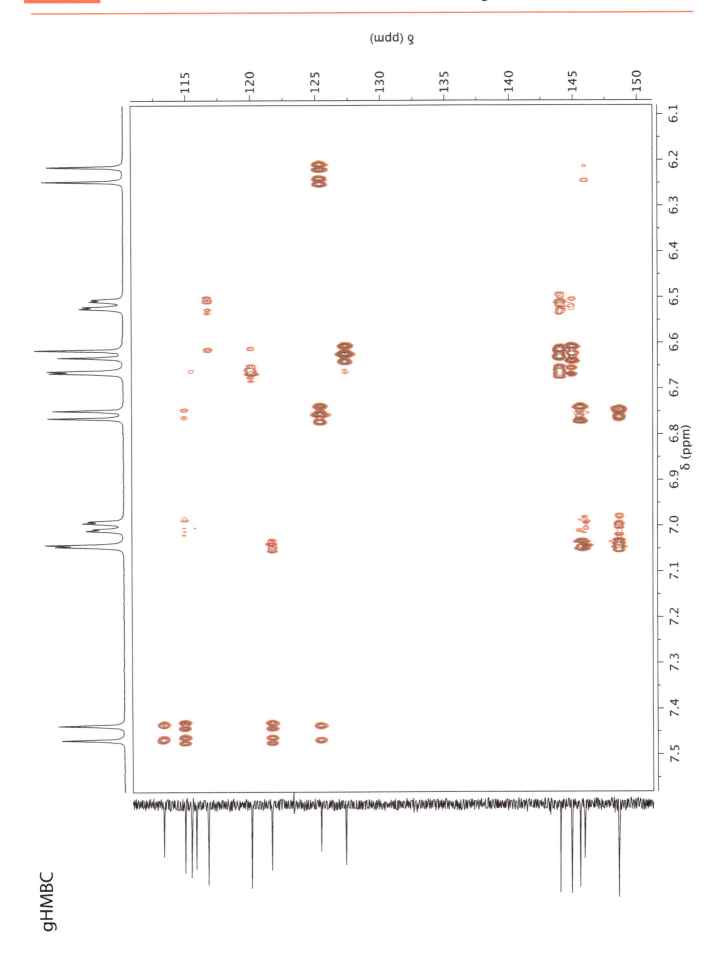

gHMBC

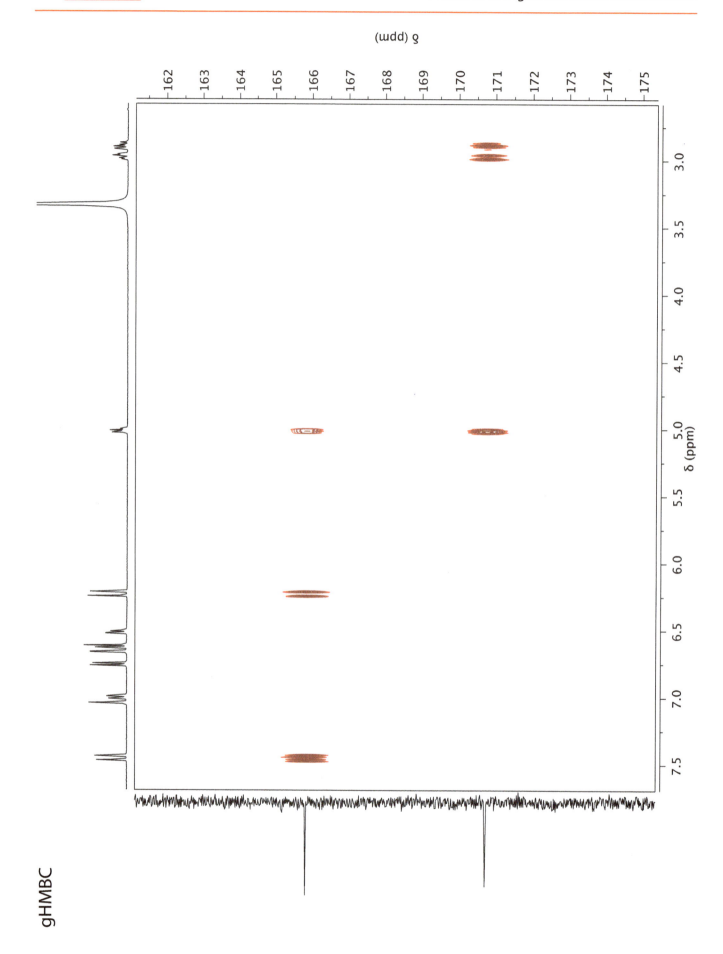

SECTION 4 Problem 90

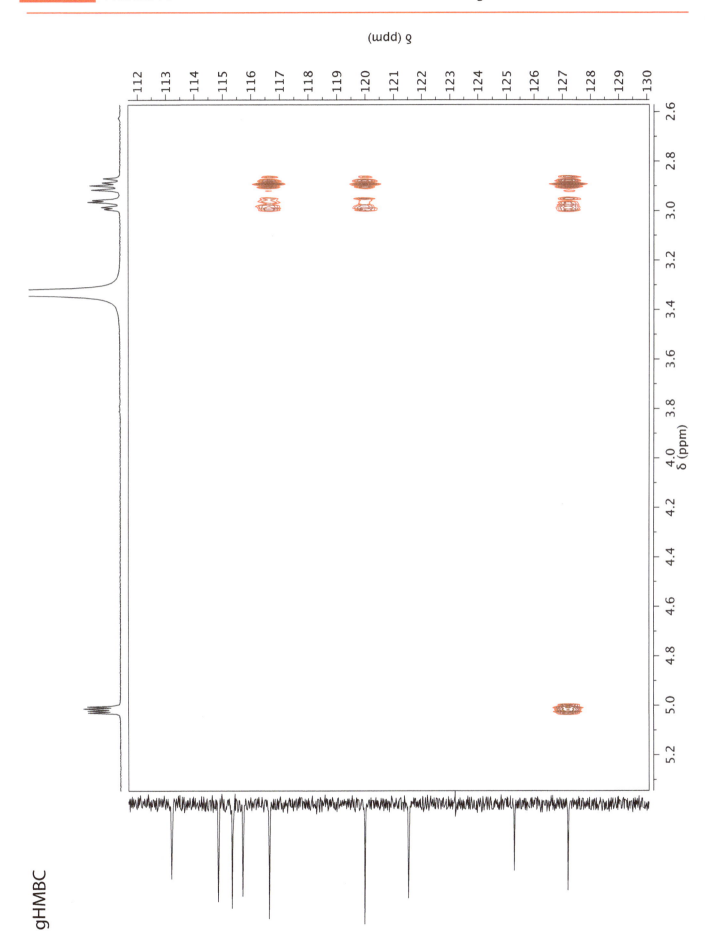

gHMBC

Problem 90

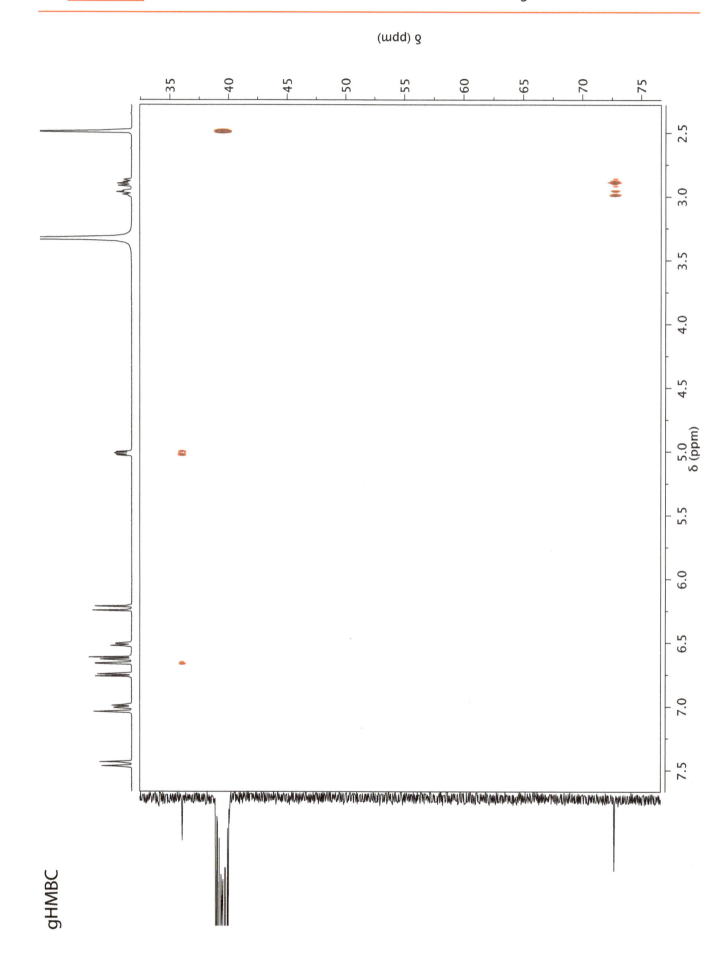

gHMBC

SECTION 4 Problem 91 — Problems in Organic Structure Determination

Determine the structure of the unknown compound below with molecular formula $C_{15}H_{19}NO_3 \cdot HCl$.

Spectra acquired in $CDCl_3$ at 500 MHz (^1H) or 125 MHz (^{13}C).

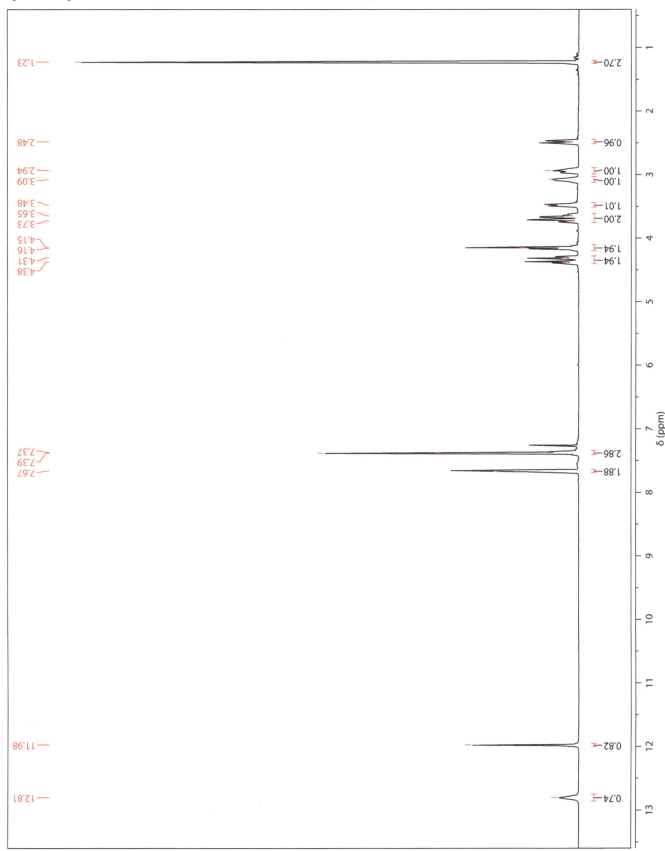

SECTION 4 Problem 91

Frequencies (Hz) near 4.15 ppm region (integration 1.94):
2056.7, 2063.8, 2067.5, 2069.8, 2074.6, 2076.9, 2081.7, 2084.0, 2088.8, 2091.2, 2094.8, 2102.0

Frequencies (Hz) near 4.35 ppm region (integration 1.94):
2145.8, 2151.3, 2158.8, 2164.2, 2181.8, 2186.5, 2194.6, 2199.5

Frequencies (Hz) near 7.38 ppm region (integration 2.86):
3678.0, 3682.5, 3687.1, 3692.5, 3694.1

Frequencies (Hz) near 7.67 ppm region (integration 1.88):
3829.5, 3832.1, 3837.1, 3838.9

SECTION 4 Problem 91

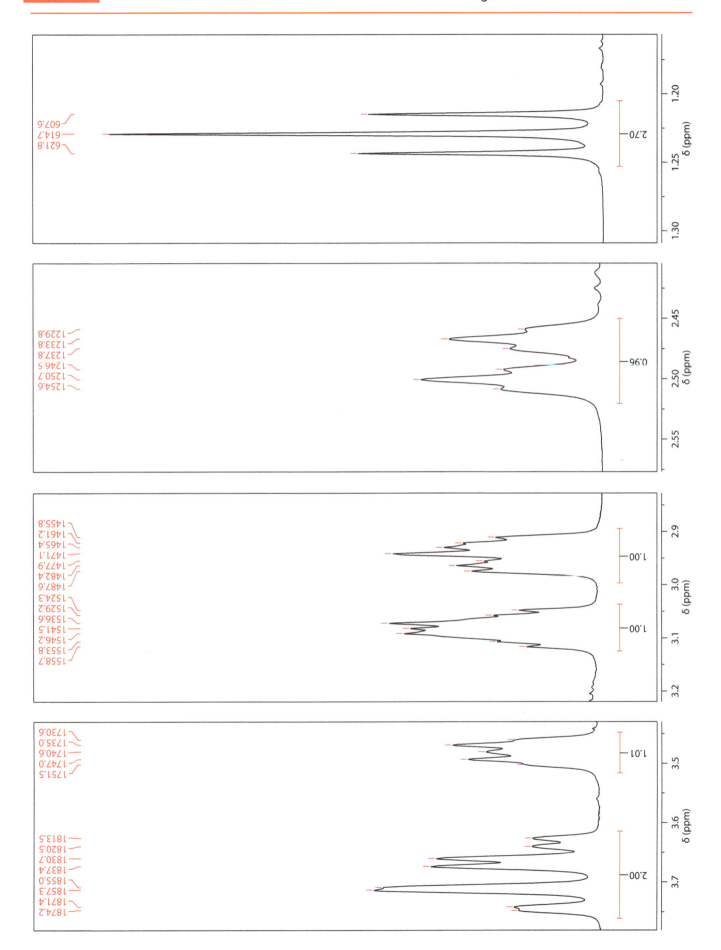

SECTION 4 Problem 91

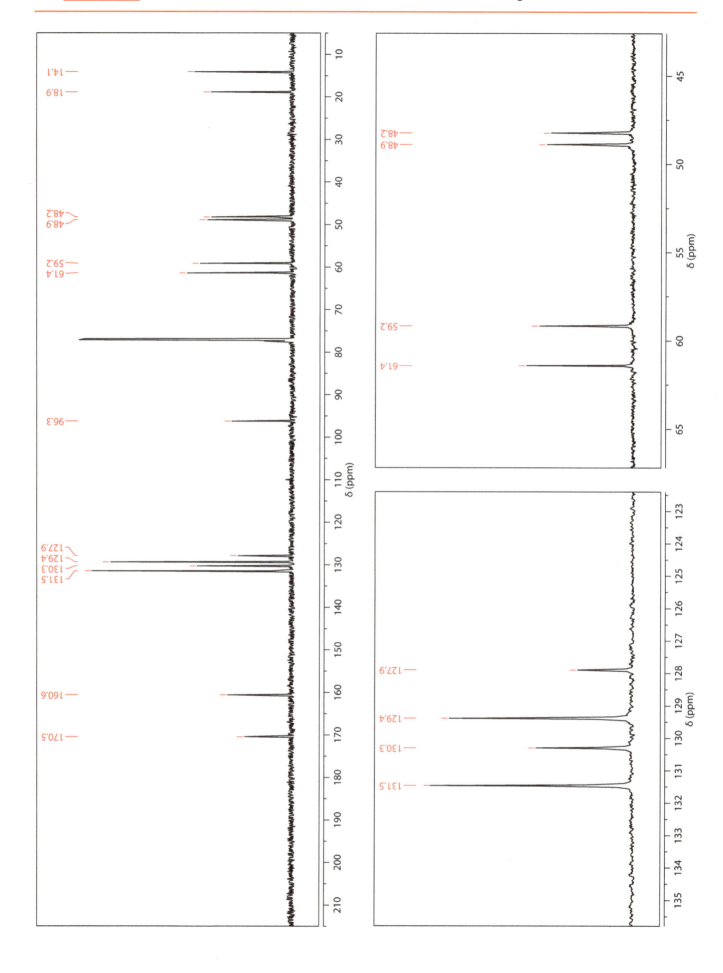

SECTION 4 Problem 91

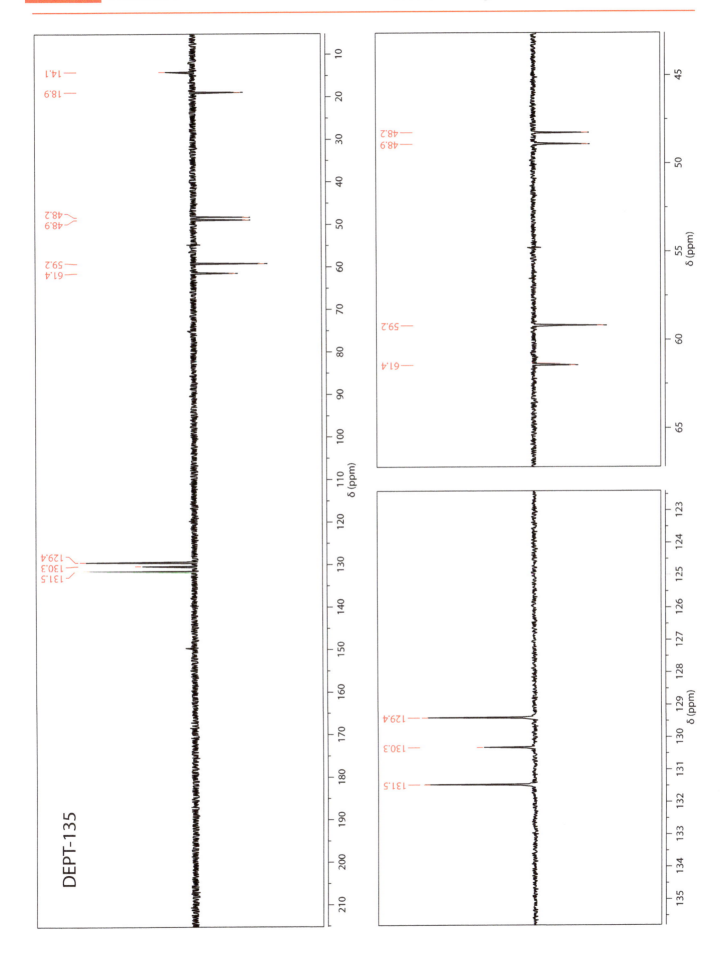

DEPT-135

Problem 91

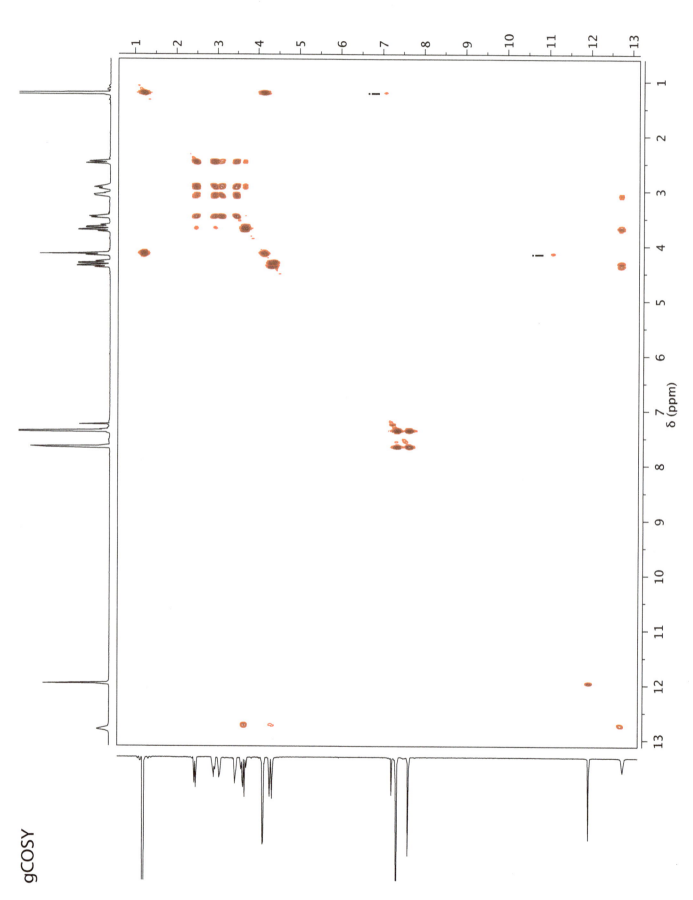

gCOSY

SECTION 4 Problem 91

Problems in Organic Structure Determination

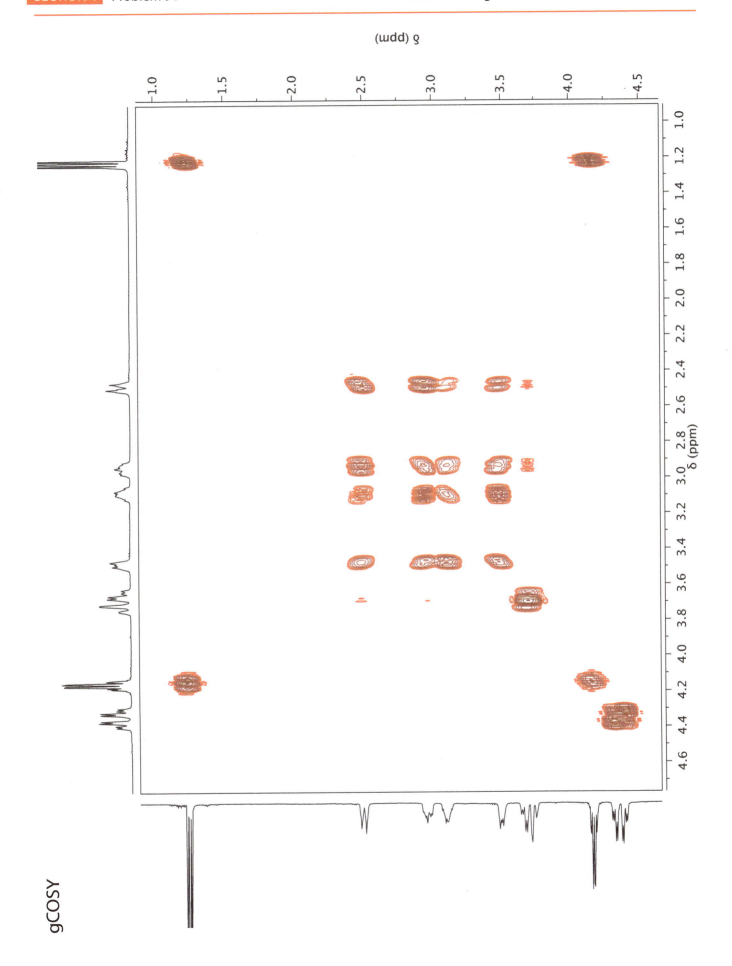

gCOSY

SECTION 4 Problem 91

Problems in Organic Structure Determination

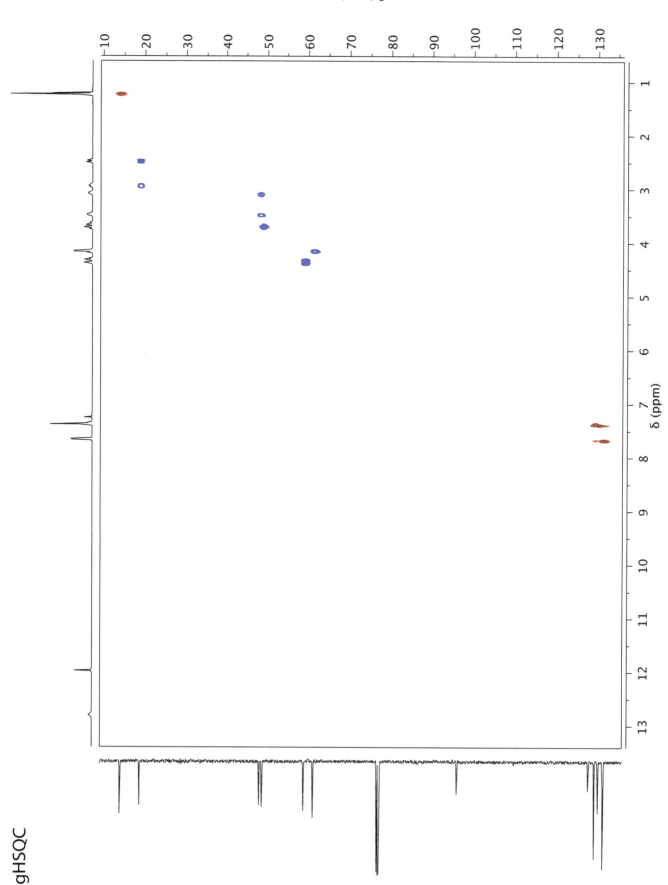

SECTION 4 Problem 91

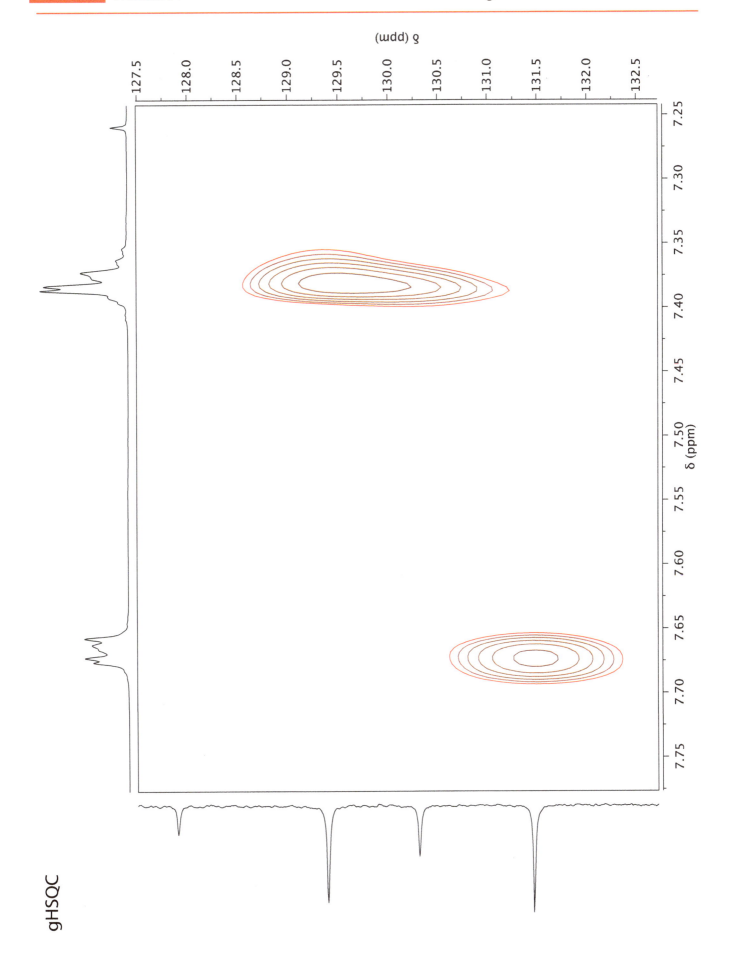

gHSQC

Problem 91

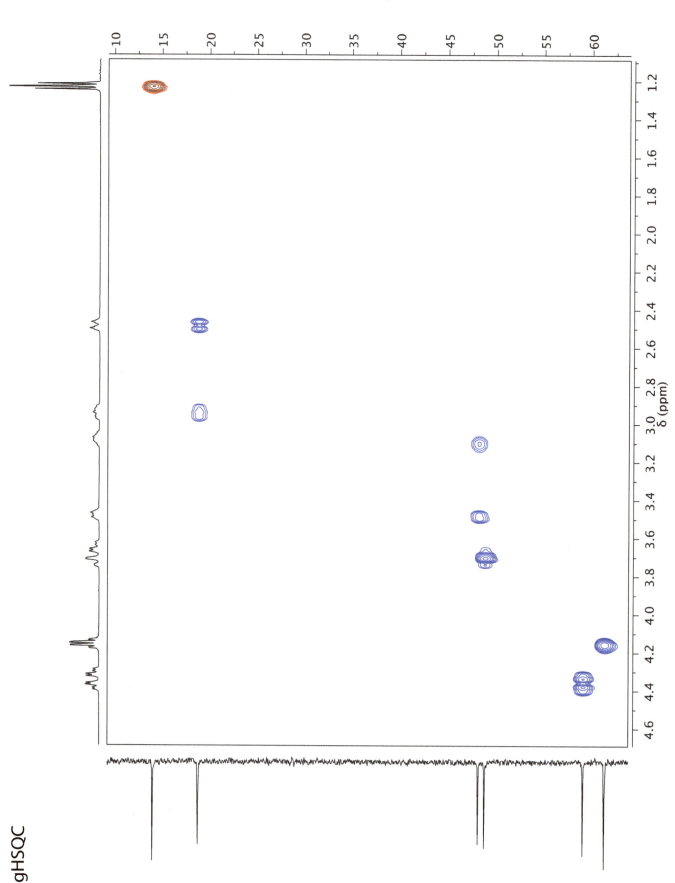

gHSQC

SECTION 4 Problem 91

gHMBC

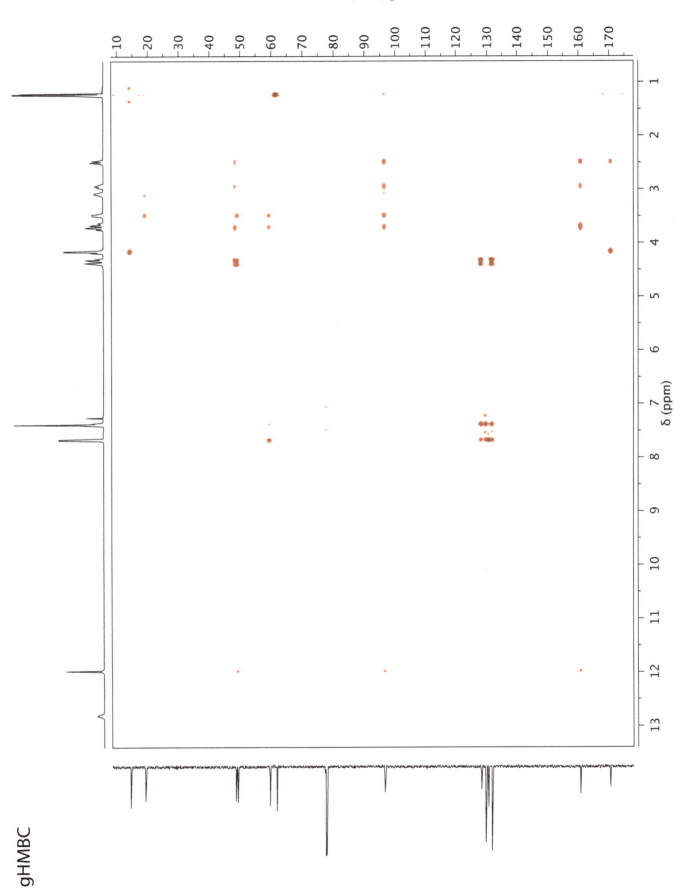

SECTION 4 Problem 91

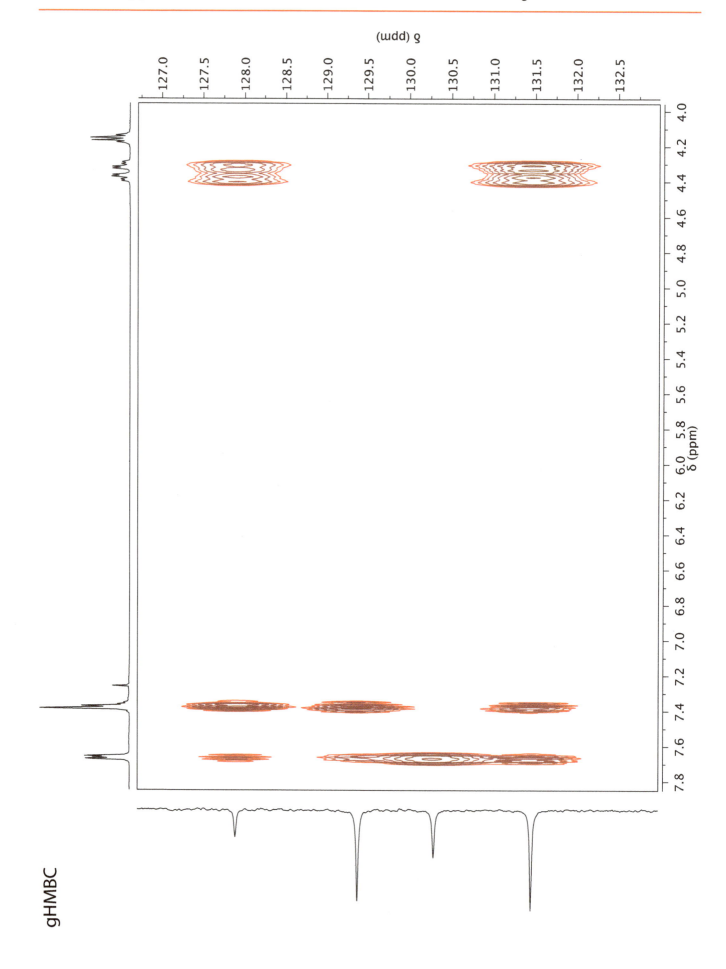

gHMBC

SECTION 4 Problem 91

gHMBC

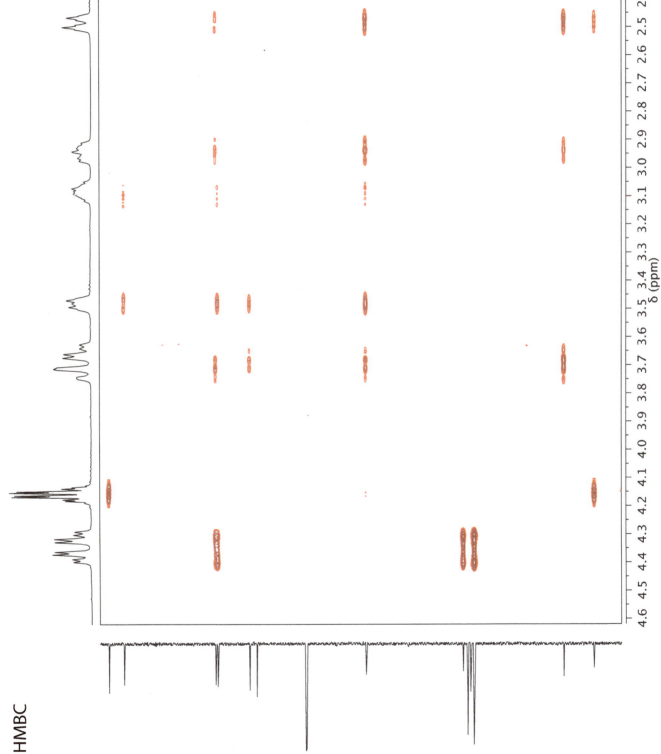

SECTION 4 Problem 91

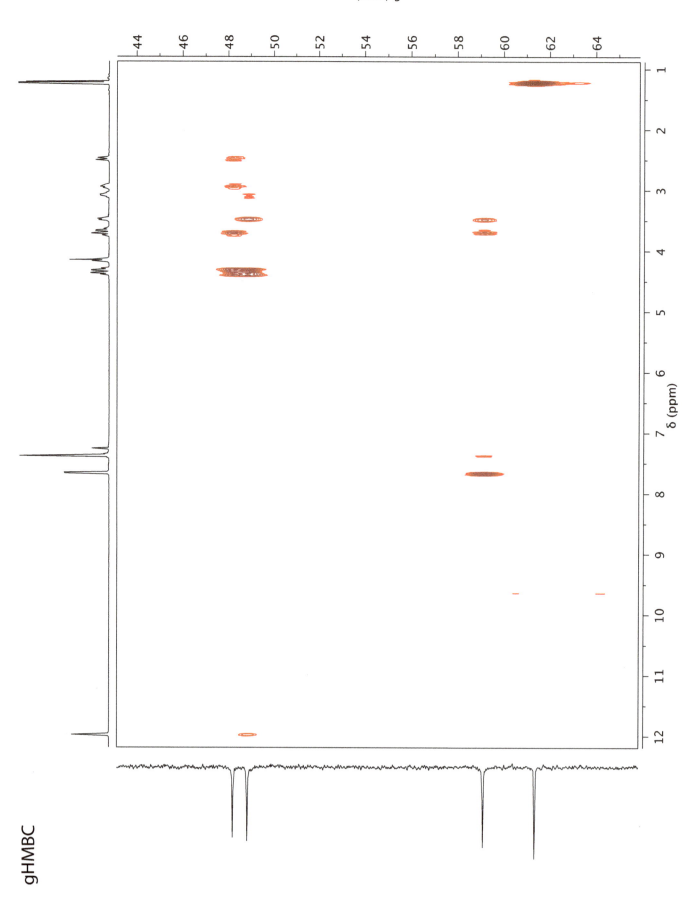

SECTION 4 Problem 92 — Problems in Organic Structure Determination

Determine the structure of the unknown compound below with molecular formula $C_{11}H_{13}FN_4O_5$.
NOTE: Some ^{13}C peak picking provided in Hz, to allow extraction of C-F coupling constants.
Spectra acquired in CD_3OD at 500 MHz (^1H) or 125 MHz (^{13}C).

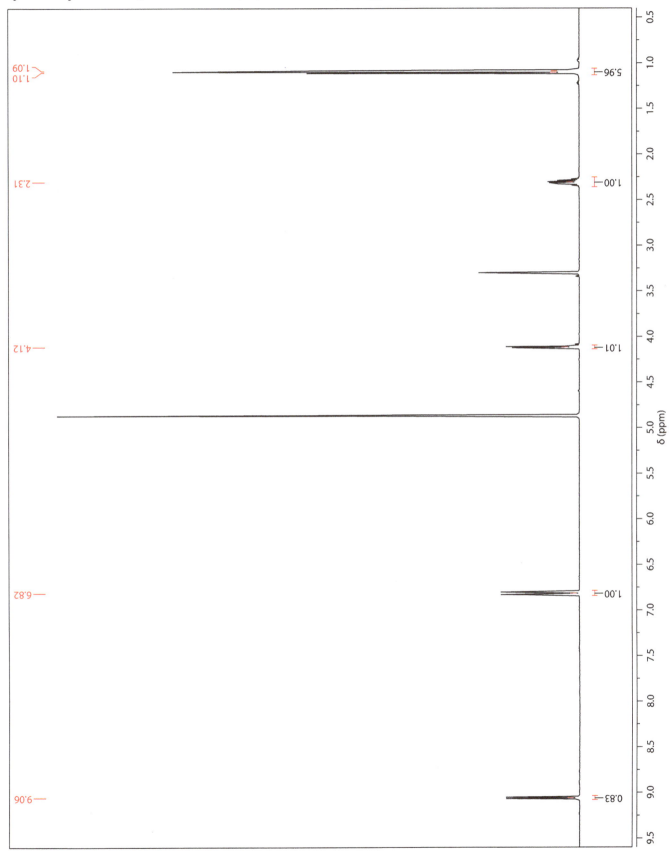

Problem 92

SECTION 4 Problem 92

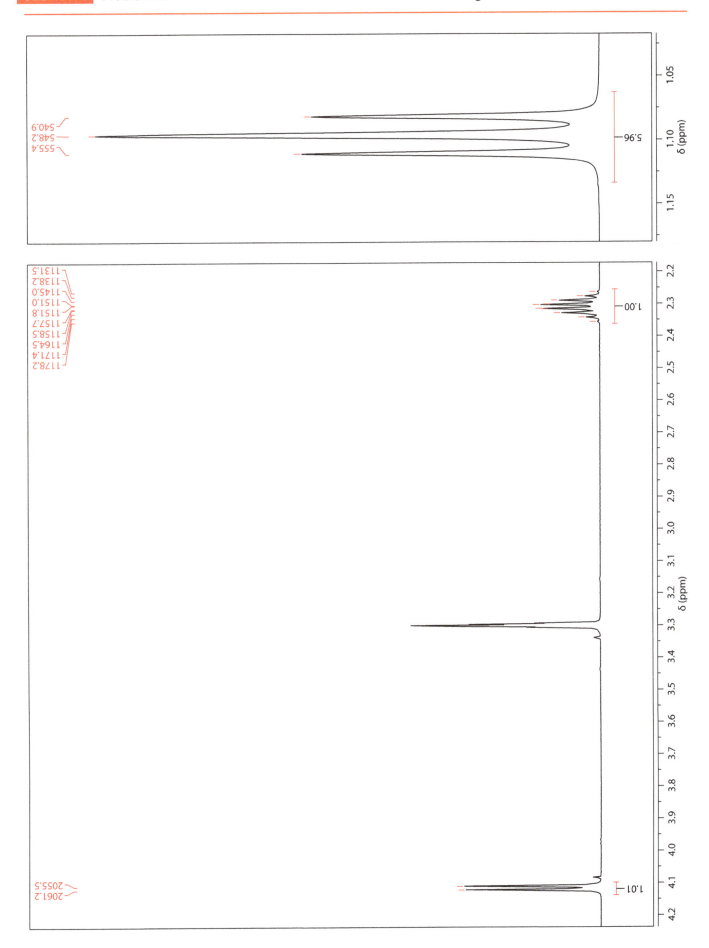

SECTION 4 Problem 92

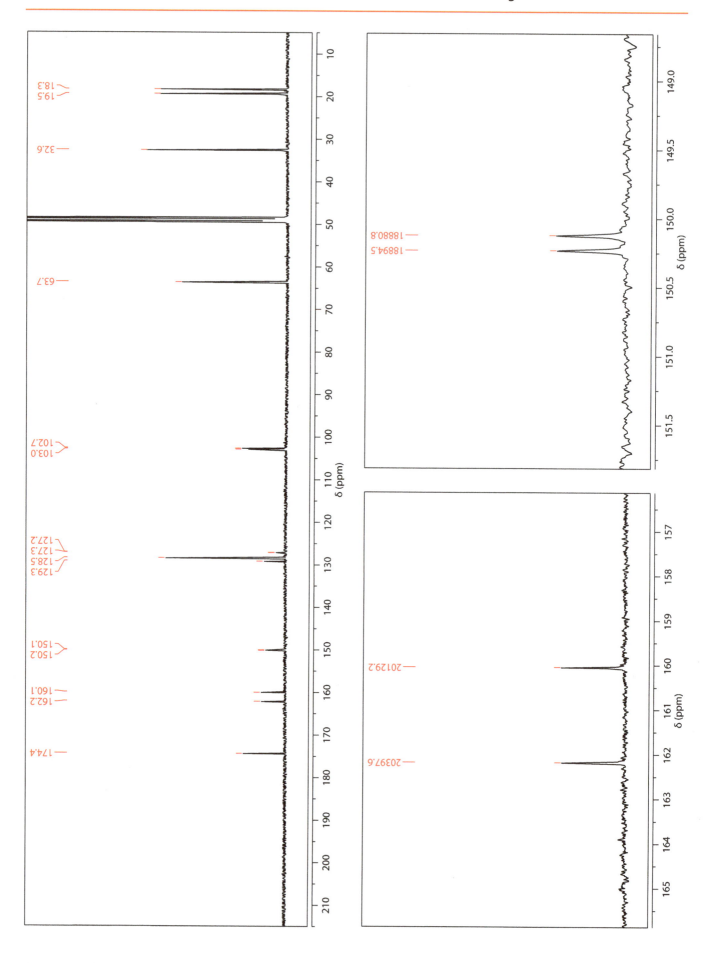

SECTION 4 Problem 92

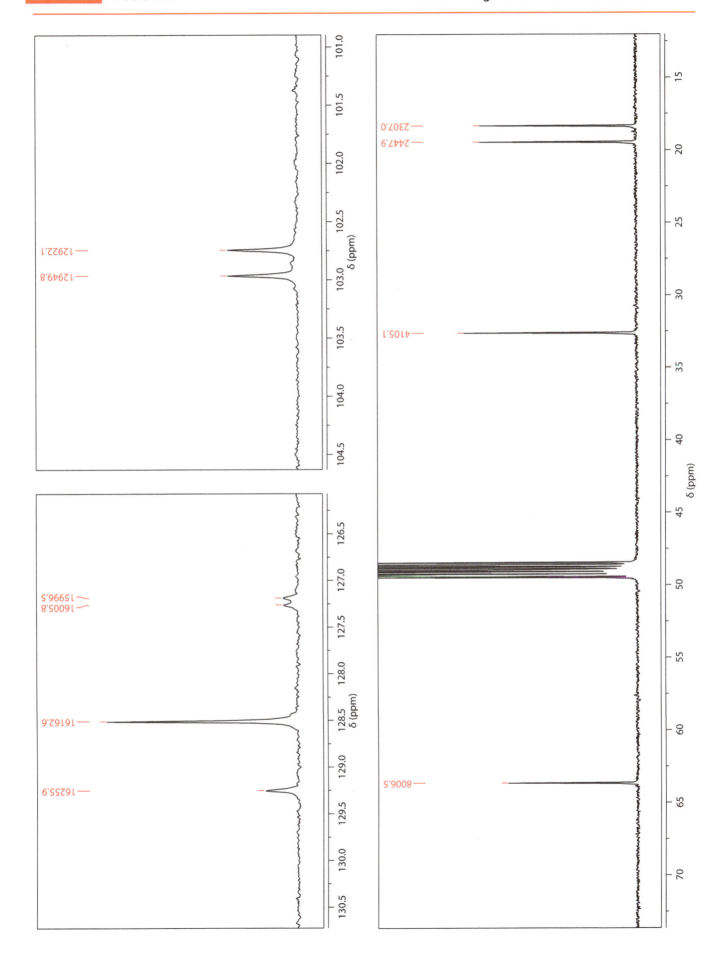

SECTION 4 Problem 92

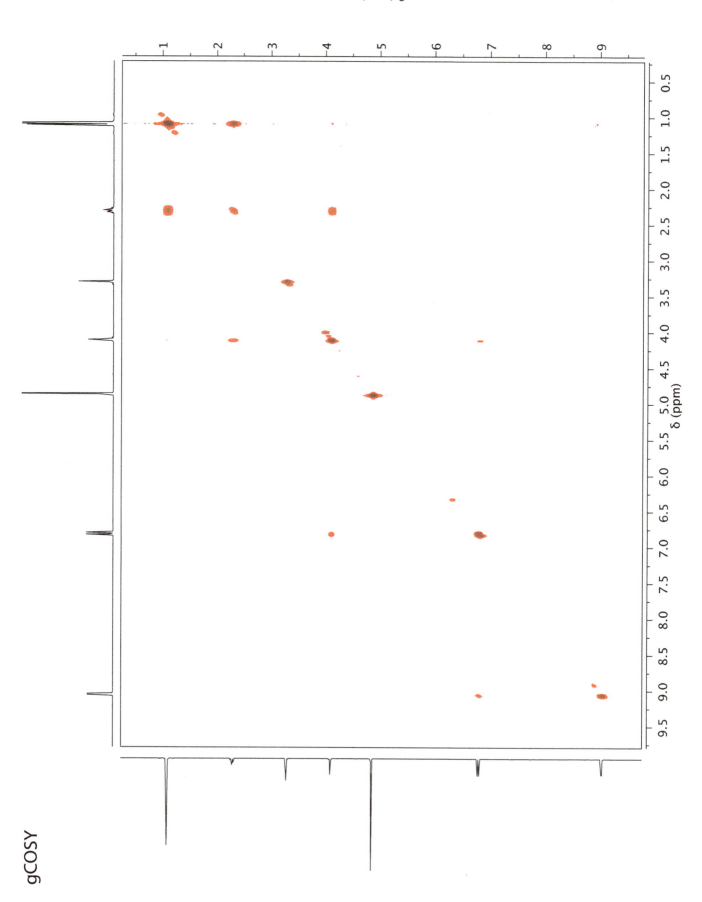

gCOSY

SECTION 4 Problem 92

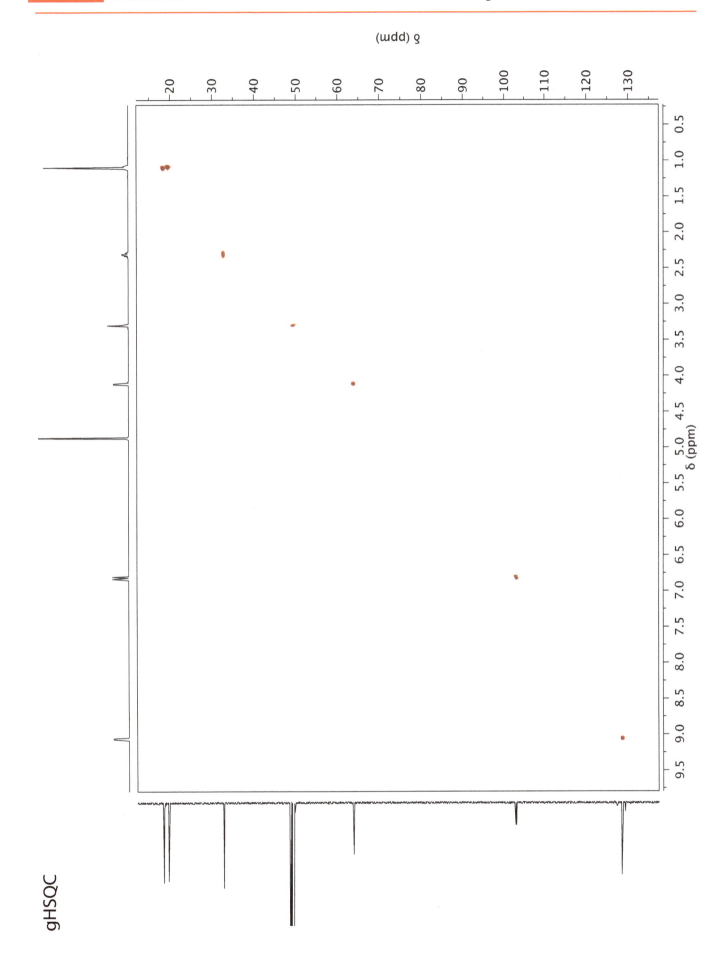

gHSQC

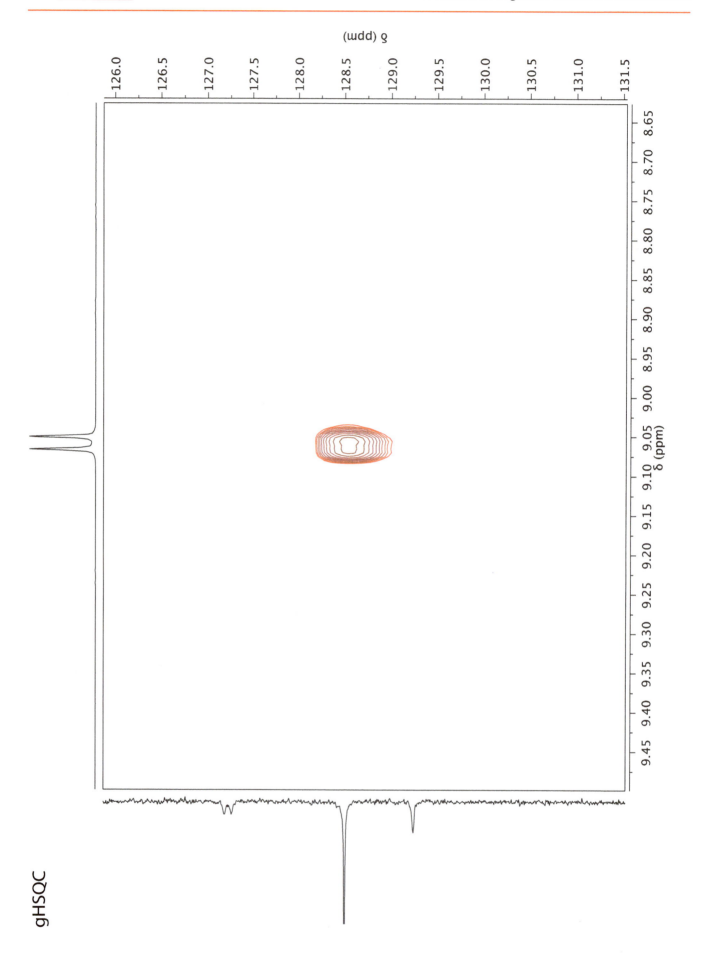

SECTION 4 Problem 92

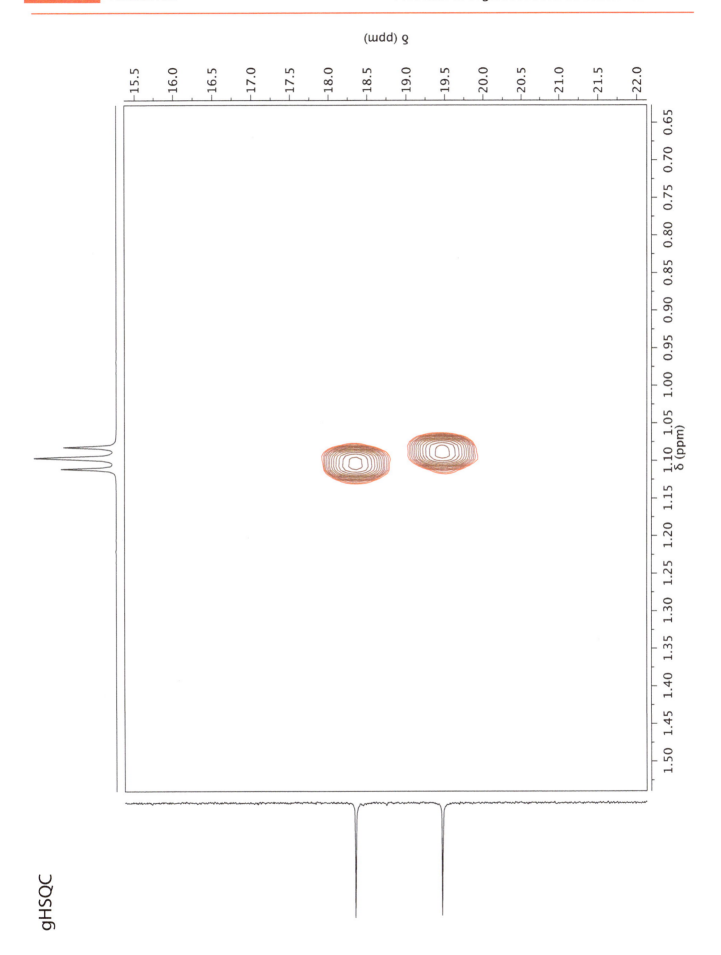

gHSQC

SECTION 4 Problem 92

Problems in Organic Structure Determination

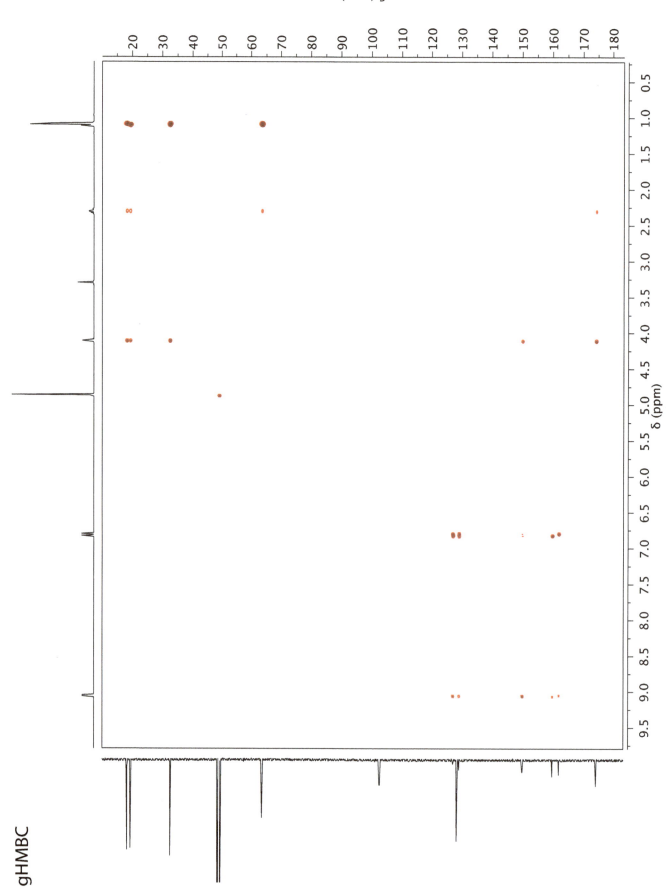

gHMBC

SECTION 4 Problem 92

Problems in Organic Structure Determination 599

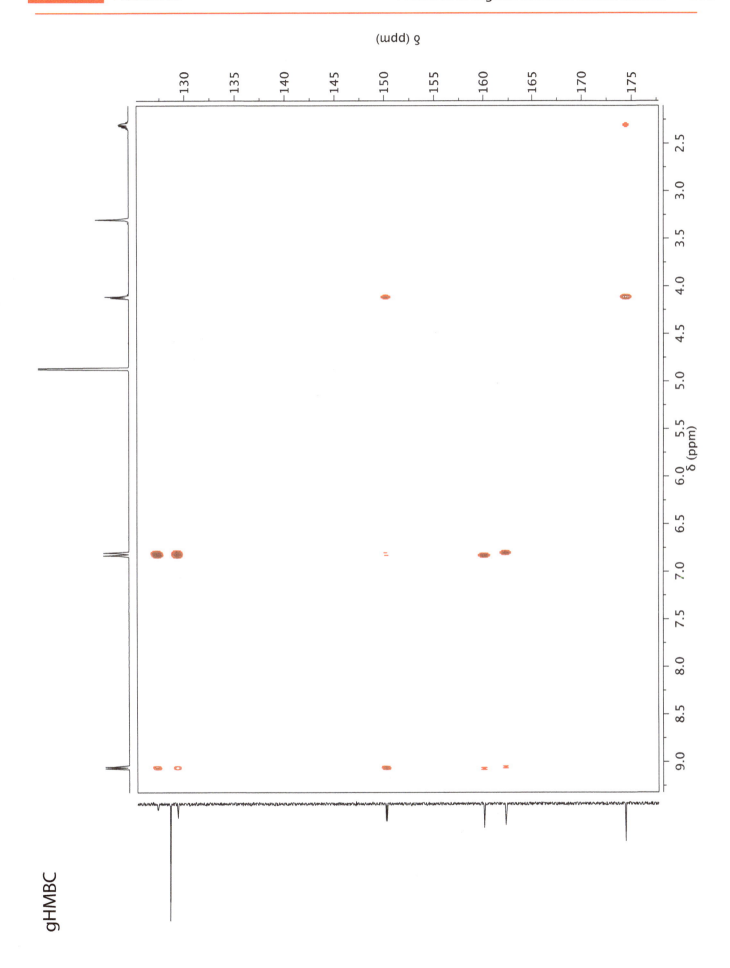

gHMBC

Problem 92

gHMBC

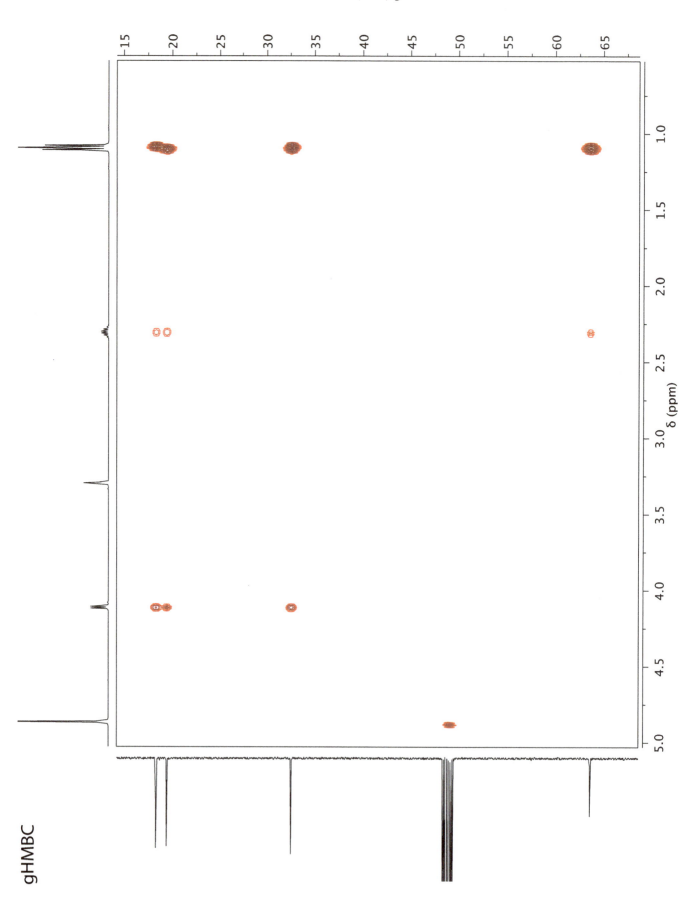

SECTION 4 Problem 93

Determine the structure of the unknown compound below with molecular formula $C_{11}H_{19}NO_5$.

Spectra acquired in $CDCl_3$ at 500 MHz (^{1}H) or 125 MHz (^{13}C). (N) (H)

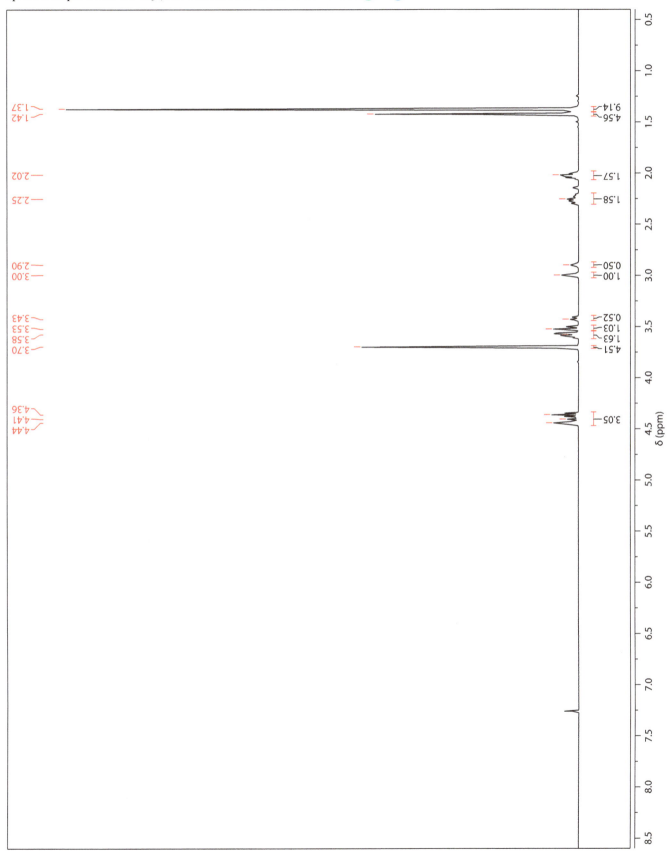

SECTION 4 Problem 93

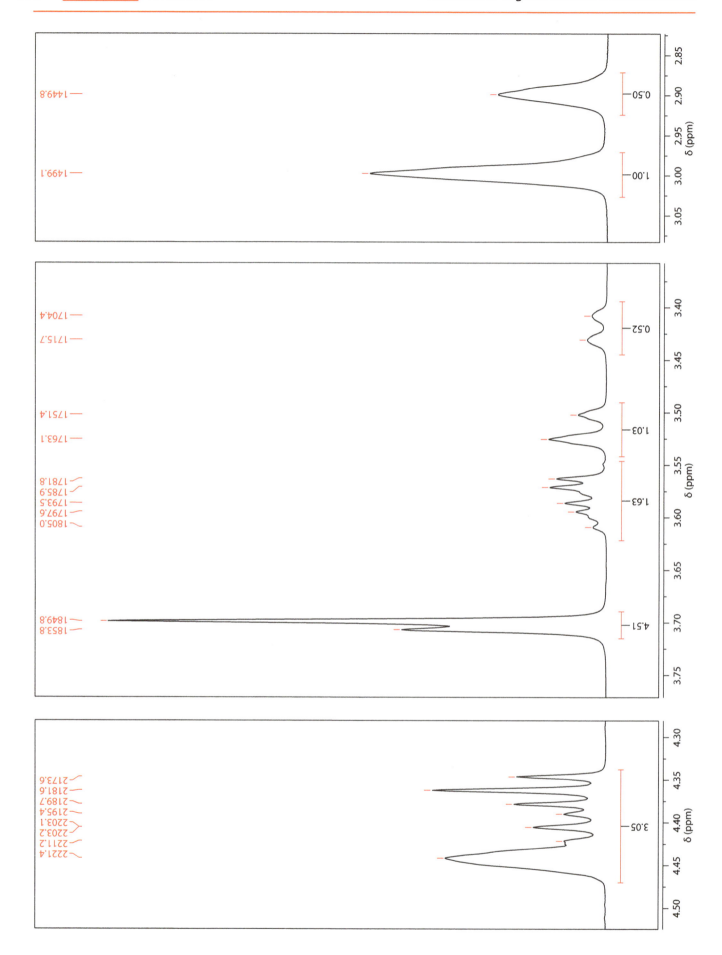

SECTION 4 Problem 93

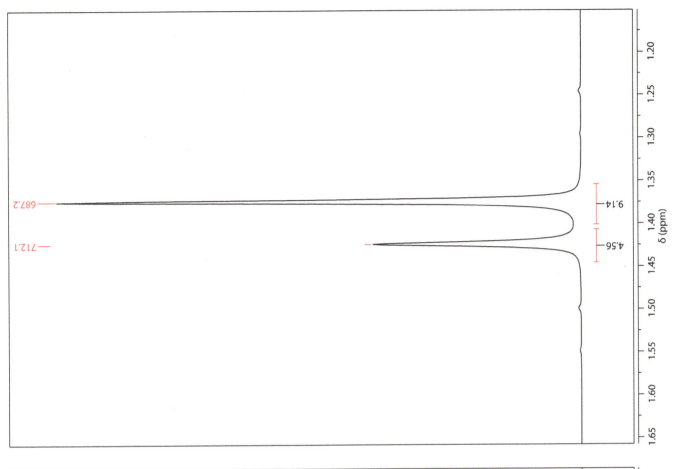

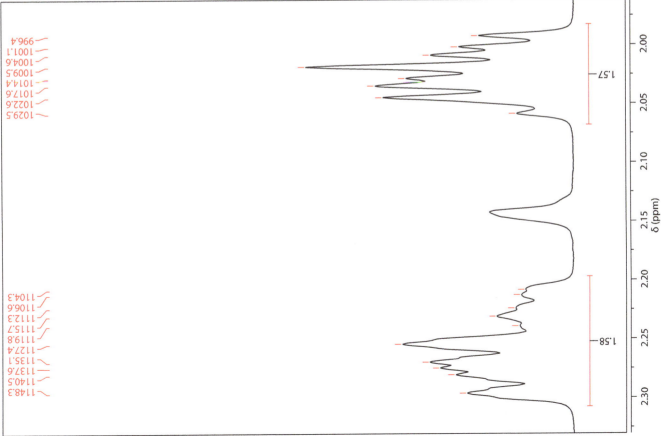

Problem 93

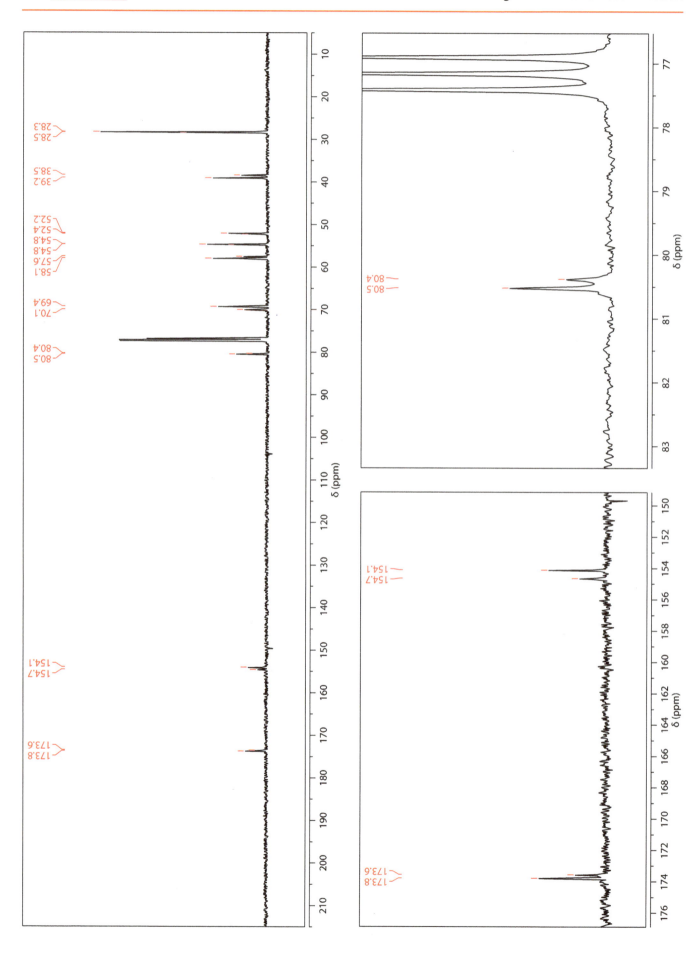

SECTION 4 Problem 93

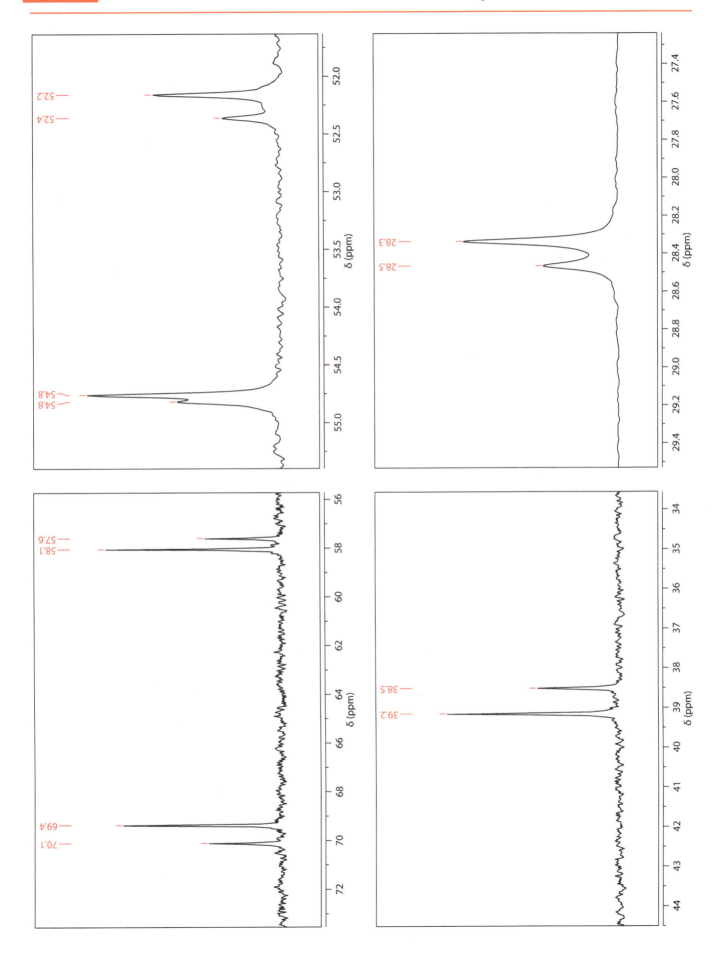

Problem 93

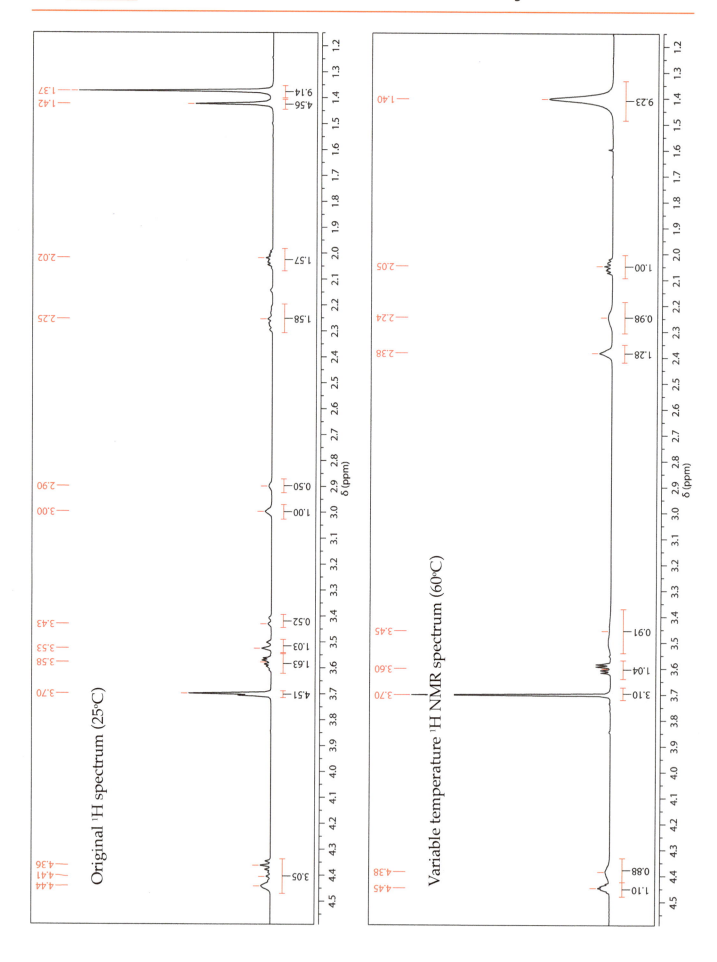

SECTION 4 Problem 93

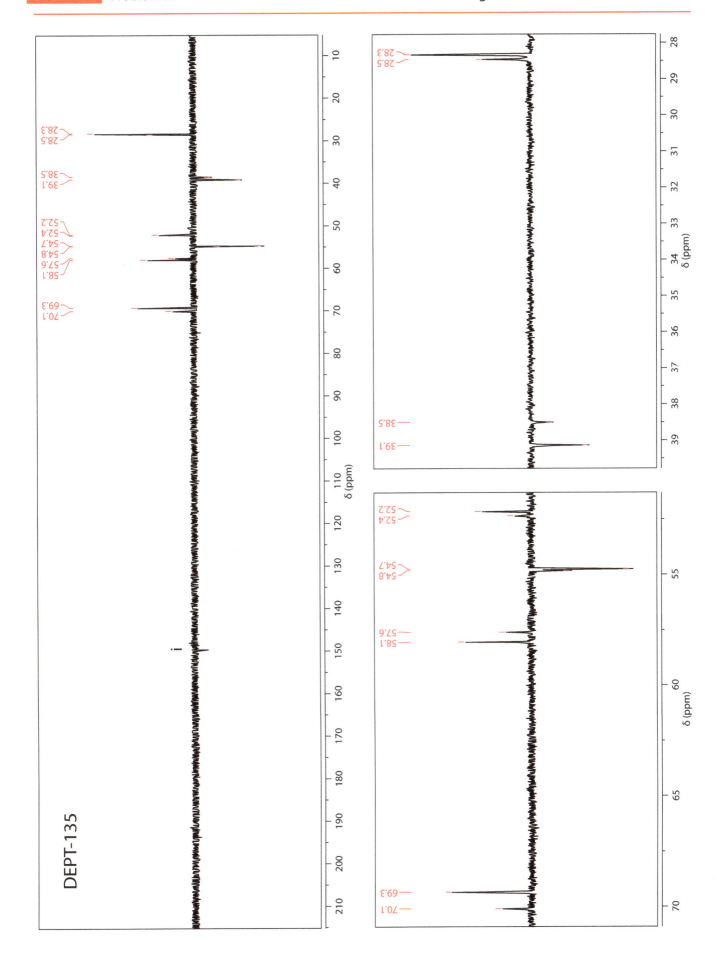

SECTION 4 Problem 93

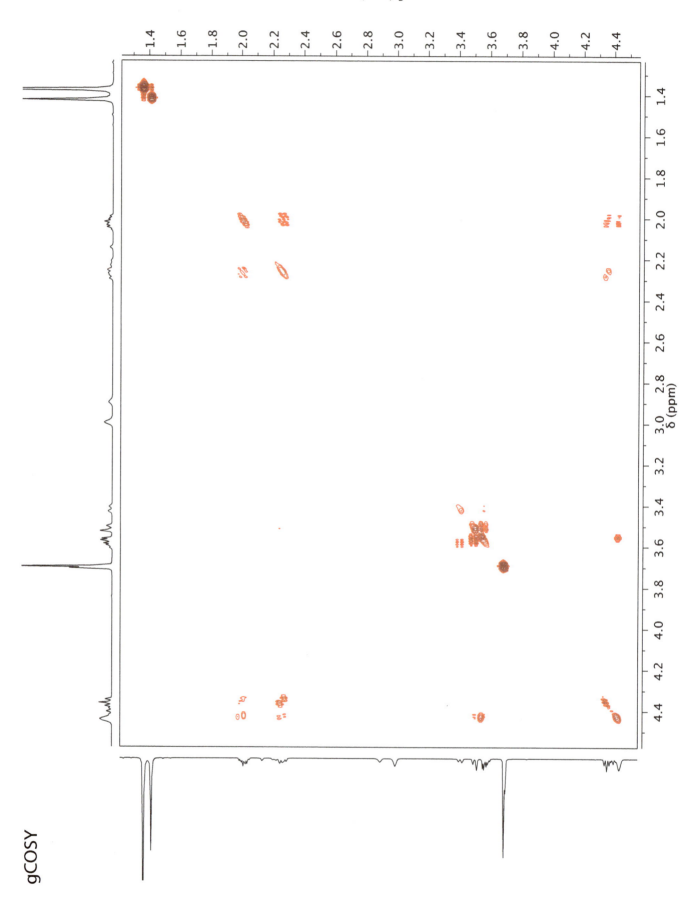

gCOSY

SECTION 4 Problem 93

gCOSY

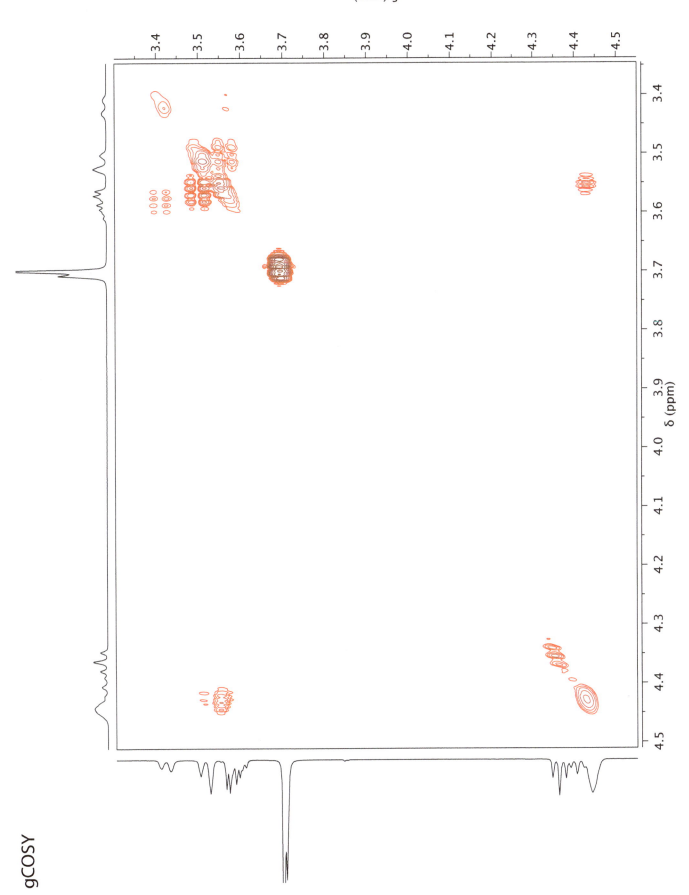

SECTION 4 Problem 93

Problems in Organic Structure Determination

gCOSY

SECTION 4 Problem 93

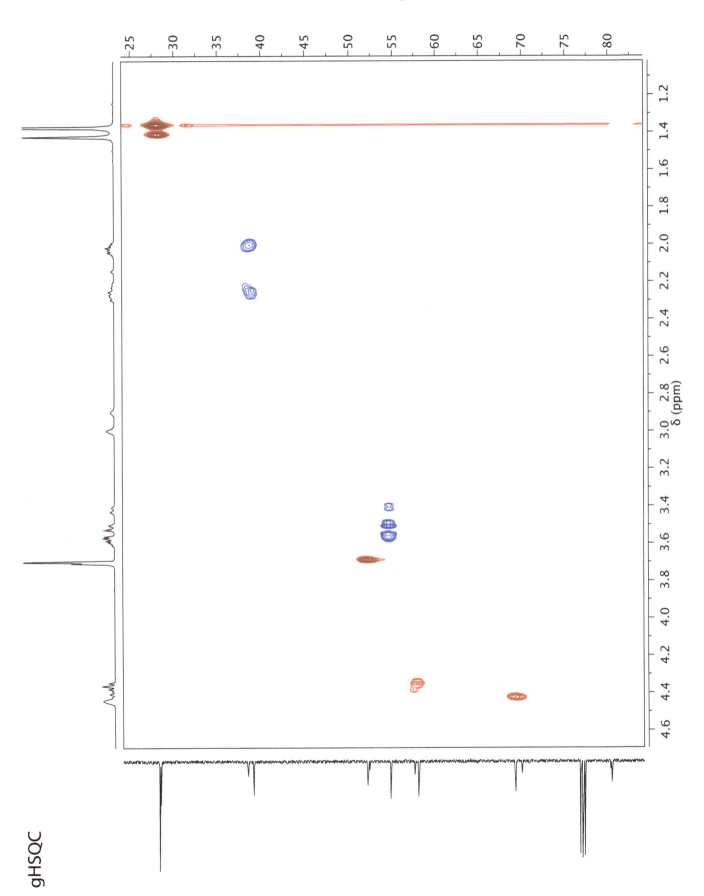

gHSQC

Problem 93

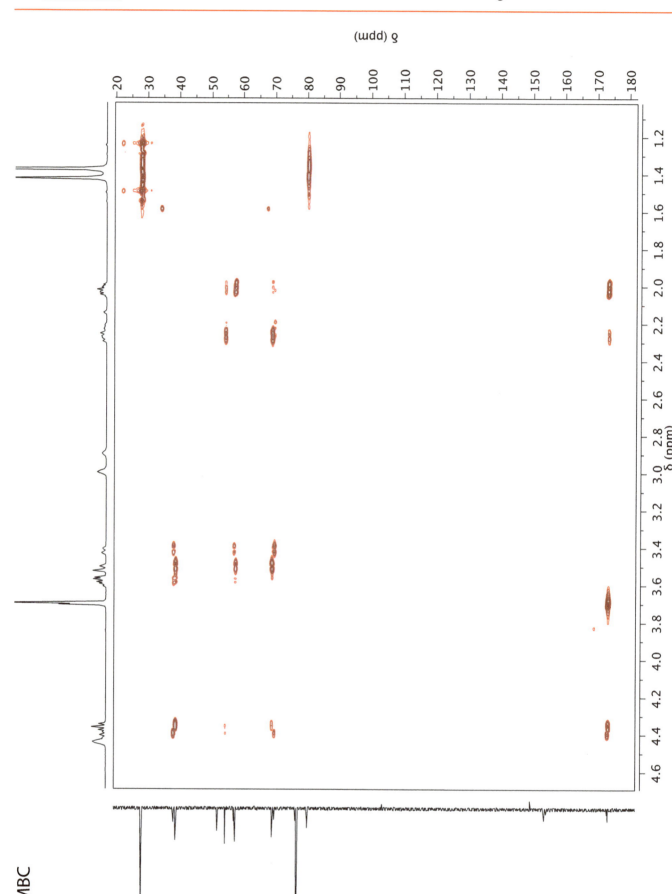

gHMBC

SECTION 4 Problem 93

gHMBC

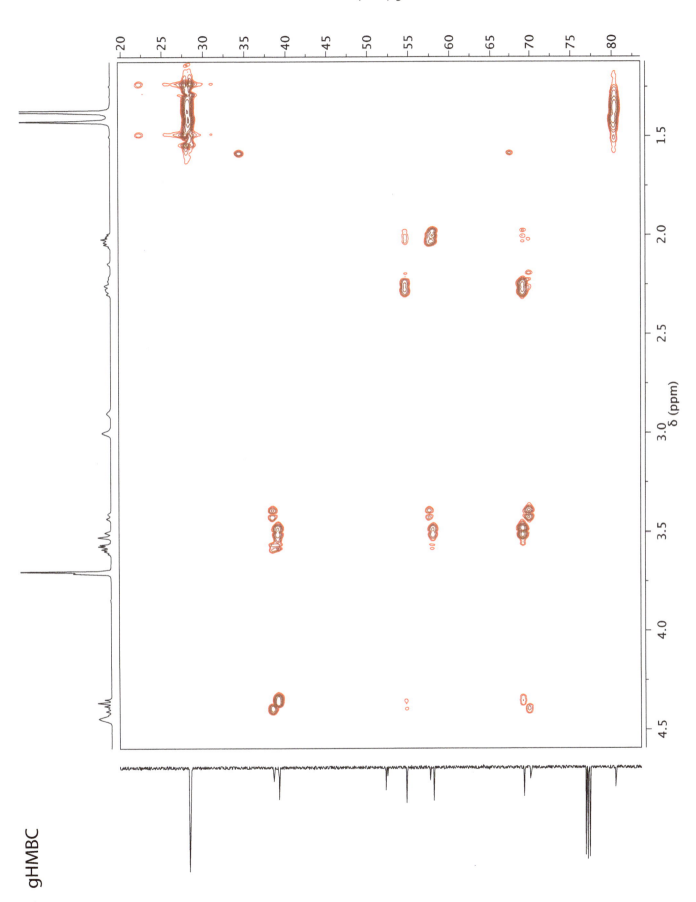

SECTION 4 Problem 94

Determine the structure of the unknown compound below with molecular formula $C_{11}H_{12}Cl_2N_2O_5$.

Spectra acquired in $(CD_3)_2SO$ at 500 MHz (^1H) or 125 MHz (^{13}C). N H

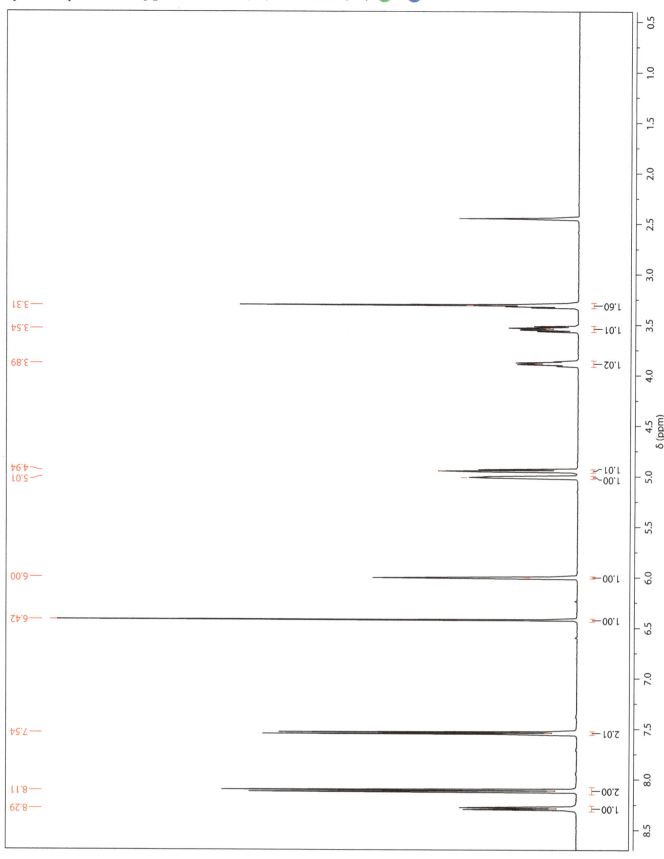

SECTION 4 Problem 94

SECTION 4 Problem 94

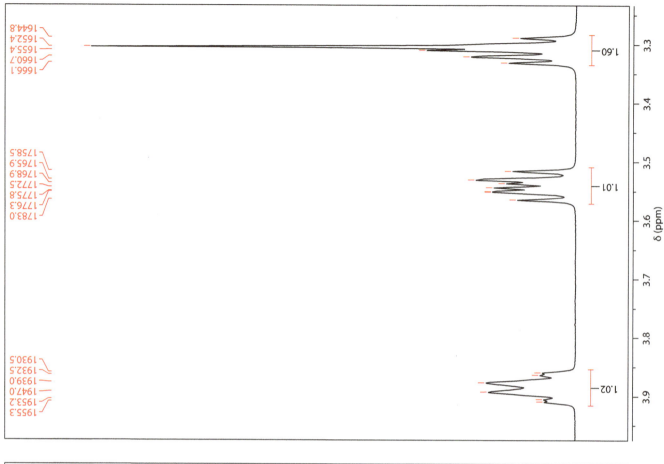

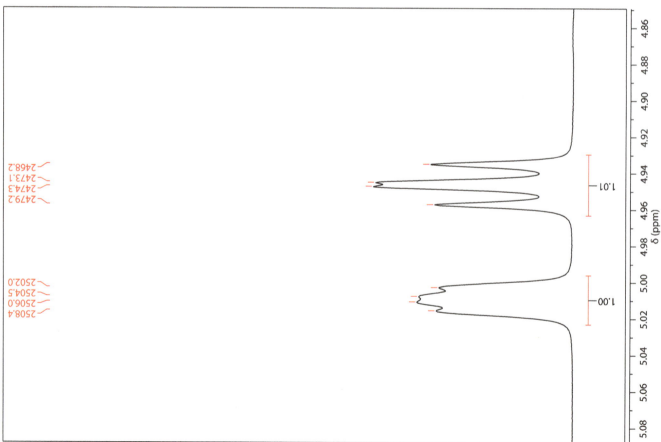

SECTION 4 Problem 94

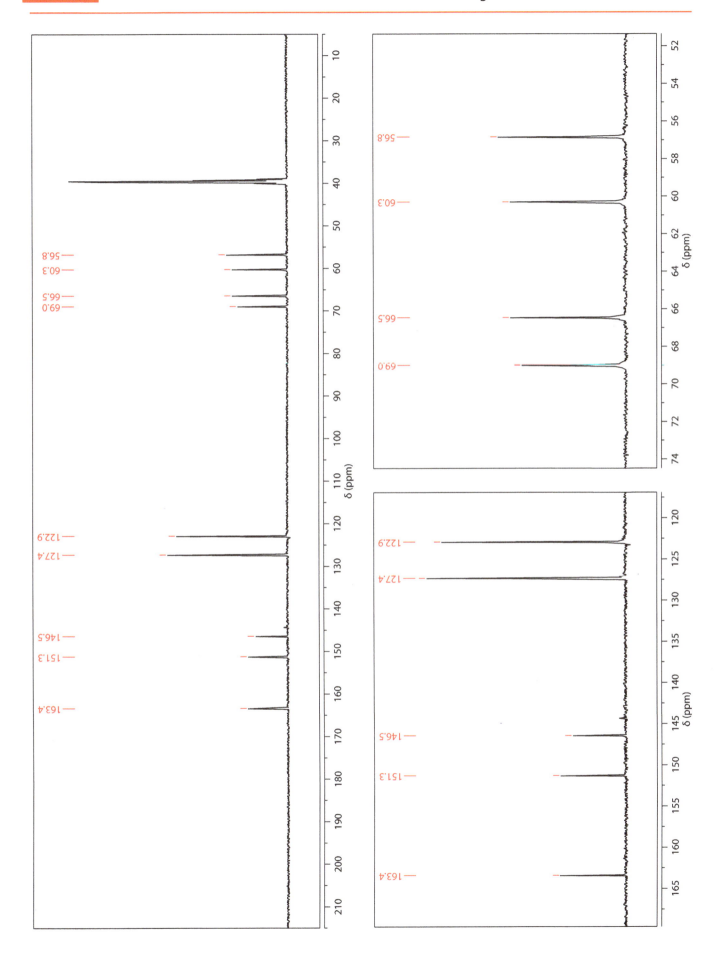

Problem 94

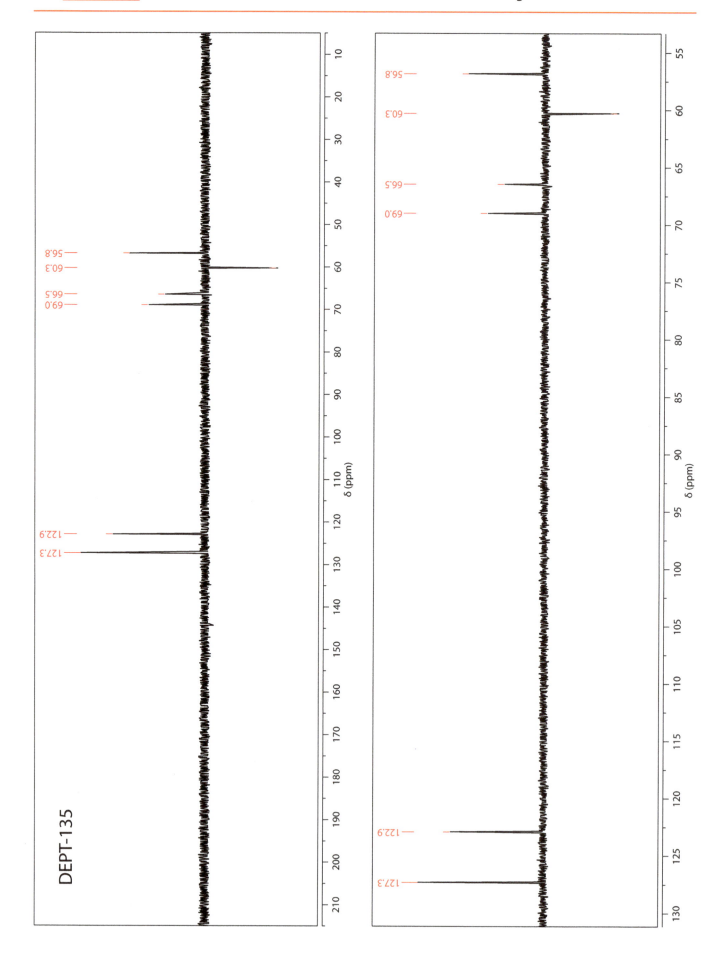

DEPT-135

SECTION 4 Problem 94

Problems in Organic Structure Determination

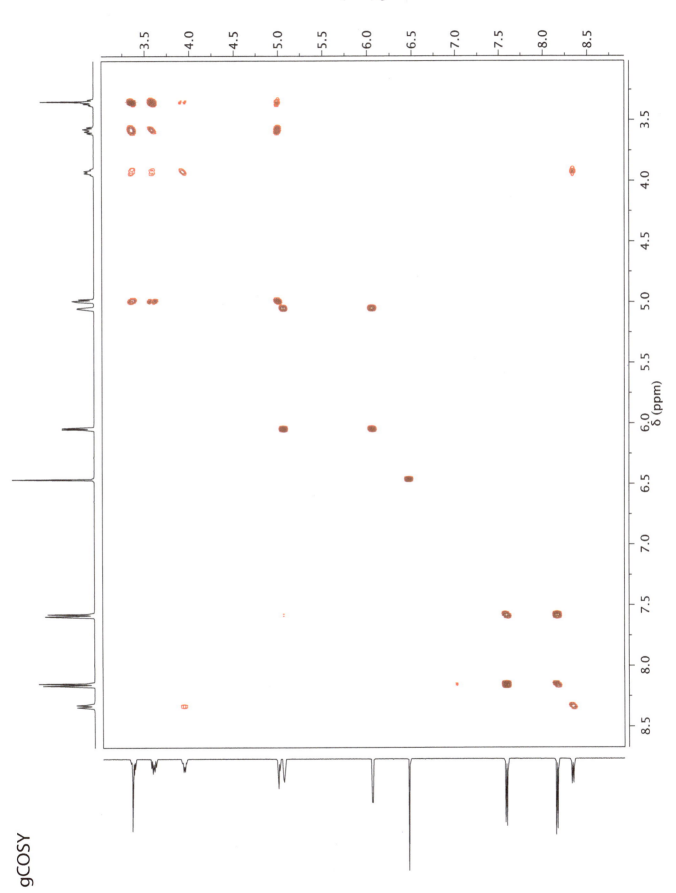

gCOSY

SECTION 4 Problem 94

Problems in Organic Structure Determination

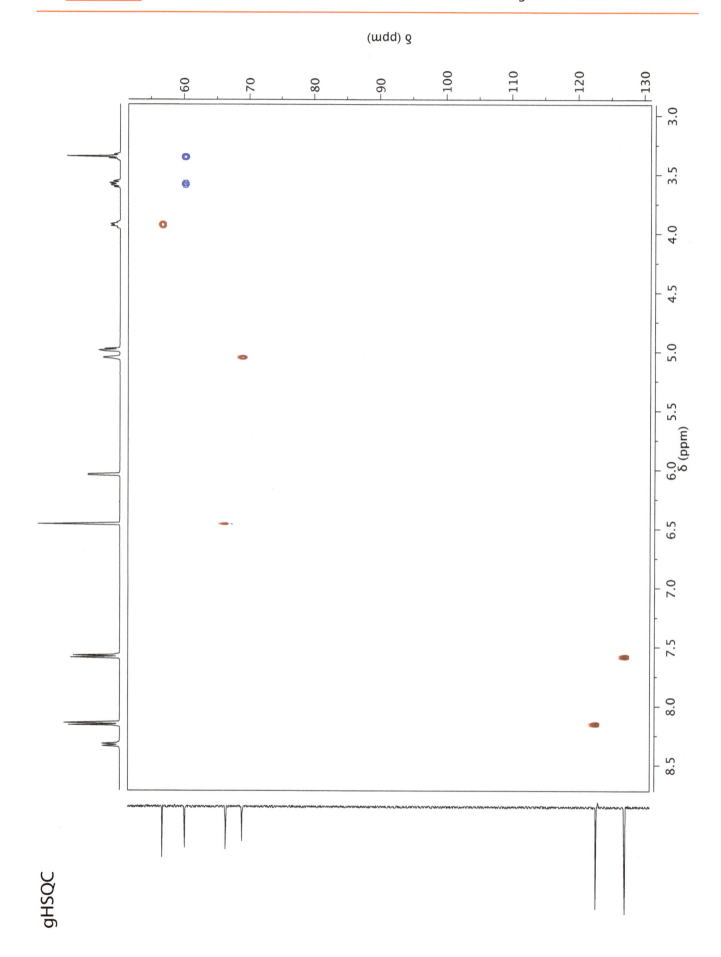

gHSQC

SECTION 4 Problem 94

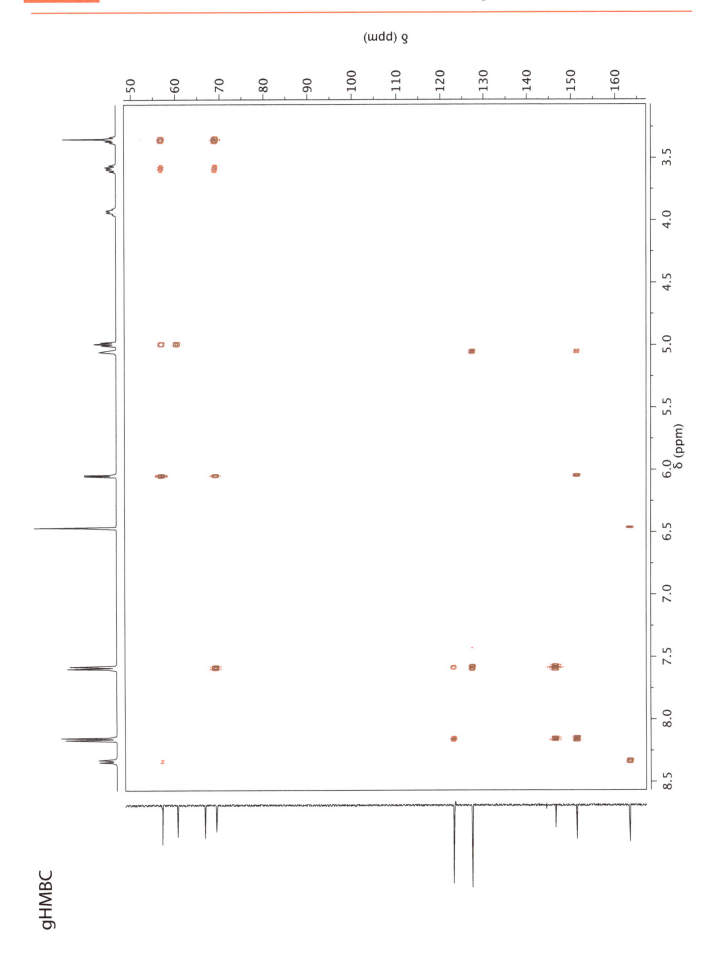

gHMBC

Problem 94

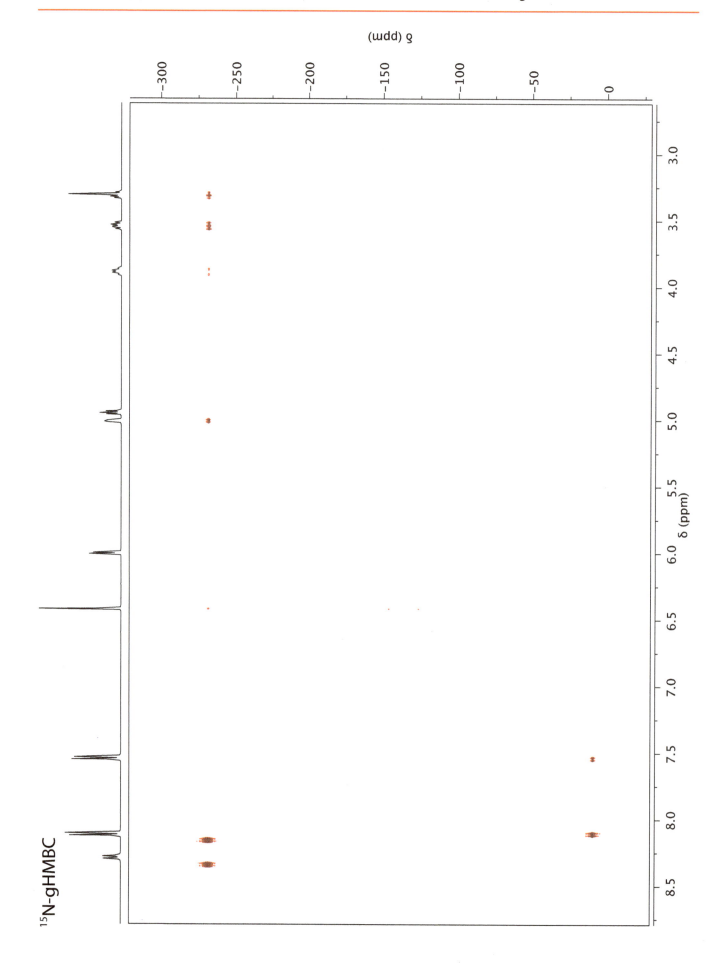

15N-gHMBC

SECTION 4 Problem 95

Determine the structure of the unknown compound below with molecular formula C₂₁H₂₆N₂O₃.

Spectra acquired in (CD₃)₂SO with trifluoroacetic acid (TFA) at 500 MHz (¹H) or 125 MHz (¹³C).

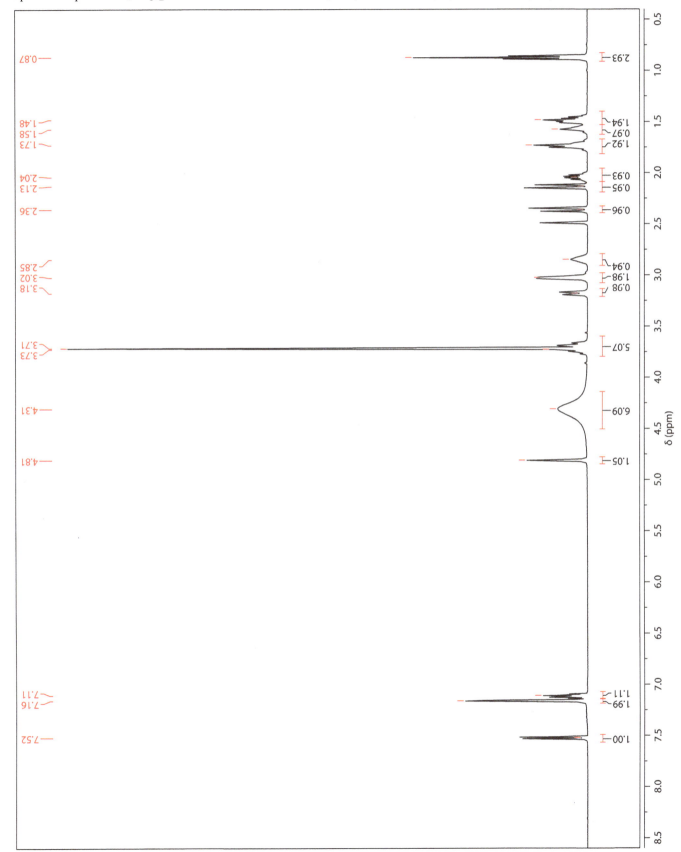

Problem 95

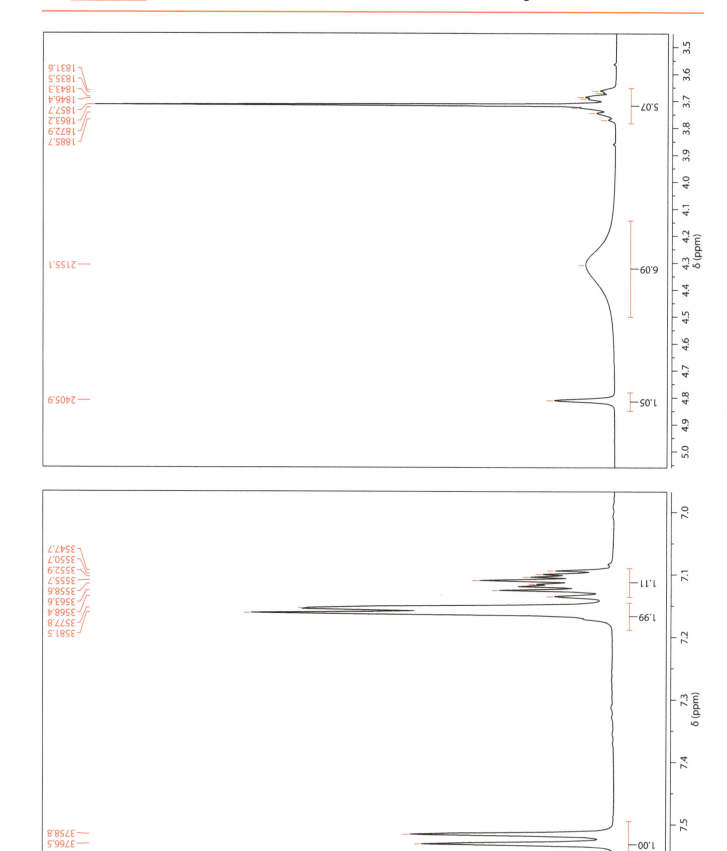

SECTION 4 Problem 95

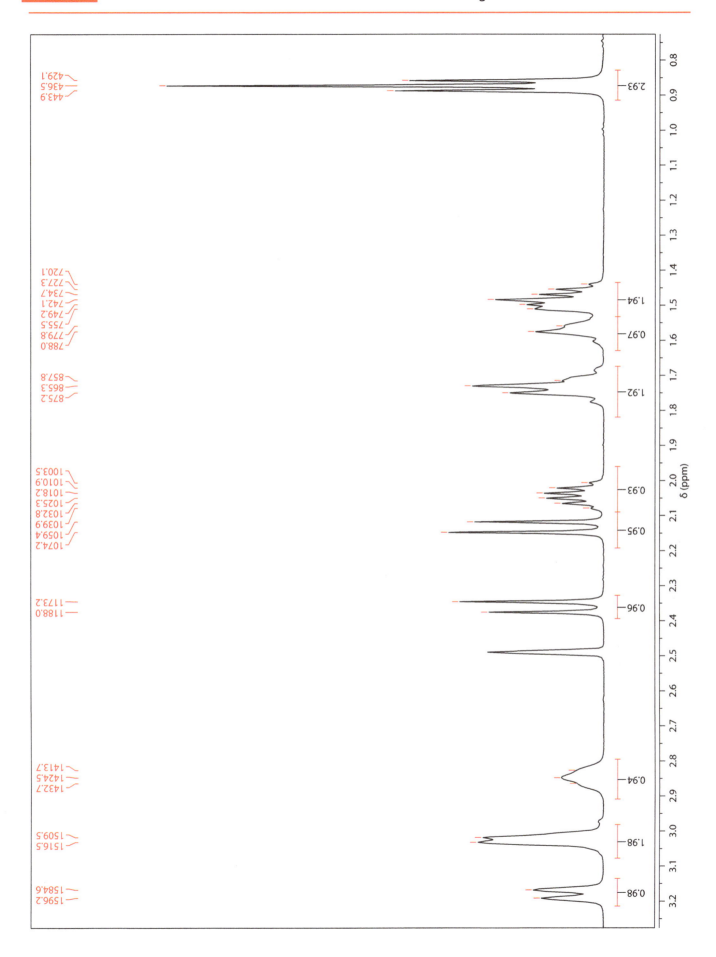

Problem 95

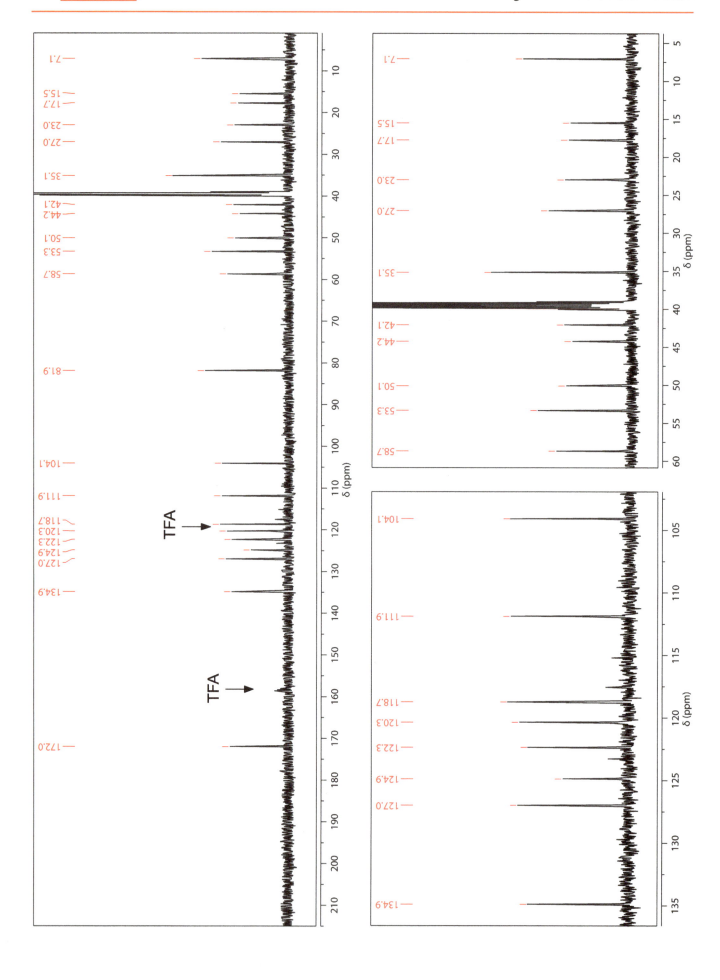

SECTION 4 Problem 95

gCOSY

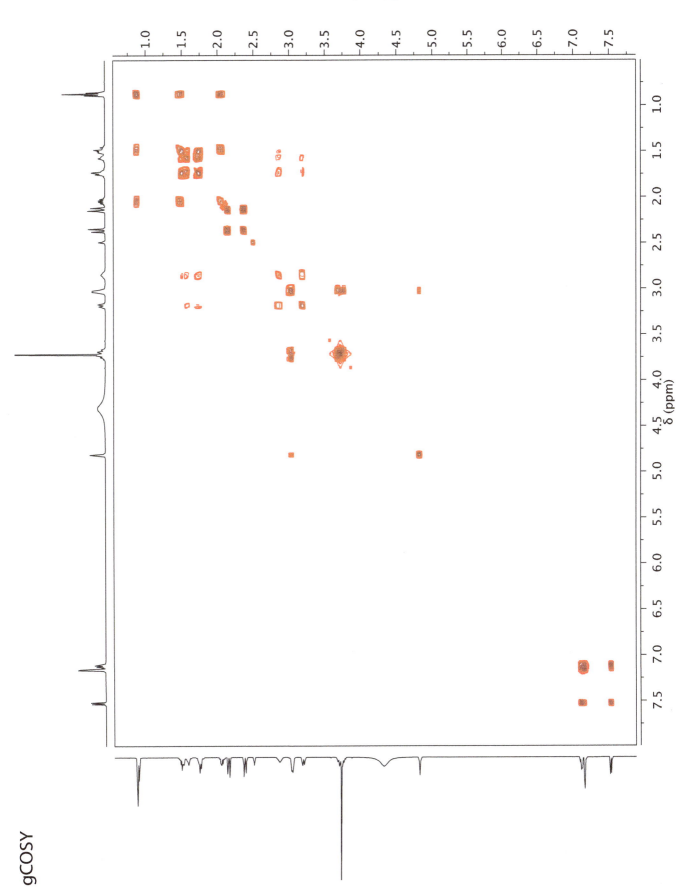

Problem 95

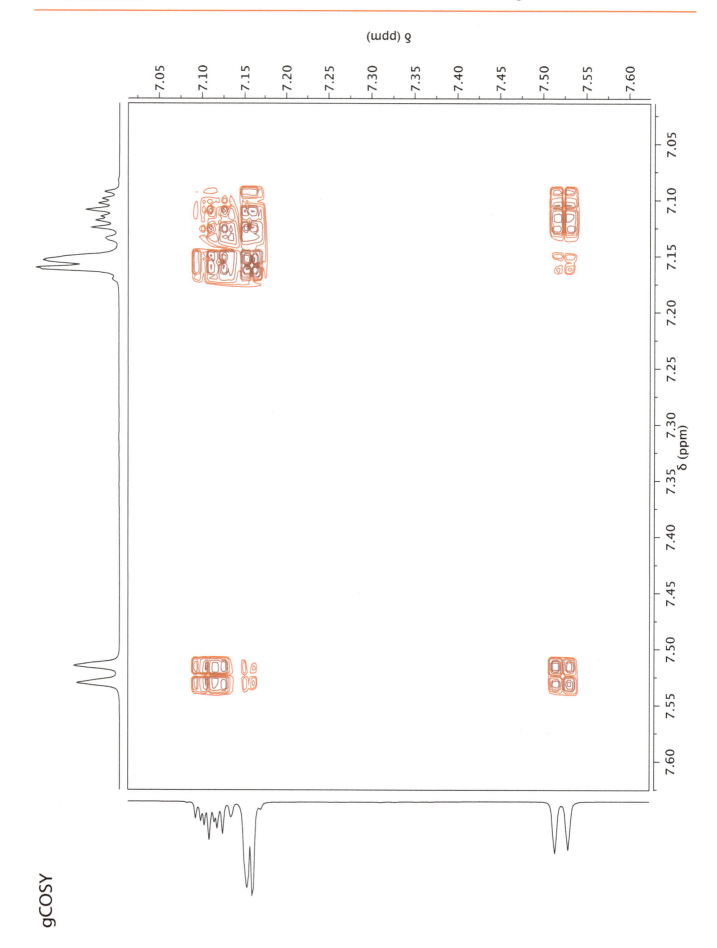

gCOSY

SECTION 4 Problem 95 — Problems in Organic Structure Determination

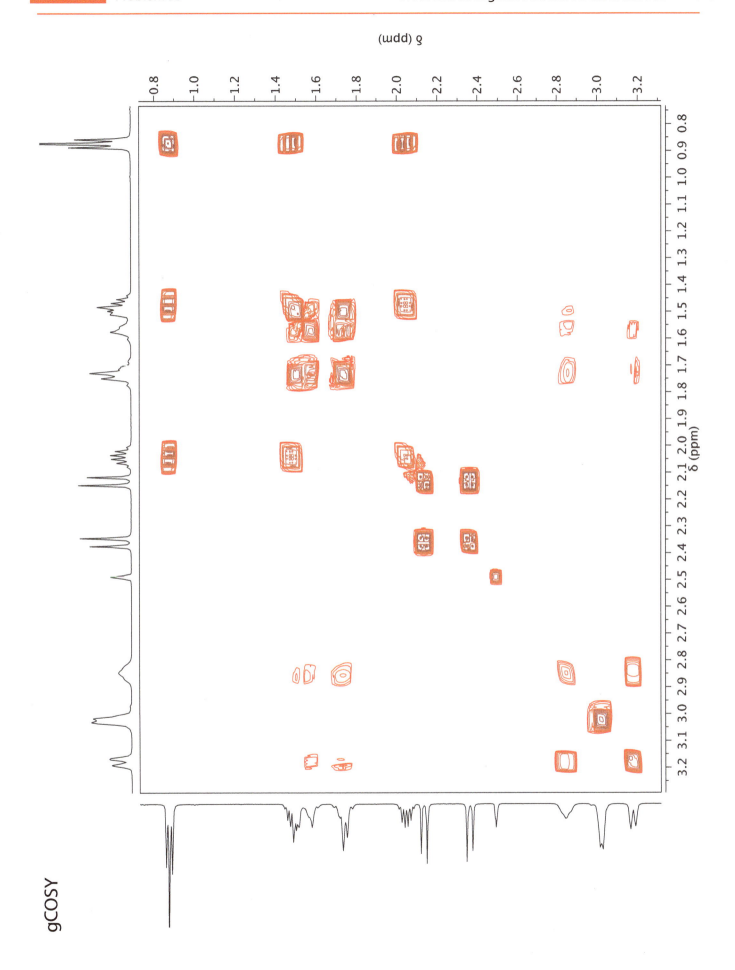

gCOSY

Problem 95

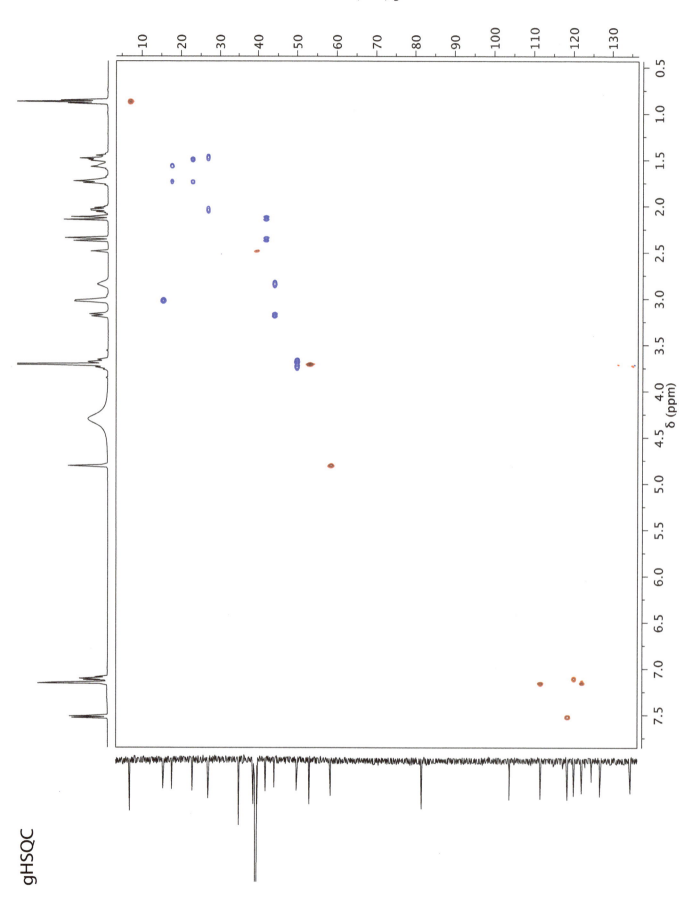

SECTION 4 Problem 95

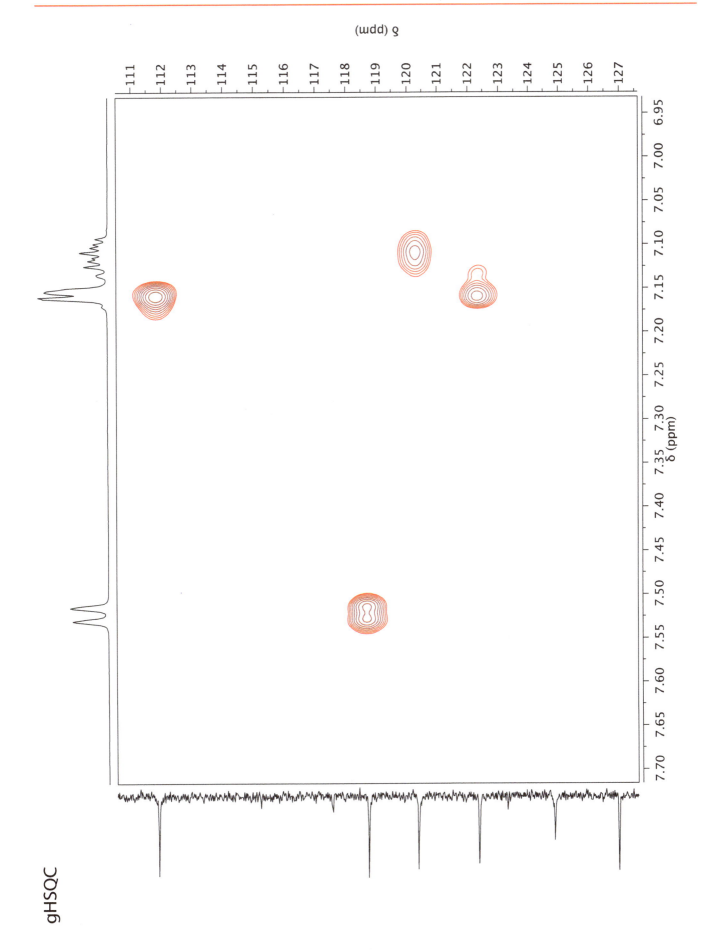

gHSQC

Problem 95

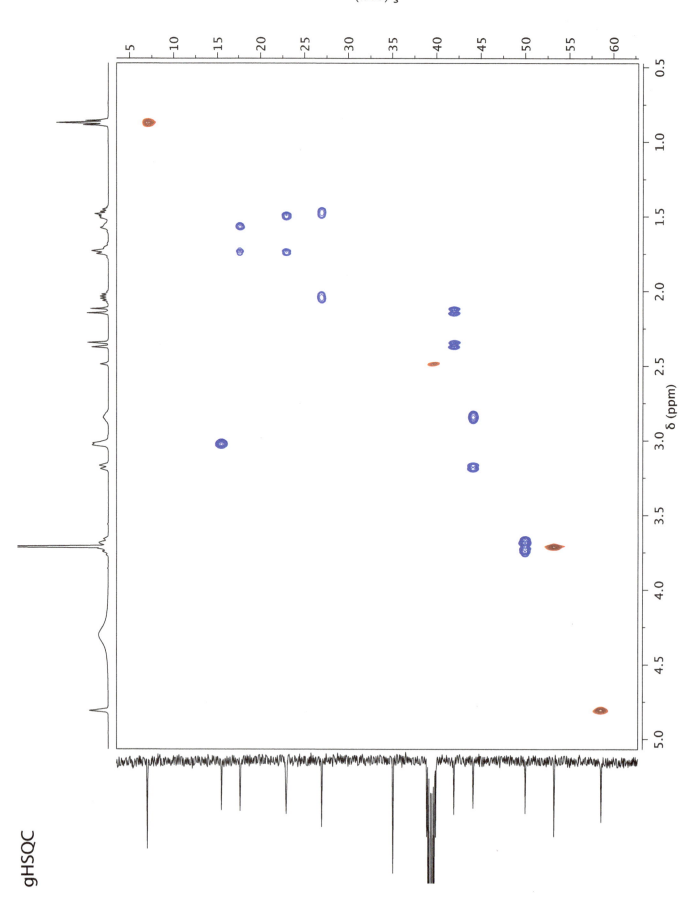

gHSQC

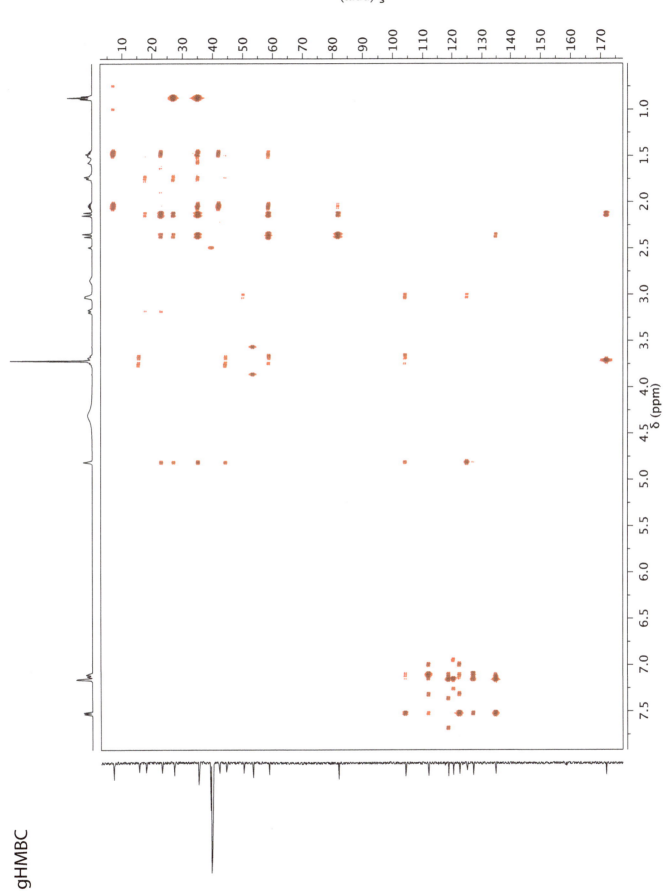

gHMBC

Problem 95

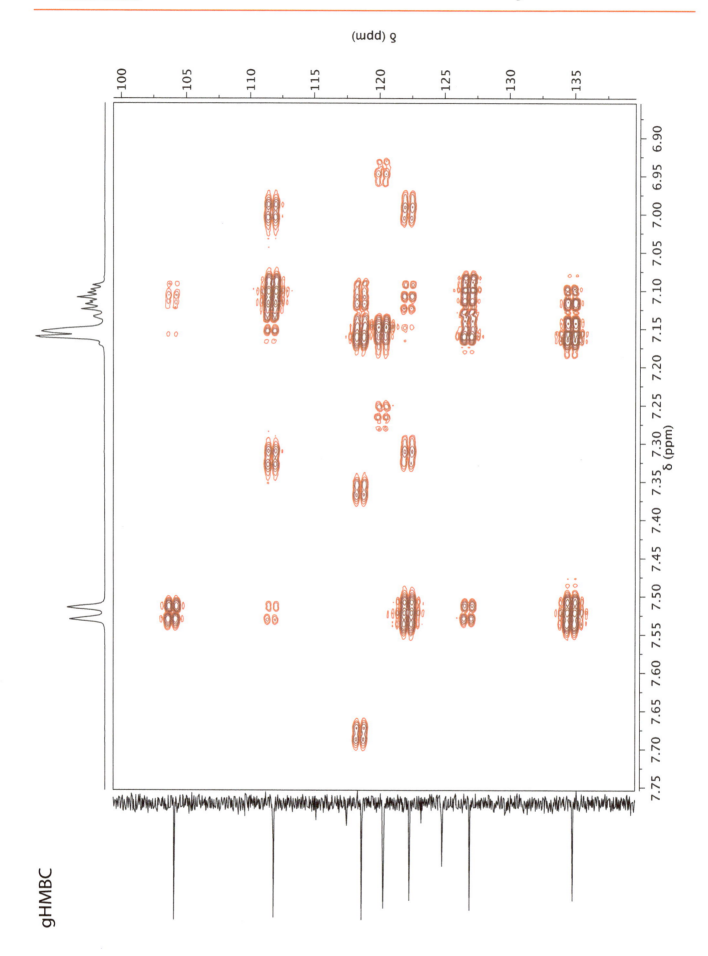

gHMBC

SECTION 4 Problem 95

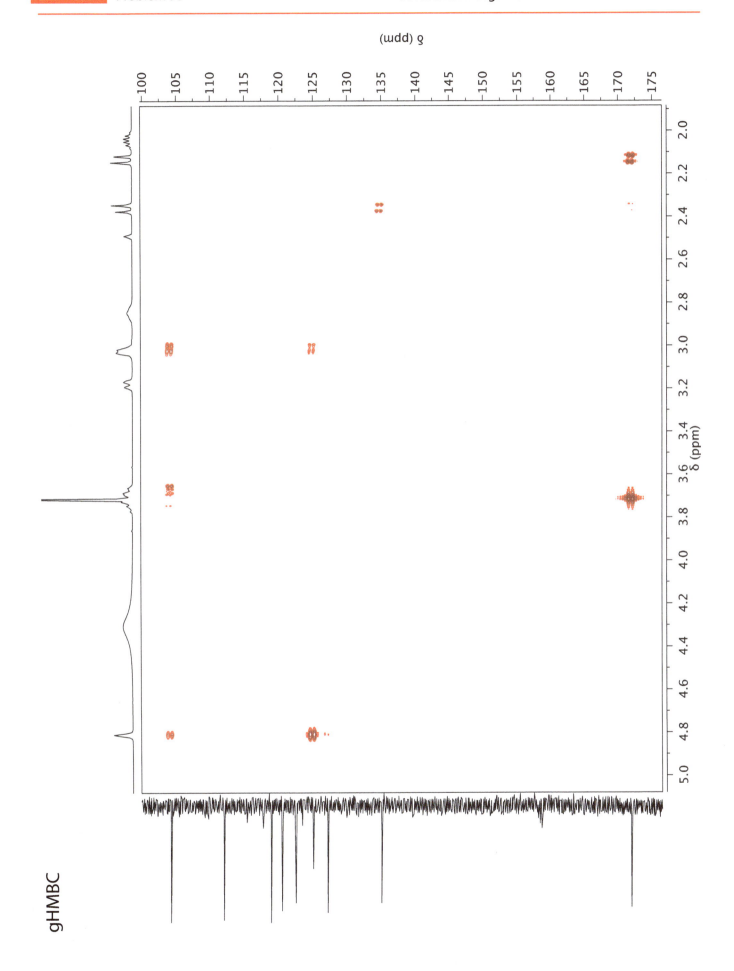

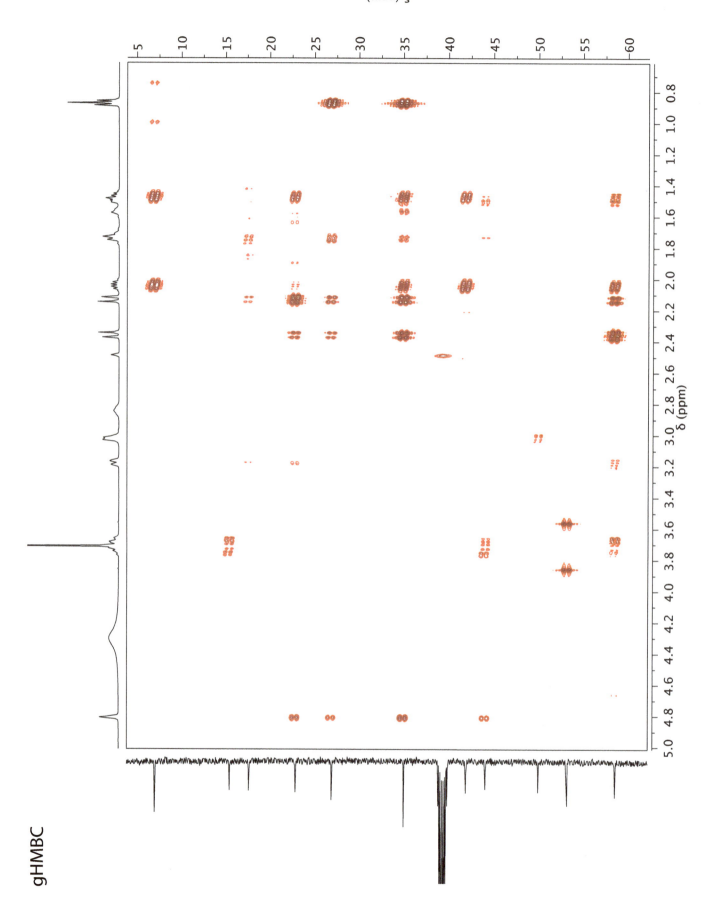

Section 5

Section 5 introduces the use of NMR spectroscopy to investigate relative configuration and conformational analysis in small molecules. The problems are designed to utilize a combination of coupling constant analysis and through-space interactions. Most of the compounds in this section are presented elsewhere in the book as planar structure elucidation problems. The locations of these planar structure determination questions are listed on the first page of each problem in this section. For those who do not wish to start by assigning the planar structures, chemical shift assignments are available for each problem in the hints section.

LEARNING OBJECTIVES

- Use coupling constant analysis to determine double bond geometry.
- Use coupling constants to determine relative configuration in simple and bicyclic ring systems.
- Use 1D or 2D NOE correlations to assign relative configuration.

EXPERIMENTS INCLUDED

^{1}H, 1D TOCSY, 1D selective NOESY, ^{1}H selective decoupling, gCOSY, gHSQC and gROESY.

Note: All ROESY experiments are multiplicity edited with cross peaks deriving from *bona fide* ROESY correlations depicted in blue, and cross peaks deriving from the diagonal and TOCSY breakthrough depicted in red.

TYPES OF MOLECULES

This section contains a variety of molecules that contain olefins, polycyclic ring systems and multiple chiral centers. In most cases these compounds also contain small-to-medium ring systems that are suitable for relative configurational analysis using NMR techniques.

STRATEGIES FOR SUCCESS

Success in this section requires a very different approach than the methods used for planar structural assignment. Users must carefully consider both the through-space correlations and the coupling constants and dihedral angles between atoms in order to deduce the conformation and configuration for each problem. It is important to remember here that the absence of through-space correlations should not be used as evidence that atoms are distantly related in space. Instead, users should rely only on observed through-space correlations as positive evidence that two atoms are in close physical proximity.

LEGEND

Spectrum annotations:

i = impurity.

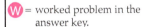 = worked problem in the answer key.

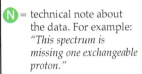 = technical note about the data. For example: *"This spectrum is missing one exchangeable proton."*

H = hint to assist in solving the problem. For example: *"This molecule would have an IR stretch at 2240 cm^{-1}."*

SECTION 5 Problem 96

Using the ¹H, gCOSY and 1D selective NOESY spectra, identify the regioisomer of 2-hexene that provides the following spectra. For additional questions of this type, see problems 12, 13, 18, 28, 33.
Spectra acquired in CDCl₃ at 500 MHz.

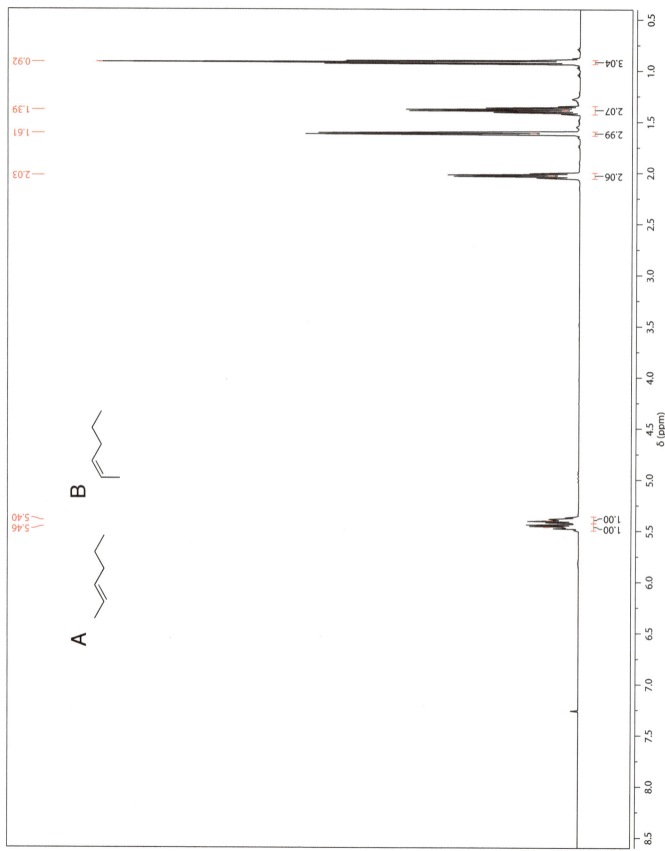

SECTION 5 Problem 96

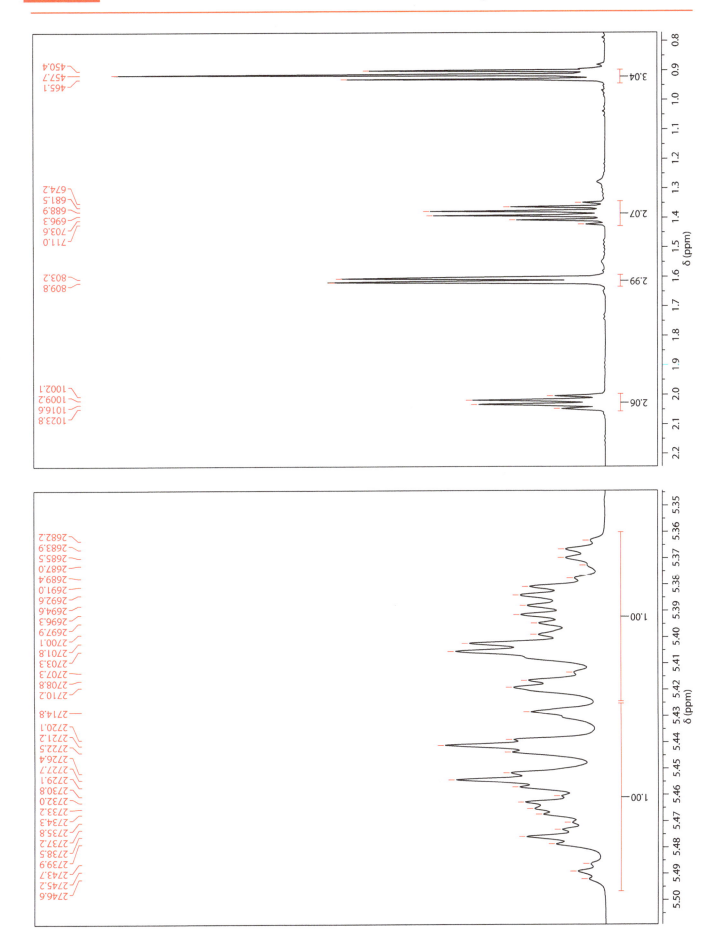

SECTION 5 Problem 96

Problems in Organic Structure Determination

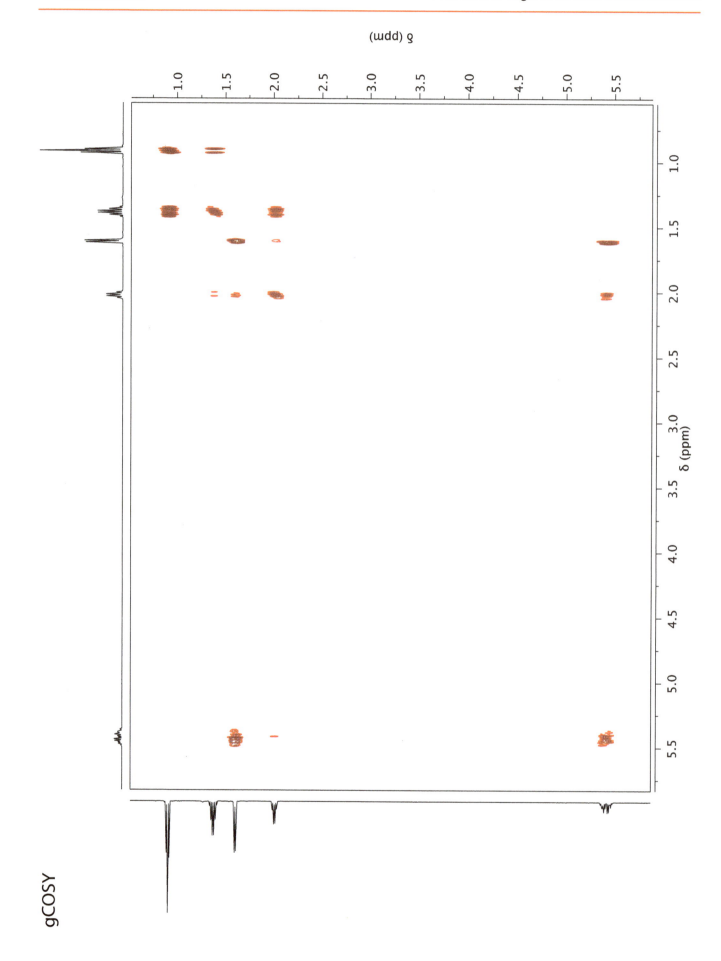

gCOSY

SECTION 5 Problem 96

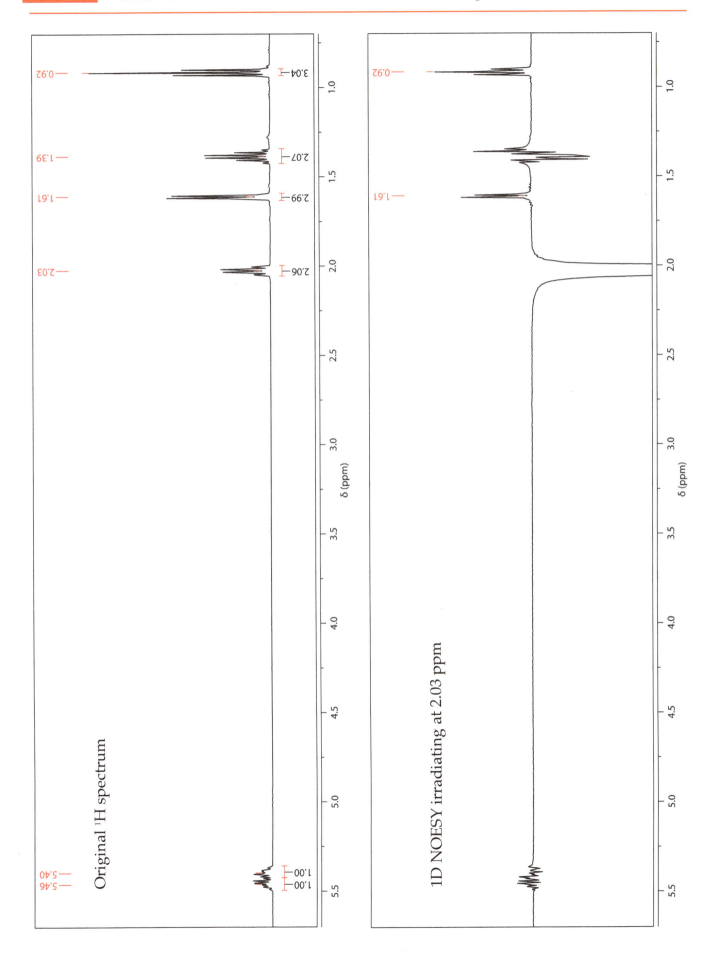

Using the ¹H, 1D TOCSY and 1D selective NOESY spectra, determine the relative configuration and conformation of the molecule below. For the assignment of proton and carbon resonances see problem 42.
Spectra acquired in CDCl₃ at 500 MHz.

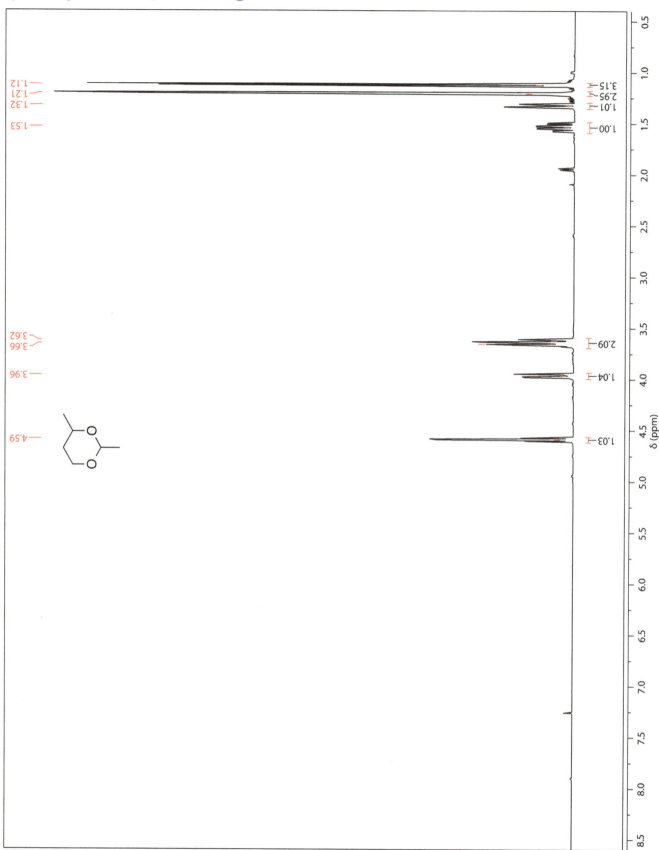

SECTION 5 Problem 97

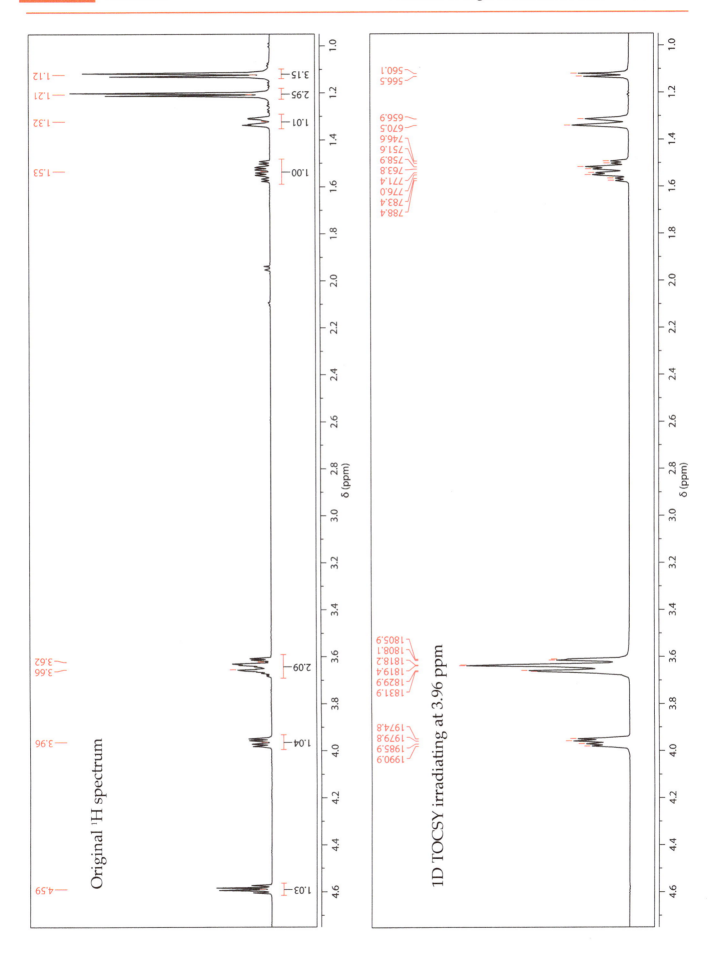

SECTION 5 Problem 97

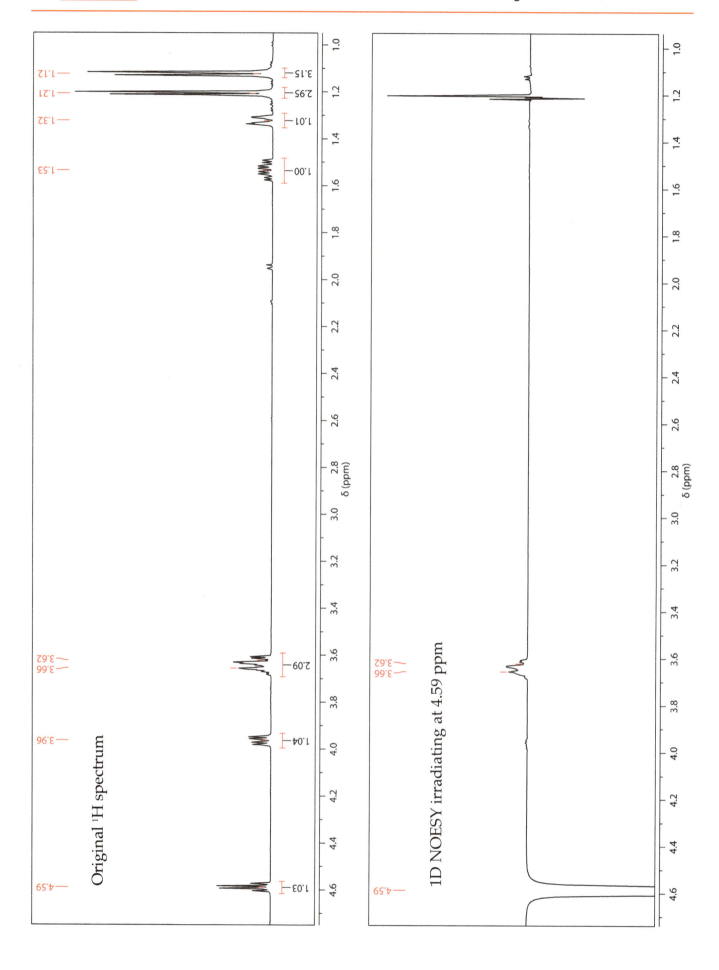

SECTION 5 Problem 98

Using the ¹H and gROESY spectra, determine the relative configuration and conformation of the molecule below. For the assignment of proton and carbon resonances see problem 31.
Spectra acquired in CD₃OD at 500 MHz.

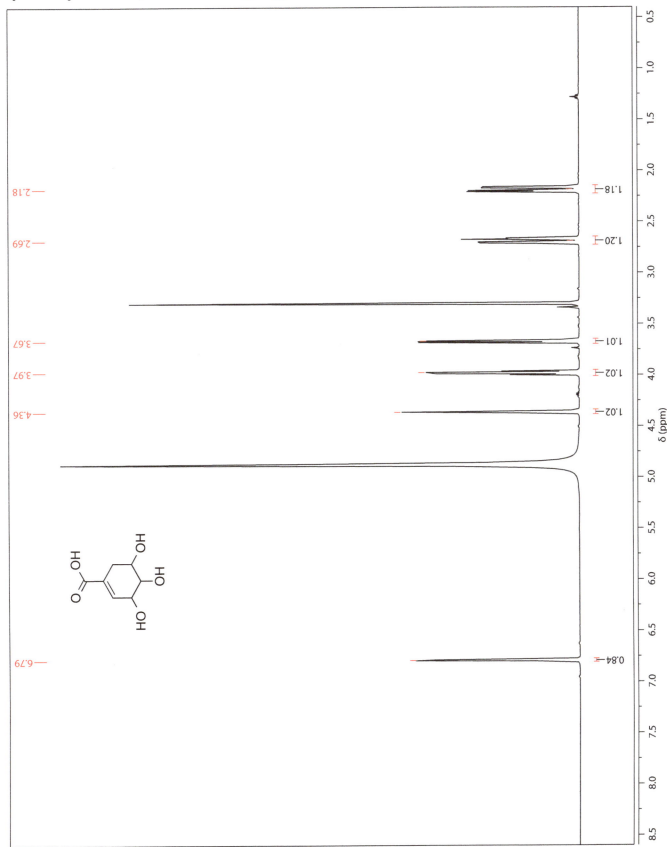

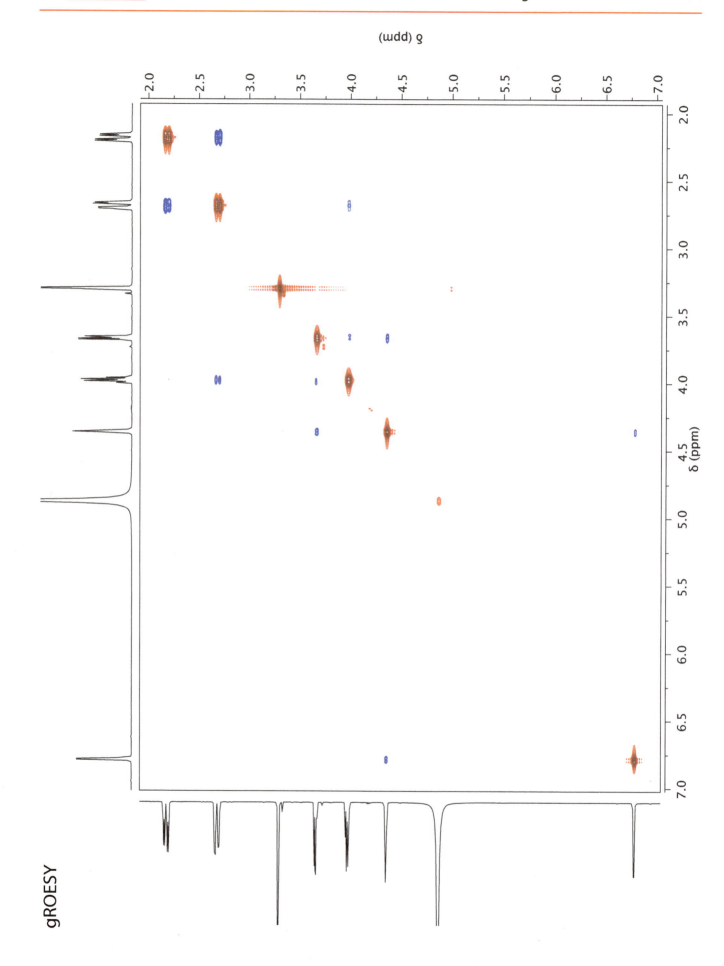

gROESY

SECTION 5 Problem 99

Using the ¹H, gCOSY and gHSQC spectra, identify the conformer of 4-piperidinemethanol that provides the following spectra. Assign all proton resonances and coupling constants.
Spectra acquired in CDCl₃ at 500 MHz.

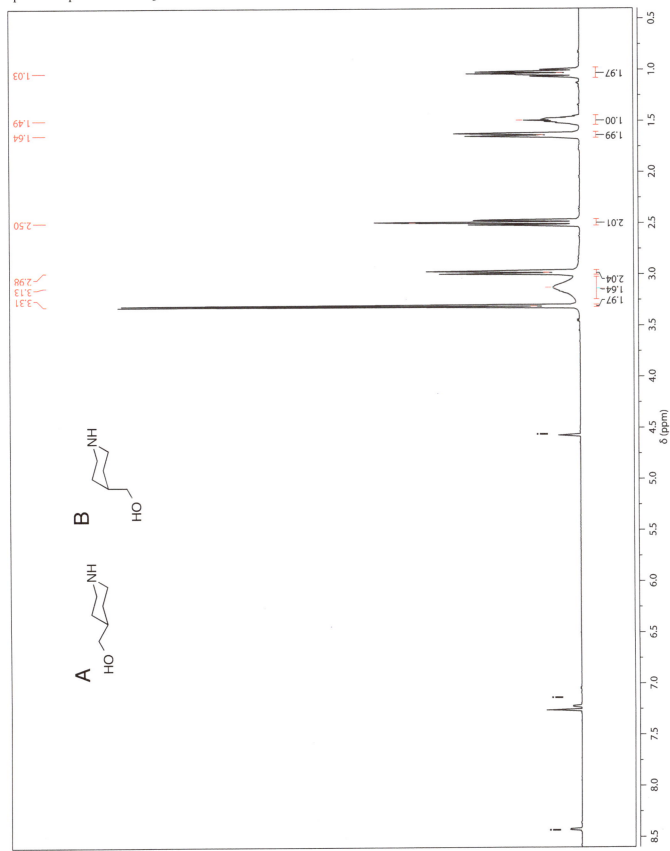

Problem 99

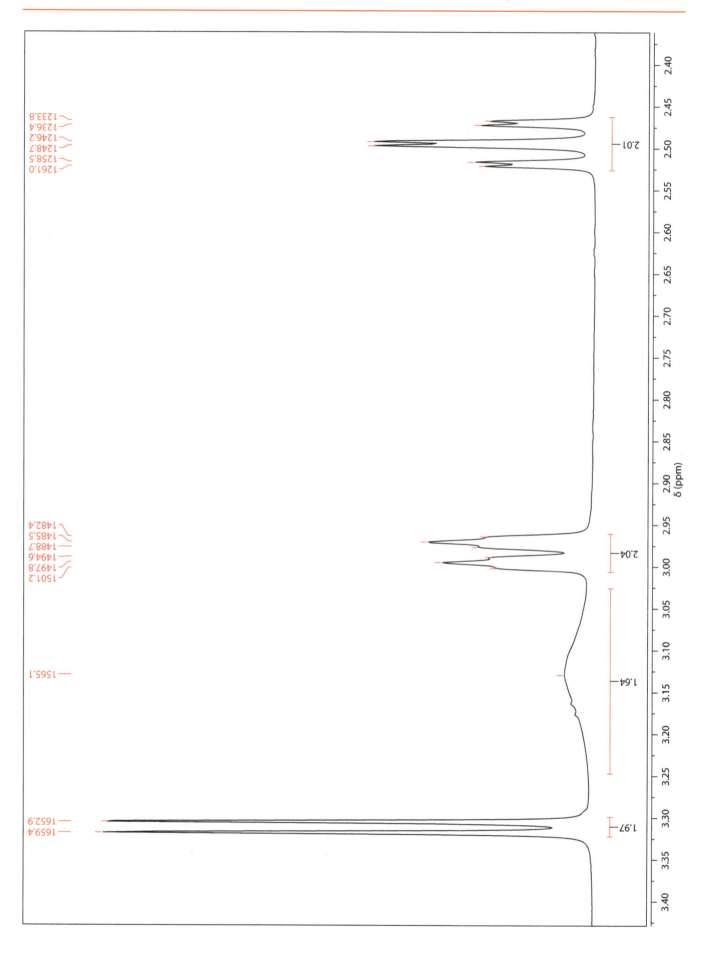

SECTION 5 Problem 99 — Problems in Organic Structure Determination

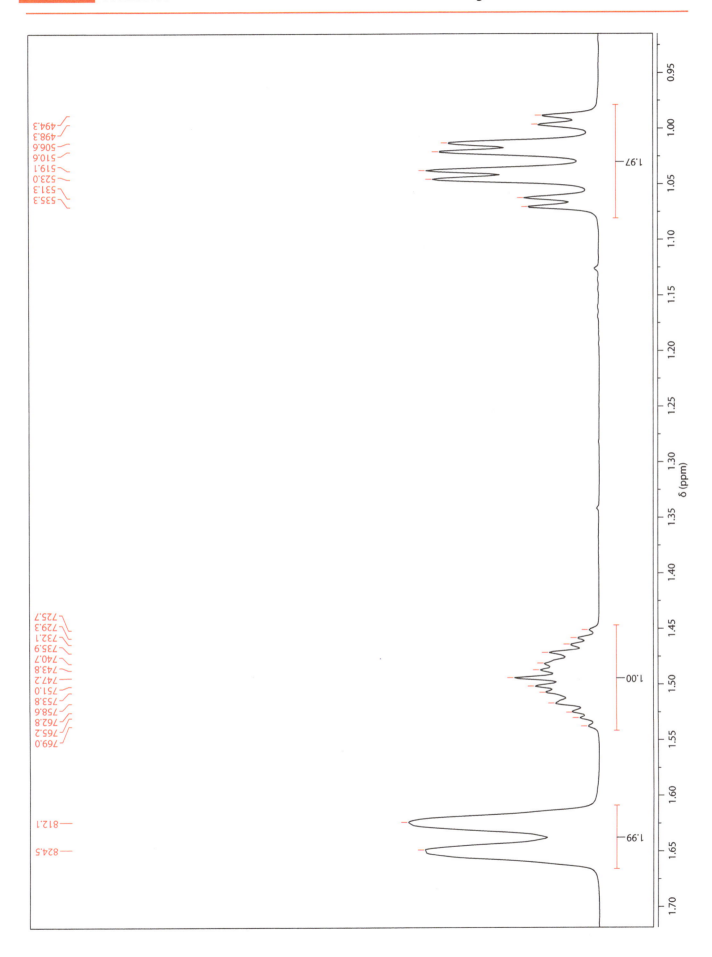

Problem 99

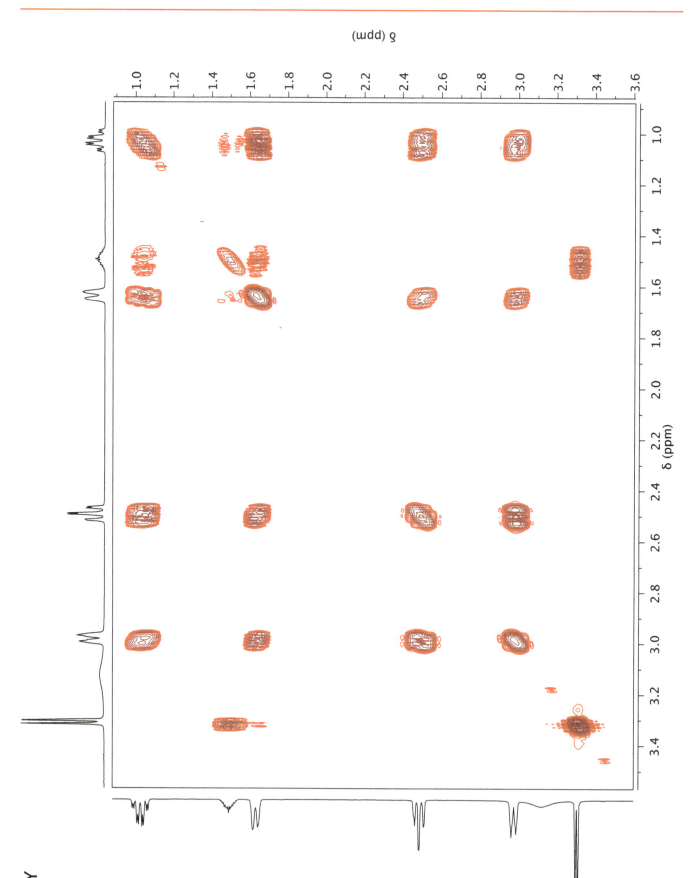

gCOSY

SECTION 5 Problem 99

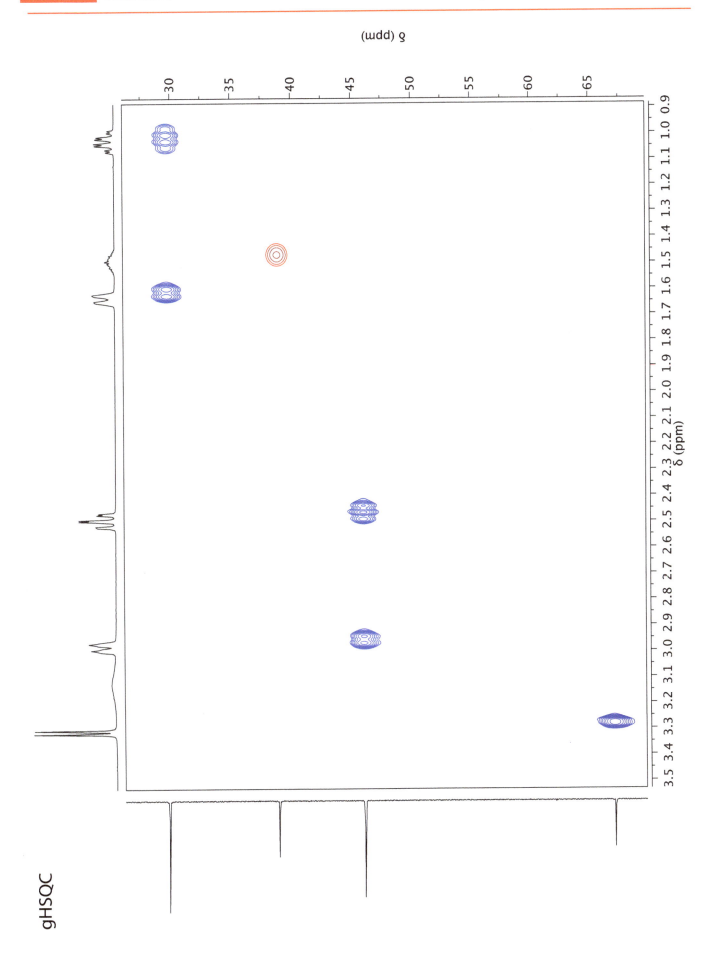

gHSQC

Problem 100

Using the ¹H and selective 1D NOESY spectra, determine the relative configuration and conformation of the molecule below. For the assignment of proton and carbon resonances see problem 111.
Spectra acquired in CDCl₃ at 500 MHz.

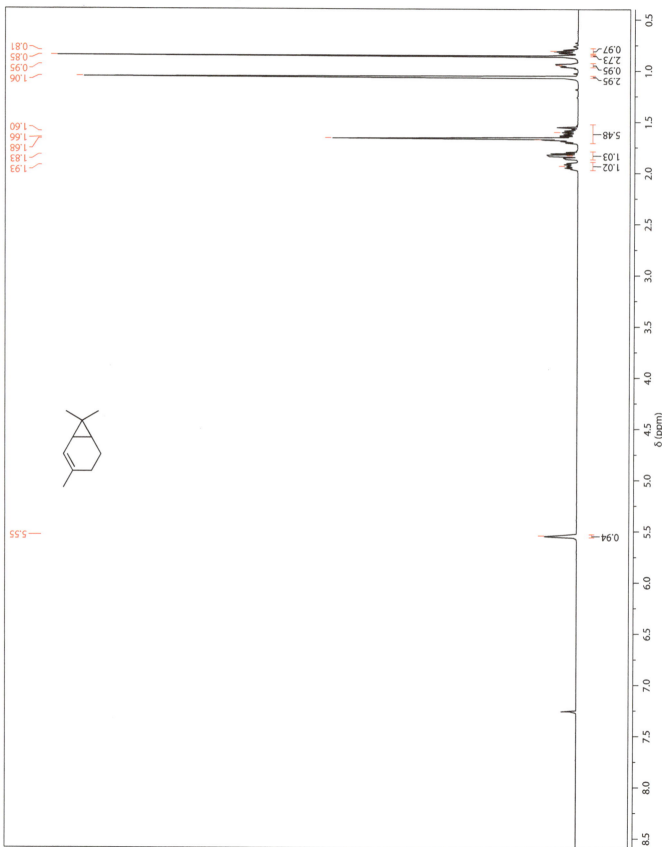

SECTION 5 Problem 100

Problems in Organic Structure Determination

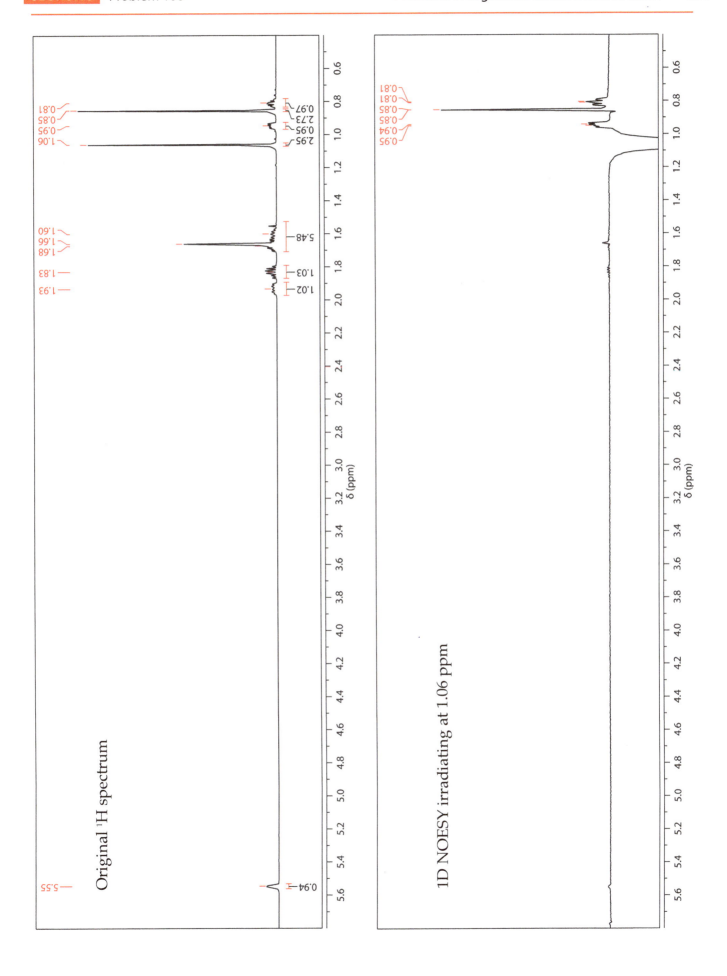

SECTION 5 Problem 100

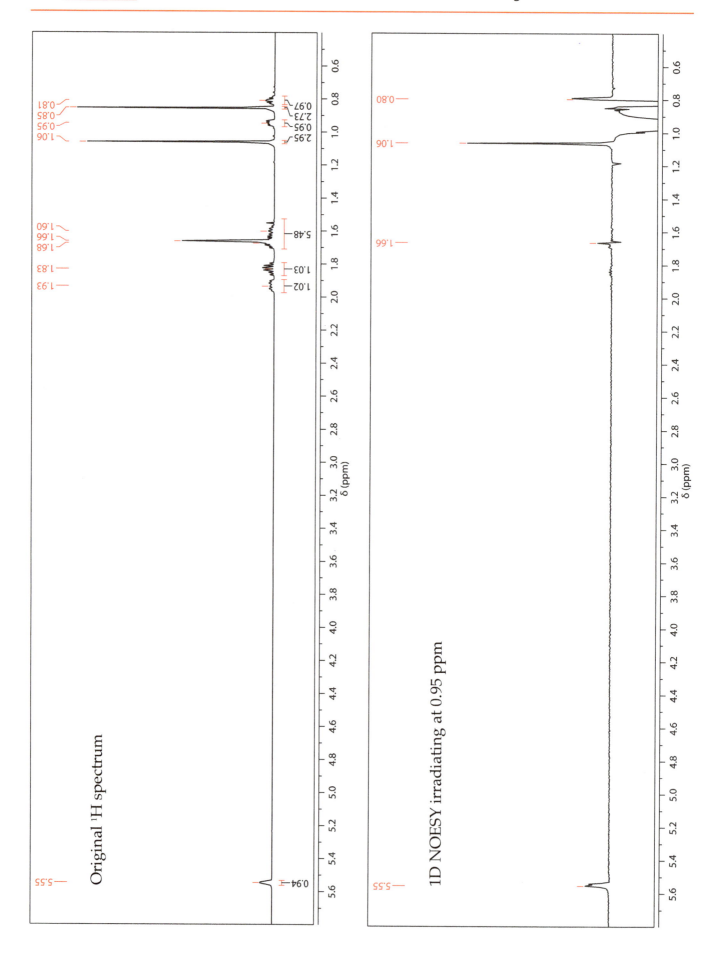

SECTION 5 Problem 100

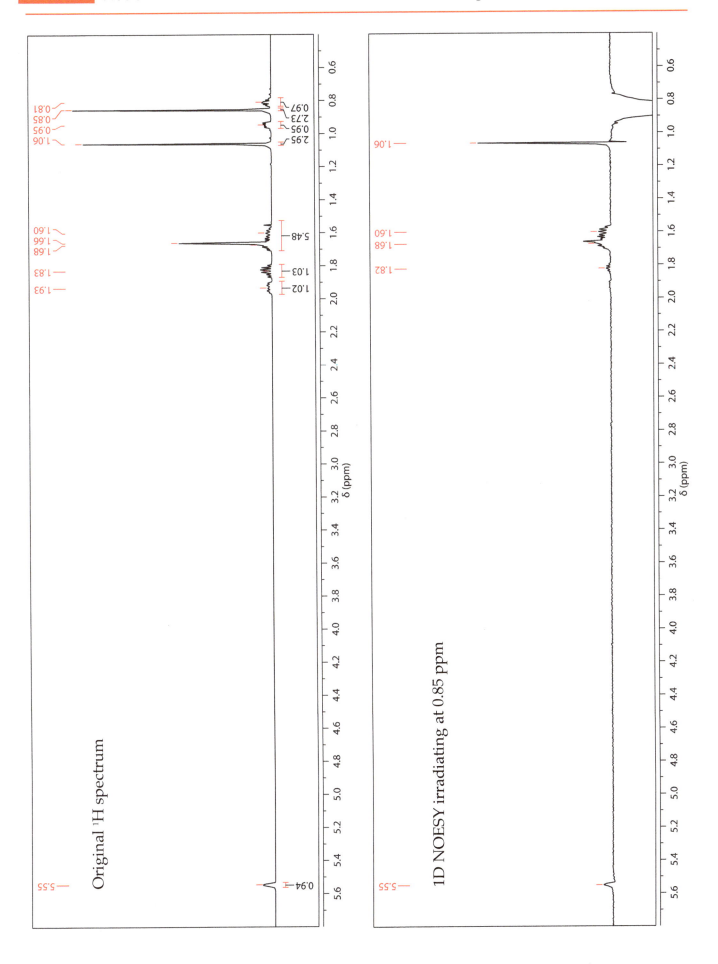

SECTION 5 Problem 100

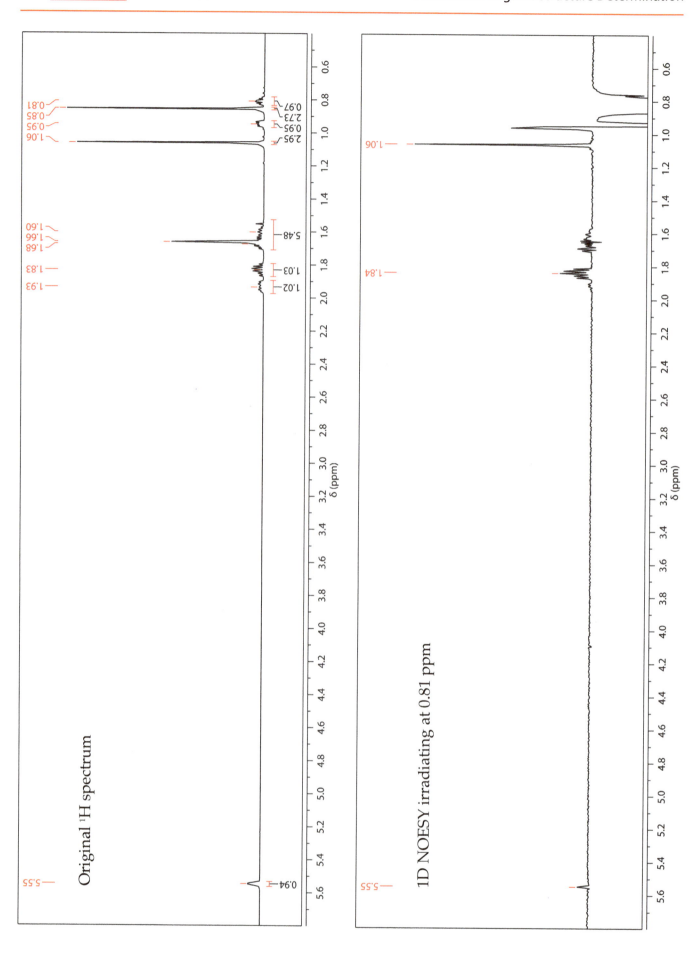

SECTION 5 Problem 101 Problems in Organic Structure Determination

Using the ¹H, gROESY and selective ¹H decoupling spectra, determine the relative configuration and conformation of 4-(aminomethyl)cyclohexanecarboxylic acid below.
Spectra acquired in D₂O at 500 MHz.

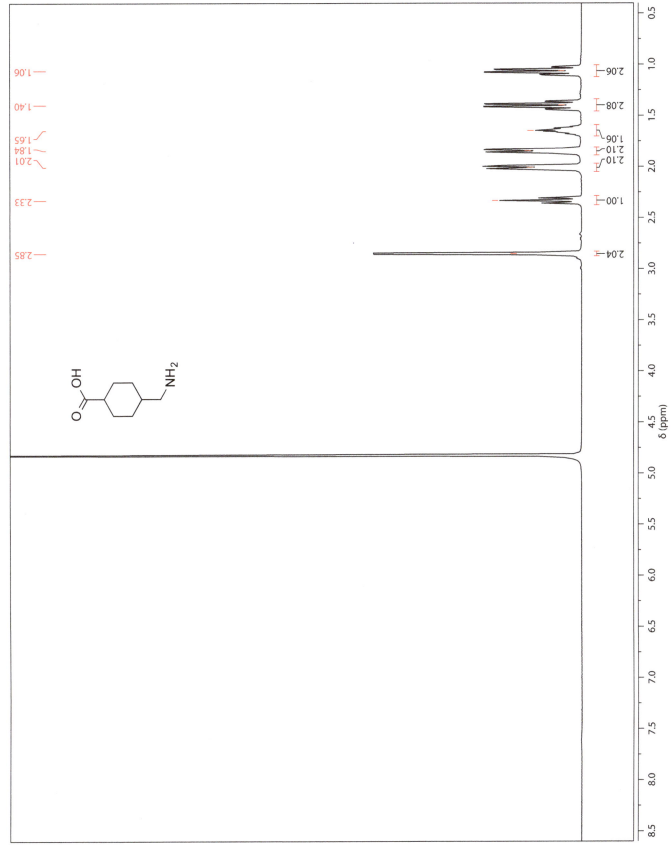

Problem 101

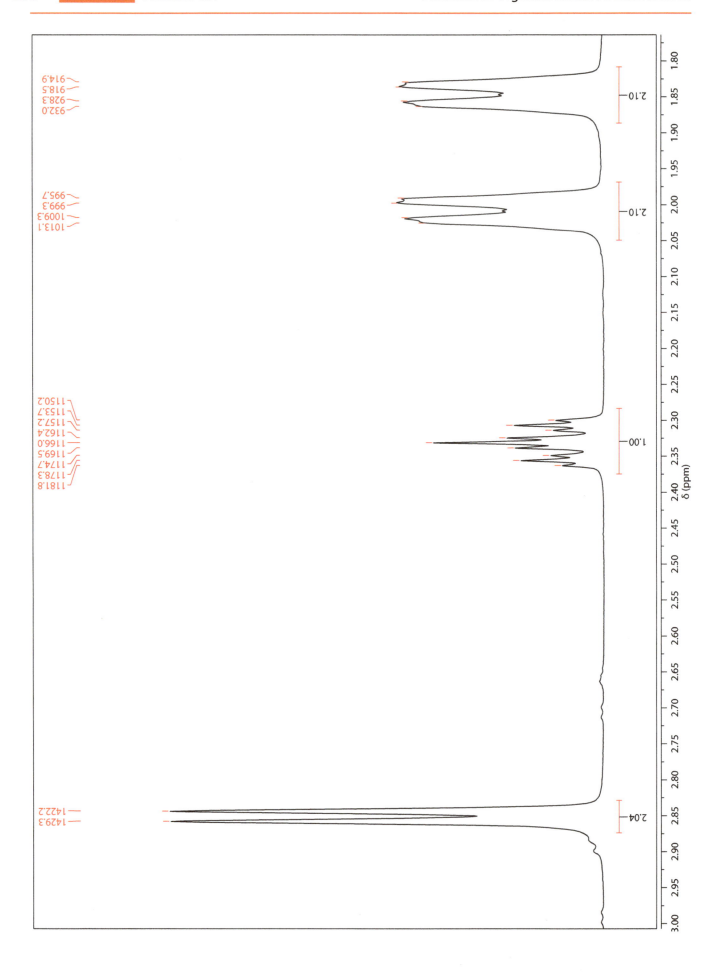

SECTION 5 Problem 101

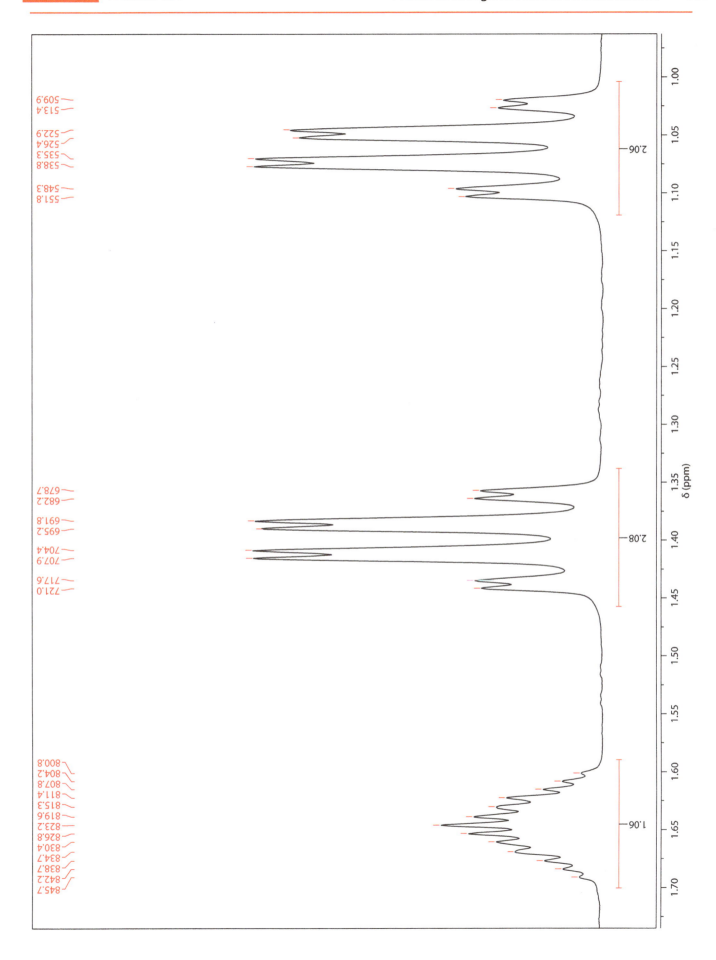

SECTION 5   Problem 101

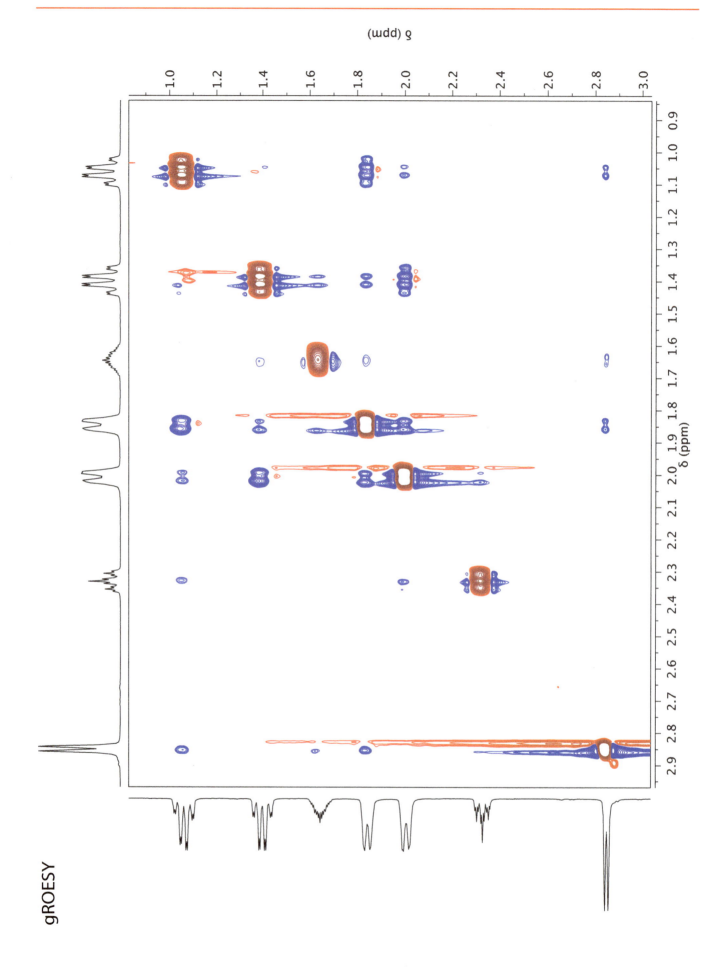

gROESY

Problem 101

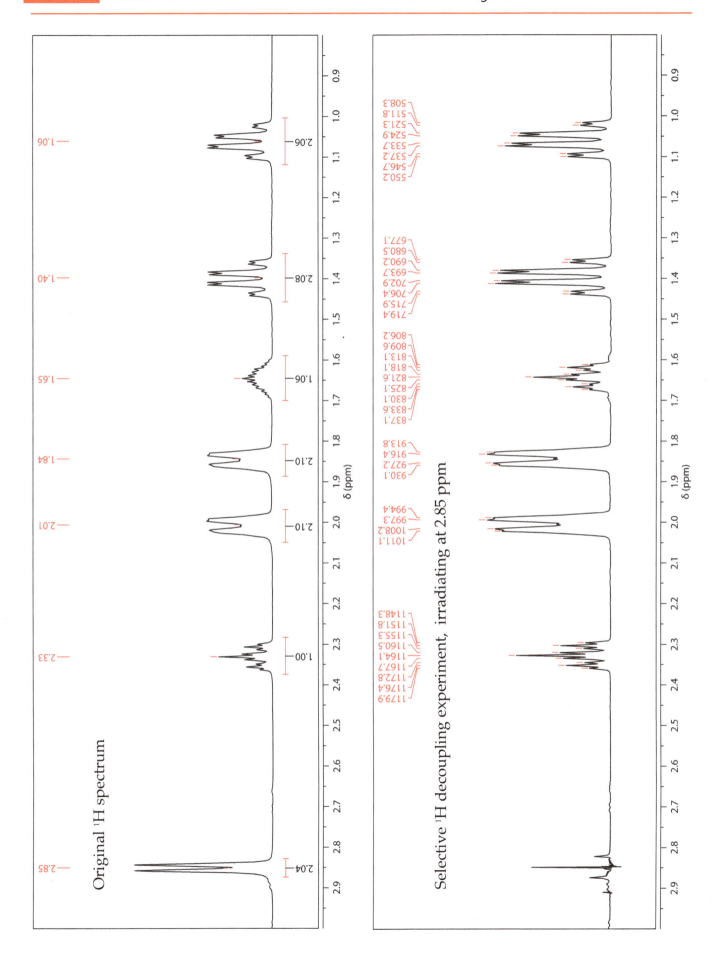

Problem 101

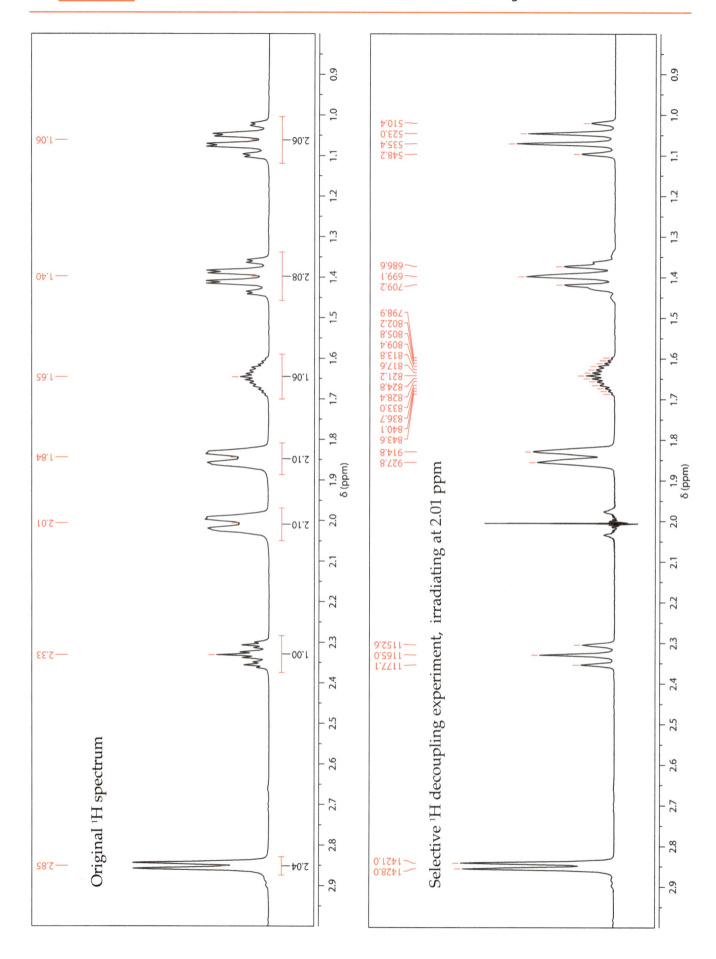

SECTION 5 Problem 101

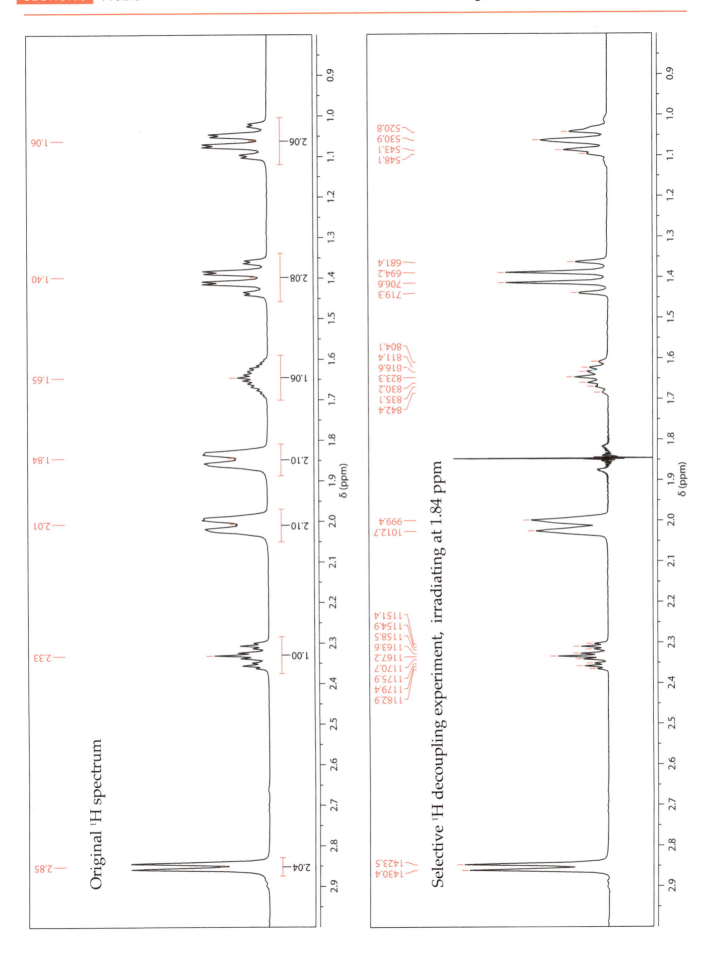

SECTION 5 Problem 102

Quinic acid typically exists as an equilibrium mixture of two conformers in solution. Determine the predominant conformer in the spectra below. For the assignment of proton and carbon resonances see problem 38. Spectra acquired in (CD$_3$)$_2$SO at 500 MHz.

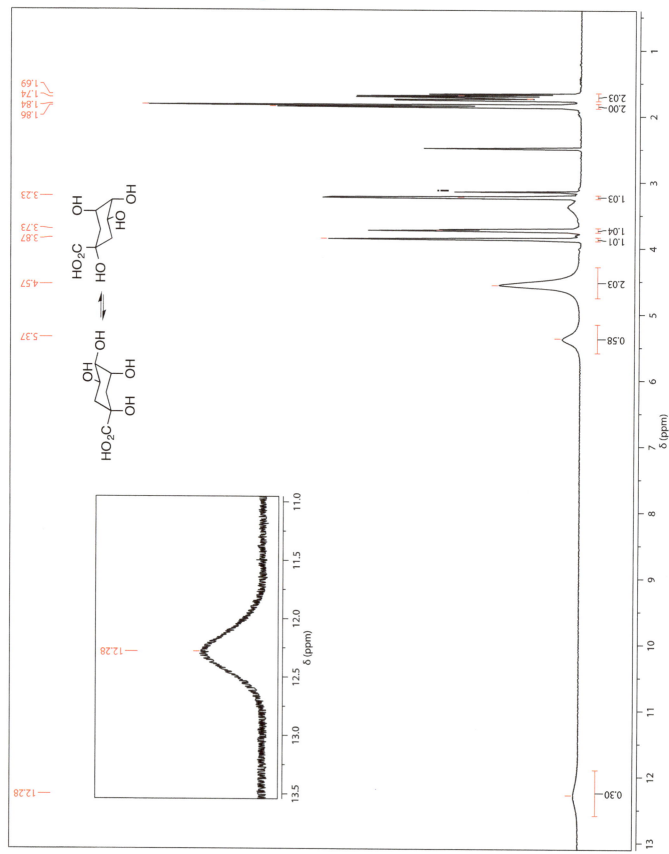

SECTION 5 Problem 102

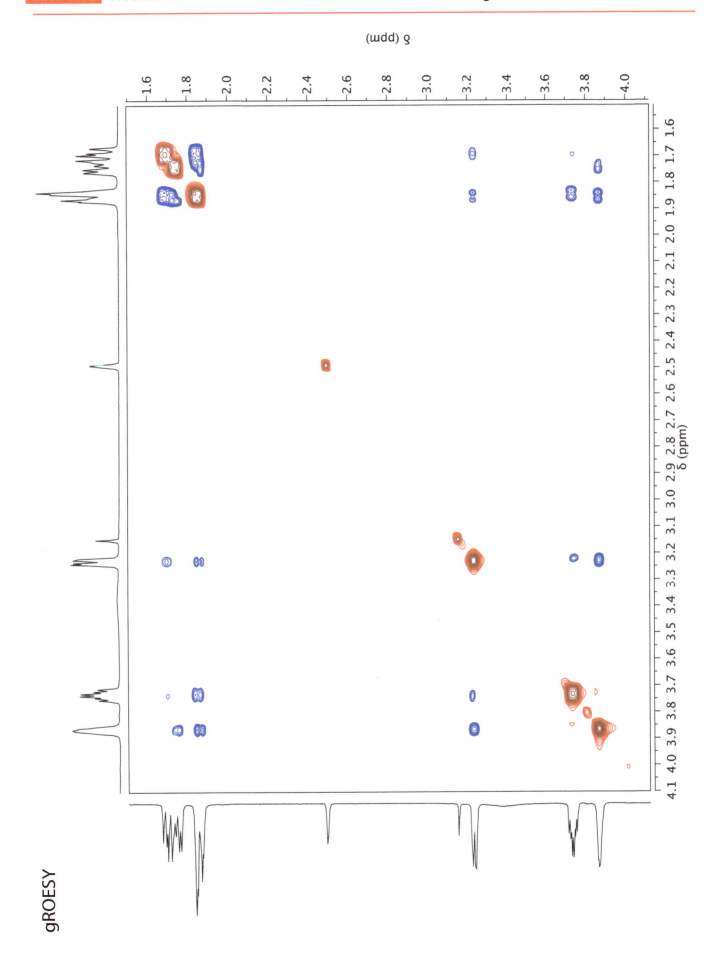

gROESY

SECTION 5 Problem 103

Using the ¹H and selective 1D NOESY spectra, determine the relative configuration and conformation of 4-methyl-5-phenyl-2-oxazolidinone below. For the assignment of proton and carbon resonances see problem 75. Spectra acquired in CDCl₃ at 500 MHz.

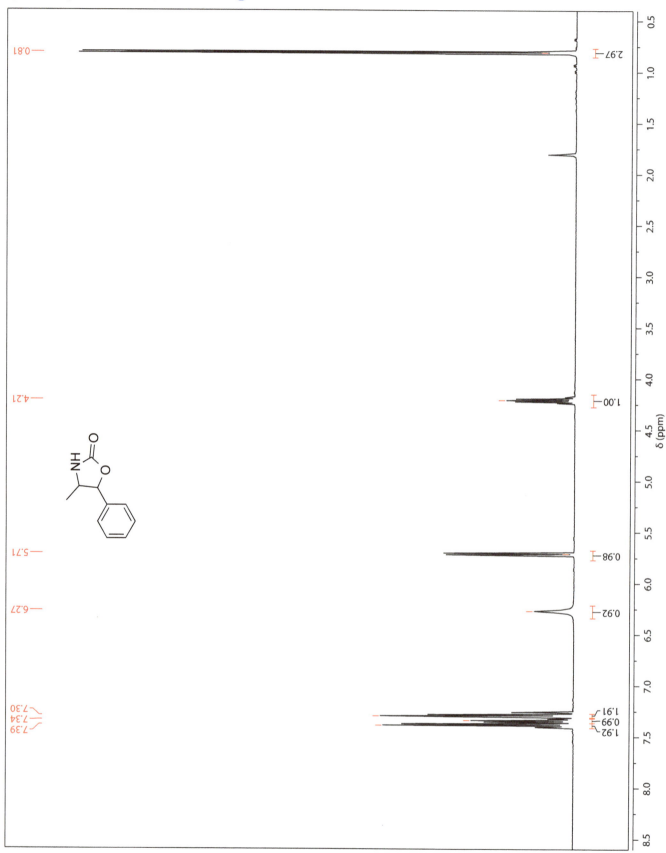

Problem 103

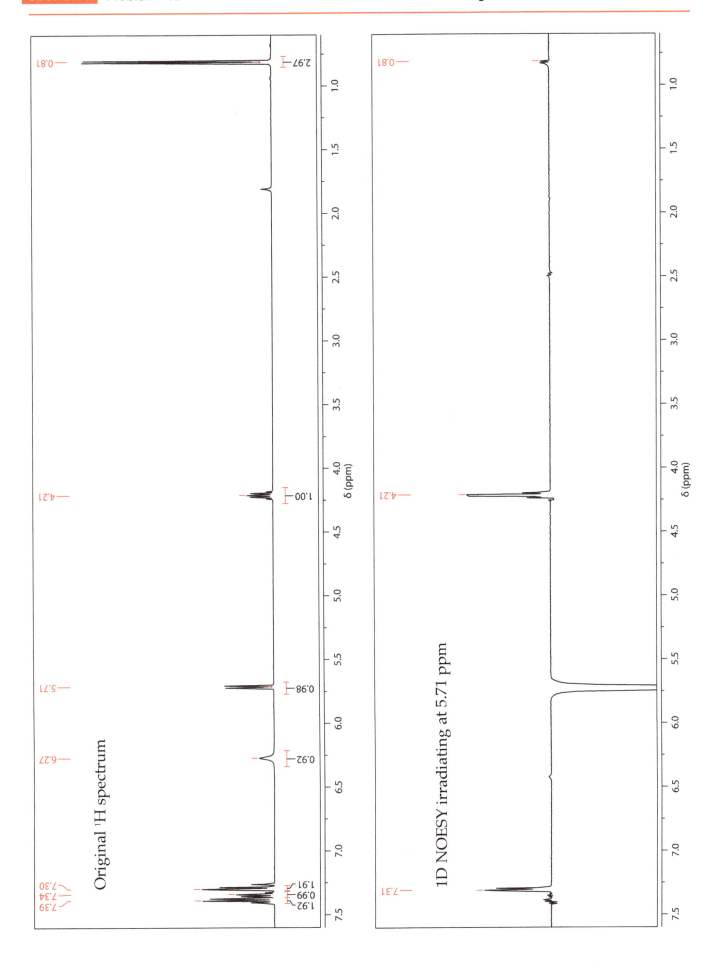

Problem 103

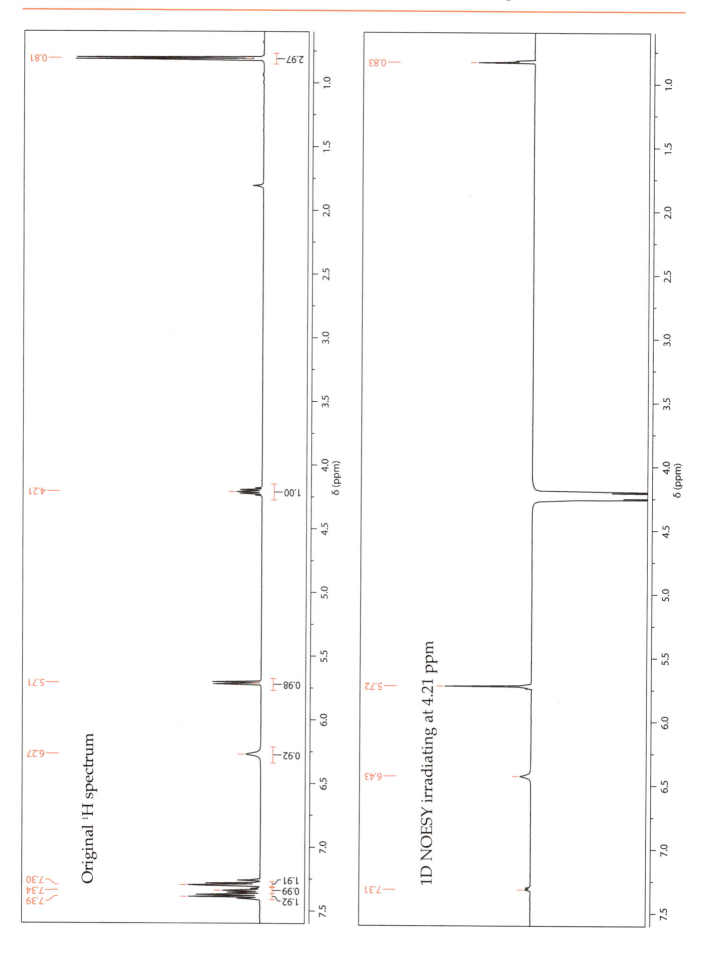

SECTION 5 Problem 103 Problems in Organic Structure Determination

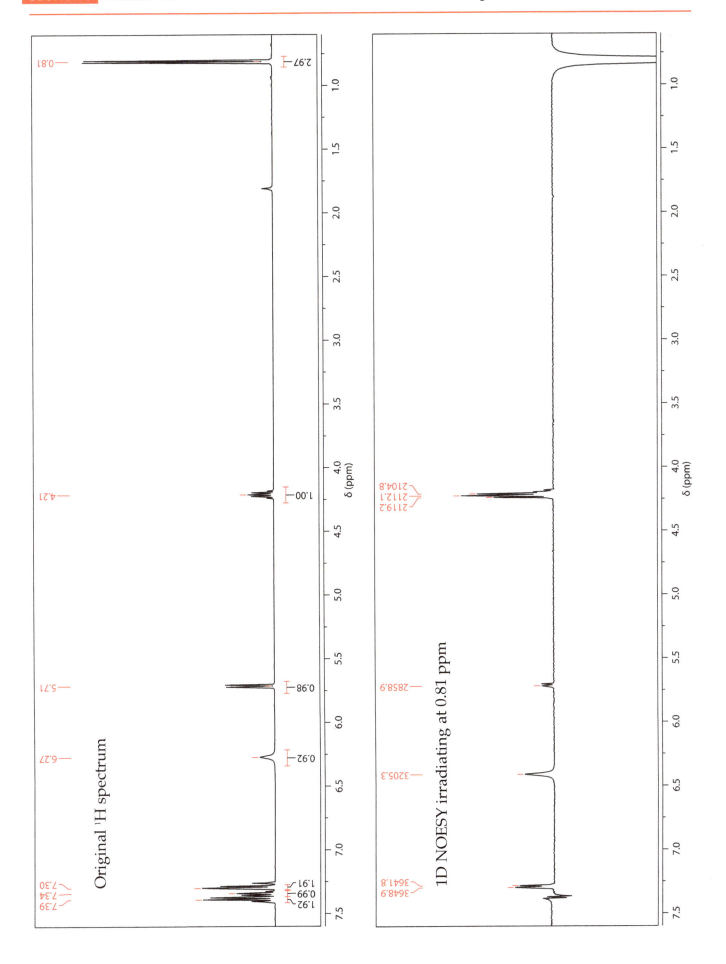

Using the ¹H, gROESY and selective 1D NOESY spectra, determine the relative configuration of the molecule below. For the assignment of proton and carbon resonances see problem 115.
Spectra acquired in CDCl₃ at 500 MHz.

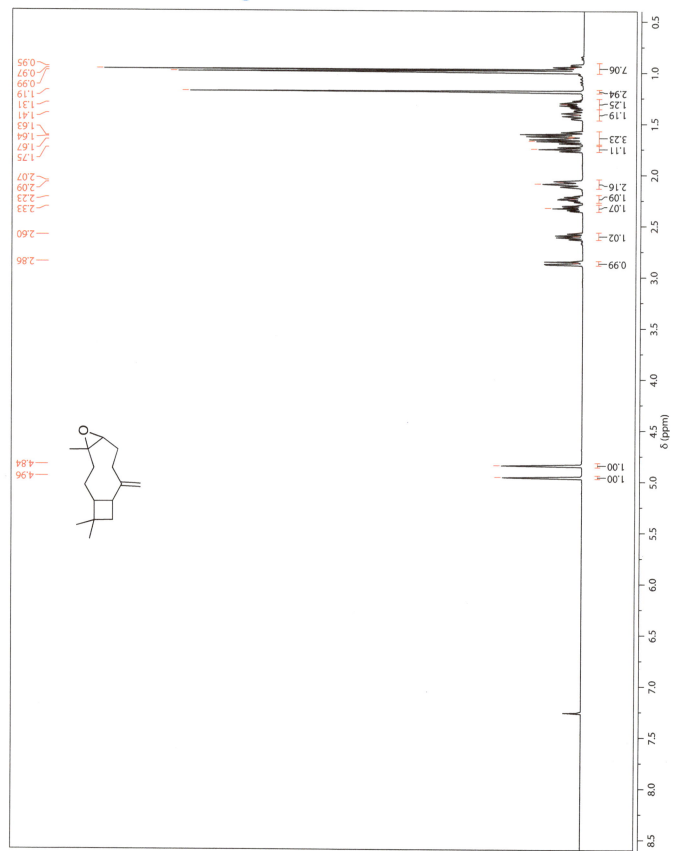

SECTION 5 Problem 104

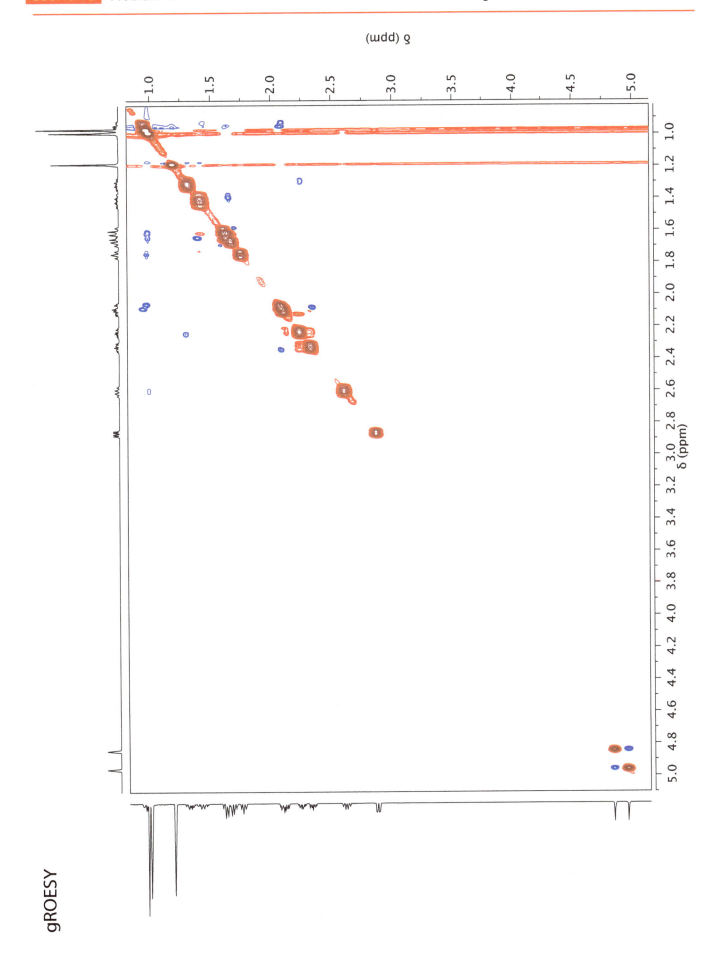

gROESY

Problem 104

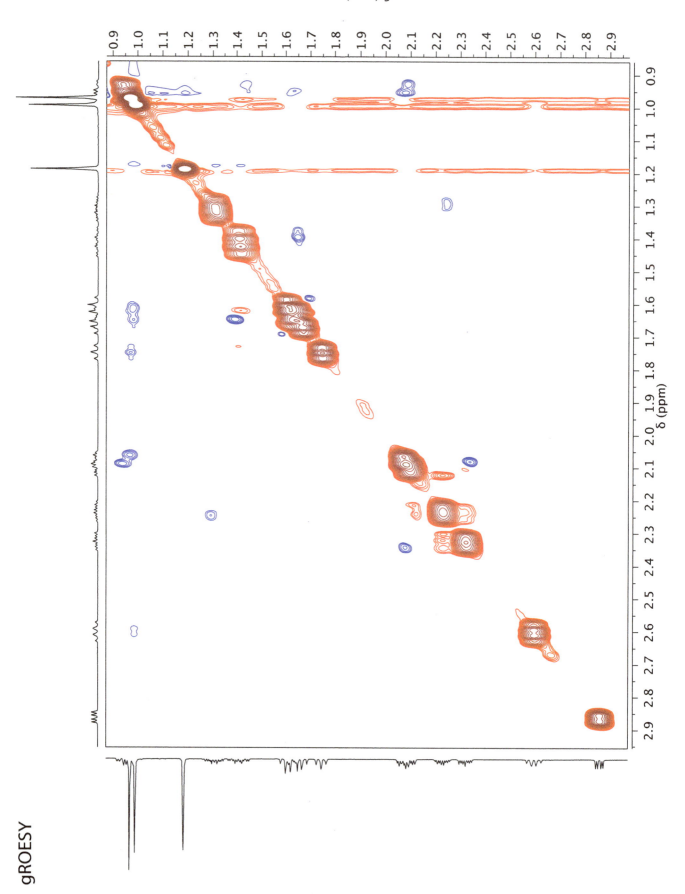

gROESY

SECTION 5 Problem 104

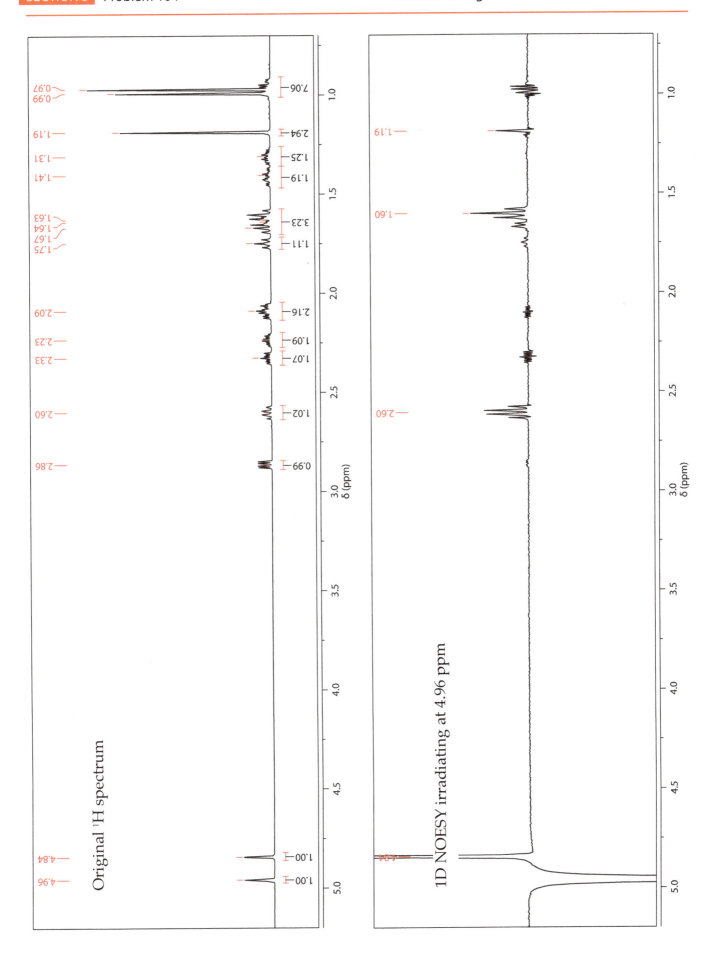

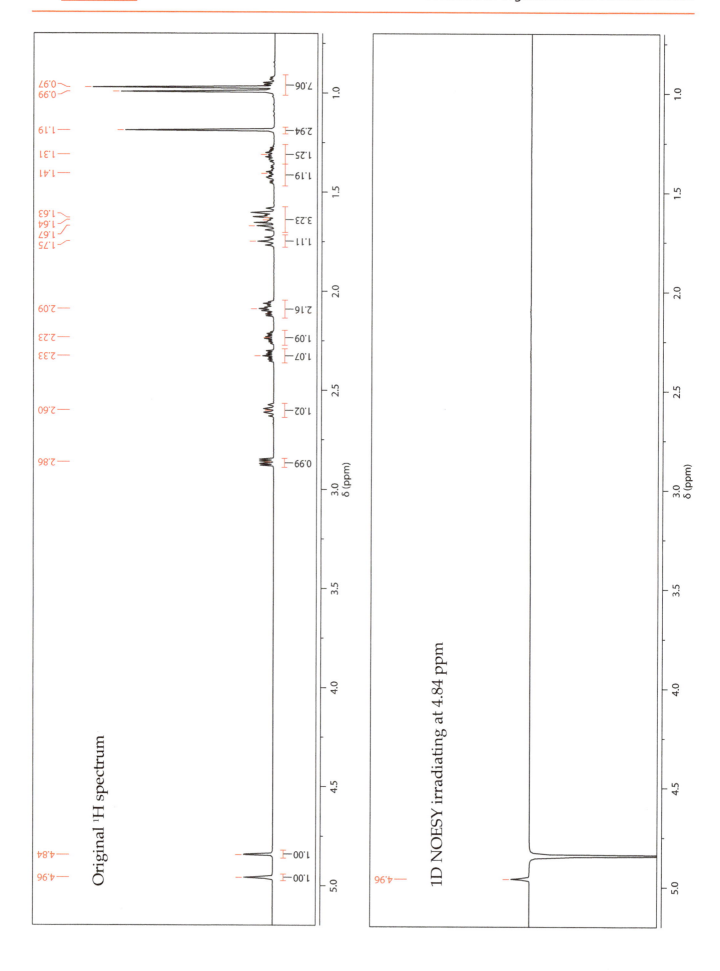

SECTION 5 Problem 104
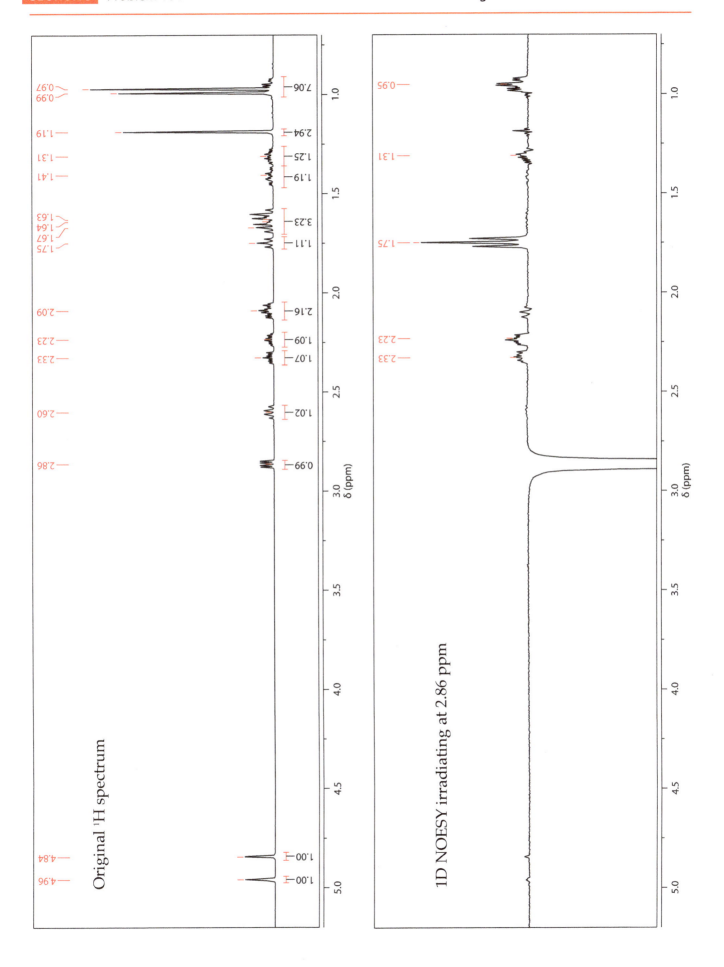

SECTION 5 Problem 104

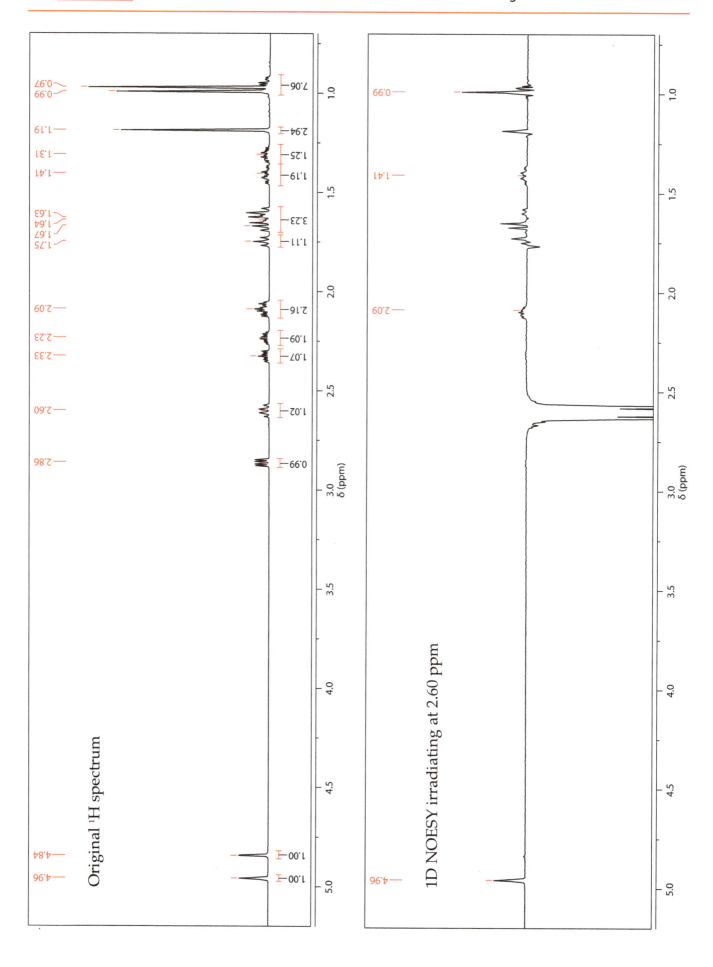

SECTION 5 Problem 104

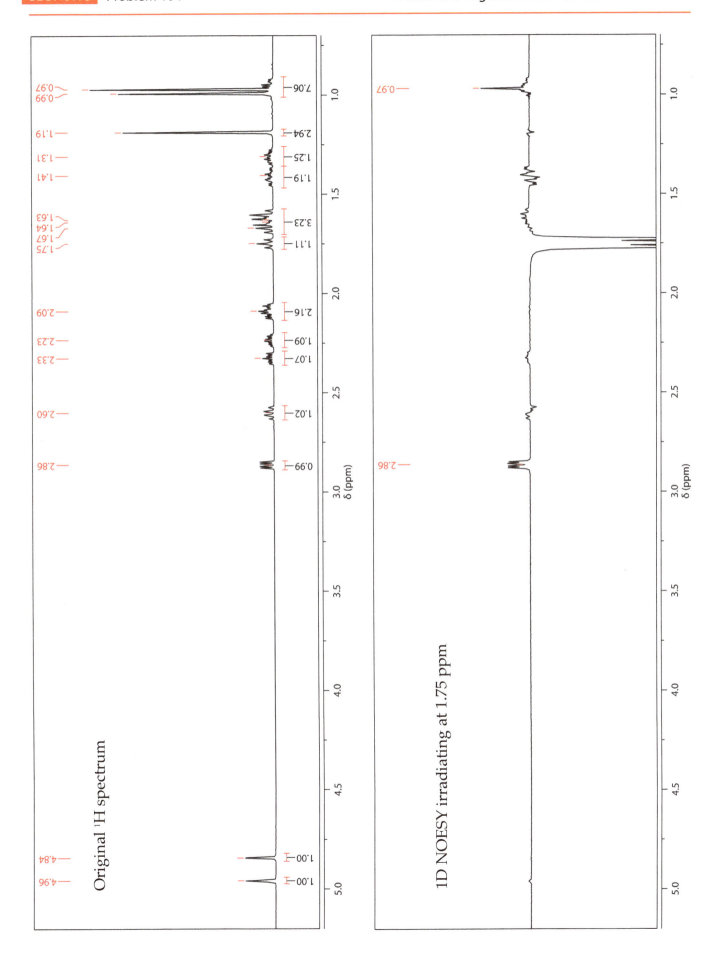

Problem 104

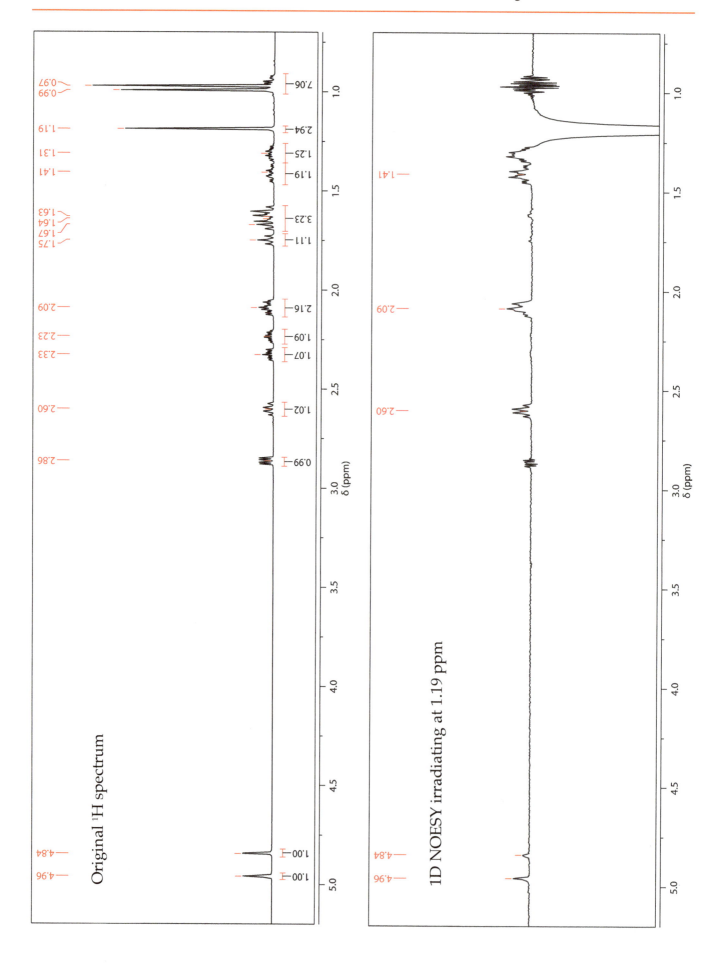

SECTION 5 Problem 104

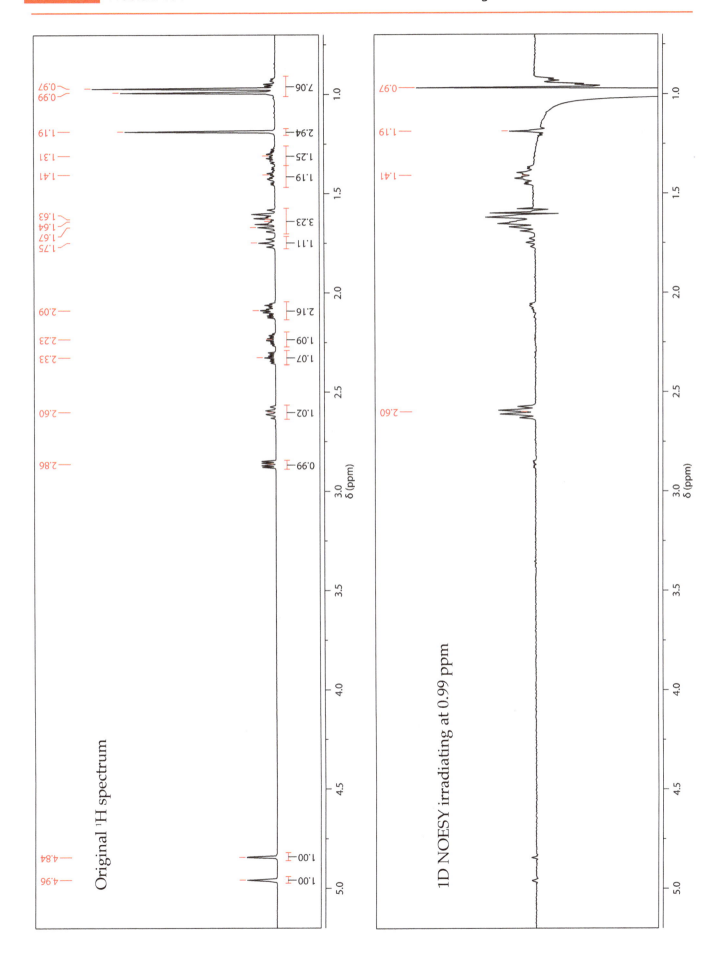

Problem 104

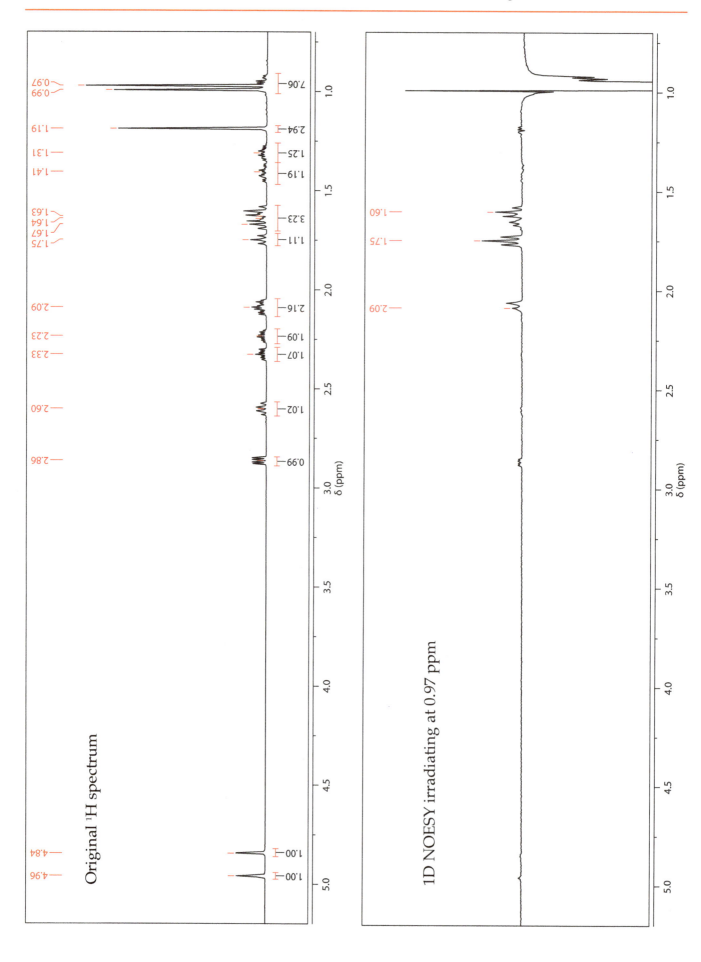

SECTION 5   Problem 105

Using the ¹H, gROESY and selective 1D NOESY spectra, determine the relative configuration and conformation of the molecule below. For the assignment of proton and carbon resonances see problem 113.
Spectra acquired in CDCl₃ at 500 MHz.

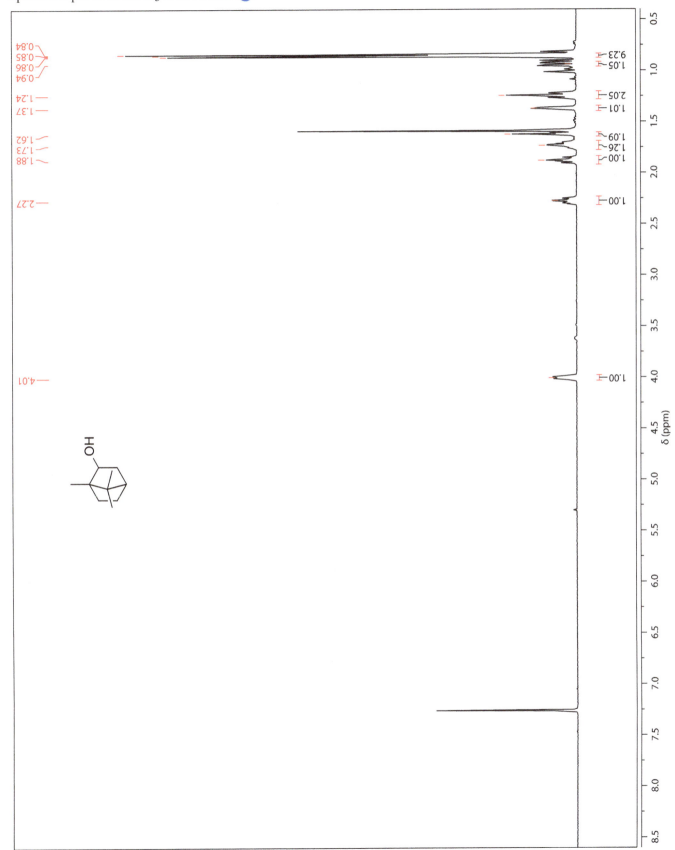

SECTION 5 Problem 105

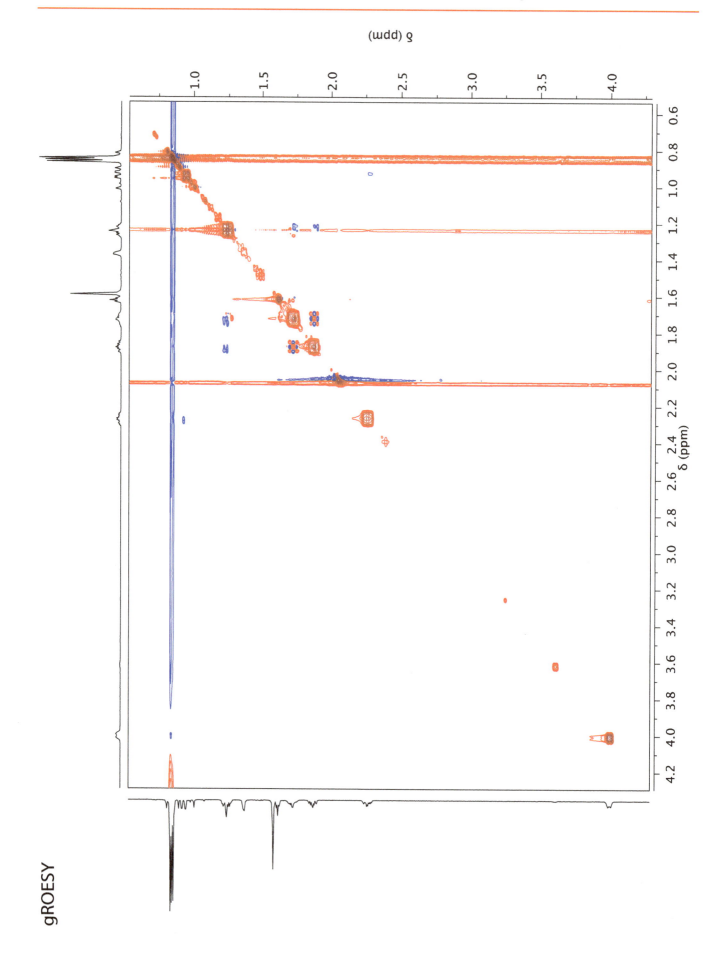

gROESY

SECTION 5 Problem 105

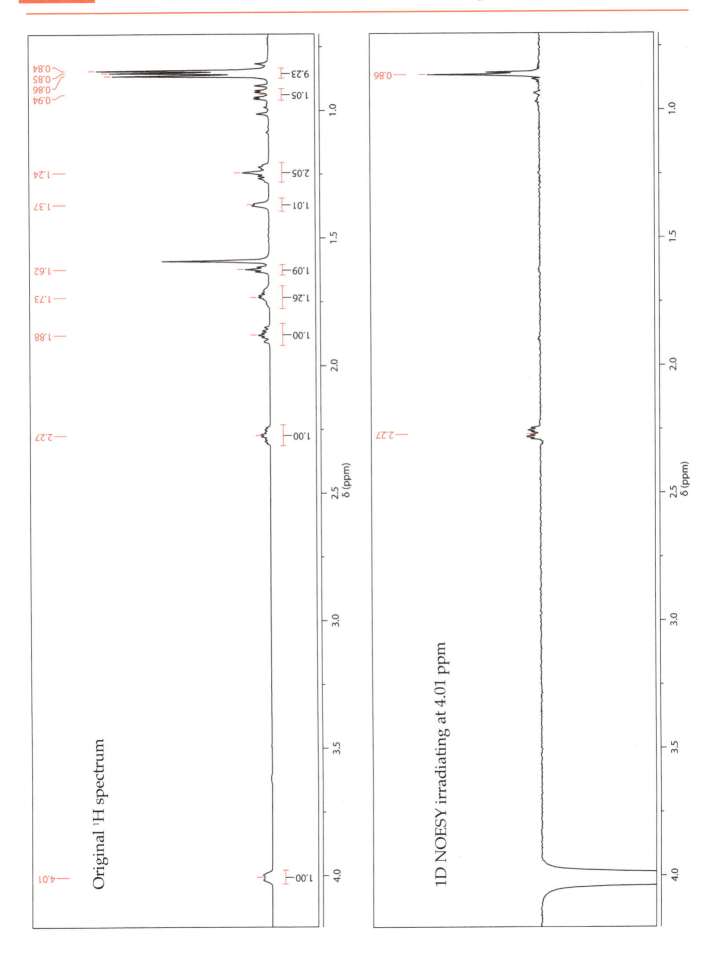

Original ¹H spectrum

1D NOESY irradiating at 4.01 ppm

Section 5 Problem 105

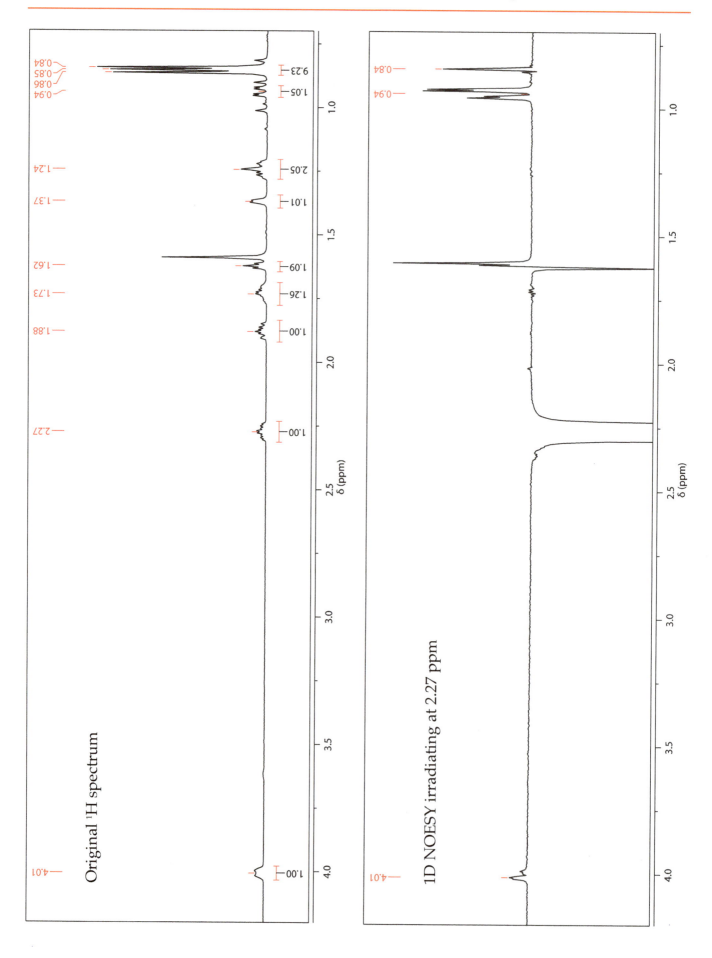

SECTION 5 Problem 105

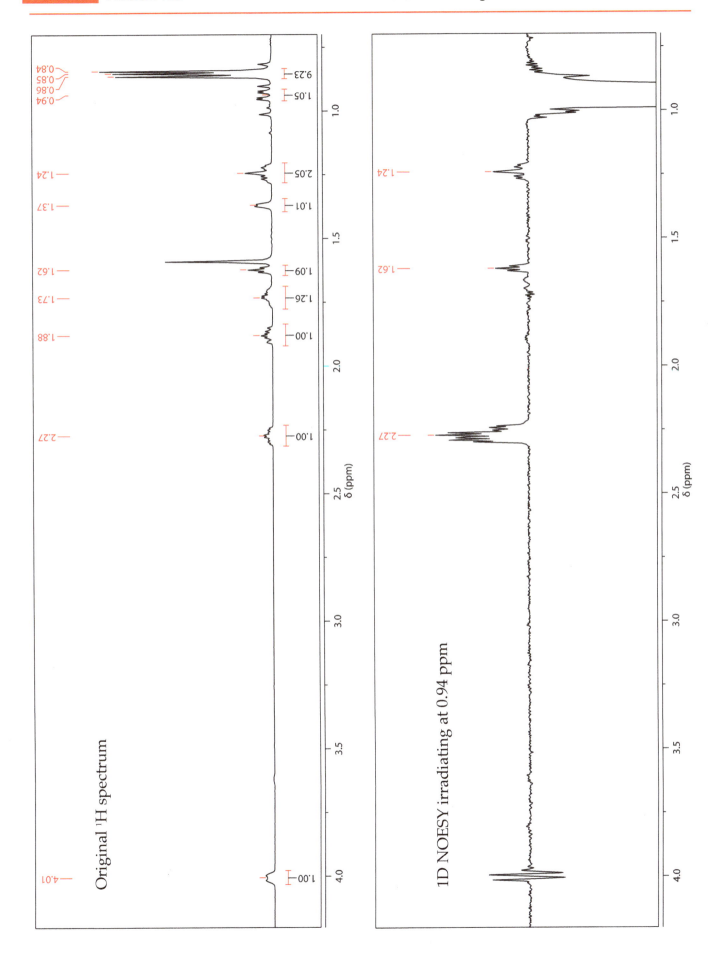

Problem 106

Using the ¹H and gROESY spectra, determine the relative configuration and conformation for the disaccharide below. For the assignment of proton and carbon resonances see problem 35.
Spectra acquired in (CD₃)₂SO at 500 MHz.

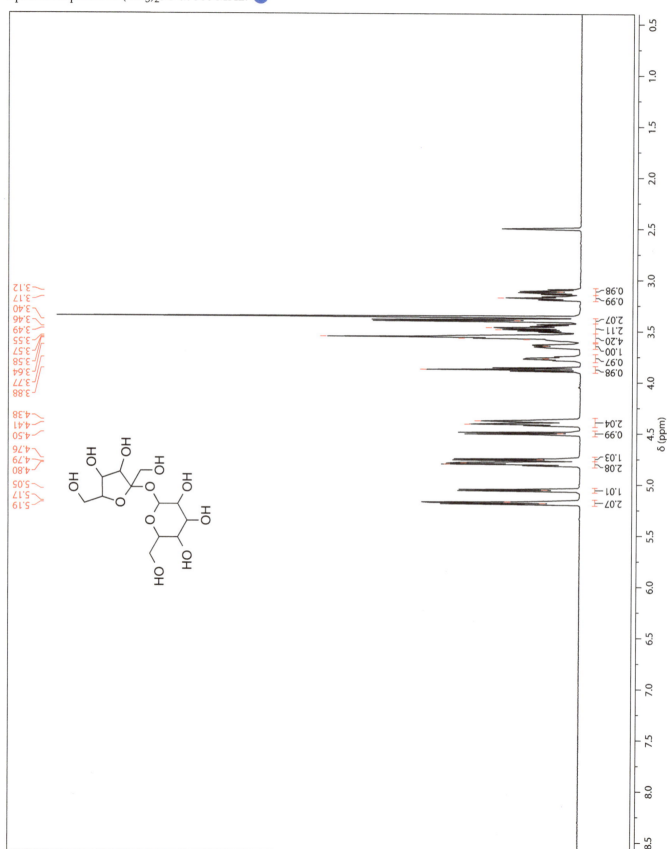

SECTION 5 Problem 106

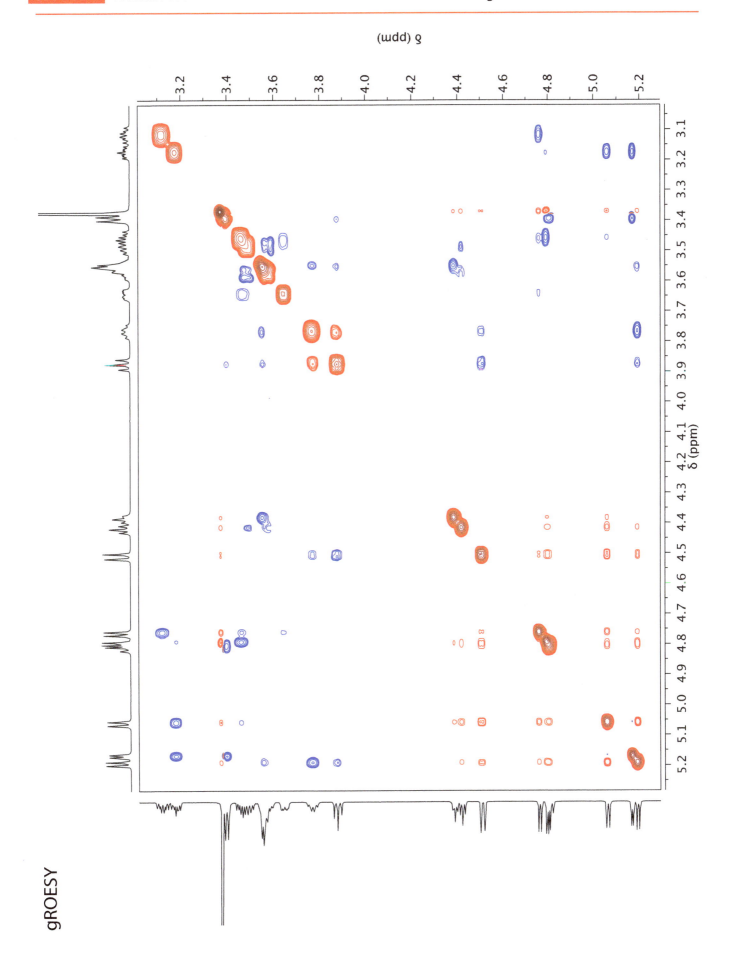

gROESY

Problem 106

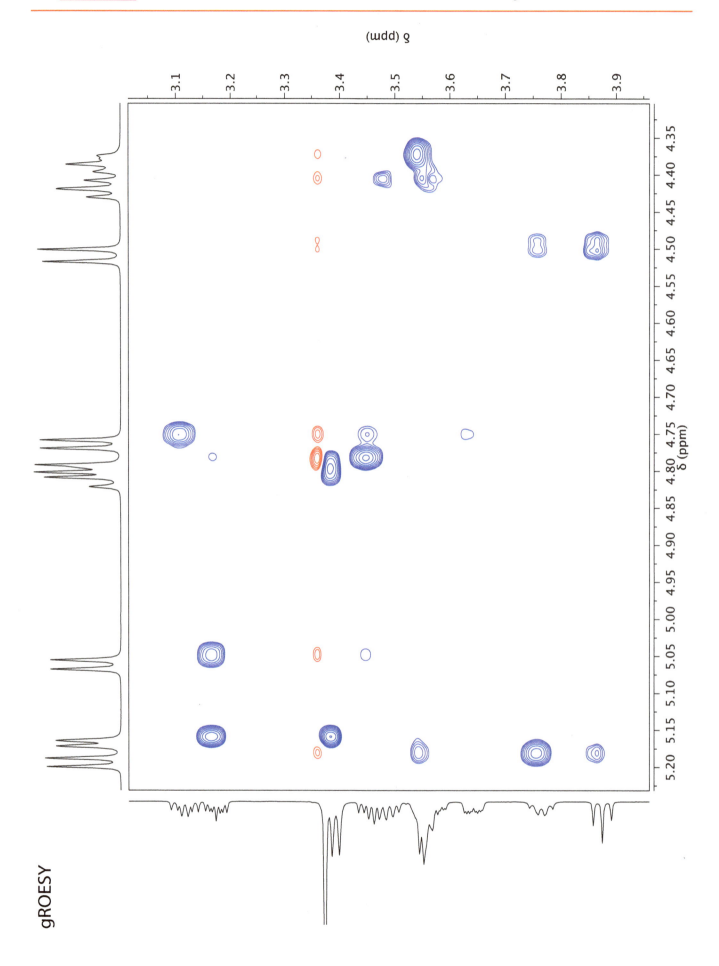

gROESY

SECTION 5 Problem 106

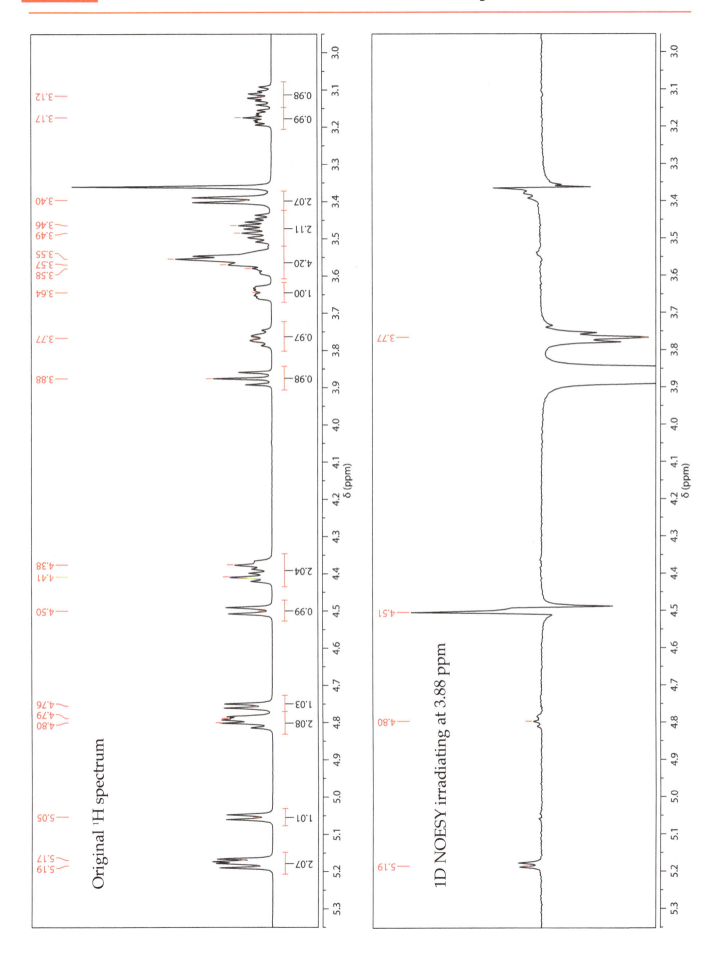

Problem 106

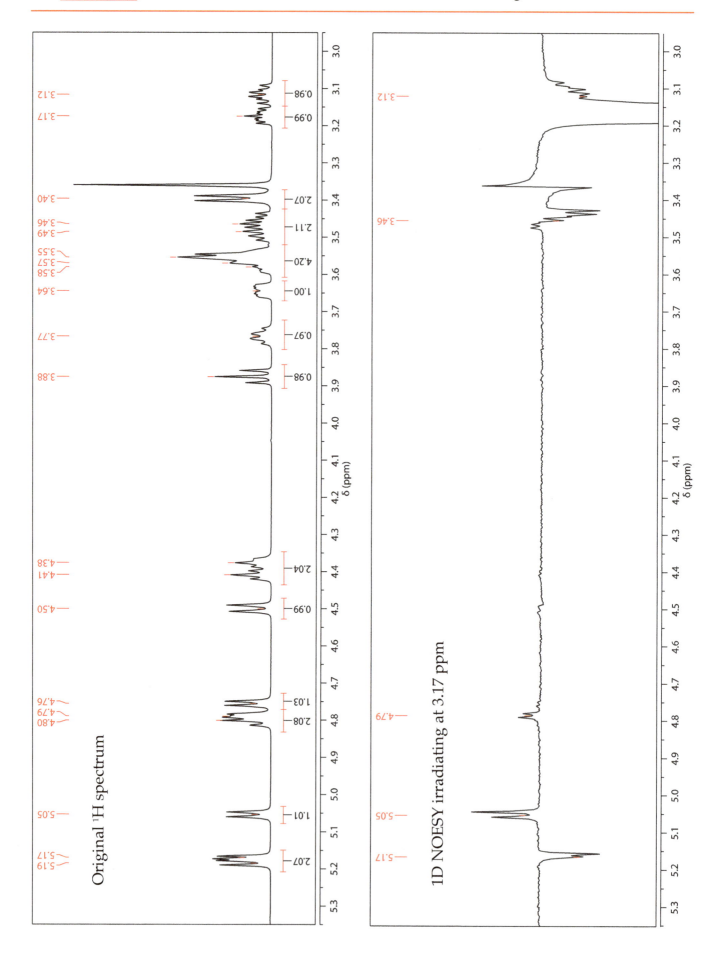

SECTION 5 Problem 107

Using the ¹H and selective 1D NOESY spectra, determine the relative configuration of the molecule below, including assignment of the two gem-dimethyl signals. For the assignment of proton and carbon resonances see problem 48. Spectra acquired in CDCl₃ at 500 MHz.

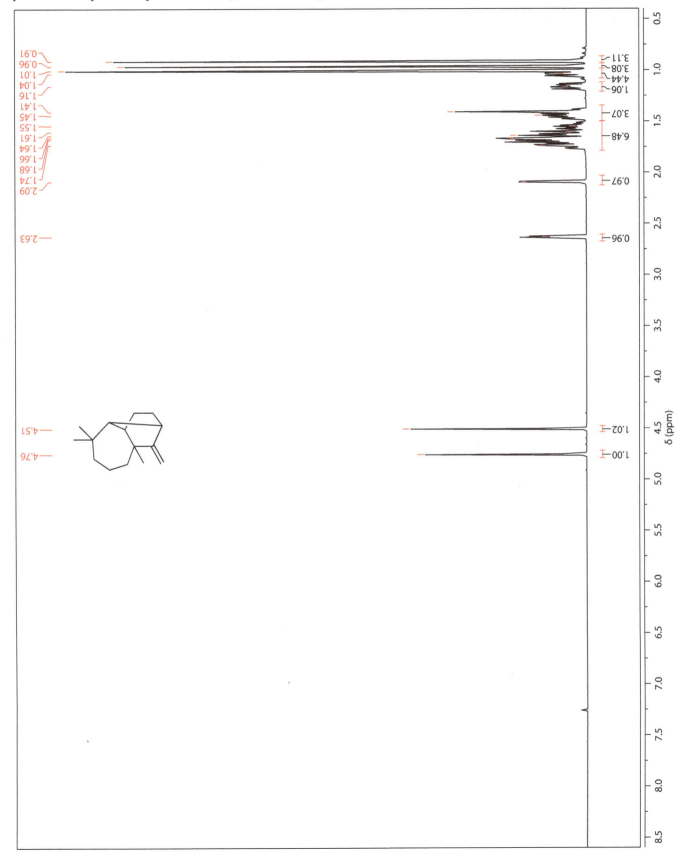

Problem 107

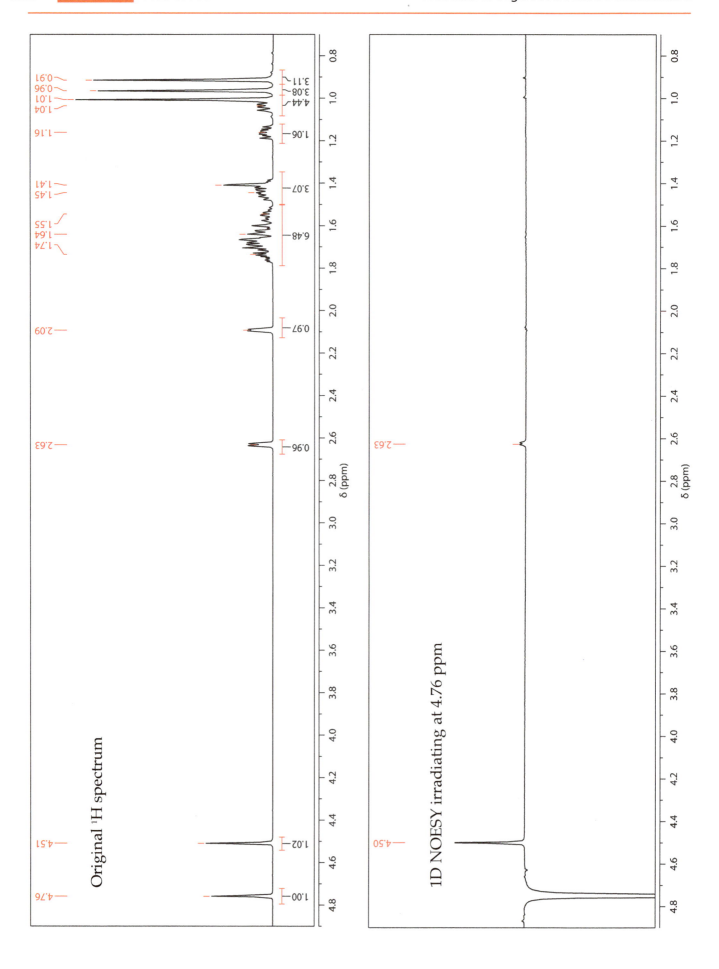

SECTION 5 Problem 107

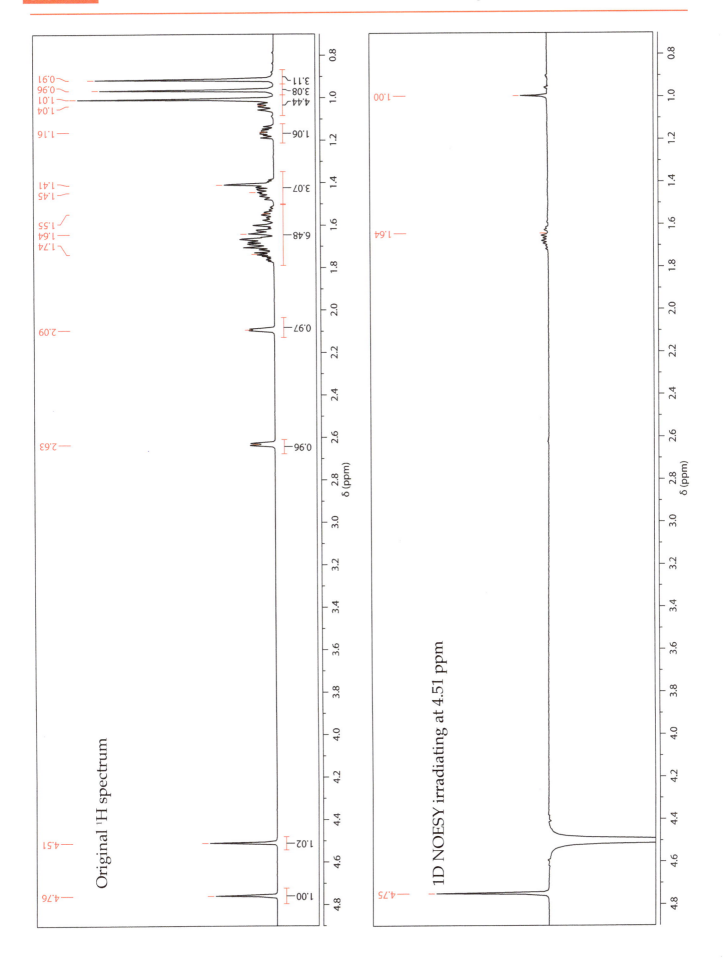

SECTION 5 Problem 107

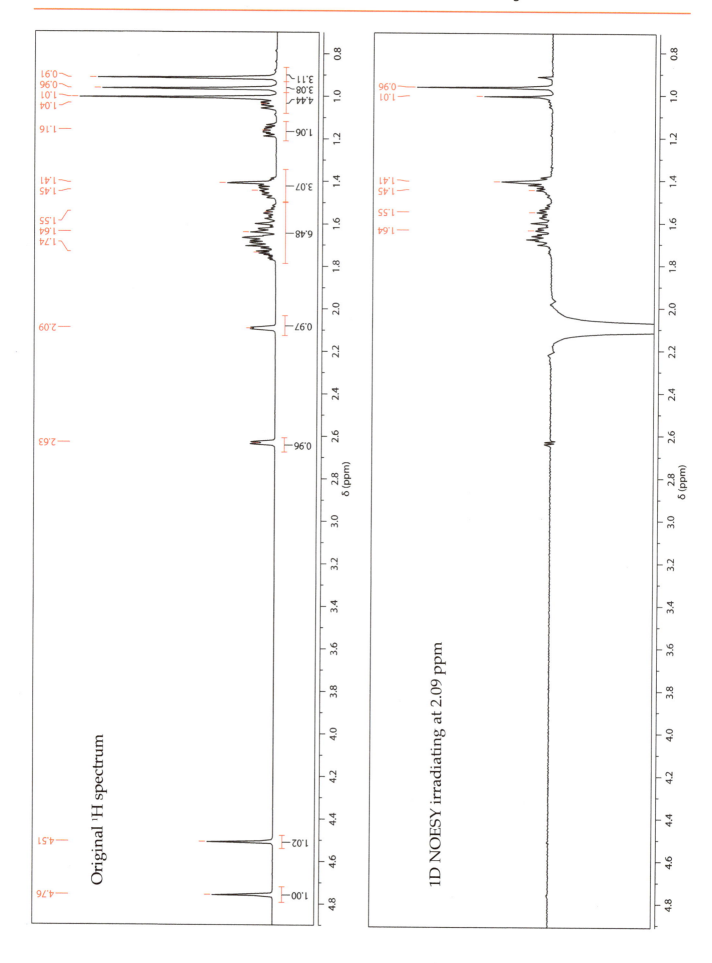

SECTION 5 Problem 107

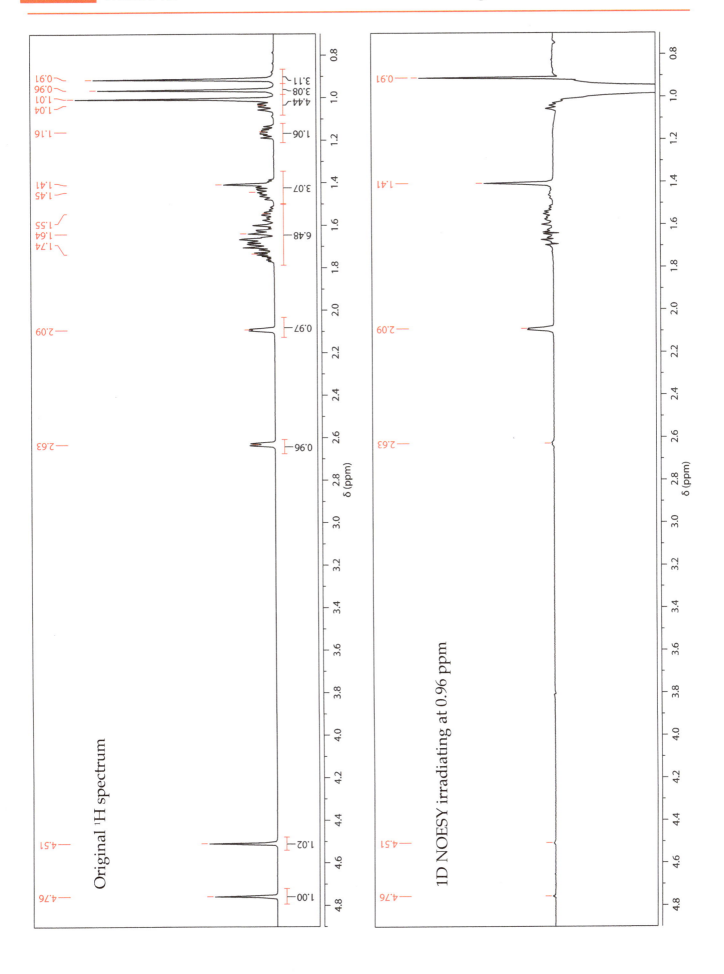

SECTION 5 Problem 107

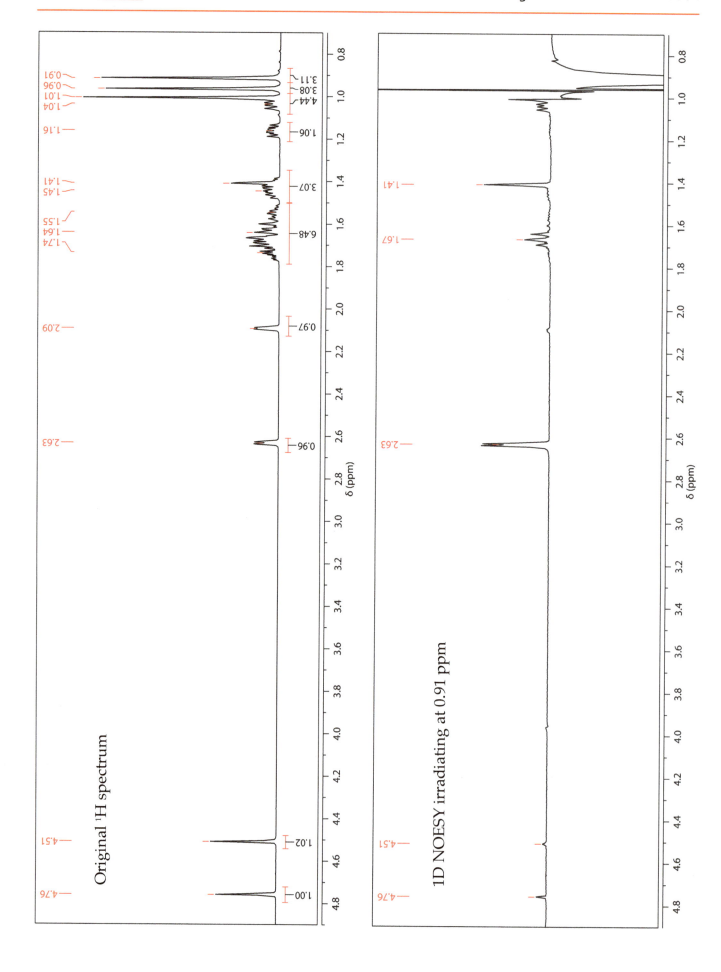

SECTION 5 Problem 108

Problems in Organic Structure Determination

Using the ¹H, gROESY and selective 1D spectra, determine the relative configuration and conformation of the molecule below. For the assignment of proton and carbon resonances see problem 116.
Spectra acquired in (CD$_3$)$_2$SO at 500 MHz.

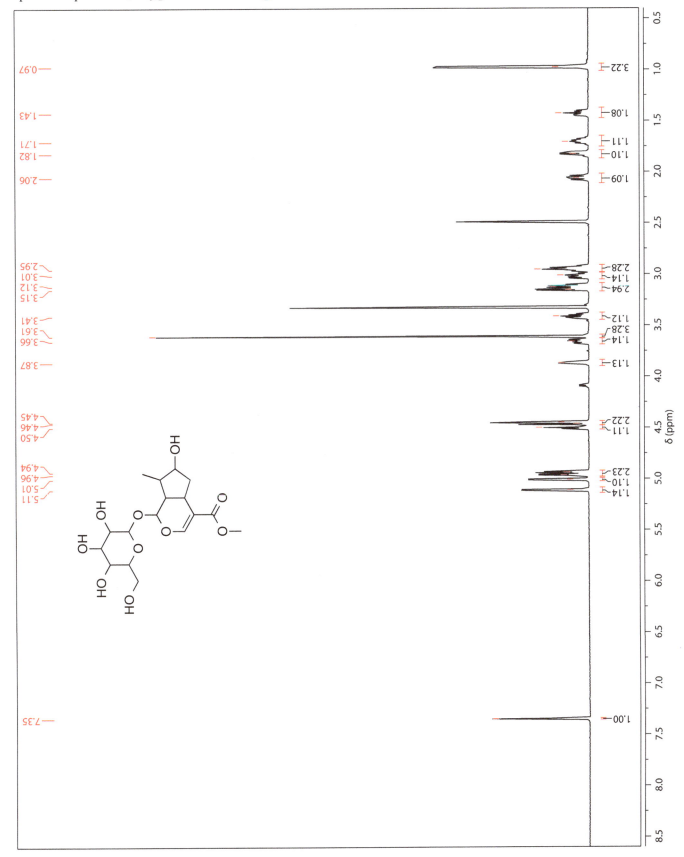

Problem 108

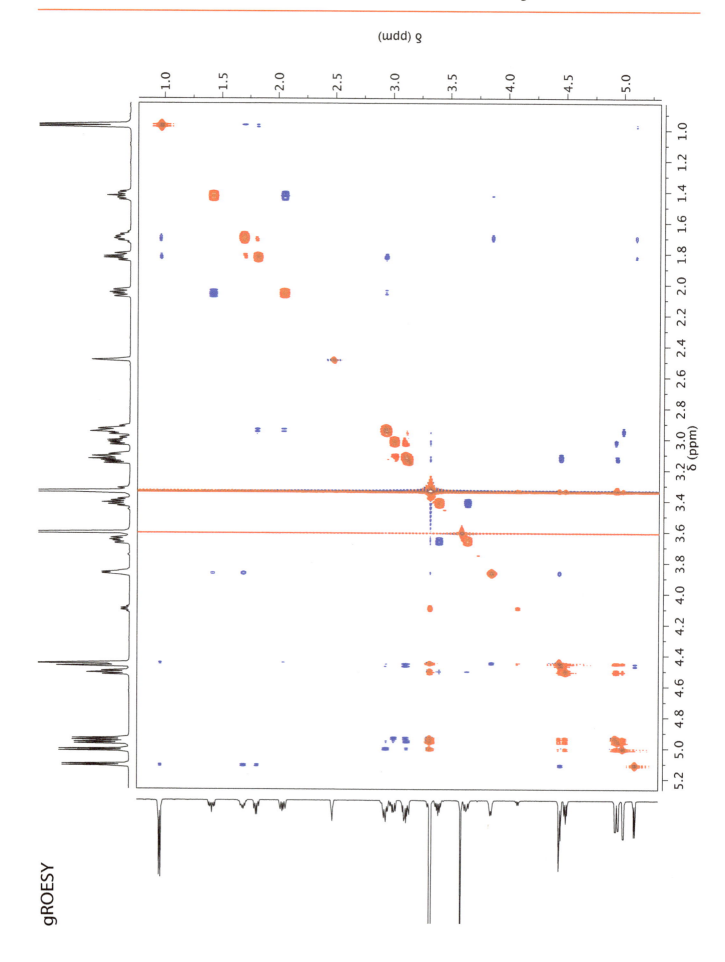

gROESY

SECTION 5 Problem 108

Problems in Organic Structure Determination

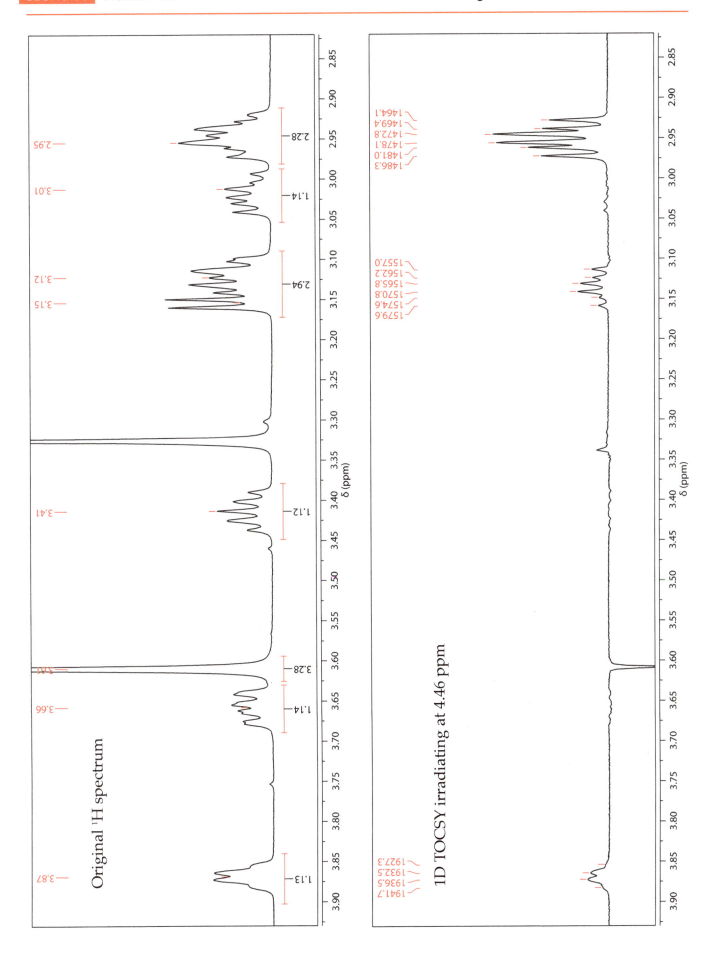

Problem 108

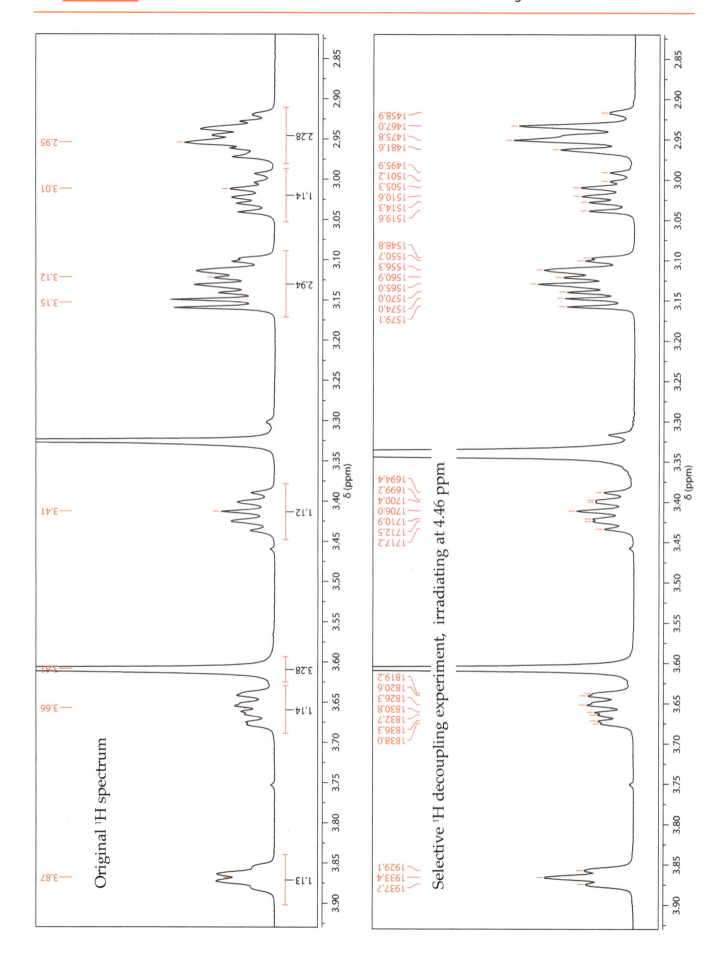

Problem 108

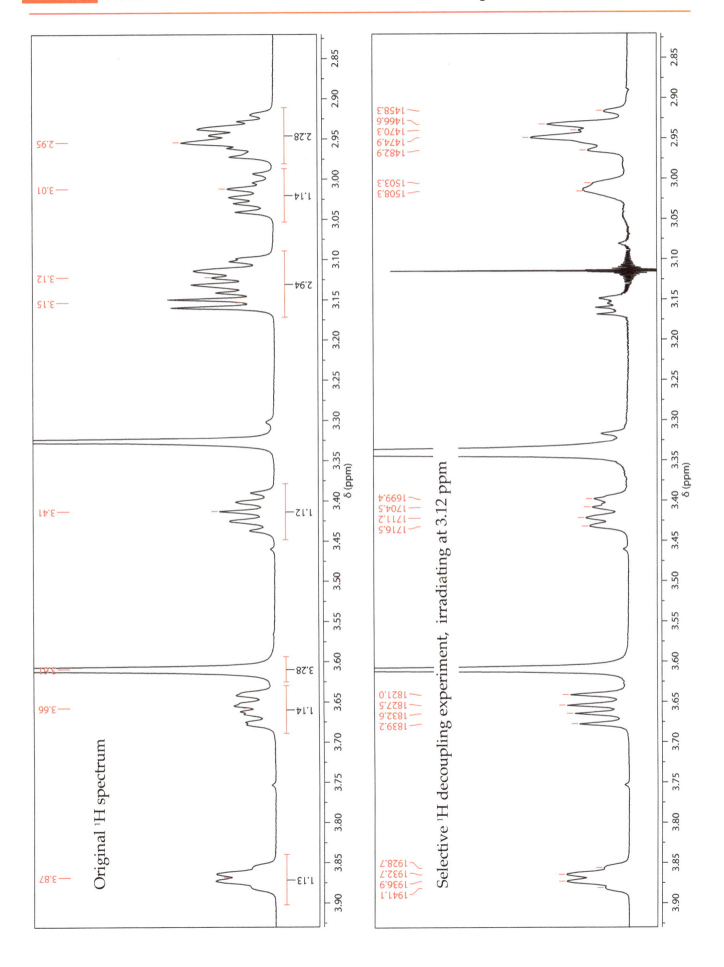

SECTION 5 Problem 108

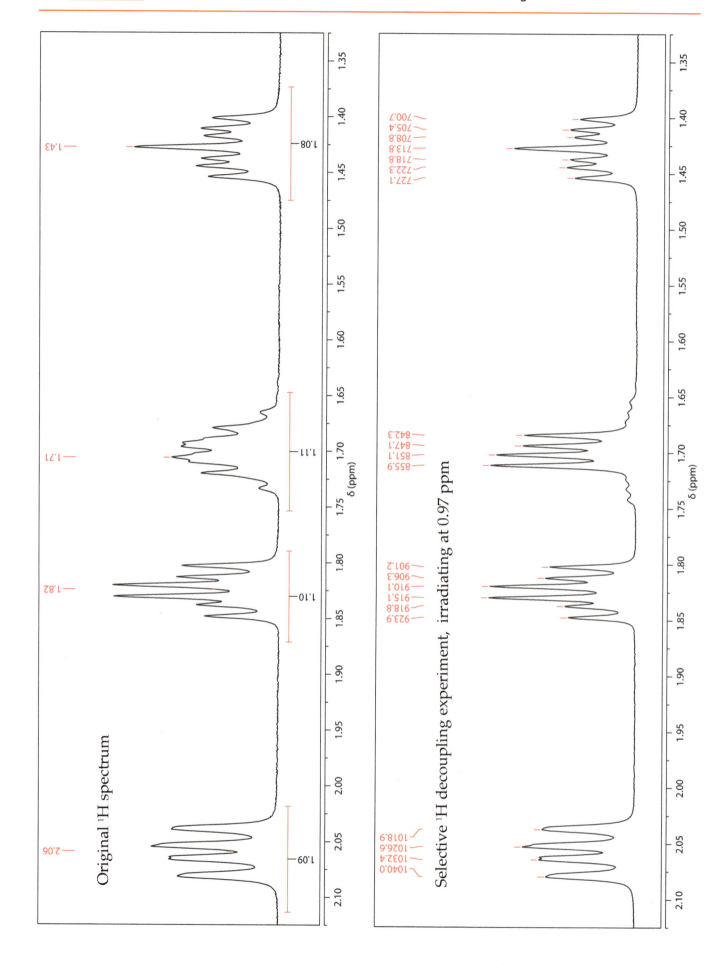

SECTION 5 Problem 109

Using the ¹H and gROESY spectra, determine the relative configuration of the molecule below.
For the assignment of proton and carbon resonances see problem 121.
Spectra acquired in (CD₃)₂SO at 500 MHz.

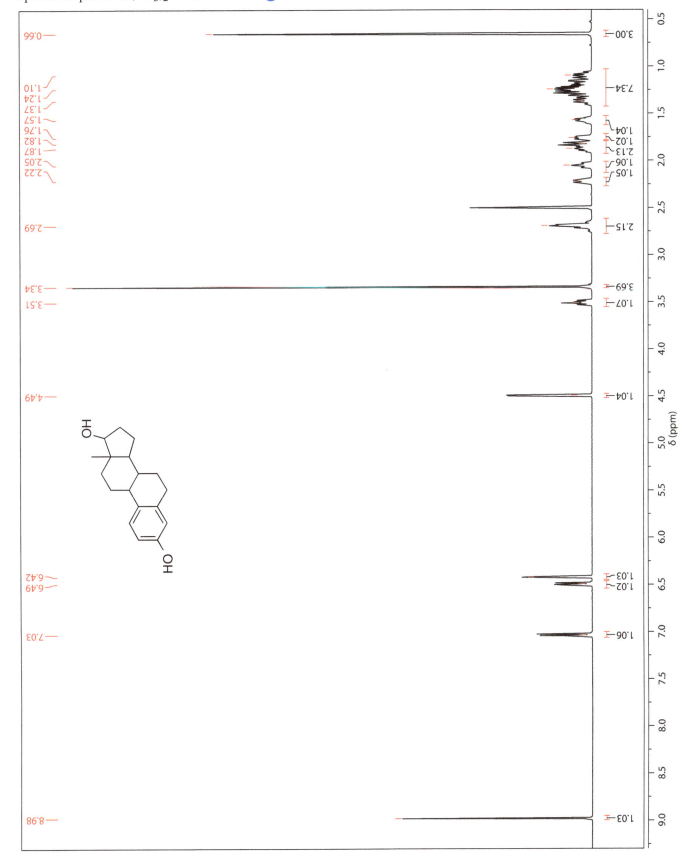

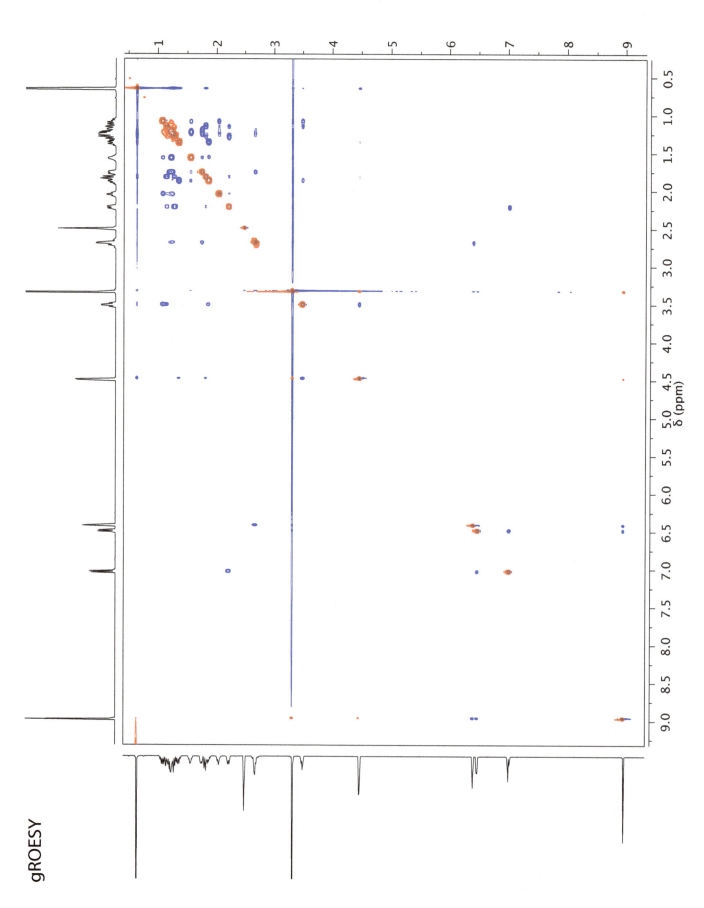

SECTION 5   Problem 109

Problems in Organic Structure Determination

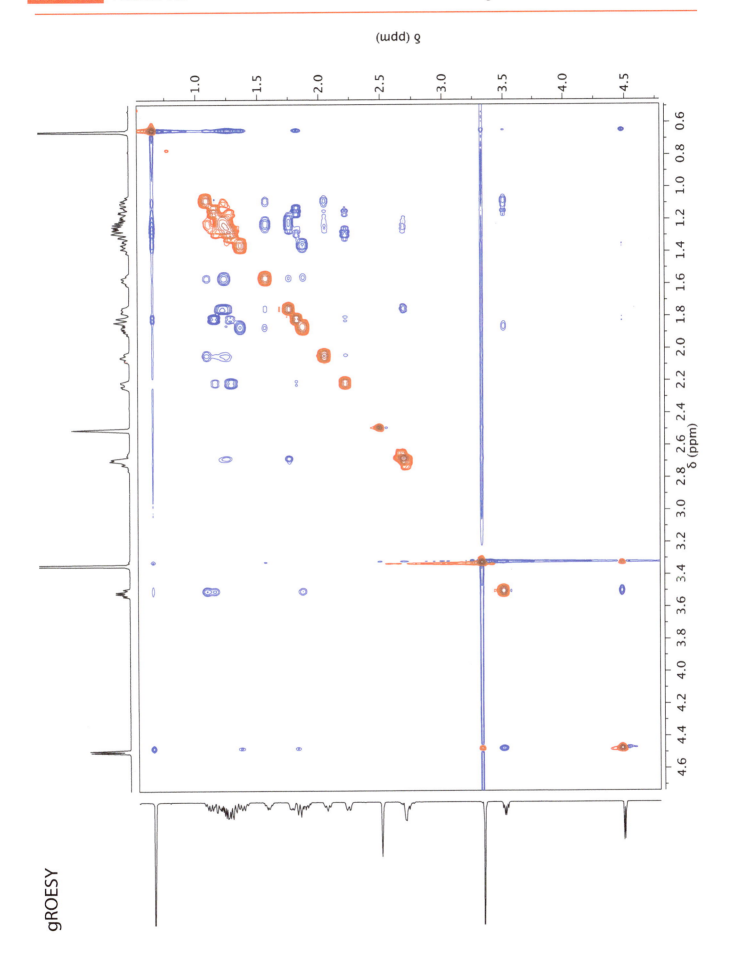

gROESY

SECTION 5 Problem 110

Problems in Organic Structure Determination

Using the ¹H and gROESY spectra, determine the relative configuration and conformation of the molecule below. For the assignment of proton and carbon resonances see problem 95.
Spectra acquired in (CD₃)₂SO at 500 MHz.

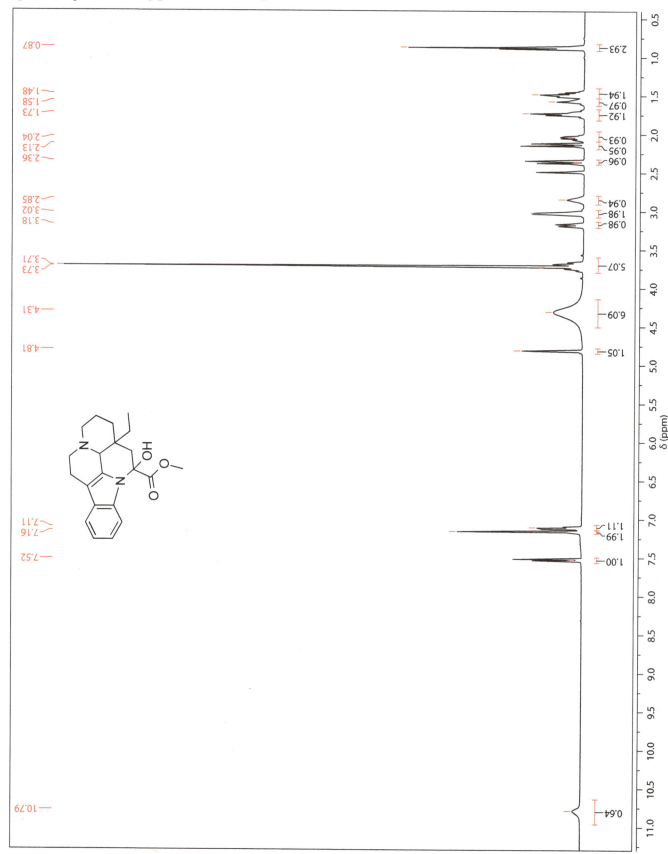

SECTION 5 Problem 110

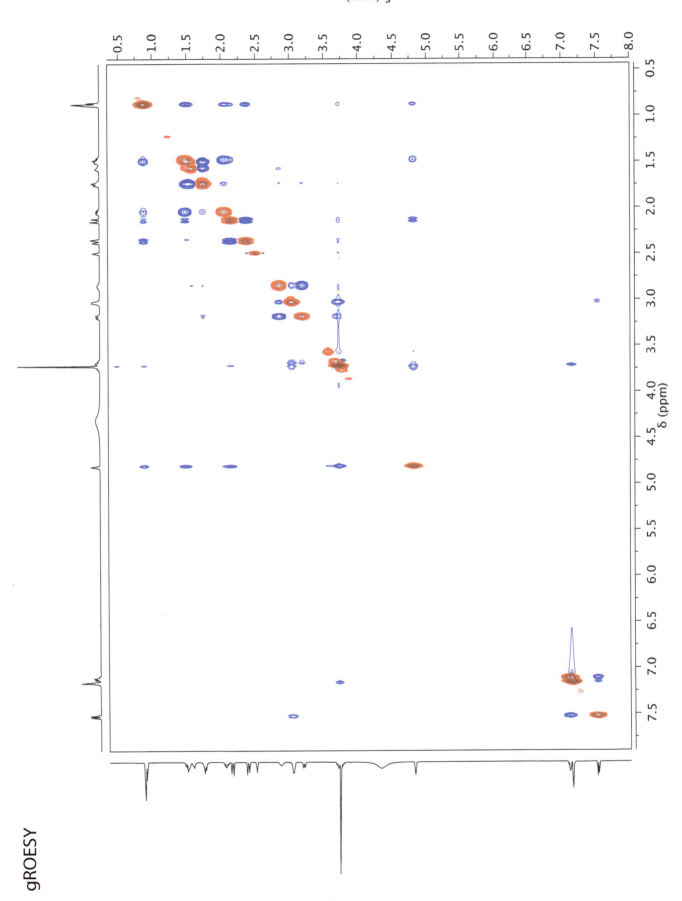

gROESY

SECTION 5 Problem 110

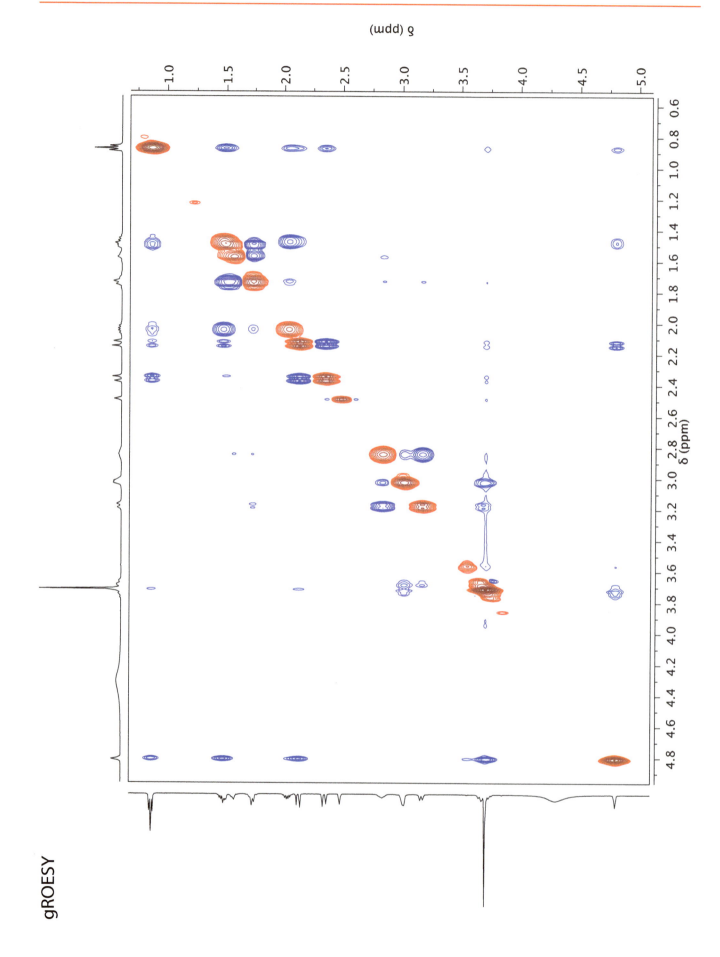

gROESY

Section 6

Section 6 contains problems at the pinnacle of NMR structure elucidation complexity, including natural products and natural product derivatives. These problems require careful use of all the fundamental and advanced techniques and strategies learned from the previous five sections, and provide challenges of the highest difficulty for aspiring structure elucidation masters!

LEARNING OBJECTIVES

- Manage large complex structure elucidation problems.
- Determine structures for compounds containing large numbers of stereogenic centers.
- Incorporate mass spectrometry data into the structure elucidation workflow.

EXPERIMENTS INCLUDED

^{1}H, ^{13}C, DEPT-135, gCOSY, gHSQC, gHMBC, gTOCSY, gROESY, ^{15}N-gHMBC, HSQC-TOCSY, 1D TOCSY, 1D NOESY, ^{1}H selective decoupling spectra. In addition, these problems provide accurate mass measurements in lieu of molecular formulae. Each of these m/z values is accurate within 2 ppm. For those not wishing to incorporate mass spectral data, the molecular formulae for all compounds are provided in the hints section.

LEGEND

Spectrum annotations:

i = impurity.

W = worked problem in the answer key.

N = technical note about the data. For example: *"This spectrum is missing one exchangeable proton."*

H = hint to assist in solving the problem. For example: *"This molecule would have an IR stretch at 2240 cm^{-1}."*

TYPES OF MOLECULES

This section contains molecules from all of the major biosynthetic classes, including polyketides, terpenes, alkaloids, steroids and peptides. Many of these molecules contain multiple heteroatoms, stereogenic centers and quaternary carbons, making them highly challenging structure elucidation objectives.

ACCESS TO SPECTRA

This section differs from the previous five sections in the book because only the ^{1}H and ^{13}C spectra are displayed in the main text. The reason for this is that these problems require very large numbers of expansions to solve from paper copies, which would necessitate a second volume!

Users have two options for accessing the spectra. The first (preferred) option is to download the raw data from the book website and solve the problem by displaying and manipulating the spectra in an NMR processing software package. This is the strategy that the authors used in preparing the answers and is now the standard approach for complex molecule structure elucidation. For those without access to such software, individual pdfs containing all required spectra are also available as downloads from the book website.

SECTION 6 Problem 111

Determine the structure of the unknown compound below with an HRMS [M + H]⁺ m/z of 137.1328.

Spectra acquired in CDCl₃ at 500 MHz (¹H) or 125 MHz (¹³C).

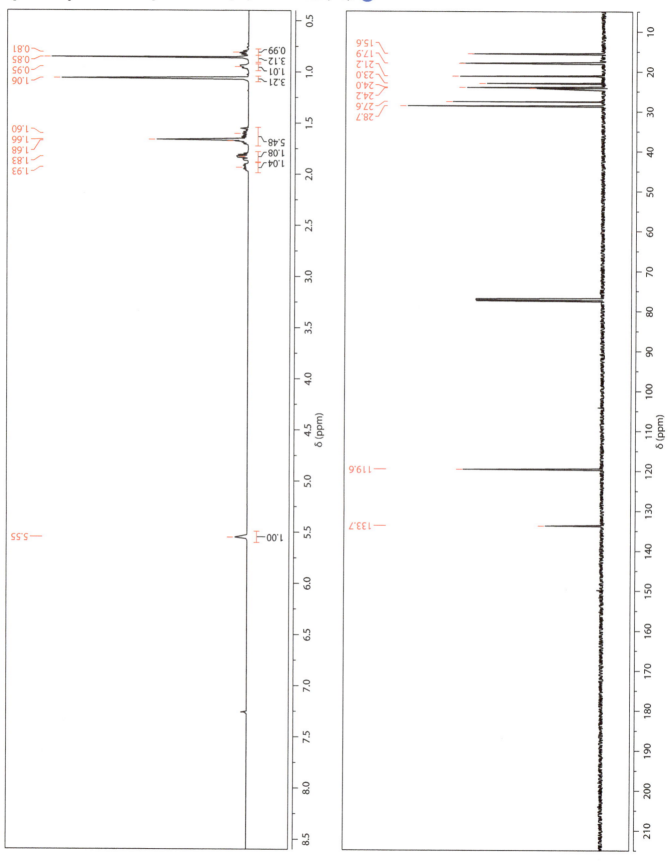

SECTION 6 Problem 112

Determine the structure of the unknown compound below with an HRMS [M + H]⁺ m/z of 340.1546.

Spectra acquired in $(CD_3)_2SO$ at 500 MHz (^{1}H) or 125 MHz (^{13}C).

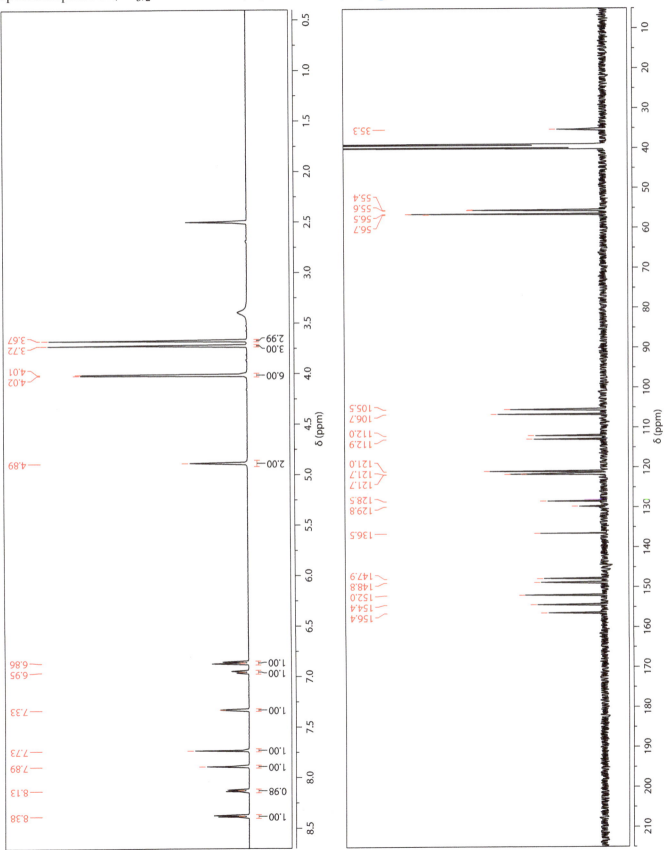

Problem 113

Determine the structure of the unknown compound below with an HRMS [M + H]⁺ m/z of 155.1438.

Spectra acquired in CDCl$_3$ at 500 MHz (^{1}H) or 125 MHz (^{13}C).

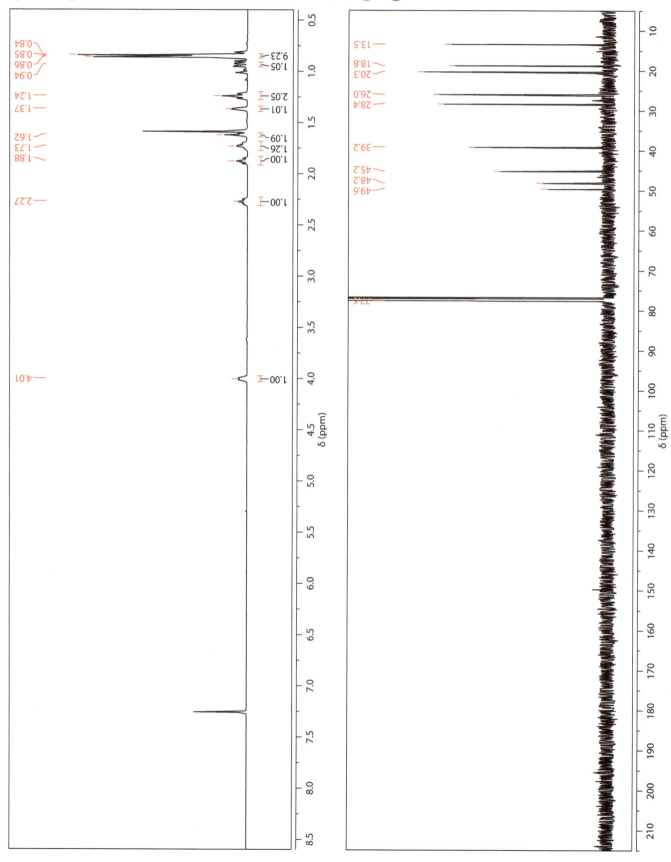

SECTION 6 Problem 114

Determine the structure of the unknown compound below with an HRMS [M + H]⁺ m/z of 338.1863.

Spectra acquired in C_6D_6 at 500 MHz (^{1}H) or 125 MHz (^{13}C).

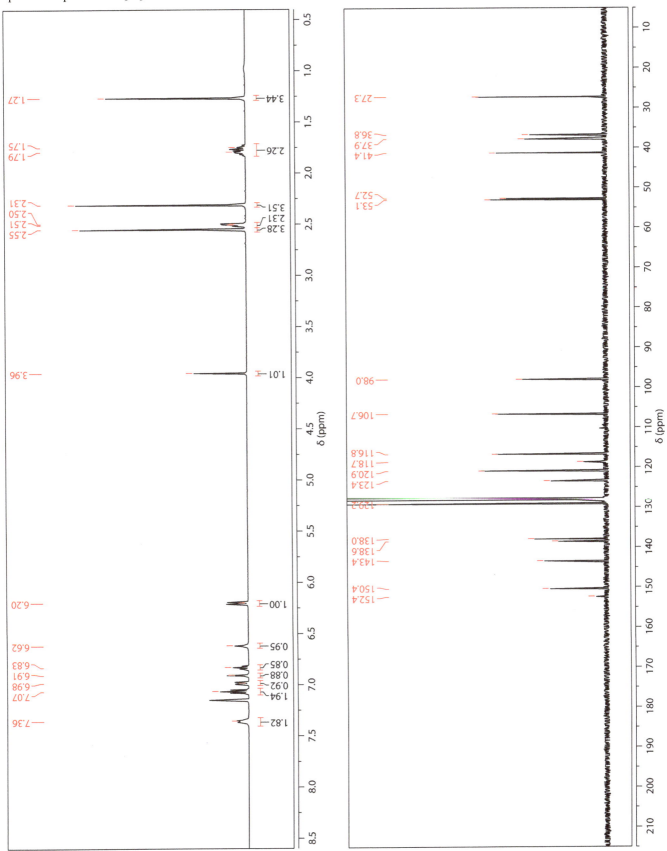

SECTION 6 Problem 115

Determine the structure of the unknown compound below with an HRMS [M + H]⁺ m/z of 221.1907.

Spectra acquired in CDCl₃ at 500 MHz (¹H) or 125 MHz (¹³C).

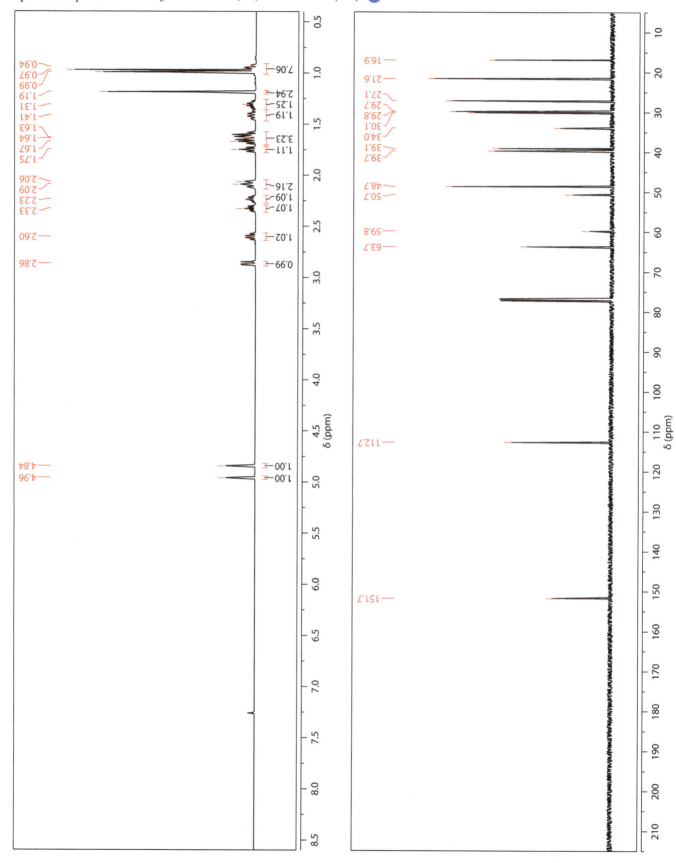

SECTION 6 Problem 116

Determine the structure of the unknown compound below with an HRMS [M + H]⁺ m/z of 391.1609.

Spectra acquired in $(CD_3)_2SO$ at 500 MHz (^{1}H) or 125 MHz (^{13}C).

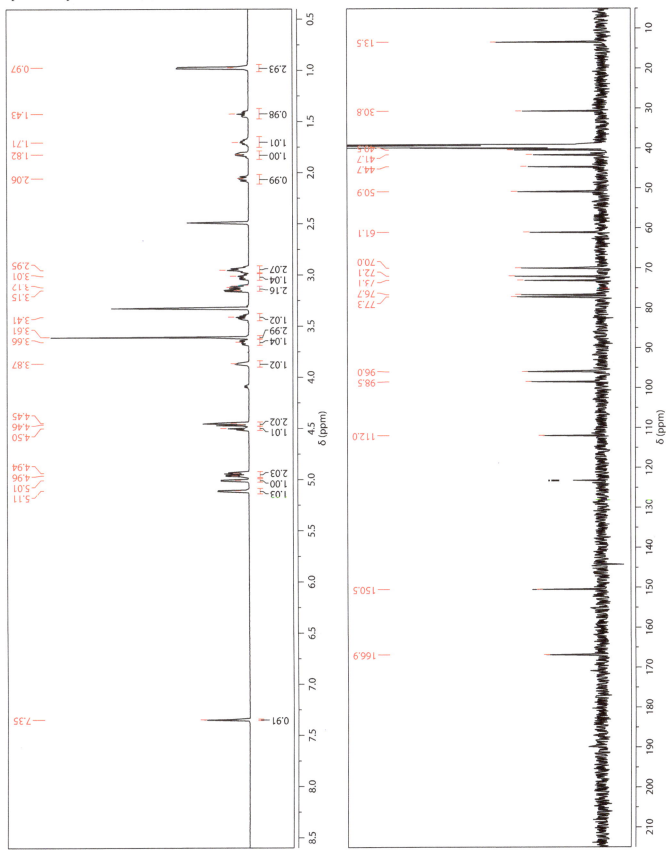

SECTION 6 Problem 117

Determine the structure of the unknown compound below with an HRMS [M + H]⁺ m/z of 489.1380.

Spectra acquired in CDCl₃ at 500 MHz (¹H) or 125 MHz (¹³C).

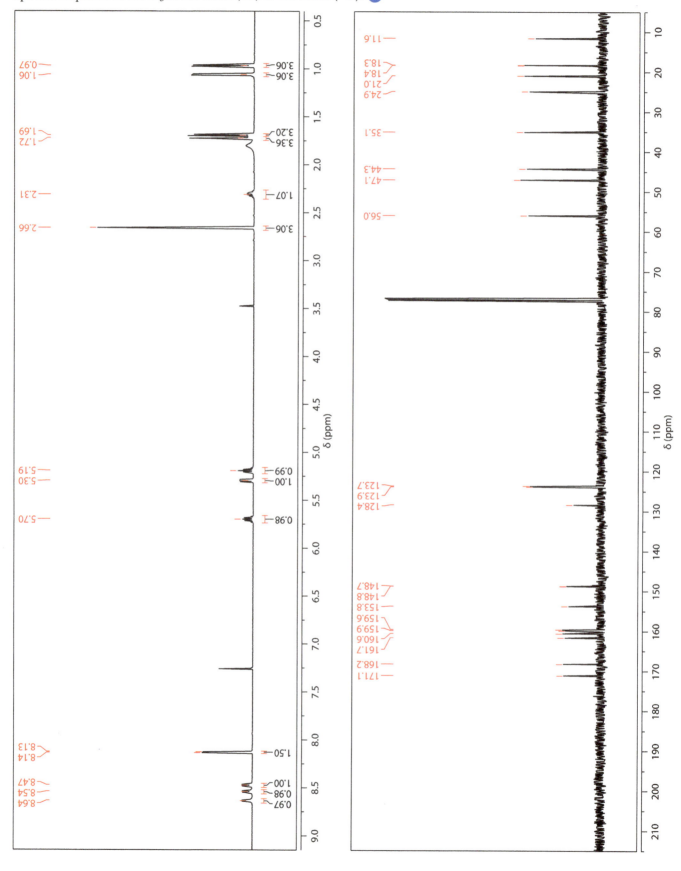

SECTION 6 Problem 118

Problems in Organic Structure Determination **717**

Determine the structure of the unknown compound below with an HRMS [M + H]⁺ *m/z* of 251.2007.

Spectra acquired in CDCl₃ at 500 MHz (¹H) or 125 MHz (¹³C).

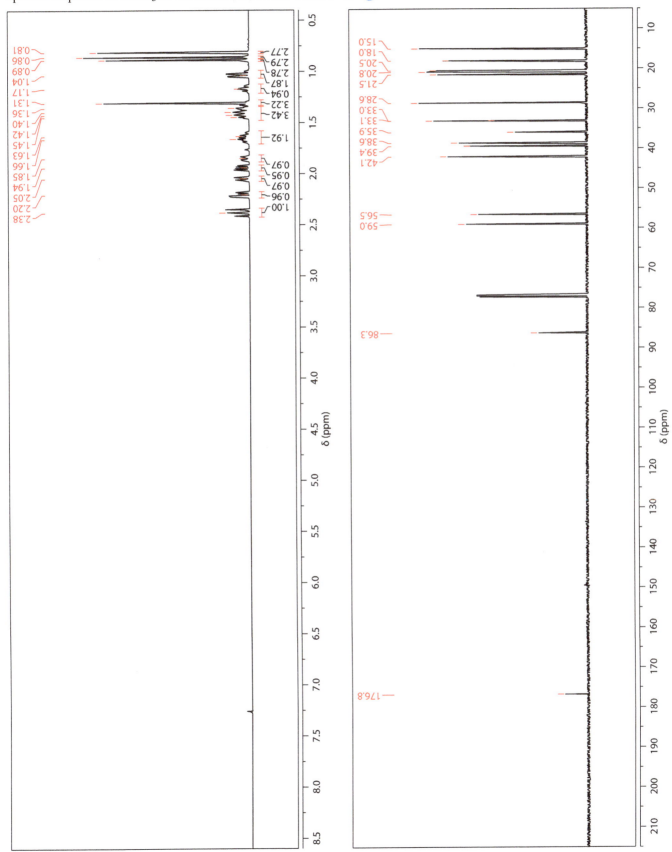

SECTION 6 Problem 119

Determine the structure of the unknown compound below with an HRMS [M + H]⁺ m/z of 327.0471.

Spectra acquired in CDCl$_3$ at 500 MHz (^{1}H) or 125 MHz (^{13}C).

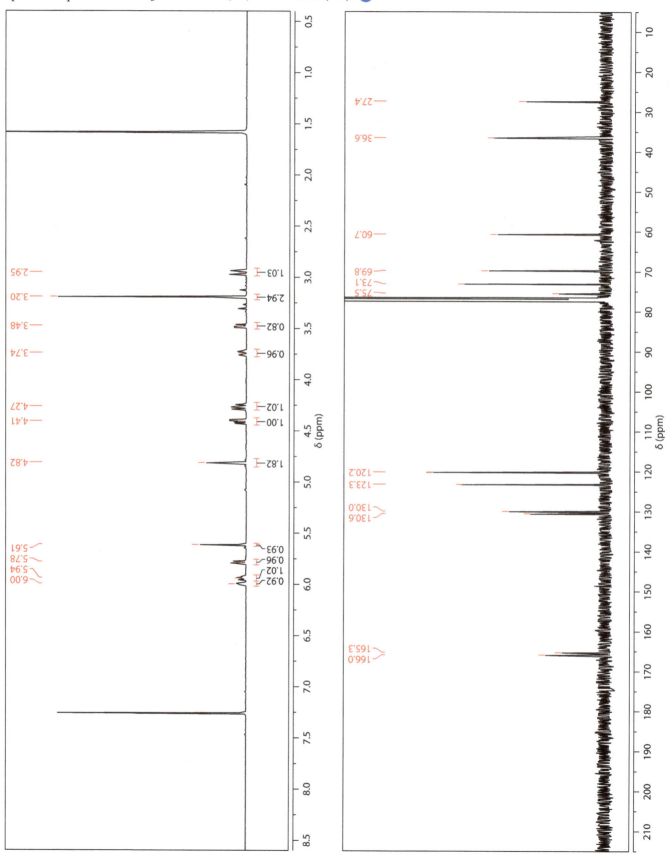

SECTION 6 Problem 120

Problems in Organic Structure Determination

Determine the structure of the unknown compound below with an HRMS [M + H]⁺ *m/z* of 407.2219.

Spectra acquired in $(CD_3)_2SO$ at 500 MHz (^1H) or 125 MHz (^{13}C). N H

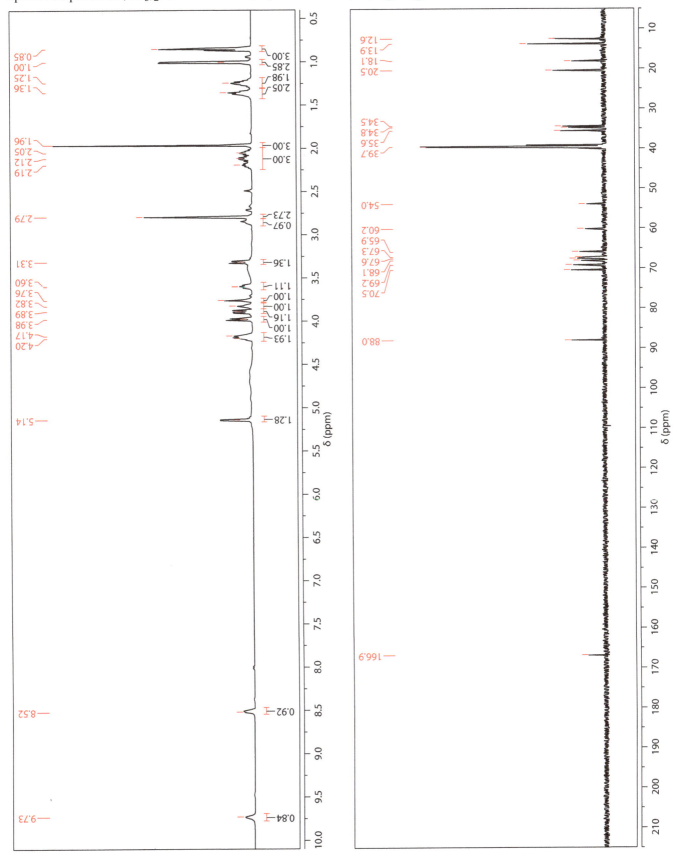

SECTION 6 Problem 121

Determine the structure of the unknown compound below with an HRMS [M + H]⁺ m/z of 273.1852.

Spectra acquired in $(CD_3)_2SO$ at 500 MHz (^{1}H) or 125 MHz (^{13}C).

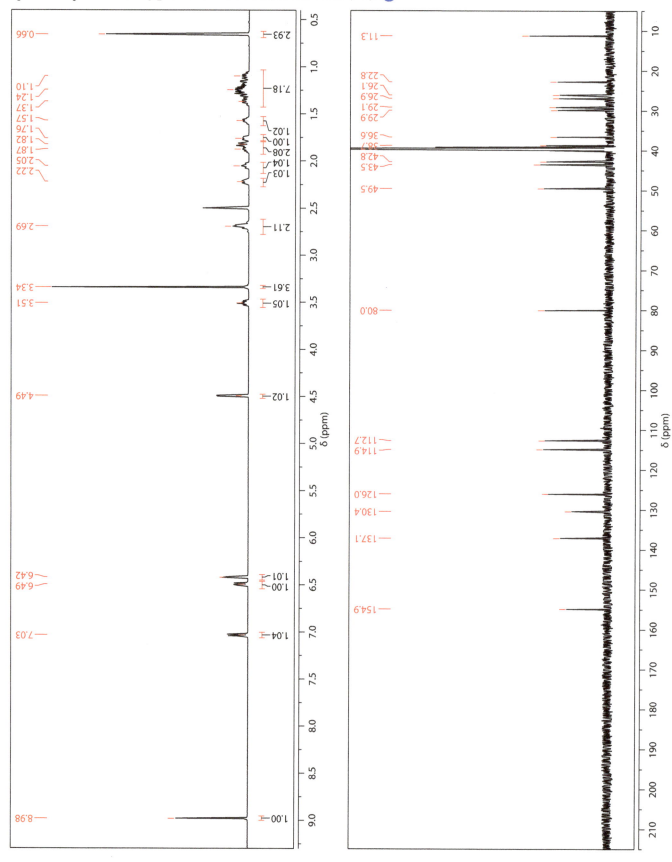

SECTION 6 Problem 122

Determine the structure of the unknown compound below with an HRMS [M + H]⁺ m/z of 351.2177.

Spectra acquired in $(CD_3)_2SO$ at 500 MHz (^{1}H) or 125 MHz (^{13}C).

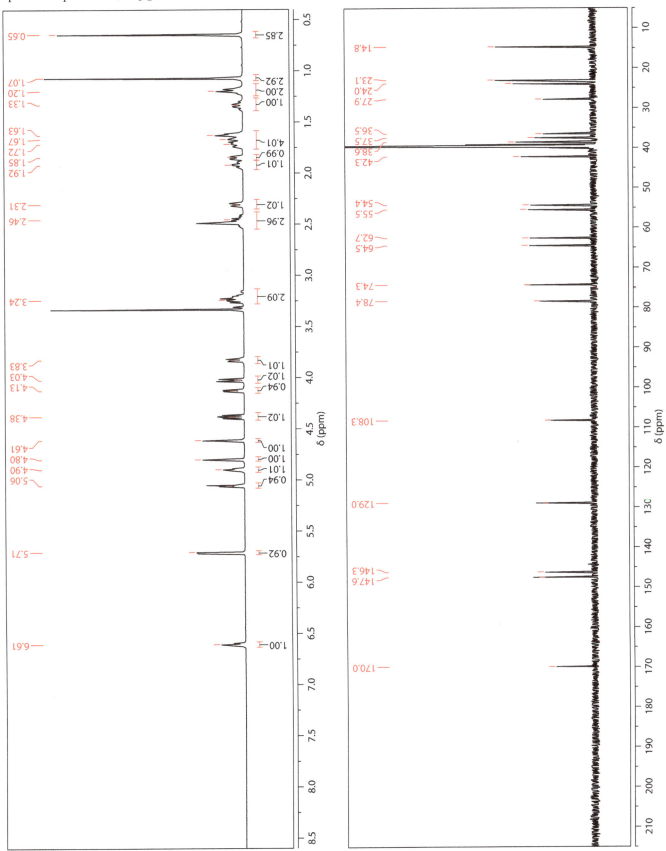

SECTION 6 Problem 123

Determine the structure of the unknown compound below with HRMS [M + H]⁺ m/z values of 423.2199 and 425.2176 in a 3:1 ratio.
Spectra acquired in CDCl₃ at 500 MHz (¹H) or 125 MHz (¹³C).

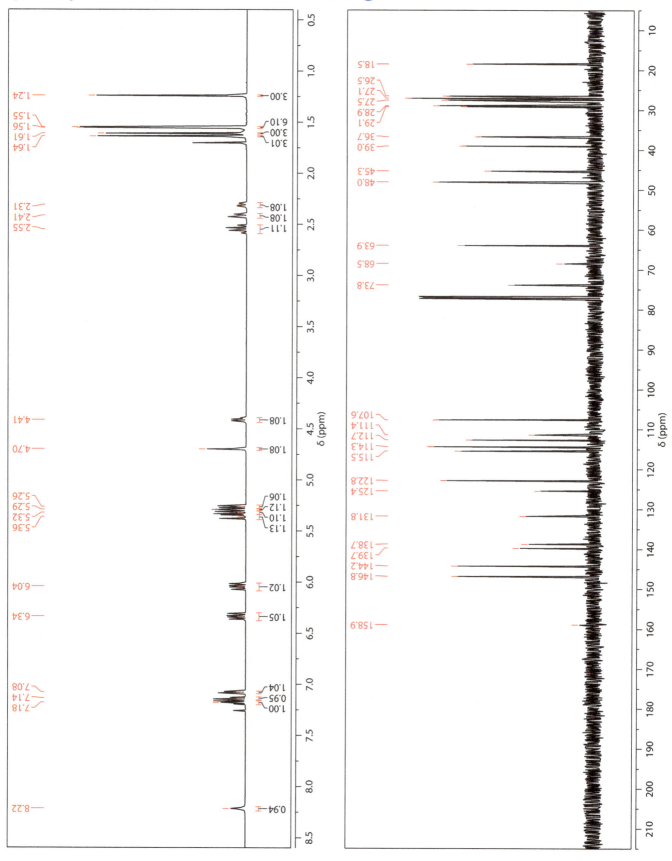

SECTION 6 Problem 124

Determine the structure of the unknown compound below with an HRMS [M + H]⁺ m/z of 501.3069.

Spectra acquired in $(CD_3)_2SO$ at 500 MHz (^1H) or 125 MHz (^{13}C).

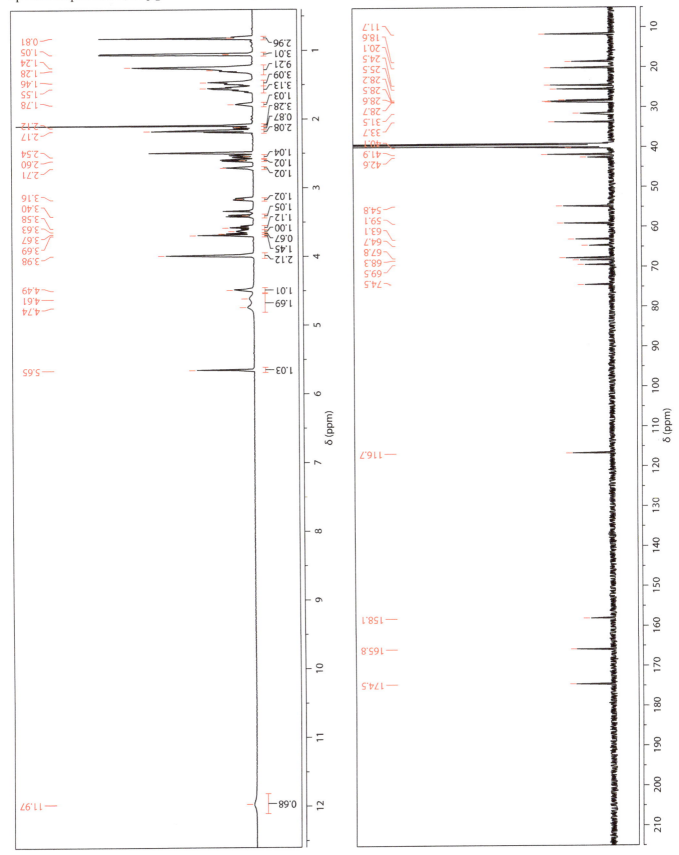

SECTION 6 Problem 125

Problems in Organic Structure Determination

Determine the structure of the unknown compound below with an HRMS [M + H]⁺ m/z of 473.3259.

Spectra acquired in CDCl₃ at 500 MHz (¹H) or 125 MHz (¹³C).

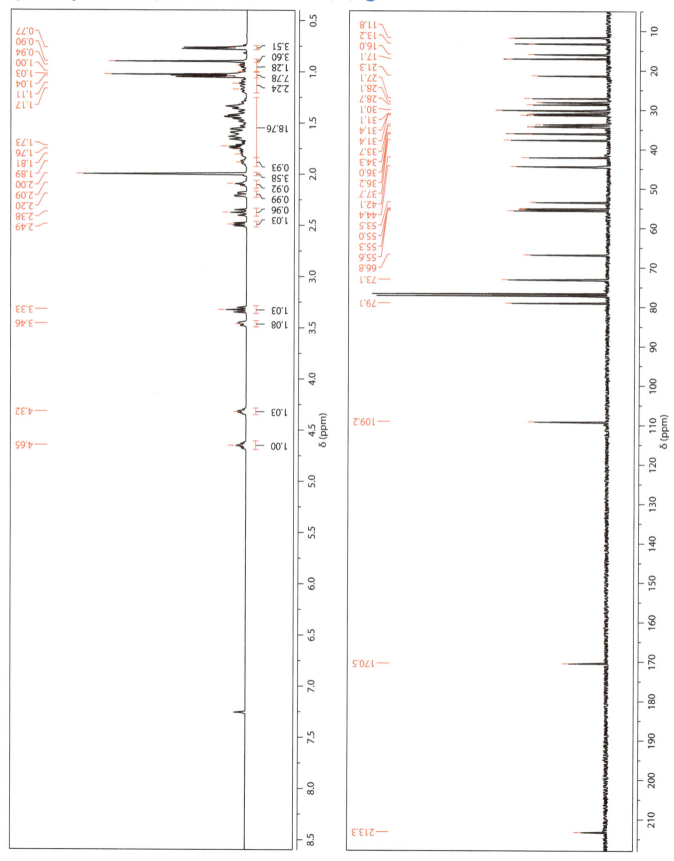

SECTION 6 Problem 126

Determine the structure of the unknown compound below with an HRMS [M + H]⁺ m/z of 494.3311.

Spectra acquired in $(CD_3)_2SO$ at 500 MHz (^{1}H) or 125 MHz (^{13}C).

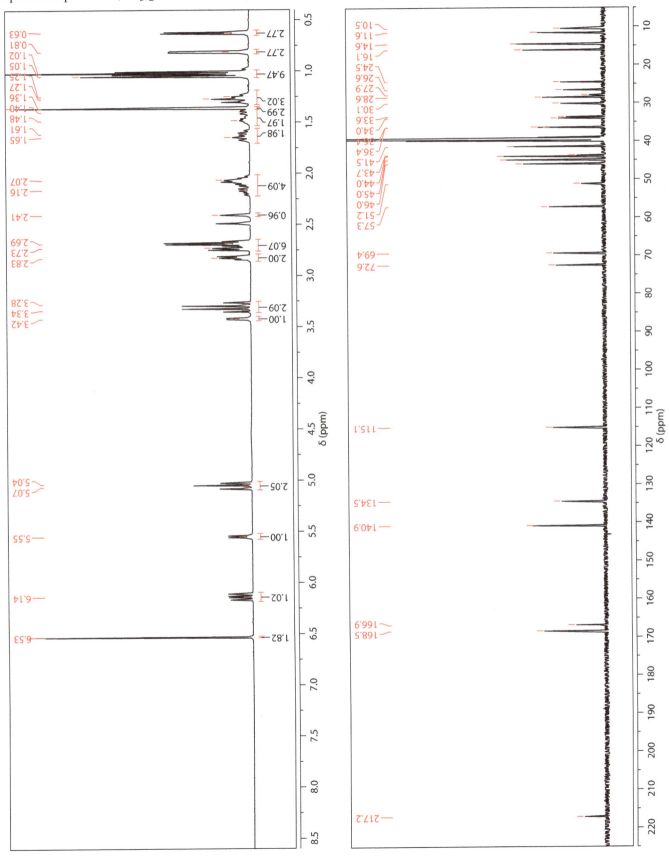

Problem 127

Determine the structure of the unknown compound below with an HRMS [M + H]+ m/z of 1111.6384.

Spectra acquired in CDCl₃ at 500 MHz (^{1}H) or 125 MHz (^{13}C).

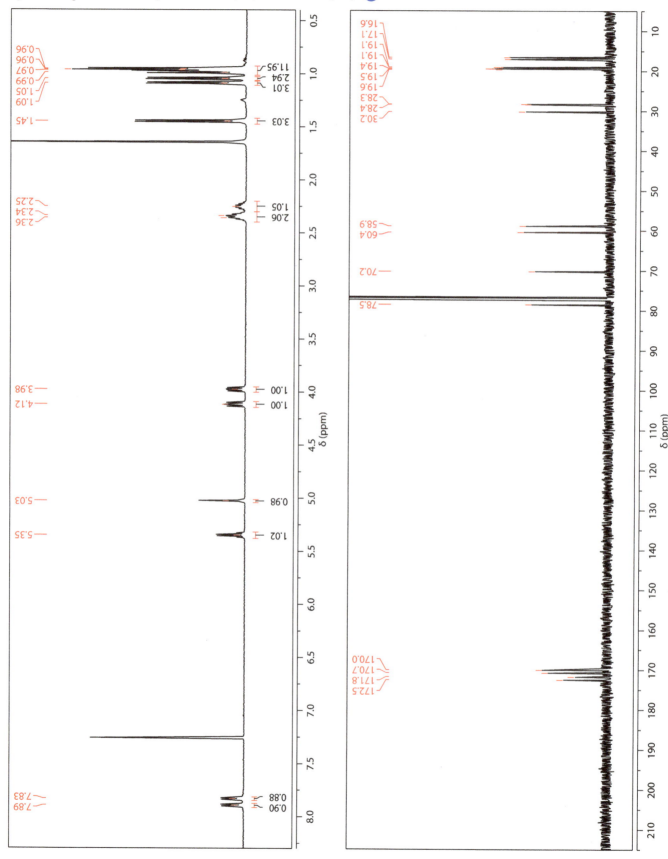

SECTION 6 Problem 128

Determine the structure of the unknown compound below with an HRMS [M + H]⁺ m/z of 749.5171.

Spectra acquired in CDCl₃ at 500 MHz (¹H) or 125 MHz (¹³C).

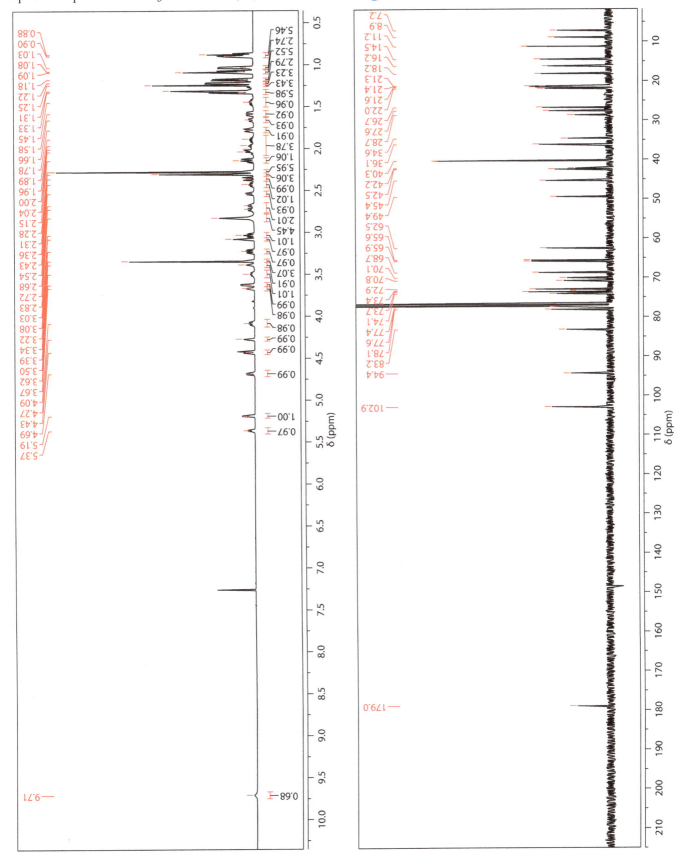

SECTION 6 Problem 129

Determine the structure of the unknown compound below with an HRMS [M + H]⁺ m/z of 791.5324.

Spectra acquired in $(CD_3)_2CO$ at 500 MHz (^{1}H) or 125 MHz (^{13}C).

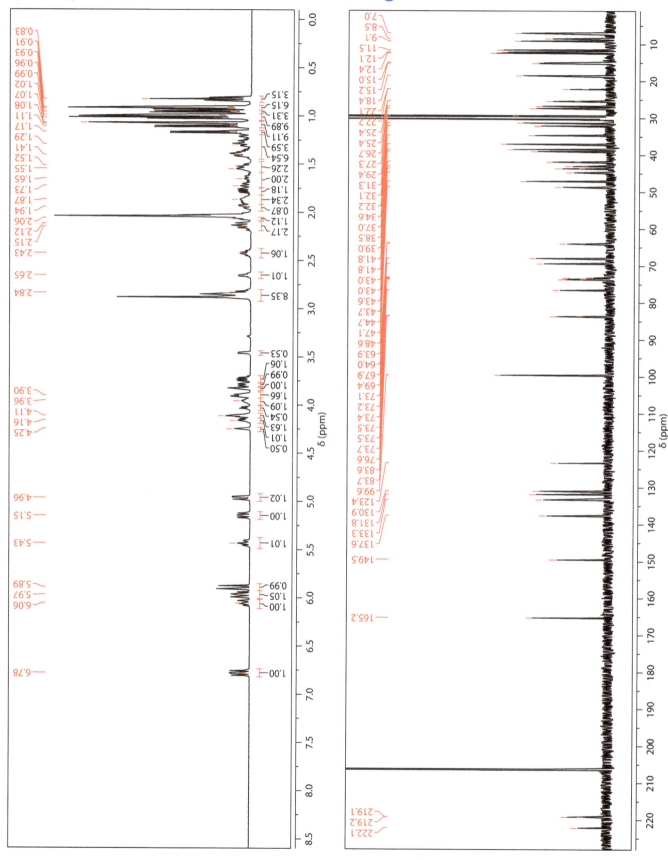

SECTION 6 Problem 130

Determine the structure of the unknown compound below with an HRMS [M + H]⁺ m/z of 850.5311.

Spectra acquired in $(CD_3)_2CO$ at 500 MHz (^{1}H) or 125 MHz (^{13}C).

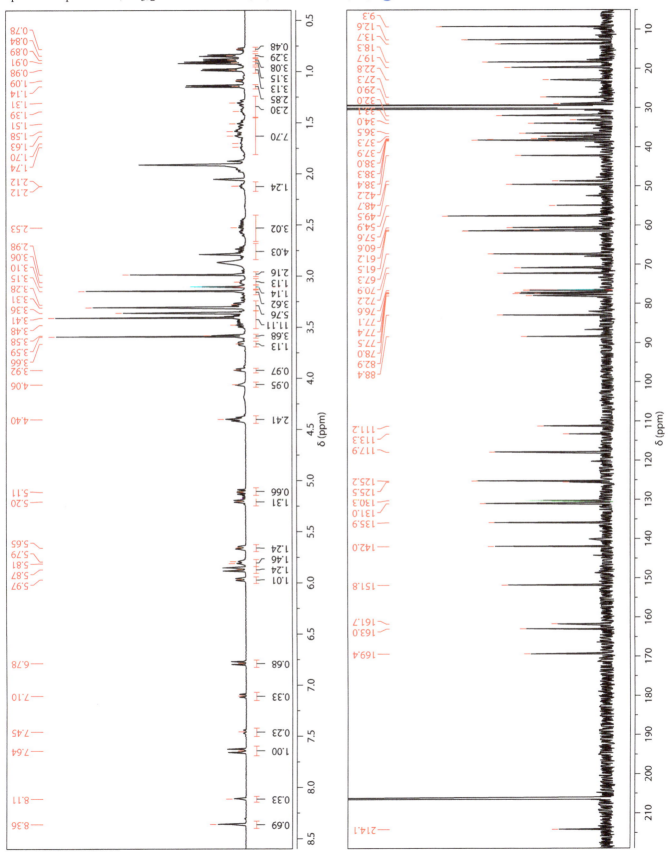

APPENDIX A

NMR SOLVENT REFERENCE CHEMICAL SHIFTS

Solvent	δ_H (ppm)	δ_C (ppm)
$(CD_3)_2CO$	2.04	29.8
C_6D_6	7.15	128.0
$CDCl_3$	7.26	77.0
D_2O	4.80	-1.90*
CD_2Cl_2	5.32	53.8
$(CD_3)_2SO$	2.49	39.5
CD_3OD	3.30	49.0

*external standard with 4,4-dimethyl-4-silapentane-1-sulfonic acid (DSS).

PROBLEM NOTES

Problem #	Note
Problem 7	Exchangeable protons are not observed in the ^{1}H NMR spectrum.
Problem 8	It is not possible to differentiate between the chemical shifts of the carbon signals designated with a "*" using the provided spectra.
Problem 17	Exchangeable protons are not observed in the ^{1}H NMR spectrum.
Problem 19	It is not necessary to make ^{1}H or ^{13}C assignments of the phenyl ring.
Problem 21	This is an HCl salt.
Problem 27	Exchangeable protons not observed in the ^{1}H NMR spectrum.
Problem 28	Exchangeable proton is not observed in the ^{1}H NMR spectrum.
Problem 31	Exchangeable proton is not observed in the ^{1}H NMR spectrum.
Problem 32	Exchangeable protons are not observed in the ^{1}H NMR spectrum.
Problem 38	Exchangeable signals at 4.57 and 5.37 ppm not assignable using data provided.
Problem 39	In the ^{13}C NMR all signals associated with J_{CF} splitting have been peak picked.
Problem 40	This compound is an HCl salt, which complicates the splitting of signals coupled to the NH.
Problem 41	You can see $^1J_{CH}$ in the gHMBC for the upfield methyl signal.
Problem 43	You can see $^1J_{CH}$ in the gHMBC for the upfield methyl signal.
Problem 45	Downfield exchangeable proton not observed in the ^{1}H NMR spectrum.
Problem 51	Not all ^{13}C signals can be assigned unequivocally.
Problem 52	Not all ^{13}C signals can be assigned unequivocally.
Problem 53	Not all ^{13}C signals can be assigned unequivocally. One exchangeable proton is missing from the ^{1}H NMR spectrum.
Problem 60	If present, exchangeable protons will not be observed in the ^{1}H NMR spectrum.
Problem 61	Not all signals can be unequivocally assigned with the data provided.
Problem 62	If present, exchangeable protons will not be observed in the ^{1}H NMR spectrum.
Problem 65	Not all signals can be unequivocally assigned with the data provided.
Problem 66	Not all signals can be unequivocally assigned with the data provided.
Problem 68	Not all signals can be unequivocally assigned with the data provided.
Problem 69	In the gHSQC, there is a signal at 68.8 ppm that is depicted in blue; however, it should be red.
Problem 72	This is an HCl salt.
Problem 76	You can see $^1J_{CH}$ in the gHMBC for the upfield methyl signal.
Problem 78	You can see $^1J_{CH}$ in the gHMBC for the upfield methyl signal.
Problem 80	If present, exchangeable protons will not be observed in the ^{1}H NMR spectrum.
Problem 84	This is an HCl salt.
Problem 86	If present, exchangeable protons will not be observed in the ^{1}H NMR spectrum.
Problem 87	This problem contains a ^{1}H-^{15}N gHMBC experiment.
Problem 88	This is an HCl salt. This problem contains a ^{1}H-^{15}N gHMBC experiment.
Problem 91	This is an HCl salt. You can see $^1J_{CH}$ in the gHMBC for the upfield signal.
Problem 92	In the ^{13}C NMR all signals associated with J_{CF} splitting have been peak picked. If present, exchangeable protons will not be observed in the ^{1}H NMR spectrum.
Problem 93	The compound for this problem exists as two different conformations.

APPENDIX B

Problem #	Note
Problem 94	This problem contains a ^{1}H-^{15}N gHMBC experiment.
Problem 98	Exchangeable protons not observed in the ^{1}H NMR spectrum.
Problem 101	Exchangeable protons not observed in the ^{1}H NMR spectrum.
Problem 113	This is an HCl salt.
Problem 120	This is an HCl salt.
Problem 126	This is a fumarate salt.

PROBLEM HINTS

Problem #	Hint
Problem 2	Multiplicity of aliphatic signals is key to the assignment.
Problem 3	Multiplicity of downfield signals will allow assignment of -CH_2Cl and -CH_2O- functional groups.
Problem 4	For aromatic protons consider resonance effects to assign signals. This molecule has an element of symmetry that influences the integrations of the methyl signals.
Problem 5	Pay attention to the ^{13}C chemical shift of the carbonyl signal and the relative ratios of the integrations. Consider whether these NMR spectra exhibit any features consistent with the presence of a symmetrical molecule.
Problem 6	Count the number of expected signals for the ^1H and ^{13}C spectra and consider the multiplicities of the ^1H signals.
Problem 7	The splitting patterns of the aromatic protons is the key to this assignment.
Problem 8	The ^1H signal at 4.17 ppm is complicated due to chemical inequivalence.
Problem 9	Think about the chemical shift of the methylene for each of the candidate structures.
Problem 10	Think about the role symmetry may play in the ^1H and ^{13}C spectra.
Problem 11	The ^{13}C chemical shifts of the carbonyl signals are diagnostic for this problem.
Problem 12	Predicting the splitting patterns for the aromatic signals will help solve this problem.
Problem 13	The ^1H NMR shows long-range allylic coupling. The gHSQC is helpful in establishing one-bond proton-carbon connections.
Problem 14	The phenolic proton signal can be identified by the lack of a cross peak in the gHSQC spectrum.
Problem 15	Analysis of the exchangeable protons is helpful in this problem.
Problem 16	You can have ortho, meta and para coupling in aromatic systems. Typically coupling through heteroatoms is smaller. These two observations will help solve this problem.
Problem 17	The methylene protons at 1.62 ppm are diastereotopic and magnetically inequivalent.
Problem 19	There is a heavy atom effect for the CH_2Br group in this problem.
Problem 20	Pay attention to the ^{13}C chemical shift of the methine carbon.
Problem 21	Predicting the number of signals in the ^1H and ^{13}C NMR spectra will help here.
Problem 22	Interpreting the multiplicities of the signals between 7.13 - 6.67 ppm in the ^1H is key to solving this problem.
Problem 23	Methylenes in cyclic structures will be diastereotopic. The ^1H chemical shift of a CH_3 group depends on whether it is next to an sp^2 or sp^3 carbon.
Problem 24	This molecule is symmetrical.
Problem 25	The following book is a great reference for chemical shift: Pretsch, E.; P. Bühlmann, et al. (2009). Structure determination of organic compounds: tables of spectral data. Berlin, Springer.
Problem 26	Protons of cyclopropane rings are shifted further upfield than normal hydrocarbons.
Problem 27	Think about what influences the ^{13}C chemical shifts of carbonyls in order to distinguish the signals at 174.1 and 176.5 ppm.
Problem 28	The multiplicities of methyl signals will help to distinguish possible structures here. The gHSQC correlations from the signals at 4.94 - 4.89 ppm are useful too.
Problem 29	Think about the relative chemical shift differences between O-CH_2 and N-CH_2 and how the gCOSY can be utilized. The 1D TOCSY experiment on the NH proton shows the entire ^1H spin system that includes this functional group. Is that data consistent with the ethyl group or one of the ring systems?

APPENDIX C — Problems in Organic Structure Determination

Problem #	Hint
Problem 30	This molecule has an element of symmetry. If the sample were treated with a drop of D_2O, the three downfield signals (4.40, 4.32, 4.13 ppm) would disappear. Do these signals show gHSQC correlations?
Problem 31	See the Hoye reference in the foreword for help interpreting complex multiplicity patterns.
Problem 32	See the Hoye reference in the foreword for help interpreting complex multiplicity patterns.
Problem 33	This relies on analyzing the coupling constant of the olefin signal on the side-chain to distinguish these cis- and trans-isomers.
Problem 34	The gHMBC is useful for assignment of the carbon signals at 145.3 and 137.1 ppm.
Problem 35	Take advantage of the gHSQC to determine which ^1H's are exchangeable.
Problem 36	The selective decoupling experiment on the 1.74 ppm methyl group is very helpful for deducing coupling constant values.
Problem 37	There are a number of CH_2's with similar chemical shifts. In depth analysis of the gHMBC will allow full assignments of these methylene signals. Remember that the integration values are relative, not absolute.
Problem 38	See the Hoye reference in the foreword for help interpreting complex multiplicity patterns.
Problem 39	Consider the effects of ^{19}F-^{13}C coupling as $^1J_{CF}$, $^2J_{CF}$ and $^3J_{CF}$ are visible in the carbon spectrum and result in overlapping signals.
Problem 42	Use the 1D TOCSY spectrum to distinguish the signals at 3.62 and 3.66 ppm.
Problem 43	These molecules contain a number of distinguishing characteristics, such as a *t*-butyl group and a tertiary alcohol. Evaluate each possible structure in the context of all aspects of the candidate structures. The gHSQC and gHMBC spectra are helpful here.
Problem 44	Analysis of gHMBC correlations from the aromatic protons is helpful in solving this problem.
Problem 45	This molecule is symmetrical.
Problem 46	Take advantage of gHMBC correlations from the hydroxy and *t*-butyl groups in solving this problem.
Problem 47	Consider the significance of the gHMBC correlations to the methine at 41.2 ppm.
Problem 49	The gHSQC-TOCSY experiment is very useful here, because it allows the carbons in each of the independent ring systems to be easily identified, despite overlapping ^1H and ^{13}C signals.
Problem 50	The J_{CF} splitting will help in the assignment of the aromatic ring carbons.
Problem 59	Think about possible heteroaromatic ring systems that contain four carbons.
Problem 60	This molecule has an IR absorption of 2270 cm^{-1}.
Problem 61	Not many functional groups have ^1H shifts at ~6.0 ppm with a corresponding ^{13}C shift at ~100 ppm.
Problem 62	This molecule has two exchangeable protons.
Problem 64	$^3J_{CH}$ values for some heteroaromatic ring systems can be extremely small. This molecule would have an IR signal at 2247 cm^{-1}.
Problem 65	This molecule would have an IR signal at 2236 cm^{-1}.
Problem 67	The methylene protons in this molecule are chemically equivalent but not magnetically equivalent.
Problem 69	This molecule would have an IR signal at 2350 cm^{-1}. Take this into consideration in combination with the gHSQC correlation for the ^1H signal at 1.95 ppm.
Problem 72	Because this molecule is an HCl salt, think about how the ^1H signal at 8.32 ppm could integrate to 3 protons.
Problem 74	This molecule has an element of symmetry.

Problem #	Hint
Problem 77	This molecule is symmetrical, but the signals are more complex due to magnetic inequivalence.
Problem 82	Note that $^3J_{CH}$ values for some heteroaromatic ring systems can be small.
Problem 84	Analysis of the coupling constants will be critical for determining the substitution pattern. Because this molecule is an HCl salt, a functional group is protonated.
Problem 88	This molecule is symmetrical.
Problem 89	The molecule contains a disubstituted amide.
Problem 90	Take advantage of the gHMBC correlations from exchangeable ^1H signals in this molecule.
Problem 92	The ^{19}F-^{13}C coupling will help establish the substitution pattern of the aromatic ring. Pay particular attention to the magnitudes of these coupling constants, and what the likely number of bonds are for each coupling (i.e. $^1J_{CF}$, $^2J_{CF}$ or $^3J_{CF}$). This molecule contains two nitro groups.
Problem 93	This molecule exists in two conformations on the NMR timescale in a roughly 2:1 ratio.
Problem 94	Use the ^1H-^{15}N gHMBC to help place the nitrogens in this molecule.
Problem 97	 1.32, 1.53 / 32.6 1.12 / 21.5 3.62, 3.96 / 66.2 3.66 / 72.4 4.59 / 98.6 1.21 / 21.0 δ 4.59, q J = 5.0 δ 3.96, ddd J = 9.5, 5.0, 1.2 δ 3.66, m δ 3.62, m δ 1.53, dddd J = 12.7, 12.7, 11.1, 5.0 δ 1.32, dddd J = 13.3, 2.5, 2.5, 1.4 δ 1.21, d J = 5.2 δ 1.12, d J = 6.2
Problem 98	 O 170.0 OH 130.7 2.18, 2.69 / 31.6 6.79 / 138.8 3.97 / 68.4 HO 4.36 / 67.3 3.67 / 72.7 OH δ 6.79, ddd J = 3.7, 1.8, 1.8 δ 4.36, dddd J = 1.7, 1.7, 1.7, 1.7 δ 3.97, ddd J = 10.5, 5.2, 5.2 δ 3.67, dd J = 7.4, 4.2 δ 2.69, dddd J = 18.1, 4.9, 1.9, 1.9 δ 2.18, dddd J = 18.2, 5.6, 1.8, 1.8
Problem 100	 1.66 / 24.0 5.55 / 119.6 0.95 / 23.0 0.85 / 15.6 133.7 24.2 1.93, 1.60 / 27.6 0.81 / 21.2 1.06 / 28.7 1.83, 1.68 / 17.9 δ 5.55, m δ 1.93, ddd J = 15.4, 6.7, 6.7 δ 1.83, dddd J = 13.6, 7.3, 6.7, 2.9 δ 1.68, dddd J = 13.6, 8.0, 6.7 δ 1.66, s δ 1.60, ddd J = 15.4, 7.3, 7.3 δ 1.06, s δ 0.95, dd J = 8.3, 4.4 δ 0.85, s δ 0.81, ddd J = 8.0, 8.0, 2.9

APPENDIX C

Problems in Organic Structure Determination

Problem #	Hint
Problem 101	Cyclohexane with COOH at C1 (2.33) and CH₂NH₂ at C4 (2.85, 1.65); ring positions: 1.40, 2.01 and 1.06, 1.84. δ 2.85, d J = 7.0 δ 2.33, dddd J = 12.2, 12.2, 3.5, 3.5 δ 2.01, brdd J = 13.6, 3.6 δ 1.84, brdd J = 13.4, 3.6 δ 1.65, m δ 1.40, dddd J = 13.1, 13.1, 13.1, 3.5 δ 1.06, dddd J = 13.0, 13.0, 13.0, 3.5
Problem 102	Quinic acid: cyclohexane with COOH (175.6, 12.28), C-OH (74.5), CH₂ (37.5: 1.86, 1.74), CH₂ (40.5: 1.84, 1.69), CH-OH (69.0: 3.87), CH-OH (66.7: 3.73), CH-OH (74.4: 3.23). δ 12.28, s δ 3.87, ddd J = 6.1, 3.2, 3.2 δ 3.73, ddd J = 8.2, 8.2, 4.2 δ 3.23, dd J = 7.6, 2.9 δ 1.86, m δ 1.84, m δ 1.74, ddd J = 13.7, 6.0, 2.0 δ 1.69, dd J = 13.1, 9.0 Note: exchangeable signals at 4.57 and 5.37 ppm not assignable using data provided.
Problem 103	Phenyl-oxazolidinone: Ph (7.39/128.4, 7.34/128.4, 7.30/125.9, ipso 134.8); CH-O (5.71, 81.0); CH-CH₃ (4.21, 52.4); CH₃ (0.81, 17.5); NH (6.27); C=O (159.5). δ 7.39, dd J = 7.6, 7.6 δ 7.34, t J = 7.4 δ 7.30, d J = 7.1 δ 6.27, brs δ 5.71, d J = 8.0 δ 4.21, dq J = 6.6, 6.6 δ 0.81, d J = 6.5
Problem 104	Terpene with gem-dimethyl cyclobutane (0.99/21.6, 0.97/29.8, 34.0), CH (1.75/50.7), CH₂ (1.67,1.63/39.7), CH (2.60/48.7), C=CH₂ (151.7, 112.7: 4.96, 4.84), CH₂ (2.33,2.09/29.7), CH₂ (2.23,1.31/30.1), epoxide C (59.8) and CH (2.86/63.7), CH₃ (1.19/16.9), CH₂ (1.64,1.41/27.1), CH₂ (2.07,0.95/39.1). δ 4.96, d J = 1.5 δ 4.84, d J = 1.5 δ 2.86, dd J = 10.5, 4.2 δ 2.60, ddd J = 9.8, 9.8, 9.8 δ 2.33, ddd J = 12.8, 8.2, 4.5 δ 2.23, dd J = 16.7, 8.2, 4.3 δ 2.09, m δ 2.07, m δ 1.75, dd J = 9.9, 9.9 δ 1.67, dd J = 10.5, 8.3 δ 1.64, m δ 1.63, dd J = 10.6, 10.6 δ 1.41, m δ 1.31, m δ 1.19, s δ 0.99, s δ 0.97, s δ 0.95, ddd J = 13.1, 13.1, 5.1

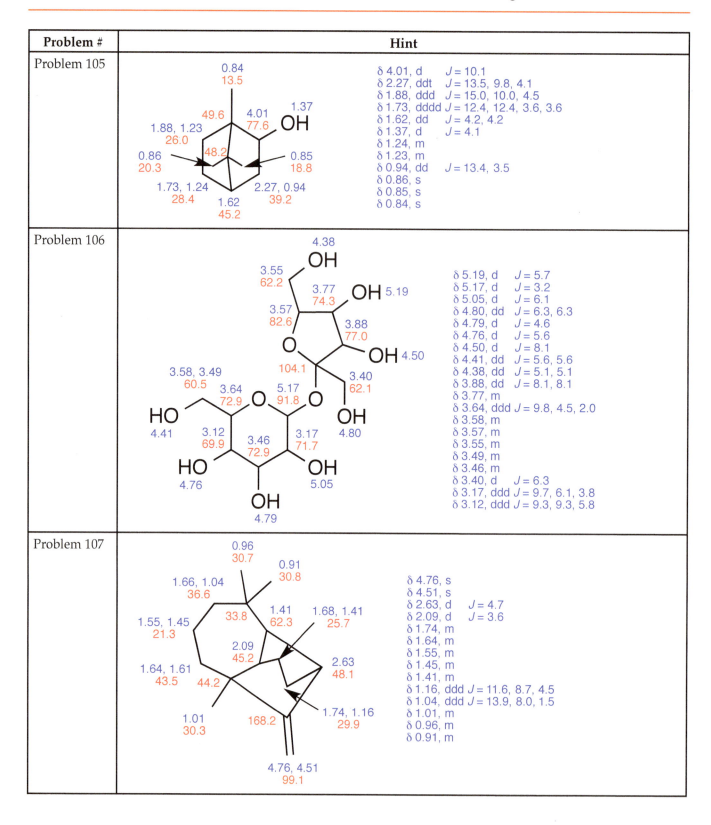

APPENDIX C

Problems in Organic Structure Determination

Problem #	Hint
Problem 108	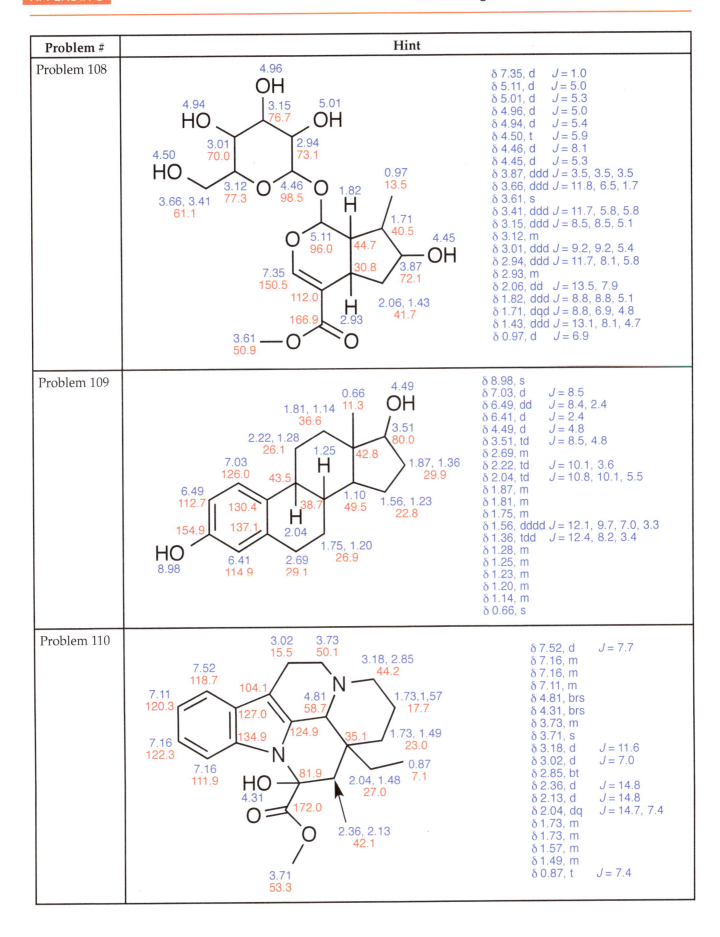

Problem #	Hint
Problem 111	Molecular formula: $C_{10}H_{16}$ This molecule is derived from isoprene biosynthesis.
Problem 112	Molecular formula: $C_{20}H_{21}NO_4 \cdot HCl$ Take advantage of the HMBC correlations from methoxy carbons to help assign the substitution patterns of the aromatic rings.
Problem 113	Molecular formula: $C_{10}H_{18}O$ This molecule is derived from isoprene biosynthesis.
Problem 114	Molecular formula: $C_{20}H_{23}N_3O_2$ This is a natural product semi-synthetic derivative and contains a carbamate function group (^{13}C shift of 155 ppm) more commonly found in synthetic compounds as part of protecting group.
Problem 115	Molecular formula: $C_{15}H_{24}O$ This molecule is derived from isoprene biosynthesis.
Problem 116	Molecular formula: $C_{17}H_{26}O_{10}$
Problem 117	Molecular formula: $C_{21}H_{24}N_6O_4S_2$ This compound contains a number of 5-membered heteroaromatic ring systems. Utilize the following: Pretsch, E.; P. Bühlmann, et al. (2009). Structure determination of organic compounds: tables of spectral data. Berlin, Springer to help identify possible rings in the molecule.
Problem 118	Molecular formula: $C_{16}H_{26}O_2$ This molecule is derived from isoprene biosynthesis.
Problem 119	Molecular formula: $C_{13}H_{14}N_2O_4S_2$ 2 vicinal methines that should display coupling are accidentally isochronous and have the same proton chemical shifts at 4.82 ppm. One carbon signal is obscured by the solvent signal. It can be detected via the HMBC spectrum.
Problem 120	Molecular formula: $C_{18}H_{34}N_2O_6S \cdot HCl$ This molecule contains a thioether moiety. Think about this with the ^{13}C signal at 88.0 ppm.
Problem 121	Molecular formula: $C_{18}H_{24}O_2$ This molecule is derived from sterol biosynthesis.
Problem 122	Molecular formula: $C_{20}H_{30}O_5$ This molecule belongs to the diterpenoid family of natural products.
Problem 123	Molecular formula: $C_{26}H_{31}ClN_2O$ This compound contains a functional group that is relatively uncommon, with an IR of 2125 cm^{-1}.
Problem 124	Molecular formula: $C_{26}H_{44}O_9$ This molecule is derived from polyketide biosynthesis. It can be divided into two components, one of which is a saturated fatty acid.
Problem 125	Molecular formula: $C_{29}H_{44}O_5$ The core of this molecule is derived from sterol biosynthesis.
Problem 126	Molecular formula: $C_{28}H_{47}NO_4S \cdot C_4H_4O_4$ This molecule is formulated as the fumarate salt and displays fumarate signals in NMR spectra.
Problem 127	Molecular formula: $C_{59}H_{90}N_6O_{18}$ This molecule is peptidic in origin.
Problem 128	Molecular formula: $C_{38}H_{72}N_2O_{12} \cdot 2H_2O$ This compound is a semi-synthetic analog of a polyketide natural product. It has two sugar moieties attached to the aglycon.
Problem 129	Molecular formula: $C_{45}H_{74}O_{11}$ There is doubling of signals due to interconversion of two equilibrium structures.
Problem 130	Molecular formula: $C_{46}H_{75}NO_{13}$ This molecule exists in multiple conformers on the NMR timescale.

APPENDIX D

Problems in Organic Structure Determination

DISTRIBUTION OF EXPERIMENT TYPES

Section 1

Problem Number	Section	Problem Type	^{1}H	^{13}C	^{1}H-^{1}H gCOSY	^{1}H-^{13}C gHSQC	^{1}H-^{13}C gHMBC	DEPT135
1	1	Predict	Predict	Predict				
2	1	Assign	√					
3	1	Assign	√					
4	1	Assign	√					
5	1	Match	√	√				
6	1	Match	√	√				
7	1	Match	√	√				
8	1	Assign	√	√				
9	1	Match	√	√				
10	1	Match	√	√				√
11	1	Match	√	√				
12	1	Match	√	√				
13	1	Assign	√	√		√		
14	1	Assign	√	√		√		
15	1	Match	√	√				
16	1	Assign	√	√	√	√		
17	1	Assign	√	√	√	√		
18	1	Assign	√	√	√	√		
19	1	Assign	√	√	√	√		
20	1	Match	√	√				
21	1	Match	√	√				
22	1	Match	√	√				
23	1	Match	√	√				
24	1	Assign	√	√		√	√	
25	1	Predict	Predict	Predict				

Section 2

Problem Number	Section	Problem Type	^{1}H	^{13}C	^{1}H-^{1}H gCOSY	^{1}H-^{13}C gHSQC	^{1}H-^{13}C gHMBC	DEPT135	1D TOCSY	Other
26	2	Assign	√	√						
27	2	Assign	√	√						
28	2	Match	√	√	√	√				
29	2	Match	√	√	√	√			√	
30	2	Assign	√	√	√	√				
31	2	Assign	√	√	√	√				
32	2	Assign	√	√	√	√				
33	2	Match	√			√				
34	2	Assign	√	√	√	√	√			
35	2	Assign	√	√	√	√			√	
36	2	Assign	√	√	√	√				^{1}H-Sel. Dec.

Problem Number	Section	Problem Type	¹H	¹³C	¹H-¹H gCOSY	¹H-¹³C gHSQC	¹H-¹³C gHMBC	DEPT135	1D TOCSY	Other
37	2	Assign	√	√	√	√	√		√	
38	2	Assign	√	√	√	√				
39	2	Assign	√	√	√	√				
40	2	Assign	√	√	√	√	√			
41	2	Assign	√	√	√	√	√			
42	2	Assign	√	√		√	√		√	
43	2	Match	√	√	√	√	√	√		
44	2	Match	√	√	√	√	√			
45	2	Assign	√	√	√	√	√			
46	2	Assign	√	√	√	√	√			
47	2	Match	√	√	√	√	√			
48	2	Assign	√	√	√	√	√			gHSQC-TOCSY
49	2	Assign	√	√	√	√	√			
50	2	Assign	√	√	√	√	√			

Section 3

Problem Number	Section	Problem Type	¹H	¹³C	¹H-¹H gCOSY	¹H-¹³C gHSQC	¹H-¹³C gHMBC	DEPT135	1D NOESY
51	3	Determine	√	√					
52	3	Determine	√	√					
53	3	Determine	√	√					
54	3	Determine	√	√					
55	3	Determine	√	√					
56	3	Determine	√	√					
57	3	Determine	√	√					
58	3	Determine	√	√					
59	3	Determine	√	√					
60	3	Determine	√	√					
61	3	Determine	√	√					
62	3	Determine	√	√					
63	3	Determine	√	√					
64	3	Determine	√	√					
65	3	Determine	√	√					
66	3	Determine	√	√					
67	3	Determine	√	√				√	
68	3	Determine	√	√				√	
69	3	Determine	√	√	√	√			
70	3	Determine	√	√		√			
71	3	Determine	√	√	√	√			
72	3	Determine	√	√	√	√	√		
73	3	Determine	√	√	√	√	√		
74	3	Determine	√	√	√	√	√		
75	3	Determine	√	√	√	√	√		

APPENDIX D
Problems in Organic Structure Determination

Problem Number	Section	Problem Type	¹H	¹³C	¹H-¹H gCOSY	¹H-¹³C gHSQC	¹H-¹³C gHMBC	DEPT135	1D NOESY
76	3	Determine	√	√	√	√	√		
77	3	Determine	√	√					
78	3	Determine	√	√	√	√	√		√
79	3	Determine	√	√	√	√	√		
80	3	Determine	√	√	√	√	√		

Section 4

Problem Number	Section	Problem Type	¹H	¹³C	¹H-¹H gCOSY	¹H-¹³C gHSQC	¹H-¹³C gHMBC	DEPT135	¹H-¹⁵N gHMBC	Other
81	4	Determine	√	√	√	√	√			
82	4	Determine	√	√	√	√	√			
83	4	Determine	√	√	√	√	√	√		
84	4	Determine	√	√	√	√	√			
85	4	Determine	√	√	√	√	√			
86	4	Determine	√	√	√	√	√	√		
87	4	Determine	√	√	√	√	√		√	
88	4	Determine	√	√	√	√	√	√	√	
89	4	Determine	√	√	√	√	√			
90	4	Determine	√	√	√	√	√			
91	4	Determine	√	√	√	√	√	√		
92	4	Determine	√	√	√	√	√			
93	4	Determine	√	√	√	√	√	√		Var. Temp
94	4	Determine	√	√	√	√	√	√	√	
95	4	Determine	√	√	√	√	√			

Section 5

Problem Number	Section	Problem Type	¹H	¹³C	¹H-¹H gCOSY	¹H-¹³C gHSQC	¹H-¹³C gHMBC	1D TOCSY	¹H Sel. Dec.	1D NOESY	ROESY
96	5	Stereochem	√		√					√	
97	5	Stereochem	√ (42)	√ (42)		√ (42)	√ (42)	√ (42)		√	
98	5	Stereochem	√ (31)	√ (31)	√ (31)	√ (31)					√
99	5	Stereochem	√		√	√					
100	5	Stereochem	√ (111)	√ (111)	√ (111)	√ (111)	√ (111)			√	
101	5	Stereochem	√						√		√
102	5	Stereochem	√ (38)	√ (38)	√ (38)	√ (38)					√
103	5	Stereochem	√ (75)	√ (75)	√ (75)	√ (75)	√ (75)			√	
104	5	Stereochem	√ (115)	√ (115)	√ (115)	√ (115)	√ (115)			√	√
105	5	Stereochem	√ (113)	√ (113)	√ (113)	√ (113)	√ (113)			√	√
106	5	Stereochem	√ (35)	√ (35)	√ (35)	√ (35)		√ (35)		√	√
107	5	Stereochem	√ (48)	√ (48)	√ (48)	√ (48)	√ (48)			√	
108	5	Stereochem	√ (116)	√ (116)	√ (116)	√ (116)	√ (116)	√	√		√
109	5	Stereochem	√ (121)	√ (121)	√ (121)	√ (121)	√ (121)				√
110	5	Stereochem	√ (95)	√ (95)	√ (95)	√ (95)	√ (95)				√

√ (Problem # where that spectra is available), e.g. √ (42) indicated that spectrum is available in problem 42.

Section 6

Problem Number	¹H	¹³C	¹H-¹H gCOSY	¹H-¹³C gHSQC	¹H-¹³C gHMBC	DEPT135	1D TOCSY	TOCSY	1D NOESY	ROESY	Other
111	√	√	√	√	√				√ (100)		
112	√	√	√	√	√	√					
113	√	√	√	√	√				√ (105)	√ (105)	
114	√	√	√	√	√				√		
115	√	√	√	√	√				√ (104)	√ (104)	
116	√	√	√	√	√		√ (108)		√ (108)		¹H-Sel. Dec. (108)
117	√	√	√	√	√		√				¹H-¹⁵N gHMBC
118	√	√	√	√	√	√			√	√	
119	√	√	√	√	√				√		
120	√	√	√	√	√	√		√			
121	√	√	√	√	√				√ (109)		
122	√	√	√	√	√	√	√	√	√	√	
123	√	√	√	√	√					√	¹H-¹⁵N gHMBC
124	√	√	√	√	√		√	√			gHSQC-TOCSY
125	√	√	√	√	√	√	√	√	√	√	gHSQC-TOCSY
126	√	√	√	√	√		√	√	√	√	
127	√	√	√	√	√		√				
128	√	√	√	√	√						
129	√	√	√	√	√				√		gHSQC-TOCSY
130	√	√	√	√	√		√				gHSQC-TOCSY

APPENDIX E

DISTRIBUTION OF FUNCTIONAL GROUPS

Problem Number	Section	Alkyl	Symmetrical Aromatic	Asymmetrical Aromatic	Heterocycle	Phenol	Olefin	Carboxylic Acid	Amine	Ester	Amide	Ketone	Aldehyde	Alcohol	Ether	Alkyne	Nitrile	Nitro	Fluorine	Bromine, Chlorine	Chirality
1	1																				
2	1	√							√												
3	1	√								√										√	
4	1		√												√						
5	1									√										√	
6	1		√									√								√	
7	1			√			√													√	
8	1	√								√											
9	1			√						√											
10	1	√					√														
11	1			√			√							√							
12	1	√		√			√								√						
13	1	√					√							√							
14	1			√		√	√													√	
15	1			√	√	√	√														
16	1			√	√															√	
17	1	√					√								√						√
18	1	√					√	√													
19	1	√	√												√					√	
20	1	√								√										√	
21	1	√		√						√											√
22	1		√	√					√						√						
23	1	√					√								√						
24	1		√		√																
25	1																				
26	2	√													√						
27	2	√					√								√						√
28	2	√					√								√						√
29	2	√								√	√										√
30	2	√												√							√
31	2	√					√	√						√							√
32	2	√												√							√
33	2		√	√		√	√					√			√						
34	2		√	√	√										√						
35	2	√													√	√					√
36	2	√					√					√									√
37	2	√					√	√													

Problem Number	Section	Alkyl	Symmetrical Aromatic	Asymmetrical Aromatic	Heterocycle	Phenol	Olefin	Carboxylic Acid	Amine	Ester	Amide	Ketone	Aldehyde	Alcohol	Ether	Alkyne	Nitrile	Nitro	Fluorine	Bromine, Chlorine	Chirality
38	2	√						√						√							√
39	2	√		√								√							√		
40	2	√	√						√												
41	2	√		√				√							√						
42	2	√													√						√
43	2	√	√											√							√
44	2		√	√	√				√		√										
45	2	√	√					√			√										
46	2	√	√	√		√		√													
47	2	√					√														√
48	2	√					√														√
49	2	√												√	√						√
50	2	√		√	√			√	√										√		
51	3	√										√		√							
52	3	√						√						√							
53	3	√							√					√							
54	3	√										√								√	
55	3	√					√			√											
56	3	√									√										
57	3	√	√									√								√	
58	3	√					√	√													
59	3	√		√	√									√							
60	3		√			√											√				
61	3		√									√			√						
62	3		√			√	√	√													
63	3	√	√									√								√	
64	3			√	√												√				
65	3			√					√								√				
66	3		√					√						√							
67	3	√													√						√
68	3	√	√						√					√							√
69	3	√						√									√				
70	3	√									√										√
71	3	√		√						√											
72	3	√						√		√											√
73	3	√		√										√					√		
74	3	√													√	√					√
75	3	√	√																		√

APPENDIX E — Problems in Organic Structure Determination

Problem Number	Section	Alkyl	Symmetrical Aromatic	Asymmetrical Aromatic	Heterocycle	Phenol	Olefin	Carboxylic Acid	Amine	Ester	Amide	Ketone	Aldehyde	Alcohol	Ether	Alkyne	Nitrile	Nitro	Fluorine	Bromine, Chlorine	Chirality
76	3	√	√					√		√	√										√
77	3	√					√			√											
78	3			√			√		√	√											
79	3	√	√					√			√									√	√
80	3			√		√	√								√			√			
81	4	√					√					√									
82	4	√		√										√	√						
83	4	√	√						√						√						
84	4	√		√					√											√	
85	4	√		√					√												
86	4	√	√					√	√		√			√							√
87	4			√	√				√												
88	4	√	√						√												
89	4	√	√							√					√					√	
90	4	√		√		√	√	√		√											√
91	4	√					√		√	√				√							
92	4	√		√					√		√							√	√		√
93	4	√							√	√				√							√
94	4	√	√							√				√				√		√	√
95	4	√		√	√			√	√					√							√
96	5	√					√														
97	5	√												√	√						√
98	5	√					√	√						√							√
99	5	√							√					√							√
100	5	√					√														√
101	5	√						√	√												√
102	5	√						√						√							√
103	5	√	√																		√
104	5	√					√								√						√
105	5	√												√							√
106	5	√												√	√						√
107	5	√				√															√
108	5	√				√					√			√	√						√
109	5	√		√		√								√							√
110	5	√		√	√			√	√					√							√
111	6	√					√														√
112	6	√		√	√										√						
113	6	√												√							√

Problem Number	Section	Alkyl	Symmetrical Aromatic	Asymmetrical Aromatic	Heterocycle	Phenol	Olefin	Carboxylic Acid	Amine	Ester	Amide	Ketone	Aldehyde	Alcohol	Ether	Alkyne	Nitrile	Nitro	Fluorine	Bromine, Chlorine	Chirality
114	6	√	√	√					√												√
115	6	√					√								√						√
116	6	√					√			√				√	√						√
117	6	√		√	√				√												√
118	6	√								√											√
119	6	√		√		√				√											√
120	6	√							√		√			√	√						√
121	6	√		√		√								√							√
122	6	√					√			√				√							√
123	6	√		√	√		√							√			√			√	√
124	6	√					√	√		√				√	√						√
125	6	√								√		√			√						√
126	6	√					√		√	√			√	√							√
127	6	√								√	√										√
128	6	√							√	√				√	√						√
129	6	√					√			√		√		√	√						√
130	6	√					√			√	√	√		√	√						√

APPENDIX F
Problems in Organic Structure Determination

SELECTED ^{15}N NMR CHEMICAL SHIFTS

Structure	Class	Shift
R–CN	Nitriles	−140 to −110 ppm
R–NC	Isonitrile	−220 to −170 ppm
R–C(=O)–NH$_2$	Amides	−280 to −230 ppm
Ar–NH$_2$	Arylamine	−330 to −270 ppm
R–NH$_2$	Alkylamine	−390 to −300 ppm

Structure	Class	Shift
1,2-Oxazole	1,2-Oxazole	ca. +2 ppm
R–NO$_2$	Nitroalkanes	0 to 40 ppm
Ar–NO$_2$	Nitroaromatics	−20 to 0 ppm
1,3-Thiazole	1,3-Thiazole	−80 to −52 ppm
1,2-Thiazole	1,2-Thiazole	−103 to −77 ppm
1,3-Oxazole	1,3-Oxazole	ca. −125 ppm

Modified from: M. Witanowski and G. A. Webb (eds.). Nitrogen NMR. Plenum Press. London and New York (1973)

Chemical shift values (δ) reported relative to nitromethane (IUPAC recommended std) and converted from σ-values reported in Witanowski (δ=−σ). Referencing to liq. NH$_3$ is also commonly encountered (δ MeNO$_2$ = δ NH$_3$ − 380.23)

ABOUT THE AUTHORS

Roger Linington received his B.Sc. in chemistry from the University of Leeds, and his Ph.D. from the University of British Columbia with Professor Raymond Andersen. His postdoctoral research with Professor William Gerwick was a joint appointment between the University of California San Diego and the Smithsonian Tropical Research Institute, which gave him the opportunity to participate in an international neglected disease drug discovery program in Panama City, Panama. He then joined the Department of Chemistry and Biochemistry at the University of California Santa Cruz as a faculty member where he remained for eight years before moving to Simon Fraser University in Canada, where he is currently a Tier II Canada Research Chair in Chemical Biology and High-Throughput Screening in the Department of Chemistry.

Philip Williams caught the spectroscopy bug as a summer research student at the University of Calgary while working on a series of alkaloids from African plants. After graduating with his B.Sc. degree, he move to warmer climates and completed his Ph.D. at the University of Hawaii at Manoa under the direction of the late Richard Moore; a pioneer in the field of cyanobacterial natural products. Postdoctoral work at the Scripps Institution of Oceanography in San Diego on microbial natural products with Bill Fenical was followed by a returned to the University of Hawaii at Manoa as a faculty member. Since returning he his interests have focused on natural products and their applications in the fields of cancer and neurological diseases.

John MacMillan received his B.Sc. from the University of Iowa, and his Ph.D. from the University of California Davis where his interests in small molecule NMR began by studying the natural products chemistry of marine organisms under the direction of Professor Tadeusz Molinski. He then went on to Postdoctoral work under the guidance of Bill Fenical at the Scripps Institution of Oceanography in San Diego where he worked on microbial natural products. In 2007 he began his independent career as a member of the faculty at the University of Texas Southwestern Medical Center as the Chilton/Bell Endowed Scholar. His research focuses on the chemical and biological characterization of natural products with therapeutic potential in the areas of oncology and infectious disease.

ABOUT THE SPECTROMETER

The NMR spectrometer used to acquire all of the data in this book is a Varian Unity Inova 500 MHz instrument housed in the Department of Chemistry at the University of Hawaii Manoa. Originally installed in 1989 (it was a General Electric GN Omega 500 back then), the instrument was upgraded in 2000 with a Varian Unity Inova console and is controlled by VNMRJ software version 4.2 running on Red Hat Linux 6.3.

In general, spectra were acquired using a switchable (broadband) 5 mm room temperature probe. All ^{15}N HMBC experiments were acquired using a 5 mm HCX (triple resonance) probe or a 3 mm HX (indirect detect) probe. All probes are equipped with pulse field gradients, which were employed for all 2D experiments.

ABOUT THE COVER

To the uninitiated, structure elucidation can sometimes feel like a maze of possibilities, with no clear view of how to traverse the gap between the starting point and the solution. In reality however, the analysis will always converge on a single answer, with the correct solution threading a logical line through the data. The cover photograph, taken over Vancouver British Columbia by Aaron O'Dea, represents both the many false avenues that can perplex the structure elucidation scientist, and the single correct route that ultimately leads to each solution.